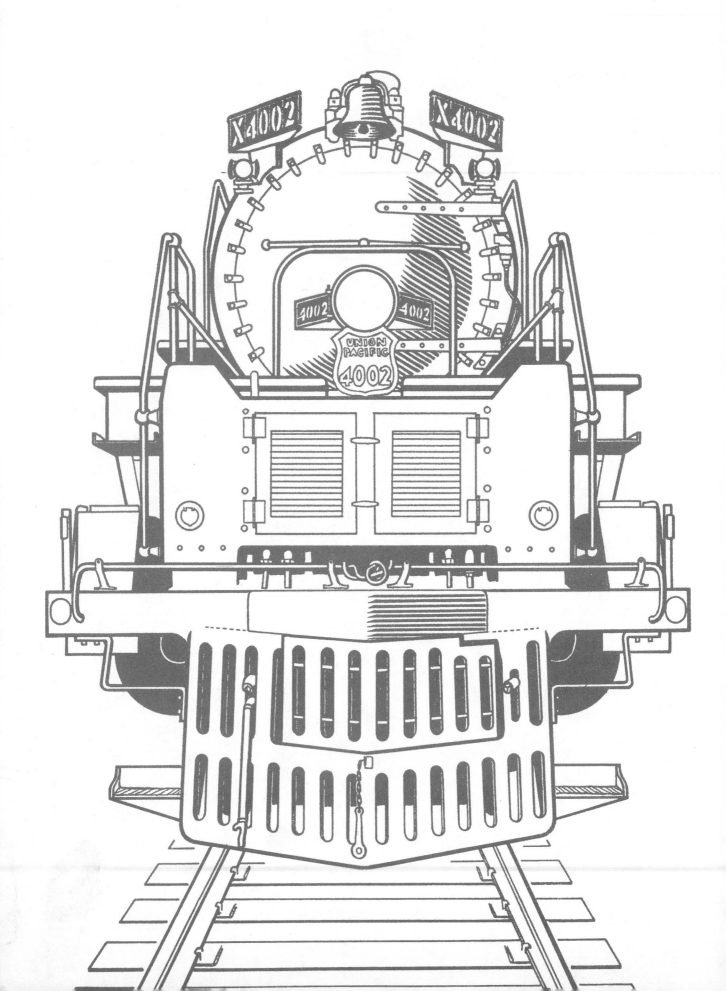

ILLUSTRATED ENCYCLOPEDIA OF
WORLD RAILWAY
LOCOMOTIVES

Edited by
P. Ransome-Wallis

DOVER PUBLICATIONS, INC.
Mineola, New York

Bibliographical Note

This Dover edition, first published in 2001, is an unabridged republication of the work originally published in the United States of America by Hawthorn Books, New York, in 1959, and simultaneously in Canada by McClelland & Stewart Ltd., Toronto, under the title *The Concise Encyclopedia of World Railway Locomotives*. All of the illustrations included in the original edition have been reproduced in the Dover edition in black and white.

Library of Congress Cataloging-in-Publication Data

Illustrated encyclopedia of world railway locomotives / edited by P. Ransome-Wallis.
 p. cm.
 Reprint. Originally published under title: Concise encyclopedia of world railway locomotives. New York : Hawthorn Books, 1959.
 Includes bibliographical references and index.
 ISBN 0-486-41247-4 (pbk.)
  1. Locomotives—Encyclopedias. I. Ransome-Wallis, P. (Patrick) II. Title: Concise encyclopedia of world railway locomotives.

TJ605 .I45 2001
625.26'03—dc21

                  00-064399

Manufactured in the United States of America
Dover Publications, Inc., 31 East 2nd Street, Mineola, N.Y. 11501

CONTENTS

CHAPTER ONE

DIESEL RAILWAY TRACTION

by J. M. DOHERTY, A.M.I. Mech.E., A.M.I. Loco.E.

CHAPTER TWO

DIESEL TRACTION IN NORTH AMERICA

by *DAVID P. MORGAN*

CHAPTER THREE

ELECTRIC MOTIVE POWER

by F.J.G.HAUT, F.R.S.A., B.Sc.(Eng.), A.M.I. Mech.E., M.I. and S.Inst.

PART V. THE ELECTRIC MOTOR COACH AND MOTOR COACH TRAIN

PART VI. UNDERGROUND RAILWAYS

CHAPTER FOUR

THE RECIPROCATING STEAM LOCOMOTIVE

by C.R.H. SIMPSON, A.M.I. Loco.E.

PART I. CONSTRUCTION AND DESIGN: A CONCISE ENCYCLOPEDIA   240

PART II. STEAM LOCOMOTIVE EXPERIMENTS

CHAPTER FIVE

ILLUSTRATED SURVEY OF MODERN STEAM LOCOMOTIVES

by H. M. LE FLEMING, M.A.(Cantab.), A.M.I. Mech.E., M.I. Loco.E., M.N.E.C. Inst.

CHAPTER SIX

THE TESTING OF LOCOMOTIVES

by S. O. ELL

CHAPTER SEVEN

THE STEAM LOCOMOTIVE IN TRAFFIC

by O. S. NOCK, B.Sc.(Eng.), M.I.C.E., M.I. Mech.E.

CHAPTER EIGHT

THE ORGANIZATION OF A STEAM MOTIVE POWER DEPOT

*by G. FREEMAN ALLEN, Parts I, II & III,
and by P. RANSOME-WALLIS, Part IV*

CHAPTER NINE

UNCONVENTIONAL FORMS OF RAILWAY MOTIVE POWER

by P. RANSOME-WALLIS, M.B., Ch.B.

CHAPTER TEN

THE GAS TURBINE IN RAILWAY SERVICE

by P. RANSOME-WALLIS, M.B., Ch.B.

CHAPTER ELEVEN

CONCISE BIOGRAPHIES OF FAMOUS LOCOMOTIVE DESIGNERS AND ENGINEERS

by H.M. LE FLEMING, M.A.(Cantab.), A.M.I. Mech.E., M.I. Loco.E., M.N.E.C. Inst.

page 494

FEATURED PLATES

LIST OF PHOTOGRAPHS

CH. 5. MODERN STEAM LOCOMOTIVES

CH. 6. TESTING OF LOCOMOTIVES

Some useful conversion factors

(to two places of decimals)

LENGTH

1 inch = 2·54 centimetres.
1 foot = 30·48 centimetres.
1 yard = 0·91 metre.
1 mile = 5,280 feet.
     = 1,760 yards.
     = 1·61 kilometres.

1 centimetre = 0·39 inch.
1 metre   = 39·37 inches.
     = 3·28 feet.
     = 1·09 yards.
1 kilometre = 0·62 mile.

AREA

1 square inch =   6·45 square centimetres.
1 square foot = 929·03 square centimetres.
1 square yard =   0·84 square metre.

1 square centimetre = 0·15 square inch.
1 square metre   = 10·76 square feet.
     = 1·20 square yards.

VOLUME

1 cubic inch = 16·39 cubic centimetres.
1 cubic foot =   0·03 cubic metre.
     = 28·32 litres.
     = 6·25 gallons.
1 cubic yard =   0·76 cubic metre.
     = 764·55 litres.
1 gallon   =   1·2 U.S.A. gallons.
     = 4·54 litres.
     = 0·16 cubic foot.
     = 277·40 cubic inches.

1 cubic centimetre =   0·06 cubic inch.
1 litre   = 61·02 cubic inches.
     =   1·76 pints.
     =   0·22 gallon.
1 cubic metre = 35·31 cubic feet.
     =   1·31 cubic yards.
     = 220·62 gallons.
1 U.S.A. gallon =   0·83 gallon.
     = 231·00 cubic inches.
     =   3·78 litres.

MASS

1 pound (avoirdupois)   =   16 ounces.
     = 453·59 grammes.
1 hundredweight (cwt)   =   50·80 kilogrammes.
1 ton (British or long ton) = 2,240 pounds.
     = 1·016 metric tonnes.
     = 1·12 U.S.A. or short tons.
     = 1,016 kilogrammes (kilos).
1 ton (U.S.A. or short ton) = 2,000 pounds.
     = 0·91 metric tonne.
     = 0·89 British or long ton.
1 tonne (metric)   = 2,204·60 pounds.
     = 1,000 kilogrammes.
     = 0·98 British or long ton.

DENSITY

1 cubic centimetre water at 4° centigrade weighs 1 gramme.
1 cubic foot water at 4° centigrade weighs 62·43 pounds.
1 litre water at 4° centigrade weighs 1 kilogramme.
1 gramme per cubic centimetre = 62·43 pounds per cubic foot.

PRESSURE AND FORCE

1 inch of mercury at 0° C.
     = 0·49 pound per square inch.
     = 0·43 foot of water.
     = 0·03 kilo. per square centimetre.
1 millimetre of mercury
     = 0·02 pound per square inch.
     = 0·02 foot of water.
     = 0·20 inch of water.
1 pound
     = 0·45 kilogramme.
     = 445,000 dynes.
1 foot of water
     = 62·43 pounds per square foot.
     = 0·43 pound per square inch.
     = 0·03 kilogramme per square centimetre.
Standard Atmospheric Pressure
     = 14·70 pounds per square inch.
     = 29·92 inches of mercury.
     = 33·93 feet of water.
     = 1·03 (rec.) kilogrammes per square centimetre.

1 pound per square inch
 = 0·07 kilogramme per square centimetre.
 = 68,971 dynes per square centimetre.
 = 27·71 inches of water.
 = 2·03 inches of mercury.
 = 51·71 millimetres of mercury.

1 pound per square foot
 = 4·88 kilogrammes per square metre.

1 kilogramme per square centimetre
 = 14·22 pounds per square inch.
 = 32·84 feet of water.

1 kilogramme per square millimetre
 = 0·63 ton per square inch.

1 ton (long) per square inch
 = 1·57 kilogrammes per square millimetre.

1 gramme per square centimetre
 = 0·01 pound per square inch.

1 hectopieze (hpz)
 = 1·02 kilogrammes per square centimetre.
 = 14·50 pounds per square inch.

C.V. (Metric horsepower)
 = 0·98 H.P.

———————

TEMPERATURE

F. indicates degrees Fahrenheit scale.
C. indicates degrees centigrade scale.
1 degree F. = $(9/5 \times C.) + 32$.
1 degree C. = $5/9 (F. - 32)$.

Abbreviations

A.C.	= Alternating current.	kW.	= Kilowatt.
Amp.	= Ampere.	lb.	= Pounds.
A.T.C.	= Automatic train control.	L.P.	= Low pressure
B.H.P.	= Brake horse power.	L.T.	= Low tension.
B.M.E.P.	= Brake mean effective pressure	M.E.P.	= Mean effective pressure.
B.S.	= British standard specification.	m.	= Metres.
B.T.U.	= British Thermal Units.	mm.	= Millimetres.
C.	= Centigrade.	m.p.h.	= Miles per hour.
C.T.C.	= Centralized train control.	M.U.	= Multiple-unit (control).
d.b.h.p.	= Drawbar horse power.	O.C.	= Oscillating cam (valve gear).
d.b.h.p.hr.	= Drawbar horse power hour.	O.P.	= Opposed piston (diesel engines).
D.B.T.E.	= Drawbar tractive effort.	p.s.i.	= Pounds per square inch.
D.C.	= Direct current.	R.C.	= Rotary cam (valve gear).
diam.	= Diameter.	r.p.m.	= Revolutions per minute.
F.	= Fahrenheit.	S.W.G.	= Standard wire gauge.
ft	= Feet.	T.E.	= Tractive effort.
H.P.	= Horse power.	t.	= Tonne
H.T.	= High tension.		(Metric ton [2,204·6 lb.]=1,000 kilos).
I.H.P.	= Indicated horse power.	v.	= Volt.
I.H.P.Hr	= Indicated horse power hour.	w.	= Watt.
in.	= Inches.		
Kg.	= Kilogrammes.		
Km/h.	= Kilometres per hour.		
kV.	= Kilovolt.		

Note: Where tons are given, these are always English long tons (2,240 lb.) except in Chapter 2, where tons refer to American short tons of 2,000 lb. Metric tons (2,204·6 lb.) are always written as *tonnes* or *t*.

Introduction

by THE EDITOR

Little more than a century and a half has passed since the invention of the steam engine started the incredible social and industrial revolution which has now reached the stage of the exploration of outer space, electronics and all that this implies, and the utilization of nuclear power in a wide range of application.

The use of steam as a means of land transportation did not really come into being until Stephenson demonstrated at the Rainhill Trials of 1829 that the multi-tubular boiler, draughted by the exhaust steam from the cylinders provided a machine which, the harder it was worked the more coal was consumed and the more steam was generated. Conversely, when the locomotive was worked lightly less coal was consumed and less steam generated. This so-called "automatic action" has been the corner-stone upon which all successful coal-burning steam locomotive practice has ever since been founded. By its very nature, therefore, the steam locomotive is a machine whose performance depends very greatly upon the skill of the men who drive it and who feed fuel to its boiler.

It is far from being a machine of precision in the accepted modern sense and men who work with steam locomotives, and indeed many who do not, have often endowed the machine with human attributes. Experienced and completely normal locomotive men may be heard talking to their engines in a manner that makes the uninformed observer question their sanity. But it is these men who, by being at one with the machine, have demonstrated time and again, that the steam locomotive is capable of almost incredible feats of power and performance when it is understood and sensitively handled. Considering the abuse to which it has often been subjected and at times the scanty maintenance which it has received, the steam locomotive has very seldom fallen down on the job and for all time it will remain as one of the staunchest friends man has ever had.

Some of the friendliness with which it has been treated must stem from the fact that unlike most other forms of power, it has never been adapted in war for the destruction of human life. Rather have its designers been at pains to produce, not only an efficient machine, but a thing of beauty, of graceful curves and perfect balance. In many countries, the fashion has been to embellish the machine with gaily coloured paint and with trimmings of polished steel, copper and brass.

In the realm of sound also, the steam locomotive possesses attributes which have the power to stir the souls of men. The rhythmic beat of the exhaust is the basis of music, its tempo denotes urgency, power, brutality when working hard, contentment, tranquillity and even lethargy, when running easily. Men have spent months of patient work to provide it with a warning cry which is at once penetrating, melodious, characteristic, and the locomotive whistle has become the most widely recognized sound in the world.

By its individualism and the fact that it demands of men a measure of conscientious hard and dirty work, the steam locomotive is becoming more and more of an anachronism in this modern age. For this is an age of impersonality, precision and remote control with the individual counting for less and less while the masses crowd ruthlessly on towards uncontrolled saturation.

In their endeavours to provide a more exact and precise instrument of transportation necessary to meet modern requirements, the engineers have somewhat belatedly turned away from steam towards electricity and the internal combustion engine. In these media they have found forms of power which depend upon no special skill in the driving and which can always be relied upon to give an exactly calculated output by the movement of a simple lever. No longer is it necessary to "nurse" the engine before making some extra demand of it, for the maximum of which the machine is capable is always and continuously available for the movement of a lever or the throwing of a switch. The human feeling towards the machine has gone and no longer do men address their charges as "old girl" or even as "old bitch"!

Such changes on a world-wide scale are taking place with great rapidity and with a seemingly inexhaustible supply of money. For modern railway motive power is expensive, even by post-war standards. In Britain one may quote the £200,000 paid for the 3,000 H.P. Deltic diesel-electric locomotive and compare it with the £8,000 paid for a main-line express steam locomotive only twenty-five years ago. Furthermore, experience on the dieselized railroads of the United States suggests that the economic life of a diesel locomotive is not more than fifteen years, about half of that of a steam locomotive, though during its life it produces as much as three times the amount of work. Also a very high degree of standardization is possible with both diesel and electric motive power and in the United States all the individual railroads are operated by no more than a total of eleven types of diesel locomotive, and this is rapidly being reduced to seven. With such opportunities for economy, it seems incredible that on the nationalized railways of Britain some forty-nine different types of diesel locomotive have been introduced since 1948.

The diesel comes really into its own, however, when operating availability is assessed. With no fire to clean and no boiler to wash out, it is capable of giving six and a half days of service out of every seven, over long periods of time. It is true that when given equal conditions of maintenance, trials on the New York Central Railroad proved the modern steam locomotive to be but very little inferior to the diesel on all counts. But these trials were of limited duration and it is certain that if the period of trial had been extended to several years, the diesel would have been greatly superior. The over-all efficiency of the diesel is without doubt much higher than that of the steam locomotive.

A deciding factor in favour of diesel traction in many countries is the ever-increasing cost of coal, and in the recent past, shortages occasioned by strikes and inefficient working. The long strike in the coal industry in the United States after the last war was a very large factor in deciding the railroads to abandon the steam locomotive in favour of the diesel. On the other hand there is very real concern that in countries such as Britain, once said to be built on coal, a major part of the transportation system should become so dependent upon oil, all of which has to be imported by sea from vulnerable and politically unstable parts of the world.

Electric traction offers even greater precision in operation than does diesel traction. Because it has no reciprocating parts, the electric locomotive has great advantages over both its rivals in simplicity, maintenance, durability and efficiency. As, however, it is not in itself a prime mover but must draw its current from wire or rail, the electric locomotive or train requires a great deal of fixed equipment before it can function at all, and the cost of such equipment is very high. None the less it would appear that for many countries, especially those of Western Europe, the ideal transportation system for the future may well be electric, using current from nuclear-powered generating stations or from hydro-electric schemes where this is practicable. This appears at present to be the best solution to the dependence of ourselves and many others on Middle East oil.

Enormous improvements have been made in electrical traction equipment in recent years, and these have enabled high-voltage, single-phase industrial alternating current to be used in comparatively small locomotives and suburban train sets.

The pattern for the future seems to be one of electrified main lines with diesel-operated branch lines and secondary lines on which the traffic offering does not warrant the high first cost of electrification. It appears unlikely that locomotives individually powered by nuclear reactors will be available for many years to come. Nuclear power on such a small scale would be extremely costly, and economically unsound in the present state of our knowledge.

In spite of these and other modern trends in means of railway transportation, it is wrong to assume that the steam locomotive is dead. It is probable that it will survive for many years and indeed at this time new steam locomotives are being built for Africa, China, India, Turkey and the U.S.S.R., and this is by no means a complete list. Very much more than 50 per cent of the railway traffic of the world is still handled by steam locomotives and is likely to be so for a long time to come.

This Concise Encyclopedia appears then at a time when great changes are occurring throughout the world in the pattern of railway motive power. This time of transition is unique because there are at work, often in a single country, the steam locomotives of the past side by side with all the other different forms of railway locomotion which are likely to be used in the foreseeable future. It is probable that there never has been and never will be again a more interesting and important period.

In this book we have given an account of the immediate past, surveyed the present and anticipated what we think may be the future.

The book has been compiled by a team of experts, each a well-known authority in the subject on which he writes. The standard of the work is such that it will be a useful book of reference for the engineer and the railwayman for many years to come. There are many outside the railway whose interest in and knowledge of locomotives is searching and profound, and to them this volume should be a mine of information.

When planning this book we paid great attention to published works, and each subject has been planned partly with a view to "filling the gaps" in existing literature. For this reason there is an unavoidable, but not, we think undesirable, lack of uniformity between chapter and chapter. For example there is an abundance of literature on the history of the steam locomotive and on its technical description. So we have omitted the former and confined our technical description to a concise encyclopedia of components. It has, however, been possible to include some interesting and little-known information in this chapter. We have given a comprehensive survey of modern practice and authoritative accounts of loco-motive testing, performance and operation. The history of electric traction has been dealt with in some detail, and a detailed survey of modern practice given. The emphasis in the chapter on diesel motive power is on engines and transmissions, which are very fully described and discussed.

A survey of the impact of the diesel-electric locomotive in the United States gives an up-to-date description of this recent form of motive power in a country which is, by now, nearly one hundred per cent dieselized.

In the chapter on unconventional motive power, a brief account is given of the efforts of engineers to find a locomotive which would improve upon the low effi-ciency of the Stephenson engine. A section on the gas turbine locomotive shows what has been and is being done to adapt this form of power to railway traction.

The bibliography has been carefully compiled to include most of the important works on railway motive power, and the appendices include several useful items which have not hitherto been published in a single volume.

In spite of our title, it has not been found convenient or rational to keep slavishly to encyclopedic form, though wherever possible this has been done, and it was thus decided that an index was not necessary if a reasonably detailed table of contents was given.

Some repetition of facts and formulae, as between chapter and chapter has been allowed for the sake of con-tinuity and completeness when dealing with each subject. In addition, the work is adequately cross-referenced.

The preparation of this book has been an enormous task of selection, reference, and research. The contribu-tors have worked as a team – I think a happy team – and have given every possible assistance to their Editor, a fact which he has deeply appreciated. For he is the only one among them who is not professionally connected in some way with world locomotives.

ACKNOWLEDGMENTS

The Editor and the Contributors wish to thank the many individuals and organizations who have so generously assisted them in the preparation of this book.

For their reading of part or all of the manuscript and for most helpful advice their thanks are due to: B. K. Cooper, Esq., P. C. Dewhurst, Esq., M.I.C.E., M.I.Mech.E., M.I.LOCO.E. and Robert G. Lewis, Esq. (U.S.A.).

For information and data willingly provided, thanks are especially due to: The Association of American Railroads; British Railways; The Curator of Historical Relics; British Transport Commission; Coras Iompair Eireann; The English Electric Co., Ltd; The General Electric Company of America; Messrs Henschel-Werke; The Institute of Locomotive Engineers, London; The Institute of Mechanical Engineers, London; A. Reidinger, Esq., M.I.LOCO.E.; Société Nationale des Chemins de Fer Français; The Society for Cultural Relations with the U.S.S.R.; A. Stephan, Esq.; Messrs The Swiss Locomotive & Machine Works; Messrs The Swiss Industrial Company; The Union Pacific Railroad Co; J. William Vigrass, Esq., and many others.

In response to our requests, many hundreds of photographs, and with them much informa-tion, have been received from all over the world. Credits are given beneath each photograph published and our thanks are due not only to those whose names appear, but to all those who have submitted illustrations for our selection. Suitable colour photographs have been very difficult to obtain. We are especially grateful to: W. A. Coons, Esq., of the Union Pacific Railroad Company, for his help in obtaining colour transparencies from the United States, and to: P. J. Bawcutt, Esq., The Norwegian State Railways and The Swiss Federal Railways, all of whom have made special colour transparencies for this book. The Canadian National Railways, The Canadian Pacific Railway Co. and Messrs. J. Stone & Co., Deptford, have been most generous in allowing us to use their colour transparencies and art work.

Frank Garnham drew the line diagrams.

CHAPTER 1

Diesel Railway Traction

by J. M. DOHERTY

Part I. Engines

BASIC REQUIREMENTS

The exacting and often conflicting nature of the demands made on diesel traction engines employed for main line railway service, present the engine builder with a number of difficult problems. Failure in service can cause severe dislocation to traffic, and in order to secure maximum availability the engine must be capable of working for long periods between overhauls with the minimum of attention. A high degree of robustness and durability is therefore required.

Service demands create wide fluctuations in speed and power output, and engines may be required to work at or near their maximum capacity for long periods. Furthermore, severe limitations of weight and space are often imposed. To facilitate overhaul and servicing, careful attention must be given to accessibility, and this is intimately linked with the general design of the locomotive or railcar in which the engine is to be installed.

For low-powered locomotives which are not subjected to severe weight limitations, a robustly constructed low-speed engine, naturally aspirated, is often preferred. An engine of this type gives exceptionally long life coupled with low maintenance costs. A more difficult problem arises in the case of engines required for intensive duty, and subjected to severe limitations with regard to space and weight, such as occur in high-powered diesel-electric locomotives. In these cases it has become necessary to adopt every available means for improving the power-weight ratio even when this entails an increase in cost and complication. The principal problem facing the engine builder is to meet these exacting requirements, without sacrificing reliability or unduly increasing operating expenses.

The following types of engines are employed for traction duty:

(i) Low-powered engines operating at 600–800 r.p.m. suitable for shunting (switching) and low-powered freight locomotives.

(ii) High-duty, low-speed engines, operating at 600–800 r.p.m. provided with pressure chargers (*see* page 31)

and sometimes with intercoolers (*see* page 31), suitable for high-powered locomotives where ample space is available.

(iii) Moderate speed engines operating at 800–1,200 r.p.m. with or without pressure charging according to requirements, suitable for both moderate and high-powered locomotives.

(iv) Moderately powered, high-speed engines used for railcars, operating at 1,500–2,000 r.p.m sometimes provided with pressure charging.

(v) High-speed engines operating at 1,200–1,600 r.p.m. provided with pressure charging and intercooling. Used in high-powered locomotives and diesel trains of advanced design.

CONSTRUCTION

Camshafts may be one or two in number, depending on the design of the engine. The drive from the crankshaft is through a train of helical gears, or by means of a duplex roller chain incorporating a device which automatically maintains the chain tension. In addition to actuating the inlet and exhaust valves, the camshaft also drives the fuel pumps and engine governor.

Connecting rods are steel stampings or forgings, the small ends having bronze bushes, press fitted, working on floating gudgeon pins, which are prevented from moving endways in the pistons by means of circlips.

The *crankcase* forms the principal structural member of the engine, and must be very rigidly constructed to resist distortion and preserve the alignment of the crankshaft bearings. The bottom part is usually made separate from the upper part, being structurally integrated with it to form the engine bed, which incorporates the lower halves of the crankshaft bearing housings (Plate IA, page 41).

Alternatively, the bottom part may act merely as an oil sump. With this type of construction the crankshaft is underslung, the upper halves of the bearing housings forming part of the upper portion of the crankcase (Plate IE, page 41). Whichever type of construction is used, a rigid assembly is secured by locating the bearing caps sideways in the crankcase. Additional security is sometimes provided by means of cross ties consisting of long bolts which pass through the crankcase and bearing caps.

The cylinder blocks may be integral with the crankcase or form separate units attached by means of studs (Plate ID, page 41). Crankcases are constructed of cast iron or aluminium alloy but for the larger type of engine an all-steel fabricated construction is often preferred, in which the transverse members are sometimes steel castings.

In the tunnel-type crankcase used both by Maybach and Saurer, the crankshaft is supported in roller bearings mounted on the crankshaft webs which are circular in shape. The crankcase is of cast iron or fabricated construction, and forms a tunnel-like structure surrounding the crankshaft, closed at the bottom by the oil sump. A short and stiff crankshaft can thus be incorporated in conjunction with a very rigid supporting system.

Another type of construction is used by Sulzer Bros, in which the fabricated crankcase is extended at one end to form a bed for the electric generator. The crankcase extends above the centre line of the crankshaft, and incorporates deep U-shaped bearing housings. Massive bearing caps are let into the housings and held firmly in position by the cylinder block, no studs being used.

Crankshafts are generally steel forgings, hardened and ground on the wearing surfaces, with separate balance weights bolted to the webs. A vibration damper is frequently mounted at the free end to damp out torsional vibrations. Four-, six- and eight-cylinder V-type engines are inherently unbalanced, and require the addition of secondary balancing systems, gear driven from the crankshaft.

Crankshaft and big-end bearings are usually of the steel-backed precision type, in which a thin layer of lead–copper bearing metal is backed by a steel shell. Such bearings, which do not require hand fitting, are non-adjustable and must be scrapped when worn. One of the crankshaft bearings is generally designed to locate the crankshaft endways, and is provided with thrust faces which bear against the webs of the adjacent cranks.

Cylinders up to eight in number may be arranged vertically (Plate 2, page 42) or horizontally in line, the latter type of construction being suitable for underfloor mounting in railcars. When more than this number of cylinders are required, the V-type of construction is generally adopted, the angle between the cylinder banks ranging from 45° to 90° (Plate 3, page 43).

The cylinders in the opposing banks may be staggered so that the two opposing connecting rods can work side by side on a common crank pin. This arrangement is used by English Electric, Mirrlees, Crossley, M.A.N., Daimler and Deutz. Alternatively, the cylinders in each bank may be in line with those in the opposite bank, thereby enabling the overall length of the engine to be reduced. When this is done the connecting rods are constructed on the fork and blade principle, or an articulated construction is adopted which causes the stroke of one piston to be slightly greater than the opposite one. The fork and blade construction is used by Paxman and Maybach, but most European builders employ the articulated arrangement.

By increasing the angle between the banks to 180° the horizontal twin bank engine is produced, which is suitable for underfloor mounting in high-powered railcars. The vertical twin bank engine developed by Sulzer has two parallel crankshafts driving the armature of the electric generator by means of step-up gearing so that it revolves at about $1\frac{1}{2}$ times the engine speed.

The Napier Deltic engine, originally developed for fast motor-boats, consists of three banks of opposed piston two-stroke engines, arranged in the form of an inverted triangle, with the three crankshafts located at the corners. The connecting rods are of the fork and blade pattern. A train of gears is used to couple the three crankshafts together, and drive the main generator. The gear train also provides drives for the auxiliary generator, centrifugal type scavenger blower, fuel pumps, etc. (Plates 4 and 12A, pages 44 and 70).

Another type of opposed piston engine has been built by Fiat, in which there are four banks arranged in the form of a square with the crankshaft at the corners. Each bank contains four cylinders, and the crankshafts are coupled together by gearing.

Cylinder heads containing the fuel injector, inlet and exhaust valves, are made of cast iron or aluminium alloy, and are attached to the cylinder blocks by means of studs. When single inlet and exhaust valves are used, the inertia of the valves and valve operating mechanism may be considerable, particularly at high speeds. Most makers, therefore, provide two inlet and two exhaust valves per cylinder, when the bore exceeds seven inches. Maybach provide six valves per cylinder. The valve rocker gear for each cylinder is mounted on the cylinder head (Plate 1C, page 41).

Cylinder liners of hard, close-grained cast iron, often specially treated to reduce wear, are inserted in the cylinder blocks, where they are held firmly in position by the cylinder heads. Wet type cylinder liners are in direct contact with the cooling water, and at the lower end, a sealing ring prevents the leakage of water into the crankcase. Dry type cylinder liners are press fitted into circular housings formed in the cylinder blocks (Plate 1B and D, page 41).

Pistons which are cooled by oil under pressure are frequently constructed of cast iron. In most other cases aluminium alloy, which possesses good heat conducting properties, is used, and effectively dissipates the heat generated by combustion.

The pistons are provided with three or more cast iron piston rings which retain the compression and prevent leakage. In addition, two or more rings with oil retaining grooves are provided to distribute the lubricant, and scrape the cylinder walls on the downward stroke, so as to prevent lubricating oil entering the combustion space. One of these rings may be located just below the compression rings and the other in the piston skirt.

DEVELOPMENT

The first internal combustion engine to use an injection system in which the fuel oil was forced into the combustion space under pressure from a pump, was constructed in 1890 in accordance with the patents of the English inventor Akroyd-Stuart, thus anticipating by many years the system of fuel injection which was ultimately generally applied to diesel engines. This engine was developed by the firm of Richard Hornsby & Sons of Grantham, under the name of the *Hornsby-Akroyd oil engine*, and in 1896 a small internal combustion locomotive was constructed incorporating an engine of this type.

The Hornsby-Akroyd engine employed a comparatively low compression ratio, so that the temperature of the air compressed in the combustion chamber at the end of the compression stroke was insufficient of itself to initiate combustion. In order to achieve this, combustion took place in an unjacketed combustion chamber, communicating with the cylinder through a passage, which prior to starting was heated by a blowlamp, and afterwards maintained at the required temperature by the heat generated during combustion.

The first compression ignition engine designed so that the temperature of the air compressed in the combustion space was sufficient to ignite the mixture of fuel and air, was completed in 1897 at the Augsburg Works of M.A.N. This engine was built in accordance with the patents of Dr. Rudolf Diesel, and gave a thermal efficiency exceeding 30 per cent compared with about 17 per cent for contemporary low-compression oil engines.

Air was sucked into the cylinder as the piston descended, and compressed to about 480 p.s.i. on the upward stroke. To obtain complete atomization and to ensure adequate penetration of the dense mass of air compressed in the combustion space, the fuel oil was injected through a cam-operated valve, mixed with a jet of compressed air at a pressure of about 1,000 p.s.i.

A disadvantage of this type of engine was the heavy and complicated auxiliary equipment required, consisting of a two-stage air compressor, an intercooler, storage bottles and piping. Maximum speeds also were limited to about 500 r.p.m. With the development of the light, high-speed compression ignition engine, solid injection came into general use, and a system was developed in which the fuel oil was forced into the combustion space by means of a plunger pump, through one or more minute orifices, controlled by a spring-loaded valve, which could be adjusted to give the required injection pressure.

In Great Britain, important pioneer work in connection with the development of compression ignition oil engines was undertaken by the firm of William Beardmore & Co. of Dalmuir, who constructed an engine using solid injection in 1922. The Beardmore engine employed direct injection, the fuel being sprayed through a number of fine orifices into the top of the cylinder. With this system there was very little turbulence to assist in the distribution of the fuel spray, and to obtain the necessary degree of penetration, a high injection pressure was required. It was, however, economical in fuel consumption, and the engine was easy to start from cold.

In 1928 this firm completed two engines for a large diesel-electric locomotive built by the Canadian Locomotive Co., each of which developed 1,330 B.H.P. at 800 r.p.m. with a power–weight ratio of only 20 lb./B.H.P. Much of the success achieved by the Beardmore engine was due to the work of a talented and versatile engineer, the late Alan Chorlton, a Past-President of the Institution of Mechanical Engineers. Probably Chorlton's most notable work in this connection was the invention of the *jerk fuel pump* embodying in principle the system of control now generally used, in which the quantity of fuel supplied by the pump can be varied by partial rotation of the pump plunger.

FOUR- AND TWO-STROKE CYCLES

Both the four-stroke and two-stroke cycles are employed for diesel traction engines. In the *four-stroke cycle* one power stroke occurs during two revolutions of the crankshaft. Air is sucked into the cylinder through the inlet valve and compressed to a pressure of 500–600 p.s.i. Just before the piston reaches the top of its stroke, injection of the fuel commences. The temperature of the compressed air, which is in the region of 1,000° F, is sufficiently high to cause the mixture of fuel and air to ignite accompanied by a substantial rise in pressure. The power stroke follows, during which combustion is completed and the gases expand performing useful work. During the final stroke of the cycle the exhaust valve opens, and the burnt gases are expelled (Fig. 1A).

In the *two-stroke cycle* one power stroke occurs during each revolution of the crankshaft, and the admission and exhaust phases overlap. It is therefore necessary that the air should be admitted to the cylinder under pressure in order to secure effective scavenging of the exhaust gas. A pump or blower driven from the engine is generally

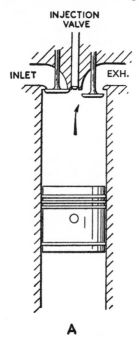

A

FIG. IA. Diagram showing four- and two-stroke cycles.

used for this purpose, supplying air at a pressure of about 2·5 p.s.i.

When the *uniflow scavenge two-stroke system* is used (Fig. IB), two or more mechanically operated exhaust valves are located in the cylinder head. Air enters the cylinder through ports in the cylinder wall uncovered by the piston at the bottom of its stroke. Compression and fuel injection take place as in the four-stroke cycle. Before the power stroke is completed, however, the exhaust valves open and the burnt gases are released at high velocity. The high exit speed of the exhaust gases induces a large volume of air to sweep through the cylinder and out through the exhaust ports, thus not only effectively scavenging the cylinder but also ensuring that the maximum quantity of air is compressed by the rising piston.

In the *loop scavenge two-stroke system* (Fig. IC), air is admitted and the burnt gases exhausted through ports in the cylinder walls controlled by the piston, and no mechanically operated valves are used. The inlet ports are inclined upwards so that the scavenge air is directed against the cone-shaped cylinder head, where it is deflected downwards into the exhaust ports.

The efficiency of the loop scavenge engine has been

improved by means of exhaust pulse pressure charging, the system used for two-stroke engines constructed by Crossley Brothers Ltd. When a cylinder is exhausting

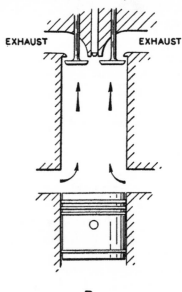

B

FIG. IB. Diagram showing four- and two-stroke cycles.

a rise in pressure occurs in the exhaust manifold, which interferes with the scavenging process in the next cylinder in sequence. If the rise in pressure is timed to occur

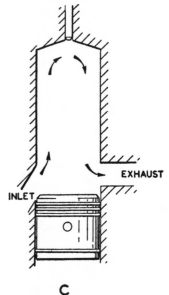

C

FIG. IC. Diagram showing four- and two-stroke cycles.

slightly later when the exhaust ports in the succeeding cylinder are closing, it is transmitted to the scavenge air which has flowed out into the exhaust manifold, causing it to be forced back into the cylinder, thus giving a certain degree of pressure charging.

In *opposed piston two-stroke engines* (Fig. 1D), where each cylinder contains two pistons coupled to separate crankshafts, the injection valve is located in the centre, and the pistons control the inlet and exhaust ports respectively. In this case also the uniflow scavenge principle is used. Effective scavenging and charging are obtained by phasing the two crankshafts so that the exhaust ports open before the inlet ports.

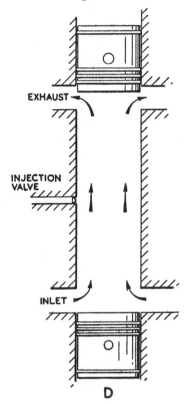

FIG. 1D. Diagram showing four- and two-stroke cycles.

INJECTION SYSTEMS

To obtain complete combustion, intimate mixing of the fuel and air is necessary. This may be achieved by imparting a high degree of turbulence to the imprisoned air and injecting the fuel into the air stream at a comparatively low pressure. Such is the system used in the *Ricardo Comet air cell type of combustion chamber* (Fig. 2A). Alternatively, *direct injection* may be used (Fig. 2B), in which the fuel is injected into the cylinder at high pressure, in the form of a spray, capable of penetrating the dense mass of compressed air to a considerable depth.

Fuel oil is supplied to the injection valve of each cylinder by the fuel pump which is timed to deliver the correct quantity of oil when required. The valve is held on its seating by means of a spring, and opens when the pressure of the fuel oil acting on the annular area formed by the enlargement of the valve stem is sufficient to over-

come the spring pressure. The valve spring can be adjusted so that the valve opens at a pre-determined pressure.

Different types of injection valve are illustrated in Fig. 3. Type A is a single-hole injector in which the valve has a conical seating with a single orifice arranged centrally. Type B is a multi-hole injector in which four or more very small orifices are used. In the pintle type shown at C, the orifice is considerably larger in diameter, and the area of opening is controlled by a cylindrical

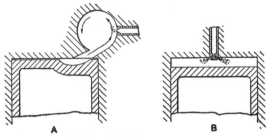

FIG. 2. Injection systems.

extension or pintle on the end of the valve. Types A and B are generally used in conjunction with high-pressure direct injection, whereas Type C is suitable for operating with a comparatively low injection pressure under conditions promoting turbulence.

The air-cell type of combustion chamber enables a rather lower grade of fuel to be used, and the injection system is subjected to less wear and tear but heat losses are greater than with direct injection, causing increased fuel

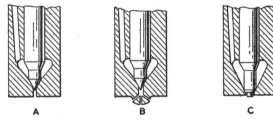

FIG. 3. Types of injection valve.

consumption. The Ricardo Comet type of combustion chamber gives exceptionally good combustion conditions at high speeds resulting in increased power output. A pintle-type injector is generally used operating at an injection pressure of 1,800 p.s.i.

With direct injection, a lower compression ratio can be used giving improved cold starting characteristics; heat losses are less and scavenging more effective. A multi-hole injector operating at about 3,000 p.s.i. is often employed.

By masking the inlet ports, a rotational swirl can be imparted to the air during the suction stroke, which is retained during compression. Specially shaped cavities formed in the piston crown also assist in promoting turbulence. The toroidal cavity direct injection system developed by Saurer of Arbon, Switzerland, incorporates these features, and in Great Britain is used by A.E.C. Ltd, of Southall, Middlesex.

The system of direct injection is well adapted for using the highly refined grades of fuel oil complying with British Standard Specification class A which have been developed for use in diesel engines. Lower grades of fuel oil can be used successfully in large-cylinder, slow-running diesel engines with direct injection, such as those manufactured by Sulzer Bros., and in engines with air-cell type combustion chambers.

LUBRICATION AND COOLING

The majority of rail traction engines are lubricated on the *wet sump principle*, in which the bottom of the crankcase forms an oil reservoir. The oil is circulated under pressure by means of a pump, and in addition to lubricating the working parts, also helps to carry away excess heat from the areas exposed to high temperatures.

A gear-type pump draws oil through a suction filter situated in the sump and delivers it at a pressure ranging from 40 to 70 p.s.i. to the crankshaft bearings, from where it flows through passages in the shaft to the big-end bearings. In many cases it is then conveyed to the gudgeon pins through passages passing up the connecting rods, and may also be used to cool the pistons, when it is directed in the form of a spray against the inside of the crown. The cylinder walls are lubricated by splash, and oil at a reduced pressure of about 10 p.s.i. is used to lubricate the camshaft bearings, rockers, etc.

Occasionally *dry sump lubrication* is used, when the bulk of the lubricating oil is contained in a tank separate from the engine. It is pumped through the engine in the normal manner, and after draining into the sump is collected by a suction pump and returned to the tank.

During its passage through the engine the oil becomes contaminated with minute particles of carbon and other impurities. Efficient filtration, therefore, is of vital importance, and in addition to the gauze suction strainer, external filters of various special types are provided in which the filtering elements are removable for cleaning or replacement. The lubricating oil is cooled by passing it through a section of the radiator reserved for this purpose. When starting from cold the system is primed by means of a hand-operated or electrically-driven pump.

Water is circulated through the water jackets surrounding each cylinder, and through passages formed in the cylinder heads round the valves and injectors. The circulation is maintained by means of a centrifugal pump driven from the engine which also pumps the water through the radiator for cooling.

For successful working of the engine, the cooling water must be maintained at the correct temperature. Not only must overheating be prevented, but excessive wear of the cylinder bores will occur when starting up, if the temperature of the cooling water remains too low. In order to counteract this, a thermostatically controlled by-pass valve is provided, which cuts the radiator out of the circuit until the water has attained a certain temperature. Radiator shutters, which, when closed, prevent the flow of air, are sometimes provided for the same purpose.

It is an advantage if the cooling capacity of the radiator can be varied to meet the requirements of the engine and the best method of achieving this is by varying the speed of the fan. When the fan is driven by an electric motor, the speed of the motor can be varied automatically by means of a thermostat. A method has also been developed which enables this to be done when the fan is driven from the engine. A fluid coupling is incorporated in the drive, the filling of which, and hence the amount of slip, can be varied. The filling pump is thermostatically controlled.

For power outputs up to about 250 B.H.P. a method of cooling the engine by means of a current of air induced by a fan is sometimes used.

POWER OUTPUT AND SPEED CONTROL

The *power output* of the engine is controlled by varying the amount of fuel oil delivered by the *injection pumps*. Sometimes the pump stroke is made variable, but in the jerk pump system which is most generally used, the stroke remains constant. The opening and closing of the suction ports through which the oil enters the pump barrel is controlled by the pump plunger, and the timing is such that delivery occurs when the plunger is near the middle of its stroke, and moving at maximum velocity. Before the stroke is completed, a vertical channel communicating with a helical groove cut in the periphery of the plunger, establishes communication between the suction and delivery sides. This causes a sudden collapse in pressure, and the injection valve closes smartly, thereby preventing the oil from dribbling. This action is further assisted by a cylindrical extension forming part of the delivery valve which increases the volume of the delivery pipe just before closure occurs. The point at which communication between suction and delivery is established, is termed the *spill point*, and can be varied by partially rotating the plunger. Each plunger is provided with a pinion mounted on a sleeve, which meshes with the control rack. The pumps are usually grouped together in units of six or eight, and driven from the engine camshaft.

The control rack is actuated by a *centrifugal governor*, generally through the medium of a *servo-mechanism*. In some types, lubricating oil under pressure is used as the operating medium, thereby safeguarding the engine in the event of a failure of the lubricating system. A governor of this type (Fig. 4) is used for engines constructed by Davey, Paxman & Co. Ltd. It is mounted at one end of the camshaft and driven by bevel gearing. When the governor weights move outwards under the action of centrifugal force, the pivoted levers to which they are attached move upwards, actuating a pilot valve which controls the admission and release of lubricating oil under pressure, to a cylinder containing a piston which operates the fuel pump control rack.

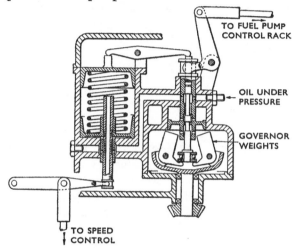

FIG. 4. Diagram showing principle of servo-governor.

The pilot valve is kept closed by the pressure exerted by a nest of coil springs, termed the *speeder springs*, on one end of a pivoted lever, the other end of which bears on the end of the pilot valve. By varying the compression of the springs the governor can be set to operate at different speeds ranging from idling up to maximum. This is done by means of a similar type of servo mechanism to that already described. The springs are seated on a piston encased in a cylinder, and movement of the piston is effected by lubricating oil under pressure. In this case, however, the pilot valve controlling the admission and release of oil is operated by an external control mechanism. To prevent the governor responding too rapidly, a delayed action spring and dashpot are included, which limit the acceleration of the engine from idling up to maximum, to a period of 5–6 seconds.

This type of governor, which is extremely sensitive, is well suited for use in conjunction with load control applied to diesel-electric working, and can also be used for remote control with multiple unit operation.

Control of the engine output by setting the governor to operate at a number of fixed speeds is the system generally applied to diesel traction engines, and for any set speed the torque developed by the engine adjusts itself automatically to suit the load. This is the *speed control system*.

The smaller type of engine used for railcars and locomotives sometimes incorporates a different method of control. The governor regulates the maximum and minimum speeds only. The driver controls the power output between these limits by adjusting the fuel pump control rack so that the fuel pumps deliver a larger or smaller quantity of fuel to the cylinders. Movement of the control rack may be effected manually by means of a suitable linkage or by a compressed air motor. With this system, therefore, the engine may be set to develop a certain torque, and the speed automatically adjusts itself to suit the load. This is known as the *torque control system*.

PRESSURE CHARGING AND INTERCOOLING

If insufficient air is drawn into the cylinder to ensure complete combustion of the fuel, black smoke will be emitted from the exhaust, indicating that the engine is loaded beyond its economic limit. By means of *pressure charging*, the density of the air is increased, more air can be delivered, and a greater quantity of fuel burned, with a resultant increase in power output.

Pressure charging is widely employed for railway traction engines working on the four-stroke cycle. The system generally used was invented by Dr. Buchi, and consists of a turbo-blower unit driven by exhaust gas from the engine. By this means the power output may be increased by 50 per cent or more.

A single-stage turbine and centrifugal blower are mounted on the same shaft, which is carried in bearings housed in the surrounding casing. The blower portion of the casing is insulated against the heat of the exhaust gas, and the turbine portion incorporates a water jacket connected to the engine cooling system. Engines with twelve or more cylinders are frequently provided with two pressure-charging units. Air is delivered to the cylinders at a pressure of from 4 to 6 p.s.i. For the smaller type of engine mechanically driven blowers of the Roots type are used.

By cooling the air delivered by the blower its density can be still further increased, and an even greater increase in power output obtained. *Intercooling*, as this process is termed, is applied to rail traction engines where exceptionally high performance characteristics are required, and enables the power to be increased by 80 per cent above that of the naturally aspirated engine. The engine cooling water is generally used as the cooling medium for the intercooler.

STARTING EQUIPMENT AND AUXILIARIES

When electric transmission is used, it is generally more convenient due to exigencies of the design, to start the engine by *motoring the main generator* with current from a storage battery, special starting windings being provided for this purpose. In other cases, either one or two *electric starter motors* may be employed, driving a toothed ring fixed to the flywheel by means of a Bendix pinion, which disengages automatically when the engine commences to fire. Occasionally compressed air starting is used.

The first system, where the generator functioning as a motor drives the engine direct, has the disadvantage that a heavy and expensive storage battery is required. With the second system a comparatively small storage battery can be used, since the torque developed by the starter motor is multiplied many times by the gearing. Current for the starter motor and other purposes is supplied by a small generator driven by the engine, usually at a pressure of twenty-four volts.

Compressed air starter motors require air at a pressure of about 450 p.s.i. supplied by a two-stage air compressor. This may be driven from the main engine or separately driven by a small auxiliary engine.

Many different types of *air filters* are available for the important task of filtering the air required for combustion. The filter may be mounted on the engine air inlet or be of panel form built into the engine housing.

Devices to safeguard the engine may include electrically operated alarm switches which shut down the engine if the lubricating oil pressure becomes too low, or the temperature of the cooling water becomes excessively high, and an engine shut-down device which functions in the event of overspeeding.

A *fuel lift pump* driven from the engine camshaft is generally used to raise the oil from the fuel tank, which in many cases is mounted beneath the underframe. The fuel oil is carefully filtered before it enters in the fuel pumps, and often de-aerated to prevent air locks forming in the fuel pipe.

TABULATED PARTICULARS

The table on the opposite page gives particulars of typical diesel traction engines used for locomotives and railcars. The examples given are representative in many cases of a standard range which may include engines with four up to sixteen or twenty cylinders. The outputs given are the maximum continuous outputs, but in some applications the engines are rated to deliver less than this.

High-powered two-stroke engines are represented by the Crossley loop scavenge engine with exhaust pulse pressure charging, and the Deltic opposed piston engine. Deutz also manufacture a range of loop scavenge two-stroke engines, some of which are equipped with exhaust

gas pressure charging in addition to scavenge blowers driven from the engine.

Air-cell type combustion chambers are used by a large number of European manufacturers, including Werkspoor who use the Ricardo Comet head in conjunction with pressure charging and intercooling. This type of combustion chamber is also used by Paxman for their naturally aspirated models, but for the pressure charged engines, direct injection is used.

TORQUE AND POWER CURVES

Brake horsepower is measured at the engine output flange, and is less than the indicated or cylinder horsepower by the amount of power required to overcome internal friction and drive the injection pumps, water circulating pumps, etc. It is given by the formula:

$$\text{B.H.P.} = \frac{apln}{33,000} \times (\text{number of cylinders}) \text{ where}$$

a = area of piston in sq. in.,
p = brake mean effective pressure in p.s.i.,
l = stroke of piston in ft,
n = number of working cycles per minute.

The *mean effective pressure* is the mean or average pressure acting on the piston during one cycle of operations. The brake mean effective pressure, generally expressed as B.M.E.P., will be less than this, and forms a convenient basis of comparison for different engines,

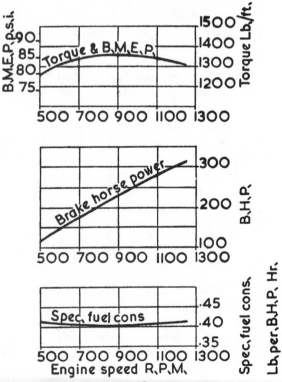

FIG. 5. Engine performance curves.

DIESEL ENGINES

Maker	Model	Cylinders No. Arrgt.	Bore × Stroke in.	Stroke per cycle	Type of injection	System of Press charging	Max. continuous output B.H.P.	B.M.E.P.	Max. speed r.p.m.	Piston speed ft per min.	Dry weight lb. per B.H.P.
LOCOMOTIVES											
Sulzer	12 LDA 28	12 twin bank	11×14·3	Four	Direct	Exhaust gas	2,300	150	750	1,760	20
English Electric	16 SVT	16 Vee	10×12	,,	,,	,,	2,000	124	850	1,700	20·2
Paxman	16 YLX	,,	9·75×10·5	,,	,,	,,	2,000	126	1,000	1,750	18·3
M.A.N.	VV 22/30	,,	8·6×11·8	,,	Air cell	Exhaust gas & intercooler	2,080	147	1,000	1,965	9·8
Maybach	MD 870	,,	7·28×7·87	,,	,,	,,	1,800	170	1,500	1,970	7
Werkspoor	RUHB 215	,,	8·5×10·3	,,	Direct	Scavenge blower	1,728	150	1,000	1,720	14·5
Napier	Deltic	18 three bank	5·125×14·5	Two	,,	,,	1,650	80·9	1,500	1,815	5·5
Crossley	HSTV 8	8 Vee	10·5×13·5	Four	Air cell	Exhaust gas & intercooler	1,200	81	625	1,406	25·5
Daimler	MB 820 Db	12 ,,	6·9×8	,,	,,	,,	1,250	146	1,500	2,020	5·4
M.G.O.	V16 SHR	16 ,,	6·89×7·32	,,	Direct	Exhaust gas	1,130	137	1,500	1,830	11·75
Fiat	288 E.S.	8 vertical	11×14·2	,,	,,	Exhaust gas & scavenge blower	1,050	108	700	1,655	27·5
Deutz	T12 M 625	12 Vee	7·88×8·07	Two	,,		1,050	95	750	1,225	13
Paxman	12 RPHL	6 vert.	7×7·75	Four	Air cell	None	562	83	1,500	1,935	17·75
English Electric	6 KT	,,	10×12	,,	Direct	,,	400	82·5	680	1,360	48·5
RAILCARS											
Maybach	MD 650	12 Vee	7·28×7·87	Four	Air cell	Exhaust gas	1,200	159	1,500	1,970	8·25
M.A.N.	L12V17.521A	16 ,,	6·9×8·25	,,	,,	None	950	135	1,500	2,065	8·25
Ganz	XVI JV	12 ,,	6·7×9·5	,,	,,	Exhaust gas intercooled	600	81	1,100	1,730	18
Renault	577	,,	5·5×7·08	,,	Direct	None	500	132	1,500	1,680	12
O.M. Saurer	SBD	12 hor. 12 opp.	6·3×7·9	,,	,,	,,	480	92	1,400	1,840	15
Breda	D 195	6 horiz.	5·3×7·45	,,	Air cell	,,	460	122	1,500	1,875	15·75
Paxman	6 ZHX	,,	7×7·75	,,	,,	,,	450	132	1,500	1,935	13
Rolls-Royce	C 8 NFH	8 ,,	5·125×6	,,	Direct	Blower	238		1,800	1,800	12·5
Fiat	700	6 ,,	5·9×7·48	,,	,,	None	210	87	1,550	1,935	13·25
B.U.T.	11.3 H	,,	5·5×6·5	,,	,,	,,	150	95	1,800	1,675	11·5
Bussing	U 9	,,	4·5×5·5	,,	Air cell	,,	110	91	1,800	1,655	15·5
M.A.N.	D1246 MTU	,,	4·4×5·5	,,	Direct	Exhaust gas	150	118	2,000	1,835	

being easily derived from the above formula when the brake horsepower is known. The B.H.P. is measured when the engine is on the test bed, and is much more easy to obtain than the indicated horsepower.

Torque is measured in pounds inches or pounds feet, and is the turning moment exerted by the crankshaft. It is given by the formula:

$$\text{Torque in pounds feet} = 63{,}025 \times \frac{\text{B.H.P.}}{N \times 12} \text{ where}$$

N = revolutions per minute of crankshaft.

Typical performance curves, with the fuel pumps set for the engine to develop its maximum continuous output, are shown in Fig. 5. The torque remains approximately constant throughout the speed range, and since the engine is not self starting, it is incapable of developing any torque until firing commences. The minimum speed

at which the engine will operate continuously is controlled by the governor and is termed the idling speed.

The torque curve is also the curve for B.M.E.P. which may be indicated by means of a suitable scale. If the drive to the wheels is direct or through mechanical gears, it may also be used to indicate the tractive effort. The B.H.P. increases in almost exact proportion to the speed.

The power output of a diesel engine will be affected by the pressure of the atmosphere, and must be adjusted for altitude and temperature as follows:

The power output must be reduced by 4 per cent for every increase of 1,000 ft above datum, which is taken as 500 ft above sea level.

The power output must be reduced by 2 per cent for every 10° F. above 85° F. ambient or engine room temperature. The power output is also reduced when much moisture is present in the atmosphere.

Part II. Transmissions

AUTOMATIC CONTROL

Automatic control refers to those systems in which a single control device such as a handwheel or lever is used to regulate the speed and power output. This feature is inherent in the various systems of automatic load control applied to diesel-electric locomotives and railcars, which by simplifying the task of the driver, gives a greater measure of security, and enables him to give his practically undivided attention to the road.

A similar principle is embodied in the *hydraulic torque converter*, in which the rail speed and power output are regulated by varying the speed of the engine. In this case the problem is often complicated by the need to incorporate more than one hydraulic circuit, or complementary change speed gearing. Nevertheless, the majority of hydraulic transmissions in use incorporate a device by means of which the hydraulic circuits can be filled or emptied, and the gear ratios changed automatically in accordance with changes in rail speed.

The simplest type of automatic control consists of a centrifugal governor, or other device sensitive to changes in speed, driven from the output side of the transmission, which operates an appropriate control device by means of oil under pressure. A device of this type is used for the *Voith L. 36 hydraulic transmission* comprising one torque converter and two fluid couplings operating in sequence.

When one torque converter operating through several sets of change speed gearing, or two or three torque converters operating in sequence are used, it is found necessary to introduce an additional source of control, namely the engine speed. The reason for this is that when

these types of transmission operate at less than full load, the changeover points between different circuits or different gear ratios must occur at lower rail speeds if the efficiency is to be maintained. Control devices are used, therefore, which are sensitive to both changes in rail speed and changes in engine speed.

Automatic gear changing is sometimes used with mechanical transmission, and can be employed in conjunction with the Wilson gearbox when electro-pneumatic valves are used for controlling the flow of air to the compressed-air cylinders which operate the gear-changing mechanism. A small generator, designed so that the voltage increases with the speed, is driven from the output side of the gearbox. The electro-pneumatic valves are operated by voltage-sensitive relays fed from the generator, which operate in sequence as the voltage rises. A device is also incorporated by means of which the speed at which gear changes occur is modified in accordance with the speed of the engine.

In another system, control is effected by a small pump driven from the output end of the transmission, which supplies oil at varying pressures dependent on the speed, to a control device. This regulates the flow of oil under pressure to the appropriate clutch or other gear-changing mechanism. Hunting at the change-over point is avoided by making upward changes occur at higher speeds than downward changes.

FACTORS GOVERNING CHOICE OF ENGINE AND TRANSMISSION

In view of the high state of development which has

been attained, it would be difficult to assign special merits to any particular type of engine or system of transmission used for diesel locomotives or railcars. It is possible, however, to indicate some of the factors which influence selection.

When maximum power output is required, and weight saving is of paramount importance, hydraulic transmission used in conjunction with a high-speed lightweight engine is probably the most suitable choice. A combination of this kind will also occupy comparatively little space. Favourable results can also be achieved with electric transmission when a similar type of engine is used, but in those cases where axle loads are restricted, and the heavier type of slow-speed engine is preferred, the employment of hydraulic transmission may make it possible to use an engine of this type.

With regard to performance characteristics, there is, generally speaking, little to choose between either electric or hydraulic transmission, and the efficiencies when operating at full or part load differ only slightly. Capital costs also would appear to be about the same. It will be apparent, therefore, that apart from those cases where the power–weight ratio is the decisive factor, the merits of the two types of transmission are roughly equal, and in the choice of a particular system other considerations may arise.

In some countries, for example, the existence of a highly developed electrical industry together with extensive railway electrification, may tip the balance in favour of electric transmission. Allowance must also be made for personal preference with regard to the mechanical design of the locomotive or railcar in which the engine and transmission are to be installed. Hydraulic locomotives mounted on bogies (trucks), require massive cardan shafts with final reduction and reverse gearboxes to transmit the drive to the axles, thus making it difficult or impossible to incorporate an orthodox type of bogie. When electric transmission is used, it may be possible for the same motor bogie to be used for both straight-electric and diesel-electric stock.

Mechanical transmission is used largely for locomotives in the lower powered range developing under 300 H.P. Although lacking the flexibility of electric or hydraulic transmission and requiring somewhat greater driving skill, it is both cheap in first cost and economical in operation. It has also found a wide field of application for railcars and diesel trains.

IDEAL PERFORMANCE AND TRANSMISSION EFFICIENCY

If there were no transmission or other losses, the horsepower measured at the wheel tread would be equal to that developed by the engine. Assuming the horsepower to remain constant from zero to maximum rail speed, the curve of tractive effort will be a hyperbola as shown in Fig. 6. The diesel engine is capable of delivering a constant power output independent of the rail speed. How closely the curve of tractive effort actually obtained approaches the ideal curve is a measure of the efficiency and effectiveness of the type of transmission employed.

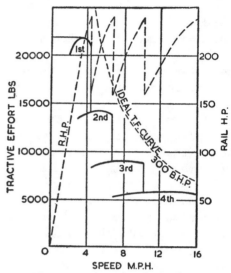

FIG. 6. Performance curves (mechanical transmission).

Part of the engine power output is absorbed in driving the radiator cooling fan, and other auxiliaries such as the auxiliary generator, air compressors, vacuum exhausters, traction motor blowers, etc. The amount of power required for this purpose ranges from about 5 per cent of the total engine output in the case of small diesel mechanical locomotives, not equipped for hauling continuously braked trains, up to 7 or 8 per cent for main line diesel-hydraulic or diesel-electric locomotives.

Transmission losses between the engine and rails are least with mechanical transmission when the engine speed and rail speed are correctly related to each other, and there is no appreciable slip in the clutch or fluid coupling. Under these conditions they may amount to 8 per cent or even less. The power utilization factor, however, is comparatively poor, since the full engine output can only be used at a limited number of points in the speed range, depending on the number of gear ratios available.

With electric transmission the losses in the traction motors, gears, main generator and leads, range between 15 and 20 per cent, but are somewhat greater at low power outputs. The losses encountered with hydraulic transmission in the torque converter and mechanical reduction gearing are in the region of 20 per cent, and are also greater at low power outputs.

The power utilization factor for both electric and hydraulic transmission varies considerably, but is usually better than that obtained with mechanical transmission. Although the peak values are less, a high proportion of the engine output is available over a wide range of rail speed.

With all types of transmission, transmission losses in the form of heat are considerably greater when operating in the lower part of the speed range, and under these conditions the sustained power output is limited by the capacity of the equipment to dissipate the heat generated. There is, therefore, a minimum speed at which the maximum tractive effort developed at that speed can be maintained continuously. This is termed the continuous rating.

MULTIPLE-UNIT OPERATION

Multiple-unit operation is intimately linked with the economical aspects of diesel traction, and enables a high degree of operational flexibility to be achieved. When applied to locomotives the maximum use can be made of the available motive power, and train loadings can be adapted to meet traffic requirements in the most effective manner. It also makes possible self-propelled diesel trains which are being increasingly used to replace locomotive-hauled trains for certain classes of work.

A simple type of multiple-unit control can be applied to small diesel-mechanical locomotives, when compressed air is used for gear changing and reversing. The compressed air pipes leading from the driver's application valve to the gearbox, are extended to both ends of the locomotive, where they can be coupled by means of flexible hoses to corresponding pipes on a second locomotive, equipped in a similar manner. In addition to the gearbox controls, the speed control mechanism of the engines must also be adapted for compressed air operation.

To avoid the multiplicity of compressed air pipes which this system would entail when applied to more complex layouts, *electro-pneumatic control* has been developed, which is widely used for all classes of multiple-unit working. In diesel trains with mechanical transmission, the operations of gear changing and reversing are effected by means of electro-pneumatic valves which control the admission and release of compressed air used in the operating cylinders. The driver's controls consist of switching devices, whereby the electro-pneumatic valves on each power car, performing a particular operation, can be energized instantaneously by means of wires passing throughout the train.

When the engine governor is of the servo type which can be set to operate at different engine speeds, electro-pneumatic control can be used as described under Load Control – Variable Speed (page 50). In other cases where the engine output is controlled by the fuel pump rack, and the governor regulates maximum and minimum speeds only, an *air-operated throttle motor* may be used. This can be of the type developed by A.E.C. for railcars in which several different torque settings are obtained by means of a series of interlocking pistons, each having a fixed travel. Admission and release of compressed air to the throttle motor is controlled electro-pneumatically.

Another system has been developed by Westinghouse in which the speed or torque of any number of engines can be controlled by varying the air pressure in a pipe which runs throughout the train. Each engine is provided with a control cylinder, communicating with the pipe, which actuates the engine control. Simultaneous operation is secured by admitting or releasing air at several different points by means of electro-pneumatic valves, which establish communication with an adjacent air reservoir or with the atmosphere. These in turn are controlled by an air-operated contact unit under the control of the driver.

Electric current for operating the electro-pneumatic valves and engine starter motors, is obtained from storage batteries, usually at a pressure of twenty-four volts. In long trains the voltage drop may be considerable, and the voltage available at the rear of the train may be insufficient either to start the engines or operate the reversing gear. This difficulty has been overcome by adopting a system in which the electrical equipment on each power car in the train is supplied with current from its own storage battery, and controlled through relays. These are operated by a comparatively small current fed through the train wires from the battery on the leading car.

An important aspect of multiple-unit control is the *supervision and safeguarding of the equipment*. With multiple-unit trains including as many as twelve separate power units, this problem becomes extremely intricate. Each engine is provided with an indicator lamp which goes out if the engine stops for any reason. The lamps are arranged on a panel in the driving compartment to show the location of the engine in the train.

Much damage can be caused if the train starts before all the direction change gears are properly engaged, and it is most desirable to give the driver a positive indication that this has taken place. To satisfy this requirement, a device has been developed consisting of two micro-switches and two air pressure switches for each reversing unit. The micro-switches are actuated by the direction change pistons, one of which is closed when the full travel of the piston has been made to engage forward or reverse. The air pressure switches close when the minimum operating air pressure has been attained. When the micro-switch and air pressure switch for either forward or reverse are closed, an indicator lamp in the driving compartment is illuminated.

To operate the control and protection equipment, up to seventy-six separate wires, running throughout the train, may be required, with jumper plug connections between the cars. Two to four jumper plugs may be used, each one making twenty to thirty separate connections. In Germany and other European countries, vehicles intended to operate in multiple unit are equipped with the Scharfenberg automatic coupler, which not only couples the vehicles together, but also makes all the necessary electrical connections.

The engine speed or output control, gear change and direction controls are conveniently arranged on an operating desk in the driver's compartment. Frequently the gear change and direction controls are mechanically interlocked to prevent faulty operation. An isolating switch with a removable key is often provided, so that the controls can only be operated when the isolating switch is closed.

Similar type of control gear is provided for diesel hydraulic locomotives and railcars adapted for multiple-unit working, but simplified as necessary.

When electric transmission is used, the control wires which connect the master controller with the various contactors and electro-pneumatic valves must be extended to both ends of the locomotive or throughout the train in the case of multiple unit train sets. Ten or more separate wires may be required for this purpose, and these are enclosed in a conduit with jumper plug connections at each end of the vehicle. Additional control lines are also required for operating electro-pneumatic valves included in the brake equipment, together with electro-pneumatic sanding valves and dead-man's equipment. Control lines are also required for the warning devices and safeguards.

The majority of diesel locomotives and railcars are provided with *dead man's equipment*, which in normal circumstances is prevented from functioning by the downward pressure exerted by the driver on the control handle, or on a treadle conveniently placed for the feet. If, due to illness, the pressure is removed, a valve opens and allows air to escape through a choke from a small timing reservoir. When the pressure in the timing reservoir has fallen to a predetermined figure, the emergency valve opens and causes the brakes to be applied. At the same time a pneumatically operated switch shuts off the power supply. The purpose of the timing reservoir is to allow for a delay period of six seconds, thus permitting the driver to move across the cab. When mechanical or hydraulic transmission is used the speed of the engine is reduced to idling. When locomotives are required to haul unbraked or partially braked trains an additional feature is provided which causes a light brake application to be made in order to bunch the train before the full brake power is applied.

Part III. Transmissions: electric

BASIC PRINCIPLES

The torque developed by a D.C. series motor has a maximum value at starting, and is approximately proportional to the current input. If the power supply remains constant, the voltage–current characteristic for the speed range through which the motor operates will be a hyperbola, resembling the speed–tractive effort curve for constant power. In order to maintain the engine output constant, the generator must have a characteristic of this form when operating at constant speed.

Electric generators intended for traction purposes are designed so that the voltage varies inversely as the current when operating at constant speed. This may be achieved by means of differential compounding in which the series field acts in opposition to the shunt. Fig. 7 shows the characteristic obtained from a generator of this type designed for 375 H.P. input.

If the hyperbola representing constant input H.P. is superimposed on the appropriate generator characteristic, it will intersect at points A and B. At starting when the current input to the traction motors is high, the engine can develop more power than the generator can absorb, so that until point A is reached, the engine develops less than its full output. Between points A and B, the engine is incapable of developing sufficient power to meet the

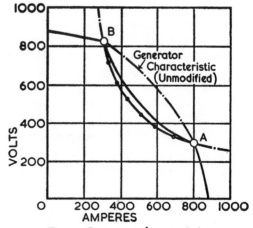

FIG. 7. Generator characteristic.

demand of the generator. Under these conditions the overload imposed on the engine would cause its speed to fall, and to prevent stalling, the generator characteristic must be modified so that it conforms approximately to the hyperbola for constant power. This process is termed *load control*.

The tractive effort curve is shown by full lines in Fig. 8 where the points corresponding to points A and B on the generator characteristic are indicated. The point B where the tractive effort curve commences to fall sharply is termed the *generator unloading point*, because from this point until the maximum rail speed is attained, the load on the engine is progressively diminished.

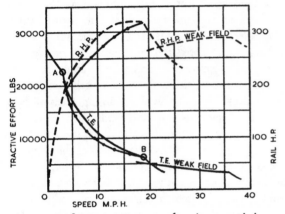

FIG. 8. Performance curves – electric transmission.

Two classes of load control are applied to diesel-electric traction generators. The first class includes all those systems in which the modified generator characteristic follows the hyperbola for constant power between points A and B (Fig. 7 full line), the engine speed remaining constant. This class of load control is applied to locomotives and railcars intended for main line service, where it is essential that the maximum power output of the engine should be available over most of the speed range.

In the second class, a balance is obtained between the power developed by the engine and that absorbed by the generator with a reduction in engine speed. The modified generator characteristic is indicated in Fig. 7 as a full line with black dots. The tractive effort and rail H.P. curves will be correspondingly modified (Fig. 8). When this class of load control is used, the maximum engine output is available at points A and B only. It is, however, less complicated and expensive than other types, and applied mainly to shunting locomotives where simplicity is an important factor.

Assuming that no alteration is made to the engine output, there are two ways in which the tractive effort and rail horse-power characteristics can be modified.

Firstly, by altering the gear ratio of the traction motors. If the gear ratio is reduced, the maximum tractive effort will also be reduced, but the speed range will be extended, with a corresponding improvement in performance at the higher speeds. Secondly, by the application of field weakening to the traction motors. This has the effect of increasing the torque developed by the motors at low current inputs. When this method is used the useful operating range can be increased, and there is no sacrifice of performance at starting or when operating at low speeds, but some increase in cost is entailed.

DEVELOPMENT

During the course of World War I, electric transmission, which had already been used for railcars (*see* Diesel Railcars and Diesel Trains – Development), was applied to a number of small petrol locomotives built by Dick Kerr Ltd, now part of the English Electric Company, for the 60-cm. gauge light railways serving the western front. These were four-wheeled machines, each weighing seven tons and provided with two axle-hung traction motors; the petrol engines developed 45 B.H.P.

The success achieved by the Swedish diesel-electric railcars led to the construction in 1921 of a Bo–Bo type diesel-electric locomotive for the metre gauge, provided with an engine developing 120 B.H.P. This was afterwards sold to the Tunisian Railways. A year later the first Russian diesel-electric locomotive was completed at the Esslingen Machinenfabric. This was a rigid frame, I–Eo–I type locomotive built to suit the five-ft gauge, and weighing $122\frac{1}{2}$ tons, A.M.A.N. diesel engine with air injection was used developing 1,200 B.H.P. at 450 r.p.m.

In these early machines *load control* of the electric generator was carried out by the driver, who had to exert considerable care to avoid stalling the engine. The introduction of automatic load control based on the work of Hermann Lemp in America, together with the application of solid injection to diesel engines, gave a considerable impetus to the development of diesel-electric locomotives and railcars. William Beardmore & Co., in Britain patented a system of servo field control in 1927. About two years later Armstrong Whitworth & Co. introduced a system in which the rheostat controlling the strength of the generator field was operated by an oil pressure motor; the motor being supplied with oil under pressure from the engine lubricating system, through a valve controlled by the engine governor.

Among the earliest successful diesel-electric locomotives to be placed in service were those supplied in 1931 to the Thailand State Railways, and constructed by the Danish firm of Frichs. They were of two types, a rigid frame, 2–Do–2 passenger locomotive, and an articulated

2–Do+Do–2 freight locomotive; both were provided with twin engines and generators. The gauge was one metre. The passenger locomotives had a maximum tractive effort of 32,000 lb., and weighed 85 tons. The combined engine output was 1,000 B.H.P. at 600 r.p.m. The freight locomotive had a maximum tractive effort of 63,000 lb. and weighed 126 tons with a combined engine output of 1,600 B.H.P. at 600 r.p.m. Frichs also constructed a number of diesel-electric railcars.

Forced ventilation was first applied mainly to the traction motors of freight and shunting locomotives to prevent overheating when operating at low speeds. The first Armstrong Whitworth diesel-electric shunter (switcher) constructed in 1932 was provided with a single force-ventilated, frame-mounted traction motor, with double reduction gearing and rod drive to the wheels. Later, however, forced ventilation was applied to most types of locomotive in order to keep the size and weight of the traction motors to a minimum. The accompanying table giving particulars of representative locomotives, shows a steady increase in the power developed per ton of weight.

Railway	Year built	Total wt. (tons)	B.H.P./r.p.m.	B.H.P. per ton
U.S.S.R.	1922	122½	1,200/450	9·8
Thailand	1931	126	1,600/600	12·7
L.M.S.	1947	121	1,760/750	14·5
British Rlwys (Deltic)	1955	106	3,300/1,500	31·1
S.N.C.F.	1956	110	1,800/1,500	16·4

CONTROL EQUIPMENT

The control equipment required for electric transmission comprises the traction motor contactors, and reversers, carrying the full load current, together with various contactors, rheostats and switches incorporated in the control and auxiliary circuits (Plates 5A and 5B, page 45).

The *auxiliary generator* is provided with a voltage regulator which maintains the voltage constant independent of the speed. It is used for charging the battery, supplying power to the electrically driven auxiliaries, for lighting, and sometimes for supplying current to the separately excited field of the main generator. The voltage used varies with different makers, ranging from 90 to 110 volts.

The *battery* is required for starting the engine by motoring the generator, and for supplying power to the auxiliaries when the engine is either stopped or rotating at idling speed. The battery charging contactor is controlled by a reverse current relay, and charging commences as soon as the auxiliary generator is generating sufficient voltage for this purpose. Provision is often made for starting the engine by towing the locomotive if the battery is inoperative. This is effected by altering the connections of one of the traction motors, so that it functions as a generator, and generates current for motoring the main generator.

When field weakening of the traction motors is incorporated, the motor field shunting contactors close automatically at the correct speed, through the medium of a potential relay.

Blowers for the forced ventilation of the traction motors may be electrically driven, or belt-driven from the main generator. The former method has the advantage of allowing the blower to be located in close proximity to the traction motors, which it is required to supply, thereby keeping the air ducts short and direct.

Contactors are of the electro-pneumatic or electro-magnetic type, and reversers may be electro-pneumatically or manually operated. It is generally found convenient to mount the remotely controlled apparatus in a frame housed in a dust proof compartment, the complete assembly being removable as a unit. Directly operated switch gear, together with various instruments and gauges, can be conveniently located in, or on, a control desk in the driver's compartment.

The *driver's controls* consist of a handle operating the power controller, which operates the traction motor contactors, regulates the engine speed, and varies the generator excitation when this is applicable, together with a handle operating the reversing switch, and a master switch. The latter, which is sometimes operated by the same handle which controls the reversing switch, is used for starting and stopping the engine, and also enables the engine to be run at any desired speed by means of the power controller, when the vehicle is stationary. The handles are mechanically interlocked to prevent incorrect operation, and the reversing switch can only be operated when the power supply to the motors is cut off (Plates 6A, 6B and 7A, pages 46 and 47).

GENERATORS

In small diesel-electric locomotives, the *engine and generator are often independently mounted*, and the drive from the engine is transmitted through a flexible coupling or short propeller shaft. When this arrangement is used the engine does not require to be specially adapted to accommodate the generator, and may be started by means of an electric starter motor driving a toothed ring fixed to the engine flywheel.

Generally, however, the *engine and generator are mounted on a common foundation*, and the armature which also serves as the flywheel, is carried on what is virtually an extension of the engine crankshaft. It is supported partly by the crankshaft bearings and partly by a single bearing located in the generator casing. Generators of this type are known as *single bearing generators* (Plates 2 and 3, pages 42 and 43).

The smaller type of single bearing generator may be overhung from the engine crankcase, to which it is attached by means of a spigoted flange. When the generator is too heavy for this method of attachment to be used, it may be supported by feet resting on an extension of the engine bed, the spigoted connection being generally retained in order to preserve the alignment. In some cases, the generator casing itself, which is rigidly attached to the engine crankcase, forms part of the engine bed. In other cases, the crankshaft extension forms the only direct connection, and the engine together with the generator are mounted on a fabricated steel underframe.

In addition to the main generator, an *exciter and auxiliary generator* are usually required, but sometimes the functions of both are combined in one machine. The exciter and auxiliary generator may be mounted on top of the main generator and driven by means of belts. This arrangement permits them to be driven at a higher speed than the engine, thus enabling smaller and lighter machines to be used. Alternatively, the auxiliary generator may be overhung from the main generator, and sometimes forms an integral part, being recessed into it to reduce the overall length.

Generators are always *self-ventilated*, and since they rotate at comparatively high speed when the maximum current is being generated, this method is effective. The fan is carried on the armature shaft at the driving end, and draws air over the commutator, armature and field coils, and through ducts in the core before being discharged to the atmosphere.

The mechanical construction of traction generators must combine lightness and strength. Frames and rotors are of cast steel or fabricated construction. To facilitate adjustment and inspection, the brush gear is sometimes mounted on a frame which can be rotated by means of a rack and pinion. Armature and field coils must be carefully insulated with well-made joints, and must be fixed rigidly in position, so that they are capable of withstanding the mechanical and thermal stresses to which they are subjected. When the single-bearing type generator is used, the driving end of the rotor carries a spigoted flange for coupling to the engine crankshaft. A self-aligning roller bearing is often used to support the rotor at the non-driving end.

LOAD CONTROL: CONSTANT SPEED

Of the load control systems which, over the greater part of the speed range, maintain the power output constant with no appreciable reduction in the engine speed, those embodying *servo field regulation* are the most widely used. The generator field strength is regulated by the engine governor through the medium of a rheostat operated by a servo motor, and since the first system was patented by Lemp in 1914, many types of load control have been evolved employing this principle. One of these systems (Fig. 9), employed by the British Thomson-Houston Co. Ltd, is used in conjunction with the Paxman servo-governor (*see* page 31, Fig. 4). The rheostat which regulates the strength of the generator field is operated by a vane motor, supplied with oil under pressure from the engine lubricating system. The flow of oil to the vane motor is controlled by the *torque control valve*, consisting of a cylindrical casing containing two concentric members, which can be rotated through an angle. Each member contains ports, which in certain angular positions allow oil under pressure to pass to the vane motor, or to be exhausted into the sump. The inner member, or stem, is connected to the governor linkage which operates the fuel pump rack. The outer member, or sleeve, is connected to the lever which regulates the speed of the engine.

The function of the stem is to modify the generator characteristic by reducing the field strength when the speed of the engine tends to fall due to overloading. Under these conditions the action of the governor tends to increase the amount of fuel supplied by the pumps and also moves the stem from the neutral position, thus allowing oil under pressure to pass to the vane motor. The angular position taken by the sleeve for various settings of the speed control mechanism insures that the stem initially occupies the neutral position at whichever speed the engine may be operating.

For supplying excitation current to the main generator, a separate exciter is used with a differential series field carrying the main traction circuit current, and a separately excited field fed from the auxiliary generator. The latter circuit, which carries a comparatively small current, contains the load control rheostat, together with a variable resistance for external regulation of the generator field (Fig. 10).

The power output is controlled by varying the engine speed by means of electro-pneumatic valves in the manner described in the next section (Load Control: Variable Speed), the speed settings being interspersed with settings giving variations in the generator field strength. These are obtained by varying the resistance in the separate excitation circuit of the exciter, by means of contactors operated from the power controller. Starting procedure is also described on page 50, the system being identical in both cases.

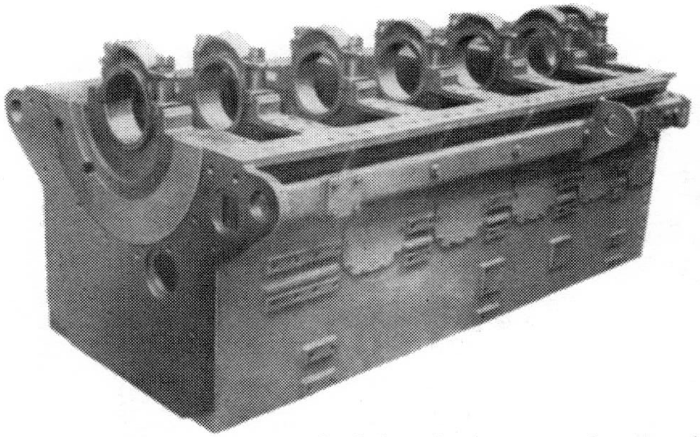

PLATE 1C. Paxman YH cylinder head complete with valves and rocker gear, showing inlet and exhaust ports.

PLATE 1B. Paxman cylinder housing showing liners and water jackets in position.

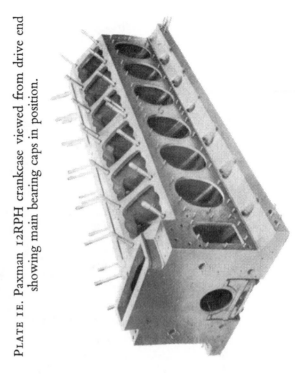

PLATE 1E. Paxman 12RPH crankcase viewed from drive end showing main bearing caps in position.

PLATE 1A. Paxman 8YL bedplate looking on drive end – showing lubricating oil manifold and alternative high and low mounting points.

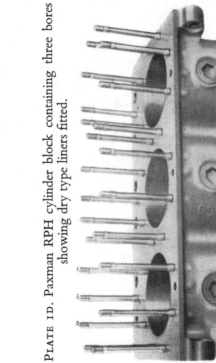

PLATE 1D. Paxman RPH cylinder block containing three bores showing dry type liners fitted.

The English Electric Co. Ltd

PLATE 2. Sectional view of a modern 6-cylinder, 4-cycle, single-acting diesel engine, and generator. Cylinders vertically in line.

The English Electric Co. Ltd

PLATE 3. Sectional view of a modern 16-cylinder, 4-cycle, single-acting diesel engine and generator. The engine is of the V-type and the cylinders are staggered.

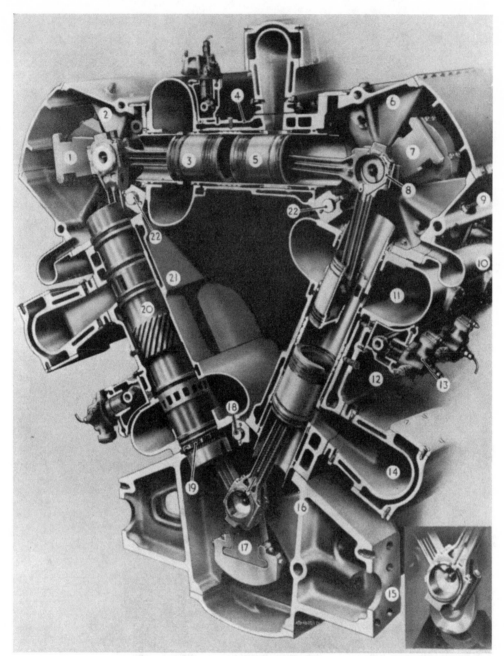

PLATE 4. Sectional view of the Deltic diesel engine. This is a 2-cycle, opposed piston engine with three crankshafts. The insert shows the arrangement of the fork and blade connecting rod big ends.

1. "BC" crankshaft	9. Crankcase tie-bolt	16. "CA" crankcase
2. "BC" crankcase	10. Drain oil manifold	17. "CA" crankshaft
3. Inlet piston	11. Air-inlet manifold	18. Pumps drive shaft
4. "B" cylinder block	12. "A" cylinder block	19. Castellated ring nut
5. Exhaust piston	13. Fuel injection pump	20. Cylinder liner
6. "AB" crankcase	14. Exhaust manifold	21. "C" cylinder block
7. "AB" crankshaft	15. Engine mounting face	22. Blower drive shafts
8. Connecting rod		

British Thomson-Houston Co. Ltd

PLATE 5A. Control frame for British Railways 800 H.P. diesel-electric locomotive – front view – showing: A. reversing switch; B. voltage regulator for auxiliary generator; C. starting contactors for motoring generator; D. motor field shunting contactor for field weakening; E. resistors for traction motor field weakening.

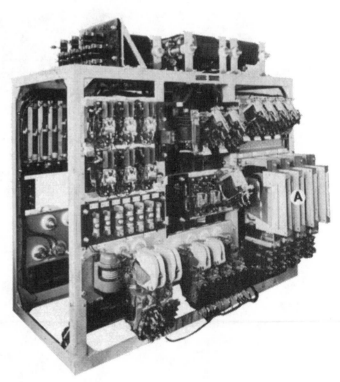

PLATE 5B. Control frame for British Railways 800 H.P. diesel-electric locomotive – rear view – showing: A. traction motor contactors; also various other contactors and relays used in the control and auxiliary circuits.

British Thomson-Houston Co. Ltd

(B)

British Thomson-Houston Co. Ltd

PLATE 6B. Driver's controls of 800 H.P. diesel-electric locomotive for British Railways. The upper handle, A, in the control desk, controls speed and power output. The lower handle, B, operates the master switch and the reversing switch.

PLATE 6A. Driver's controls of 1,700 H.P. Co–Co diesel-electric locomotive, series Bm 6/6, Swiss Federal Railways.

Swiss Locomotive and Machine Works

(A)

Waggonfabrik-Uerdingen A.G.

PLATE 7B. Driver's control desk (operating handle removed) of diesel mechanical, twin-engine, four-wheeler railcar.

Messrs. Alsthom

PLATE 7A. Driver's control desk of Co-Co diesel-electric locomotive for the S.N.C.F.

Hunslet Engine Co. Ltd

PLATE 8c. Gearbox dismantled. The lowest section is nearest the camera, the topmost section is farthest away.

Hunslet Engine Co. Ltd

PLATE 8A. Arrangement of controls.

Hunslet Engine Co. Ltd

PLATE 8B. Gearbox, main friction clutch and jack shaft.

PLATE 8. 500 H.P. Diesel Mechanical Locomotive

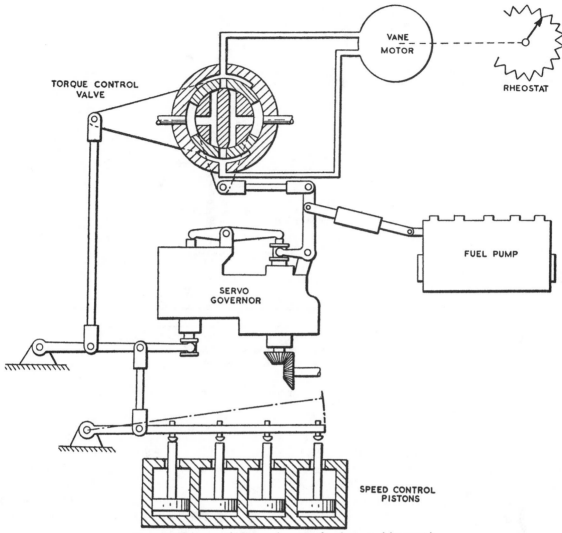

FIG. 9. Diagram of system showing load control (B.T.H.).

Servo field regulation has been used for many years, and there are several systems:

(i) Brown, Boveri & Co. Ltd, in conjunction with Sulzer engines, use a rheostat to regulate the strength of the separately excited field, operated by an oil pressure motor. The admission and release of oil is controlled by a piston valve connected by means of a mechanical linkage to the engine governor. In addition to the automatic control effected by the governor, an additional linkage is provided which enables the oil pressure motor to be operated from the driver's controller, either directly or through the medium of solenoids. The engine governor can be set to operate at a number of fixed speeds by means of electro-pneumatic valves. Movement of the controller handle, therefore, varies both the speed of the engine and the generator field strength. The main generator is separately excited from the auxiliary generator.

(ii) The English Electric system employs a small electric motor to operate the rheostat which controls the strength of the separately excited field. Rotation of the motor in a clockwise or anti-clockwise direction reduces or increases the field strength, and is controlled by the engine governor through the medium of a switch operated by a contact fixed to the rod which actuates the fuel pump control rack. The current for separate excitation is supplied by the auxiliary generator, and the driver's controller enables the engine to be run at several fixed speeds in combination with notches giving field strength variations. In this case, servo field regulation comes into operation only when the fuel pump rack is set for maximum injection, and at lower outputs the generator characteristic is unmodified.

(iii) Metropolitan-Vickers Electrical Co. Ltd, employs a system incorporating a hydraulic servo motor which regulates the strength of the separately excited field by means of resistors with cam operated contacts. A

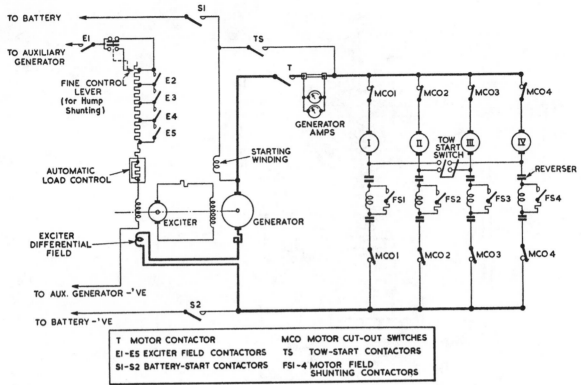

T	MOTOR CONTACTOR	MCO	MOTOR CUT-OUT SWITCHES
E1-E5	EXCITER FIELD CONTACTORS	TS	TOW-START CONTACTORS
S1-S2	BATTERY-START CONTACTORS	FS1-4	MOTOR FIELD SHUNTING CONTACTORS

FIG. 10. Simplified wiring diagram for diesel-electric locomotive (B.T.H.).

pilot valve controlled by the engine governor is used to regulate the oil supply to the servo motor. In this system, movement of the controller handle through the first four notches increases the strength of the separately excited field with the engine at idling speed. Automatic load control comes into operation on the fourth notch; thereafter the speed of the engine and the generator excitation are increased by equal increments up to a maximum. The servo motor is also used to initiate progressive field weakening of the traction motors, when the generator unloading point at full output is reached.

LOAD CONTROL: VARIABLE SPEED

The method generally used to modify the generator characteristic so as to avoid overloading and stalling the engine consists in *reducing the excitation of the separately excited field*. This can be achieved by using an exciter designed in such a manner that the voltage decreases rapidly when a slight fall in speed occurs. If, therfore, the speed of the engine falls slightly due to overloading, the strength of the generator field, and hence the output, is reduced, so that a balance is obtained between the power developed by the engine and that absorbed by the generator at a slightly reduced engine speed.

The generator is provided with self- and separately excited shunt field windings, and also with a differential series winding, producing a characteristic similar to that

shown in Fig. 7. To meet the requirements of the engine, the characteristic is modified as shown in full line with black dots by the effect of the exciter which is separately excited from the battery. In Fig. 8 the effect of the reduction in engine speed on the tractive effort and horsepower curves due to this system of load control is shown, the modifications being indicated in a similar manner to that in Fig. 7.

This type of load control is employed for moderately-powered freight and shunting locomotives. The speed and power output of the locomotive are controlled by varying the speed of the engine, which is provided with a maximum fuel stop, a *servo governor* (Fig. 4) being used for this purpose. Movement of the pilot valve which regulates the compression of the speeder springs is controlled by a lever, and by varying the setting of the lever the governor can be set to operate at a certain number of fixed speeds. Small compressed air cylinders coupled to the lever are used for this purpose, one for each speed setting, and the admission and release of compressed air is controlled by electro-pneumatic valves, energized from the driver's power controller. Multiple-unit control can thus be easily incorporated. When this is not required the layout can be simplified by omitting the electro-pneumatic equipment and operating the pilot valve by means of a mechanical linkage.

Starting and stopping. Starting is carried out by closing

the master switch which sets the electrically driven engine priming pump in motion. When the lubricating oil pressure has built up sufficiently, it operates a piston which pushes the fuel pump control rack to the starting or maximum fuel position. This action closes the starting contactors which connect the generator as a series motor across the battery, and thus starts the engine. When the engine commences to fire, the generator motoring circuit is broken by means of a potential relay, which operates when the auxiliary generator voltage reaches a pre-determined value, and the priming pump is also switched out. The servo-governor then takes over control, and moves the fuel pump control rack to the idling position. Servo-governors are also available with direct air operation, giving infinitely variable speed control. The engine is stopped by means of an electro-pneumatic valve which cuts off the supply of lubricating oil to the engine governor. This method of starting and stopping is widely used with other control systems.

In another system of load control, which has been developed for this type of service, a variable resistance is used to regulate the strength of the separately excited field. When this system is employed, the excitation current can be supplied by the auxiliary generator or the battery. It also provides an additional method of controlling the output, and enables a larger number of control notches to be incorporated in the power controller.

When the controller handle is moved through the first few notches, the strength of the separately excited field is progressively increased by cutting out the resistance, with the engine at idling speed. Movement of the handle through the next series of notches increases the speed of the engine to its maximum value, without a further increase in field strength. The final notch cuts out the resistance completely and the maximum field strength is attained, but a device is incorporated which makes this effective beyond the generator unloading point only, thus preventing the engine from being overloaded. In shunting (switching) locomotives where the starting tractive effort is high in relation to the power output, the generator unloading point occurs at a comparatively low rail speed, and any improvement in performance beyond this point will be of value. The device used to prevent overloading of the engine may consist of an *over current relay* or a *maximum voltage relay*, which operates when the controller handle has been moved to the final notch. Both these devices are used on shunting locomotives operating on British Railways.

It is thus evident that the output of the generator can be controlled by two different methods. Firstly, by varying the speed of the engine, and secondly, by inserting a variable resistance in the separately excited field circuit. The first method is termed *speed control*, and the second *torque control*. When it is necessary to obtain the maximum amount of power variation both methods are used (*see* Engines – Power Output and Speed Control, page 30).

TRACTION MOTORS

The majority of diesel-electric locomotives and railcars are equipped with *D.C. series motors of the axle-hung type* resembling those used for straight electric stock.

Rail and track stresses can be reduced by employing *fully spring-borne traction motors*, and these are being increasingly used for high-speed vehicles. A compact type of drive has been developed by Brown, Boveri, suitable for use when the traction motors are mounted in the bogies, which permits a fully spring-borne traction motor to be placed in the same position relative to the axle as the orthodox type of axle-hung motor.

The *gearcase* is a rigid axle-hung structure, flexibly connected to the bogie frame at the nose end, and incorporating bearings for the pinion which drives the axle-mounted gear. The motor is rigidly fixed to the bogie frame, and the armature shaft is hollow. A *cardan shaft*, passing through the interior of the armature shaft, is used to transmit the drive from the motor to the pinion; flexible blade couplings being used at both ends to connect the cardan shaft to the armature shaft and pinion respectively. The arrangement is similar in principle to the type of drive sometimes used, in which the motor is fixed to the frame of the vehicle, and the drive is transmitted through a propeller shaft to a gearbox mounted on the axle.

Traction motors are generally *grouped permanently in parallel*, or in groupings of two motors in series connected in parallel. *Series-parallel control* of the motor grouping, although employed extensively in North America, is little used in either Great Britain or Europe, where it is generally found that the operating range can be adequately covered by controlling the speed of the engine, varying the generator field strength, and the application of field weakening to the traction motors.

Forced ventilation is widely used, since it enables the size and weight of the motor to be kept to a minimum for a given power output.

Gear ratios range from 2·5 : 1 to 7 : 1, depending on the class of service for which the vehicle is intended. For shunting service, double reduction gearing is frequently used giving a reduction as high as 23 : 1. This arrangement enables high starting tractive efforts to be developed by locomotives having low maximum power outputs, but the maximum speed which can be attained is generally too low for main line operation. It is usual, therefore, to provide facilities for putting the final reduction gears out of mesh by moving the intermediate gear shaft. The locomotive can then be hauled as part of an ordinary freight train, without causing damage to the motors when travelling at speed.

Spur type double reduction gearing, when used, requires driving wheels of comparatively large diameter. By placing the motor with its axis parallel to the longitudinal centre line of the vehicle, and using bevel gears to turn the drive to the intermediate shaft through 90°, driving wheels of quite small diameter can be employed. This arrangement also enables larger and more powerful motors to be fitted when the rail gauge is restricted.

Part IV. Transmissions: hydraulic

DEVELOPMENT

Hydraulic transmissions originally applied to diesel locomotives during the period following World War I were of the hydrostatic type in which oil under pressure and in varying quantities was delivered by a pump to an hydraulic motor. A variable torque characteristic was obtained by making the stroke of the pump infinitely variable, thus enabling the quantity of oil supplied to the motor to be varied. The construction was complex with a number of small moving parts subjected to wear. Oil pressures were necessarily high, and slight wear caused a serious loss of efficiency.

A simpler and more successful system was developed by Lentz employing vane pumps in conjunction with a vane motor, but since the tractive effort curve was not continuous, this system offered little advantage over mechanical transmission.

The *hydraulic torque converter* originated as the hydraulic speed transformer devised by Dr. Föttinger as a speed-reducing gear for marine steam turbines. With the development of helical reduction gearing, the hydraulic speed transformer was no longer used for this purpose and other applications were investigated. In 1926 Föttinger designed a transmission for a 1,600 H.P. diesel locomotive, which was never actually constructed, comprising two torque converters, and a fluid coupling operating in sequence, on the same principle as that now used by Voith.

About this period other engineers interested themselves in the development of the hydraulic torque converter, notably Lysholm in Sweden and Coates in Great Britain. Lysholm developed the multi-stage torque converter, and Coates designed a torque converter incorporating pivoted guide vanes which adapted themselves automatically to the direction of flow. Coates' torque converters were applied to small diesel locomotives constructed by Hudswell Clarke & Co. in 1932 for the Scarborough Miniature Railway.

During the period prior to the outbreak of World War II, a number of locomotives and railcars were built, mostly in Germany, with outputs up to 1,400 H.P. and equipped with the Voith type of transmission. In 1935 a 330 H.P. locomotive provided with this type of transmission was constructed by Harland and Wolff Ltd, for the Northern Counties Committee of the L.M.S.R. in Ireland.

Lysholm torque converters were applied to road vehicles and railcars, and also to a few small locomotives. In Great Britain they were used for a 750 H.P. experimental diesel train built by the L.M.S.R. in 1938.

During recent years remarkable developments have taken place in this field, and today hydraulic transmissions are widely used for both locomotives and diesel trains. An interesting development is the revival in Italy of *hydrostatic transmission* with variable stroke pumps, which has been applied to locomotives developing up to 300 H.P. The hydraulic motors used with hydrostatic transmissions are reversible. Torque converters are not reversible and spur or bevel reverse gear must therefore be used.

HYDRAULIC TORQUE CONVERTERS

The simplest type of hydraulic torque converter consists of three elements, an impeller or pump fixed to the input shaft, a turbine or runner fixed to the output shaft, and a stationary reaction member forming part of the casing. Each member is provided with suitably shaped vanes (Fig. 11).

When the impeller rotates, the power of the engine is absorbed in giving motion to the working fluid, and when there is no reaction member, as in the case of a fluid coupling, the torque developed by the turbine is the same as the input torque. The reaction member causes a reaction torque to be applied to the turbine shaft which is additional to the input torque. The output torque, therefore, is equal to the sum of the reaction and input torques.

The magnitude of the reaction torque depends on the amount by which the fluid stream is deflected during its passage through the turbine. The deflection is greatest when the turbine is stationary. As the speed of the turbine increases, the relative movement between the turbine and reaction vanes causes the deflection of the fluid to become less. Above a certain speed the fluid stream is deflected in the opposite direction. When this occurs the reaction torque becomes negative, and the output torque is less than the input torque.

A converter of this type is known as a *single-stage torque converter*, and typical torque and efficiency curves are shown by the full lines in Fig. 12. The maximum output torque is developed when the turbine is stationary, and the load is a maximum. This is termed the stalled condition, and the ratio of output to input torque at this point

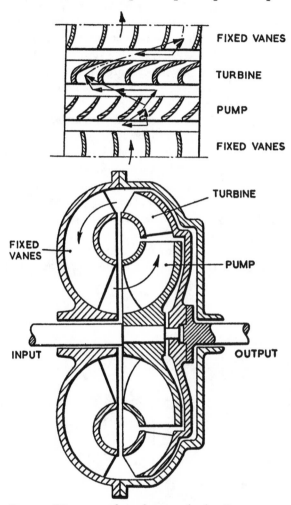

FIG. 11. Diagram of single-stage hydraulic torque converter.

is termed the *stall torque ratio*. When the output speed is about 70 per cent of the input speed the output torque becomes equal to the input torque, and with a further increase in output speed the output torque becomes less than the input torque. It will be apparent, therefore, that the useful speed range of the converter will be that part of the speed range where the torque multiplication is positive. A maximum efficiency of about 85 per cent is attained at approximately half the maximum output speed.

The losses which occur in the converter are reflected by an increase in temperature of the working fluid, which must be circulated through a heat exchanger or cooled by some other method. In order to prevent the expenditure of too much power for cooling purposes, the efficiency over that part of the speed range where the output torque must be maintained for appreciable periods of time, should not be less than 70 per cent, and the maximum output speed at 70 per cent efficiency divided by the minimum output speed at 70 per cent efficiency is termed the *utility ratio*.

Single-stage torque converters can be designed with stall torque ratios ranging from 3 to 5, but when the stall torque ratio is high the utility ratio is reduced. If, however, a three-stage torque converter is used, a stall

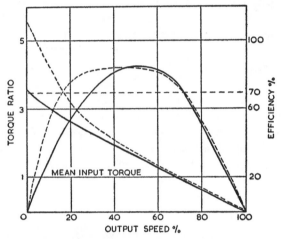

FIG. 12. Performance curves for single-stage and three-stage hydraulic torque converters.

torque ratio greater than 5 can be obtained in conjunction with a flat efficiency curve, showing that the efficiency is well maintained over the useful speed range (Fig. 12, dotted lines). A corresponding improvement is obtained with a two-stage torque converter.

In a two-stage torque converter, the path of the working fluid is as follows:

Pump→turbine first stage→reaction member→turbine second stage→pump.

And in a three-stage torque converter thus:

Pump→turbine first stage→first reaction member→turbine second stage→second reaction member→turbine third stage→pump.

The effect of the turbine stage between the reaction member and the pump is to modify the pump characteristics so that it absorbs more power at starting. At starting, therefore, the speed of the engine is reduced to a point approximately corresponding to the speed at which it develops its maximum torque. The converter output torque is thus correspondingly increased.

The characteristics of a torque converter can be

further improved by means of adjustable pump vanes. The vanes are pivoted, and can be moved from the closed to the fully open position by the aid of suitable mechanism. When the vanes are closed, and the engine is idling, the drag torque is reduced, thus facilitating the engagement of the dog clutches used for obtaining forward or reverse. When adjustable pump vanes are used in a three-stage torque converter, a stall torque ratio of 6 can be obtained, together with an improved utility ratio. It is also possible to regulate the power output within fine limits.

The power output of a torque converter varies as the cube of the rotational speed and the fifth power of the diameter: thus for a small increase in diameter, a large increase in power output is obtained. To keep the size and weight of the converter to a minimum, it is usual for the drive from the engine to be transmitted through step-up gears.

KRUPP TRANSMISSION

Cases often occur in which a single torque converter, even when it is of the multi-stage type, cannot meet the requirements with regard to starting tractive effort and maximum rail speed. In these circumstances the range may be extended by using complimentary change speed gears. The Krupp transmission makes use of a Lysholm-Smith three-stage hydraulic torque converter in conjunction with constant mesh change speed gears providing two or three different ratios according to the nature of the service for which the locomotive is intended. Air-operated bevel reversing gear is also included.

The torque converter which is driven through helical step-up gears from the engine, is situated at the input end of the gearbox. The input and output shafts are carried in cylindrical roller bearings, and deep groove ball bearings are provided for carrying the thrust loads. The vanes of the turbine and guide wheels are machined from solid and securely riveted in position.

The pump wheel is provided with adjustable vanes. Each vane carries a toothed segment which meshes with a toothed regulating wheel concentric with the input shaft, and rotating with it. By varying the angular position of the wheel, the positions of the vanes can be altered from fully open to fully closed. The position of the vanes is regulated by means of the engine speed control lever through the medium of an oil-operated servo mechanism. Labyrinth glands are provided between the stationary and moving parts, to prevent excessive leakage of the working fluid (Fig. 13).

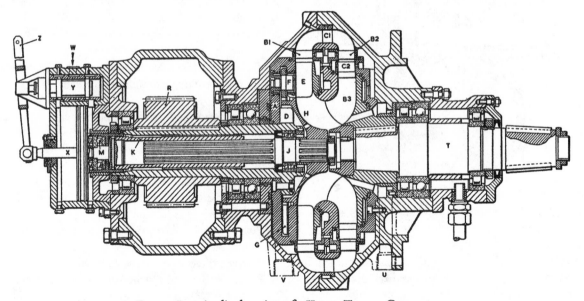

FIG. 13. Longitudinal section of a Krupp Torque Converter.

(A)	pump wheel	(F)	blade pivot	(U)	oil admission at turbine casing
(B1)	1st turbine stage	(G)	toothed segment	(V)	oil exit at turbine casing
(B2)	2nd turbine stage	(H)	regulating wheel	(W)	oil admission for the servo-motor
(B3)	3rd turbine stage	(J)	regulating shaft		
(C1)	1st guide stage	(K)	regulating sleeve	(X)	piston of servo-motor
(C2)	2nd guide stage	(M)	double bearing	(Y)	control valve
(D)	pump shaft	(R)	pinion	(Z)	reconducting lever
(E)	adjustable pump blades	(T)	turbine shaft		

At starting, the converter is filled with oil under pressure by means of a gear-driven pump, which maintains the converter in the filled condition during the period of operation. To prevent trouble due to cavitation a pressure of 56–70 p.s.i. is maintained. Besides constituting the working fluid, the oil is also used to lubricate the gears and bearings under pressure, and operate the control system.

While the converter is operating, oil is continuously bled from the working circuit and passed through an external cooler. It may be bled from a point in the circuit where the pressure is high, and returned at a point of comparatively low pressure, the difference in pressure being sufficient to maintain the oil in circulation. Alternatively, a pump may be used for this purpose. Before entering the converter the oil is passed through a filter.

The change speed gears which have helical teeth are in constant mesh, and engaged by means of multiple plate clutches, operated by oil under pressure fed through passages cut in the shafts. The loosely mounted gears are provided with plain bearings lubricated by oil under pressure. When two change speeds are provided, two sets of gears and two clutches are necessary. For three change speeds, three sets of gears arranged on three shafts, and four clutches are required, two of which are engaged simultaneously to give the required ratio (Fig. 14).

Control is carried out by means of a master controller which controls both the engine speed and the position of

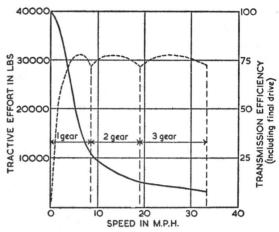

FIG. 15. Performance curves for Krupp three-speed hydraulic transmission.

the pump vanes. The change speed gears operate automatically at the correct rail speed, and the control impulse is derived from a centrifugal governor driven from the output side of the converter, and interconnected with the engine speed control. This operates a servo mechanism which controls the admission and release of oil to the appropriate clutches. There is very little loss of tractive effort when gear changing takes place. Krupp hydraulic torque converters are made in three sizes with capacities of 220 H.P., 500 H.P. and 900 H.P.

Performance curves for a torque converter operating through three sets of change speed gears at full engine output are shown in Fig. 15. The efficiency does not differ appreciably at lower engine outputs.

LYSHOLM-SMITH TRANSMISSION

The Lysholm-Smith transmission incorporates a three-stage hydraulic torque converter. The drive from the engine is transmitted through a double acting friction clutch with two driven members, either of which may be clutched to the input shaft. One of the driven members is mounted on the output shaft which passes through the centre of the torque converter, and the other is mounted on a sleeve which connects it to the pump wheel and is concentric with the output shaft. The turbine drives the output shaft through a free wheel device which allows the output shaft to overrun the turbine.

At starting the pump is clutched to the input shaft and the drive is transmitted through the torque converter. When the torque multiplication becomes zero, the clutch disconnects the pump, and clutches the input shaft to the output shaft giving a direct drive. The output shaft then overruns the turbine as the latter comes to rest. This type of transmission has undergone considerable development in America where it is known as the *twin disc transmission*.

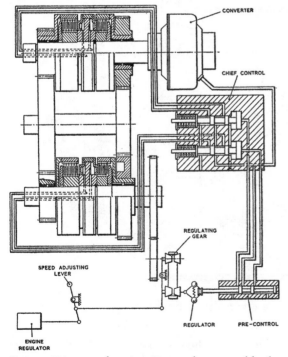

FIG. 14. Diagram showing Krupp three-speed hydraulic transmission.

MEKYDRO TRANSMISSION

The *Mekydro* transmission (Fig. 16), manufactured by Maybach Motorenbau, consists of a single-stage hydraulic torque converter, operating in association with change speed and reverse gears, four different ratios being available in both directions. By this means the converter always operates in the region of maximum efficiency except for a short period when starting from rest.

All the gears which have helical teeth are in constant mesh, and engagement is effected by means of dog clutches of special design known as Maybach interlocking claw couplings. The jaws have oblique faces, so that when clutching is attempted and one member is rotating faster than the other, the two members repulse each other. Practically no wear takes place but a certain amount of reciprocating movement in an axial direction is imparted to the selector forks and operating gear. This movement is reduced to a very small amount by means of interlocking rings, concentric with the clutch rings, which partially cover the gaps when the clutch is disengaged, and are rotated against spring pressure by the clutch jaws when engagement takes place. This can only occur when the speeds of the two members synchronize.

To enable gear changes to be made, the torque converter is designed to fulfil two additional functions, firstly to act as a clutch for disconnecting the drive from the engine, and secondly to synchronize the speeds of the members which are to be connected. These functions are performed by moving the turbine disc axially along the output shaft, thus breaking the working circuit, and at the same time introducing a supplementary ring of blades into the fluid stream. The supplementary blading is designed to impart a weak backward torque to the output shaft, thereby reducing its speed to synchronizing point when changing up. In the case of a downward change, synchronization is achieved by speeding up the output shaft, which takes place when the working circuit is restored. Axial movement of the turbine disc is effected by oil under pressure admitted to an operating cylinder concentric with, and forming part of, the output shaft.

The drive is transmitted to the converter through step-up gears, and by means of a sleeve surrounding the output shaft. The complete assembly consisting of the torque converter, change speed and reverse gears form a single compact unit. Three pairs of gears in constant mesh are required to give the four different ratios, and the necessary combinations are obtained by four interlocking claw couplings. Three additional gears and two claw couplings are required to transmit the drive in forward and reverse.

A gear-driven pump supplies oil under pressure for filling the converter, operating the control system, and

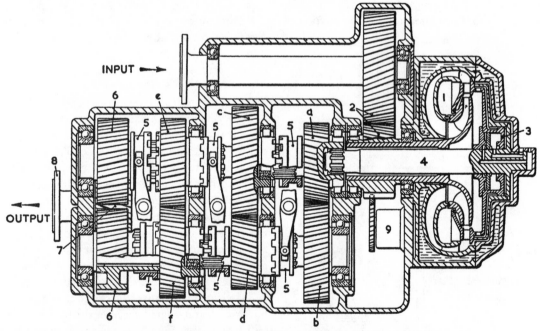

FIG. 16. Longitudinal section of Mekydro Hydraulic Transmission.

(1) Torque converter	(4) Torque converter output shaft	(8) Output flange
(2) Step-up gears	(5) Maybach claw coupling	(9) Oil pump for torque converter
(3) Operating cylinder for disengaging torque converter	(6) Reversing gear	and control gear
	(7) Output gear	(a-f) Change speed gears

J. Stone and Co., (Deptford) Ltd.

PLATE I. Victorian Railways of Australia: "The Westlander Express" in charge of two Co–Co type 1,950 H.P., diesel electric locomotives built in Australia by the Clyde General Motors Organisation.

for lubricating the gears and bearings. To prevent over-heating of the working fluid, the converter is surrounded by a water jacket through which the engine cooling water is circulated.

An automatic servo mechanism operated by oil under pressure controls the movement of the claw couplings and turbine disc. The control impulse is derived from a governor, driven from both the input and output sides. Electro-magnetic valves are used to operate the servo mechanism controlling the direction change, and also for engaging the converter at starting.

VOITH TRANSMISSIONS

Hydraulic transmission. The distinguishing feature of the Voith system of hydraulic transmission is the principle of using two or more hydraulic circuits which are filled and emptied in sequence. In its simplest form it consists of a single-stage torque converter and a fluid coupling arranged in series. At starting the torque converter is filled, and operates up to that point in the speed range where the output torque becomes equal to the input torque. When this point is reached the working fluid is transferred from the converter to the fluid coupling which continues to transmit the drive until the maximum rail speed has been attained.

Single-stage torque converters of the type used by Voith are simple in construction, but have a somewhat limited working range. This disadvantage is overcome by using up to three torque converters in sequence, or by using one torque converter and two fluid couplings, reduction gearing giving two alternative ratios being used in both cases.

The L.36 type of transmission (Fig. 17) comprising three torque converters is generally applied to high-speed locomotives and railcars, and can be used to transmit up to 1,000 B.H.P. A torque converter with a high stall torque ratio and restricted working range is used at starting. When the maximum speed for efficient operation is reached, the working fluid is pumped from the first to the second converter. The turbines of both these converters are coupled together by a sleeve shaft, and drive through the same reduction gearing, the second converter, however, has a comparatively low stall torque ratio and a much increased working range, thus enabling a higher rail speed to be attained without any serious loss of efficiency. The third converter, which is brought into operation in the same manner, and covers the highest part of the speed range, is identical with the second converter, but in order to give a different torque-speed characteristic, drives through a second set of reduction gearing having a lower gear ratio.

In the L.37 type of transmission (Fig. 18) which is generally applied to freight and shunting locomotives equipped with engines developing up to 800 B.H.P., one torque converter and two fluid couplings are used. The torque converter is employed at starting, after which the working fluid is transferred to the first fluid coupling, the same reduction gearing being used for both. To cover the

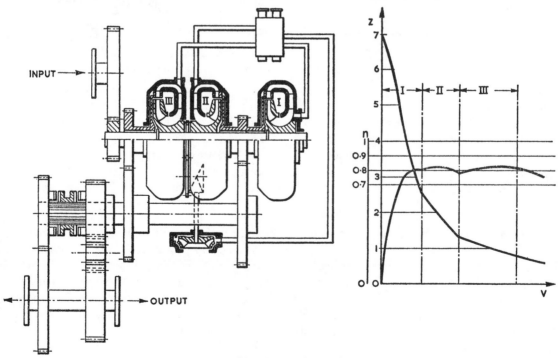

FIG. 17. Diagram of Voith L.36 hydraulic transmission.

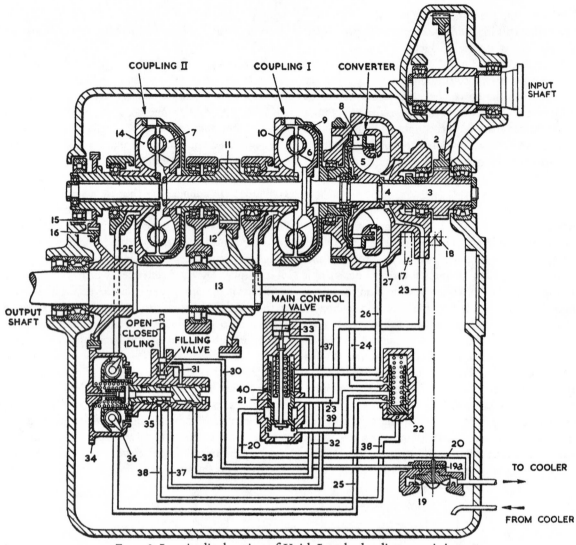

FIG. 18. Longitudinal section of Voith L.37 hydraulic transmission.

Key to major items	9 Casing connecting items 8 and 10	14 Turbine wheel coupling 2	22 Second control valve
1–2 Step-up gears			27 Converter casing
3–4 Converter input shaft	10 Turbine wheel-coupling 1	15 Second gear pinion	34 Centrifugal governor gear drive
5 Converter pump wheel		16 Second gear wheel	
6 Pump wheel coupling 1	11 First gear pinion	17–18 Drive to oil circulating pump	35 Centrifugal governor – piston valve
7 Pump wheel coupling 2	12 First gear wheel		
8 Converter turbine wheel	13 Output shaft	19 Oil circulating pump	36 Centrifugal governor – casing

upper part of the speed range the second fluid coupling is filled, and the second set of reduction gearing, giving a lower gear ratio, brought into operation. When this type of transmission is used, the final reduction gears are often designed to provide two alternative gear ratios, a low gear for use in shunting yards, and a high gear for use when operating on the main line. These gears can only be engaged when the locomotive is stationary.

In both systems *step-up gearing* is used to transmit the drive to the torque converters or fluid couplings, and filling and emptying is carried out by means of a gear-driven pump. The time taken to transfer the fluid from one circuit to the next is about one second, and there is no loss of tractive effort during the process. The pump is also used to circulate the fluid continuously through an external oil cooler, and maintain the pressure of 40–60 p.s.i. in the working circuit through which the drive is being transmitted.

The control system is automatic and embodies either a centrifugal governor driven from the output end, or two small control pumps, one driven from the input and the other from the output. The first system of control is used for the L.37 type of transmission. The governor actuates a piston valve, which supplies oil under pressure to the main control valve, which controls the filling and emptying of the various circuits. The second system is applied to the L.36 type transmission. The two pumps generate a pressure differential in the oil control circuit which is used to operate the main control valve.

For starting, a mechanically-operated filling valve is used, actuated by the handwheel which controls the engine speed. When the engine is accelerated, the filling valve opens and allows oil under pressure to pass to the torque converter. So that the engine may be speeded up when the locomotive is stationary, a mechanical lock is fitted by means of which the filling valve can be kept closed.

In Fig. 19, performance curves are shown for the two types of transmission, designed to cover the same speed range, with the same input horsepower. Curves for the L.36 type of transmission are shown in full lines, and those for the L.37 type of transmission in dotted lines. In the L.36 transmission the engine speed is maintained

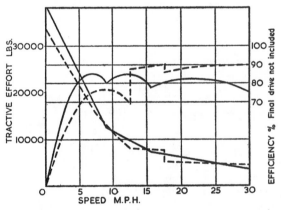

FIG. 19. Comparative performance curves L.36 and L.37 transmissions.

nearly constant throughout, so that very little power is lost when the drive is transferred from one converter to the next. In the L.37 transmission the engine speed falls considerably when the drive is transferred from converter to fluid coupling, resulting in loss of power at the rail. It follows, therefore, that the L.36 transmission is better suited for passenger locomotives and railcars, where maximum acceleration is required. When operating through the fluid couplings the efficiency of the L.37 transmission is high, the losses being considerably less than those encountered in an hydraulic torque converter, even when operating at its maximum efficiency.

Split-drive or Diwar transmission. An entirely different principle is used in the Diwar drive developed by Voith, and used for transmitting up to 200 horsepower (Fig. 20). The drive from the engine is split by means of a differential gear so that part of the engine output is transmitted through a single-stage hydraulic torque converter, and part direct. The two drives are then combined and drive the road wheels through epicyclic reduction gears. The converter drive is transmitted to the output shaft through the medium of a free wheel, so that under certain conditions the output shaft can overrun the converter.

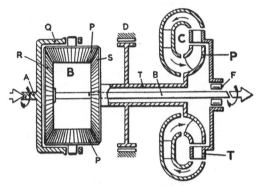

FIG. 20. Diagram showing principle of Diwar transmission.

(B) distribution gear	(D) distributor brake
(C) torque converter	(F) freewheel
(P) pump impeller	(P) bevel pinion
(T) turbine runner	(Q) planet carrier
(A) input shaft	(R) Sun wheel
(B) output shaft	(S) Sun wheel
(T) Primary shaft	

At starting when the driving wheels are stationary, the entire engine output is transmitted through the torque converter, the speed of the pump element being stepped up considerably due to the action of the differential. The load on the engine is thereby increased so that its speed falls to the region where the maximum torque is developed. As the vehicle commences to move, the action of the differential causes some of the power developed by the engine to be transmitted direct and the proportion of engine power transmitted in this manner increases with the rail speed. When the engine reaches its maximum speed, the input shaft to the converter is held stationary by means of a brake. This causes the engine speed to fall, but the direct drive shaft is speeded up due to the action of the differential and the entire engine output is then transmitted direct, the torque converter being overrun.

Due to the division of power, this type of transmission operates with a very high efficiency. When part of the power developed by the engine is being transmitted through the torque converter, the efficiency exceeds 90 per cent, and this is still further increased when the entire power output is transmitted direct. It also enables the maximum torque developed by the engine to be used at starting. Since the torque converter is not required to transmit the maximum power output which the engine develops when operating at its maximum speed its size and weight can be kept to a minimum.

In practical applications, a spur-type differential is used. The torque converter is surrounded by a water jacket through which the engine cooling water is circulated. Two sets of epicyclic change speed gearing are provided which can be brought into operation by means of band brakes, operated by compressed air. A similar type of brake, termed the distributor brake, is also used to hold the converter input shaft stationary, when direct drive is being used. This is operated automatically at the appropriate speed. The change speed gears, however, are engaged by the driver according to requirements.

ZAHNRADFABRIC: HYDROMEDIA TRANSMISSION

Hydraulic torque converters are being increasingly used in a supplementary capacity to mechanical transmission, particularly for the smaller powers. One example is the Hydromedia transmission used for railcars and manufactured by Zahnradfabric (Fig. 21).

This comprises a gearbox of the layshaft type with gears in constant mesh giving three different ratios,

engagement being effected by means of multiple plate clutches operated by oil under pressure. The Trilok hydraulic torque converter is carried in a bell housing at the input end of the gearbox and is of the single-stage type comprising three members.

The pump element is mounted on the input shaft which is extended into the gearbox, where it can be connected to the gearbox output shaft through the second and third gear trains. The turbine is fixed to a sleeve concentric with the input shaft, which is also extended into the gearbox and drives through the first gear train. The reaction member is fixed to a sleeve concentric with the turbine sleeve and input shaft, the torque reaction being transmitted to the gearbox casing through the medium of a free-wheel device.

At starting the drive is transmitted through the torque converter and first-speed gear train. When the speed of the turbine rises to that point where the torque reaction becomes negative, the fluid stream is deflected in the opposite direction and the reaction member commences to rotate, since it is no longer held stationary by the free wheel. The torque converter now functions as a fluid coupling until second gear is engaged. In second and third gear the torque converter is emptied and the drive to the gearbox is direct through the input shaft. It will be apparent, therefore, that the torque converter fulfils its specific function only part of the time during which

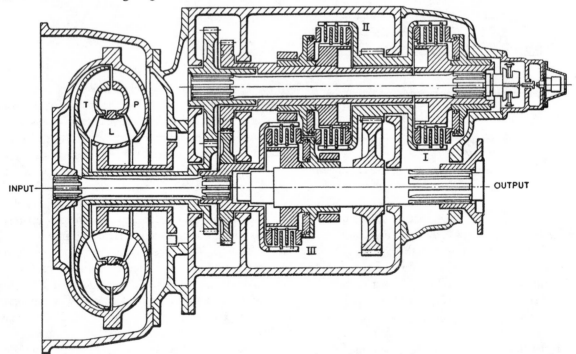

FIG. 21. Diagram showing Hydromedia transmission.

| (P) impeller | (T) turbine | (II) multiple disc clutch 2nd gear |
| (L) reactor | (I) multiple disc clutch 1st gear | (III) multiple disc clutch 3rd gear |

first gear is being used. The efficiency measured between the engine and rail ranges from about 80 per cent when the drive is transmitted through the torque converter and gears, to about 94 per cent when the gearing only is used.

A gear-driven pump supplies oil at a pressure of 30 p.s.i. to the converter circuit which includes an oil cooler. A second pump supplies oil under pressure for operating the gear change clutches. Gear changing is carried out by means of electro-pneumatic valves.

Part V. Transmissions: mechanical

BASIC PRINCIPLES

The torque developed by the engine must be multiplied by some means, so that it can be usefully applied for traction purposes. A simple method of doing this is by means of mechanical reduction gearing which makes it possible to exert a tractive effort at the rim of the driving wheels up to the limit allowed by the adhesion.

If a single train of gears is used, the tractive effort curve will duplicate the engine torque curve, but with the horizontal and vertical scales altered to indicate rail speed and tractive effort. The tractive effort curve can be derived from the engine torque curve by means of the following formula:

$$TF = \frac{Tr}{R}e$$

where TF = tractive effort in pounds.

T = torque developed by engine in pounds feet.

r = total gear multiplication between engine and wheels.

R = radius of driving wheels in feet.

e = transmission efficiency, this figure includes the power absorbed in driving the radiator fan, air compressor, etc.

By using four-gear trains giving different gear reductions, four tractive effort curves can be obtained (Fig. 6) thus considerably extending the speed range over which the locomotive can operate. The four separate tractive effort curves can be combined to give a continuous tractive effort curve, with steps where changes in ratio occur.

The horsepower exerted at the wheel tread can be derived from the tractive effort by means of the following formula:

Rail horsepower = $\dfrac{TF \times V}{375}$

where V = speed in m.p.h.

The corresponding curve of rail horsepower is shown in Fig. 6, from which it will be seen that the horsepower exerted at the wheel tread reaches a peak value, when the rail speed corresponding to the maximum engine speed in each gear is attained.

For a transmission of this type to be practicable, the following conditions must be observed:

(i) The maximum tractive effort must be available at starting.

(ii) It must be possible to change the gear ratios while the locomotive is in motion.

(iii) All the gear ratios must be available in both directions of travel.

The larger the number of gear ratios used, the more nearly will the tractive effort curve approach the ideal curve.

DEVELOPMENT

Most of the early internal combustion locomotives and railcars were provided with change speed and reverse gears resembling those used for contemporary petrol-driven road vehicles, which at that period were equipped with bevel reversing gear making all the speeds available in both directions. For low-powered units, change speed gears in sliding mesh in combination with a main friction clutch, have proved reasonably satisfactory, the main difficulty encountered being that of changing gear when hauling a load. As early as 1896 this problem was overcome by Richard Hornsby & Sons, who constructed a small locomotive of 8 B.H.P. in which independent friction clutches were used for each of the two sets of change speed gearing which were in constant mesh.

During the subsequent period many improvements in construction were introduced, including the complete enclosure of the gears and the provision of ball and roller bearings, but as long as power outputs remained low there was little incentive for notable technical advances.

In the field of large locomotives a number of attempts were made to employ direct drive, using compressed air for starting and for augmenting the tractive effort during the period of acceleration. These included the 1,000 H.P. Sulzer locomotive constructed in 1913 for the Prussian State Railways, the Ansaldo locomotive of 1926, and the 1,600 H.P. Deutz locomotive constructed in 1934. Varying degrees of success were obtained.

An ambitious effort to introduce mechanical change speed gearing for large locomotives was made in 1926, when a 1,200 H.P. locomotive was completed in Germany for service in Soviet Russia. This locomotive was provided with three sets of change speed gears in constant mesh, with electro-magnetic gear-changing clutches, and an electro-magnetic main clutch. Although reasonably successful the arrangement was not sufficiently flexible to warrant general adoption.

In 1924 a small diesel-mechanical locomotive constructed in Switzerland inaugurated an important development, being equipped with the S.L.M. gearbox in which gear changing was effected by means of friction clutches operated by oil under pressure.

From about 1930 onwards diesel-mechanical locomotives and railcars with outputs exceeding 100 B.H.P. were built in increasing numbers, and some notable developments took place in transmission systems. The first locomotive to incorporate a fluid coupling was built by Hudswell, Clarke & Co. in 1930, and was provided with an engine developing 300 B.H.P. The fluid coupling used for this locomotive was of the ring valve type which allowed the drive from the engine to be completely disengaged. This design was not perpetuated, and other methods were developed for overcoming the problem of drag torque.

In 1932 the Hydraulic Coupling & Engineering Co. Ltd introduced a traction gearbox which maintained the tractive effort during the gear-change interval. This was the precursor of the S.S.S. Powerflow transmission, but instead of synchro-clutches, dog clutches with baulking rings were used.

Constant mesh change speed gearing with dog clutches, used in conjunction with a main friction clutch for gear changing, was developed by the Hunslet Engine Co. together with a system of pre-selection using oil under pressure as the operating medium. This type of transmission was first applied to a 150 B.H.P. diesel locomotive built in 1932.

The Wilson epicyclic gearbox, originally developed for automotive vehicles, was also successfully applied to locomotives and railcars, where it was used in conjunction with a fluid coupling. Its compactness combined with its suitability for multiple-unit control have led to its wide adoption for railcars. Constant mesh transmissions suitable for multiple-unit operation underwent considerable development in Europe, including the use of several friction clutches for gear changing, two or more of which could be engaged simultaneously.

FLUID COUPLINGS AND FRICTION CLUTCHES

Fig. 6 shows that the maximum tractive effort is developed when the rail speed is approximately 3 m.p.h. and in order to make this available at starting, a friction clutch or fluid coupling must be used. While the locomotive is at rest, the engine is started and accelerated to the speed at which it develops the torque necessary to set the train in motion. When the clutch is engaged the torque is transmitted to the driving wheels, although at the moment of starting, when the driving wheels are stationary, the clutch slip is 100 per cent, and practically all the power developed by the engine is dissipated as heat. As the locomotive accelerates the slip diminishes, and disappears when a rail speed corresponding to the rotational speed of the engine is attained.

The duties imposed on friction clutches employed in railway service are often very severe, and sometimes when starting a heavy train from rest, the maximum torque developed by the engine must be transmitted when the clutch slip is 100 per cent. In the fluid coupling, fluid friction is substituted for sliding friction during the slipping stage, thereby eliminating the heavy wear of clutch linings.

A *fluid coupling* consists of an impeller or pump mounted on the input shaft, and a turbine or runner mounted on the output shaft. Each element has a number of straight vanes radiating from the centre, and when placed face to face a series of circular passages are formed. When the pump element is rotated by the engine, the fluid flows outwards gathering momentum, until it is deflected into the runner element, where it flows inwards imparting momentum to the runner, which also commences to revolve when the torque developed is sufficient to overcome the resistance.

An appreciable amount of torque is transmitted by the coupling when the engine is idling, and may be sufficient to cause the locomotive to move. This drag torque as it is termed, also makes it difficult to engage first gear when starting from rest. If, however, the working circuit is completely emptied no torque is transmitted and the drive is disengaged.

In the *traction type coupling* (Fig. 22) the runner carries a reservoir which communicates with the working circuit. When the runner is stationary, some of the fluid contained in the coupling drains into the reservoir, and the drag torque is considerably reduced. This action is assisted by the baffle fixed to the inner edge of the runner, which breaks up the fluid vortex at low speeds. At normal operating speeds the vortex shrinks and flows clear of the baffle. When the runner commences to revolve the fluid is fed back into the working circuit by the action of centrifugal force.

A further development is the *scoop-controlled coupling* which gives a completely disengaged drive. In this case the reservoir which contains the fluid is fixed to the impeller, and the fluid is collected and fed into the working circuit by means of a movable scoop. Below a certain speed the coupling automatically empties itself through centrifugally controlled quick emptying valves. The drive can be disengaged or the filling varied by

MULTI-PLATE FRICTION CLUTCH **FLUID TRACTION COUPLING**

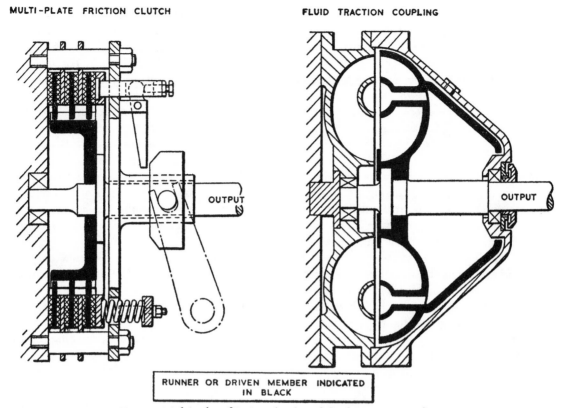

RUNNER OR DRIVEN MEMBER INDICATED IN BLACK

OUTPUT

FIG. 22. Multi-plate friction clutch and fluid traction coupling.

altering the position of the scoop tube. This is accomplished by means of suitable gearing.

When operating under conditions of maximum efficiency, the runner must always rotate at a slightly slower speed than the impeller. This is due to the pressure difference which causes the fluid to circulate. The loss from this cause amounts to about 2·5 per cent. As in the case of the friction clutch, severe slipping causes excessive heating which can result in damage to the coupling.

Like the friction clutch, the fluid coupling can transmit torque in the reverse direction, and this makes it possible to use the engine compression for braking purposes. If a locomotive fitted with a friction clutch stalls under load, the engine will also stall. This cannot occur, however, when a fluid coupling is used, since when the runner stalls no torque is transmitted, and the engine continues to rotate. The fluid coupling also acts as a useful shock absorber between the engine and transmission.

Fluid couplings also may be carried on the engine flywheel, but when the weight of the coupling is too great to be carried by the crankshaft bearings, it becomes necessary to provide additional support for the runner portion. In such cases an outboard bearing is generally used, carried in a bell housing fixed to the crankcase, or supported from the frame. To avoid trouble due to any slight misalignment which may occur, the outboard

bearing is of the self-aligning type, and the impeller is connected to a flange on the input shaft by means of a thin flexible steel driving disc which transmits the torque. The impeller is maintained central by means of a spherical spigot which fits into a block attached to the input flange. When the runner portion of the clutch or coupling is supported in accordance with one of these methods, a flexible coupling or propeller shaft may be used to transmit the drive to the gearbox. In certain cases, however, where there is very little space between the engine and gearbox, the outboard bearing may be dispensed with, and a rigid coupling or special type of flexible coupling used to transmit the weight of the runner portion to the gearbox input shaft.

In Italy, a coupling known as the *Ranzi power coupling* is used, which employs a refractory powder as the operating medium. It is claimed that this coupling, which has been applied to both locomotives and railcars, can transmit the maximum torque at high rates of slip for indefinite periods without suffering damage due to overheating. It is also claimed that when operating at maximum speed, the slip is negligible.

The *main friction clutch* is generally of the multiple-plate type, having steel plates lined with Ferodo or other friction material (Fig. 22). Single-plate clutches and occasionally cone clutches are also used. It is usually

65

mounted on the engine flywheel, the weight of the runner portion being carried by a bearing fixed to the flywheel, or forming part of a bell housing attached to the engine crankcase. For successful operation, it must be of robust construction with ample wearing surface, and adequate provision for heat dissipation and inspection. Provision should also be made for renewing clutch linings without excessive dismantling.

GEARS: CONSTANT MESH

A transmission incorporating four sets of reduction gears is shown diagrammatically in Fig. 23, in which the drive from the engine is transmitted through a friction clutch fixed to the engine flywheel. The gearbox input shaft carries four pinions of varying sizes meshing with corresponding gears fixed to the output shaft. The pinions are loosely mounted so that they are free to rotate on the shaft; usually ball or roller bearings are used for this purpose, but plain bearings may also be employed.

The pinion and gear giving the required reduction ratio are brought into operation by connecting the pinion to the input shaft so that it revolves with it. This is effected through the medium of dog clutches which can be moved axially along the splined portions of the shaft by means of selector forks actuated by the gear change lever. Instead of being provided with jaws, the dog clutches are often made with internal gear teeth which mesh with external gear teeth formed on the pinions, thus facilitating the engagement of the gears.

Reversing may be carried out by means of *bevel reversing gear* (Fig. 23). The two bevel wheels which are driven in opposite directions by the bevel pinion, are free to revolve on the cross shaft. Forward or reverse gear is obtained by connecting the appropriate bevel wheel to the cross shaft, by means of a dog clutch. The bevel reverse gear also serves the purpose of turning the drive through 90°, which is necessary in the majority of transmissions, when the engine crankshaft is parallel with the longitudinal centre line of the vehicle.

When it is unnecessary to turn the drive, spur reverse gear may be used. This consists of two trains of spur gears, one of which incorporates a third gear or idler, which reverses the direction of rotation. One gear in each train is free to revolve on its shaft, and either of the trains can be brought into operation by connecting one of the loosely mounted gears to the shaft by means of a dog clutch.

The dog clutches used for selecting forward and reverse are often moved in and out of engagement by small compressed-air pistons working in cylinders mounted on the gearbox casing. This arrangement is essential in the case of multiple unit working.

Variations of the arrangement illustrated occur in which the bevel reverse gear is situated at the input end of the gearbox, or the reverse gear and final reduction gearing, which is not shown in the diagram, may be contained in a separate casing.

To carry out an effective gear change from first to

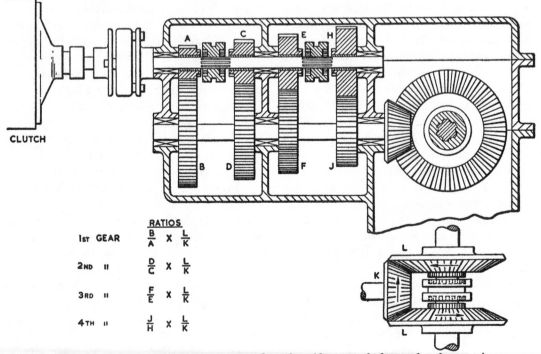

	RATIOS	
1st GEAR	$\frac{B}{A}$ × $\frac{L}{K}$	
2nd "	$\frac{D}{C}$ × $\frac{L}{K}$	
3rd "	$\frac{F}{E}$ × $\frac{L}{K}$	
4th "	$\frac{J}{H}$ × $\frac{L}{K}$	

FIG. 23. Diagrammatic arrangement of gearbox (four speeds forward and reverse).

(B)

Arn Jung

PLATE 9B. Diesel-hydraulic articulated locomotive type B + B, similar to that shown in 9A but employing rod drive.

De Dietrich et Cie

PLATE 9D. Standard gauge B type 150 H.P. diesel-electric locomotive with chain drive for S.N.C.F.

(D)

PLATE 9C. 2 ft 6 in.-gauge diesel-hydraulic articulated locomotive type B + B for the Kalha–Simla line of the Indian State Railways. This locomotive is similar to that shown in 9A but employs shaft drive.

(A)

Arn Jung

PLATE 9A. Metre gauge diesel-hydraulic, articulated locomotive type B + B with chain drive for the Emdener Kreisbahn. This locomotive has two 145 H.P. engines and is fitted with standard gauge side buffers for service on mixed gauge lines.

Arn Jung

(C)

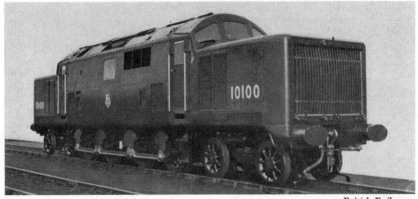

British Railways

PLATE 10A. British Railways 2–D–2 diesel mechanical locomotive employing the Fell system.

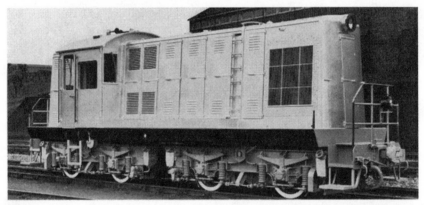

Birmingham Railway Carriage and Wagon Co. Ltd

PLATE 10B. Ghana Railways and Harbours Administration 410 H.P. Bo–Bo diesel-electric mixed traffic locomotive for 3 ft 6 in.-gauge.

British Thomson-Houston Co. Ltd

PLATE 10C. British Railways 800 H.P. Bo–Bo diesel-electric freight locomotive, standard gauge.

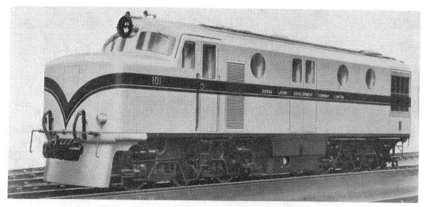

Birmingham Railway Carriage and Wagon Co. Ltd

PLATE 11A. 850 H.P. locomotive with Sulzer engine, for mineral traffic on the 3 ft 6 in. gauge lines of the Sierra Leone Development Co.

Birmingham Railway Carriage and Wagon Co. Ltd

PLATE 11B. 950 H.P. locomotive with Sulzer engine, for mixed traffic on the 5 ft 3 in.-gauge lines of Coras Iompair Eireann.

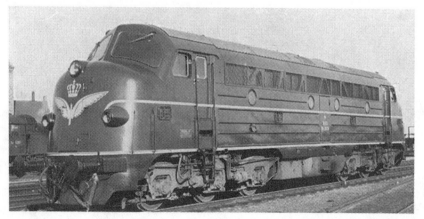

Danish State Railways

PLATE 11C. 1,665 H.P. locomotive with electro-motive V-type, 16-cylinder, 2-cycle engine, for express passenger duties on the standard gauge Danish State Railways. Built by Nohab under licence from General Motors Corporation of America.

PLATE 11. Diesel-electric Locomotives, Type A1A–A1A

69

The English Electric Co. Ltd

PLATE 12A. British Railways Deltic locomotive – the most powerful single unit diesel locomotive in the world. The two Napier Deltic engines develop a total 3,300 H.P. with a starting tractive effort of 60,000 lb. The total weight of the locomotive is only 106 tons.

P. Ransome-Wallis

PLATE 12B. Coras Iompair Eireann: 1,200 H.P. mixed traffic locomotive with Crossley engine No. A.9 at Claremorris.

PLATE 12. Diesel-Electric Locomotives, Type Co–Co

De Dietrich et Cie

PLATE 13A. Algerian Railways 960 H.P. locomotive with Sulzer engine, for passenger and freight service.

Swiss Locomotive and Machine Works

PLATE 13B. Swiss Federal Railways 1,700 H.P. locomotive with two 850 H.P. Sulzer engines for transfer freight and shunting (switching) duties.

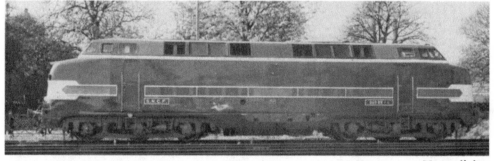

Messrs. Alsthom

PLATE 13C. S.N.C.F. 1,800 H.P. mixed traffic locomotive with two 900 H.P. M.G.O. engines.

PLATE 13. Diesel-Electric Locomotives, Type Co–Co

P. Ransome-Wallis

PLATE 14B. British Railways 1–Co–Co–1, 2,000 H.P. locomotive for mainline passenger service on the Eastern Region.

Metropolitan-Vickers Electrical Co. Ltd

PLATE 14D. West Australian Government Railways (3 ft 6 in. gauge) 2–Do–2, 1,105 H.P. locomotive with Crossley engine for mixed traffic.

British Railways

PLATE 14A. British Railways 1–Co–Co–1, 2,000 H.P. locomotive for mainline passenger service.

The English Electric Co. Ltd

PLATE 14C. New Zealand Government Railways (3 ft 6 in. gauge) 2–Co–Co–2, English Electric 1,500 H.P. locomotive for passenger and freight service, and equipped with dynamic braking.

PLATE 14. Diesel-Electric Locomotives with Guiding Wheels or Bogies

British Railways

PLATE 15B. British Railways (Western Region) 2,200 B–B express locomotive built to the same design and specification as the German locomotive Plate 15A.

North British Locomotive Co. Ltd

PLATE 15D. British Railways 2,000 H.P. A1A–A1A express locomotive with two M.A.N. 12-cylinder engines and Voith transmission.

Deutsche Bundesbahn

PLATE 15A. German Federal Railways 2,200 H.P. B–B express locomotive with two V-type 12-cylinder, 4-cycle Maybach engines and Mekydro transmission.

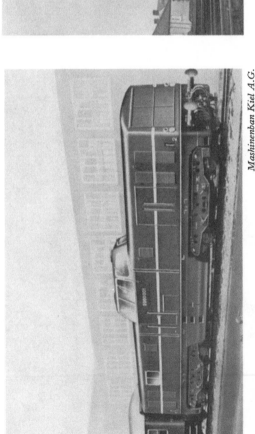

Maschinenban Kiel A.G.

PLATE 15C. German Federal Railways 2,000 H.P. B–B mixed traffic locomotive with two 1,000 H.P. M.A.K. engines and Voith transmission.

Plate 15. Diesel Locomotives with Hydraulic Transmission

Hudswell Clarke

PLATE 16B. Hudswell Clarke constant-output diesel mechanical locomotive with Paxman 6-cylinder pressure charged engine. A constant horsepower governor maintains an output of 210 B.H.P. over a speed range of from 735 to 1,250 r.p.m. Dual Fluid transmission with an S.S.S. three-speed gearbox is fitted.

P. Ransome-Wallis

PLATE 16D. British Railways standard 400 H.P. English Electric diesel-electric locomotive with two nose-suspended traction motors and double reduction gear drive. All wheels are coupled. Photograph shows No. D.3042 on gravity shunting duty.

PLATE 16. Diesel Shunting (switching) Locomotives

Mashinenbau Kiel A.G.

PLATE 16A. M.A.K. 800 H.P. locomotive with Voith hydraulic transmission and Beugniot bogies. Supplied to various of the world's railways.

P. Ransome-Wallis

PLATE 16C. L.M.S.R. (now British Railways) 350 H.P. diesel-electric locomotive with single traction motor driving through a jack-shaft and side rods.

second gear, the following sequence of operations must be performed:

(i) The main friction clutch must be released sufficiently to remove the driving pressure from the dog teeth, so that the dog clutch can be slid out of engagement with the first-gear pinion.

(ii) The dog clutch, which is revolving at engine speed, must be engaged with the second-gear pinion which is revolving considerably slower. So that the dog teeth can mesh smoothly, it is necessary to synchronize the speeds of the input shaft and pinion. This may be achieved by means of a clutch brake which reduces the speed of the runner when this is withdrawn to its fullest extent.

(iii) To avoid slipping the clutch, the engine speed must be reduced to the same speed as the input shaft, before the clutch is re-engaged.

While the clutch is released, no power can be transmitted to the wheels, and the tractive effort falls to zero. If the operation is prolonged, loss of rail speed will occur, and a severe shock will be imparted to the train when the clutch is re-engaged. It is important, therefore, that gear changing should be carried out as quickly as possible.

The process of gear changing can be much simplified if instead of dog clutches, *friction clutches* are used for connecting the pinions to the input shaft. These clutches, which are normally disengaged, may be of the cone, multiple-plate, or other suitable type, and are generally located inside the gearbox. They can be conveniently operated by means of compressed air or oil under pressure. For the smaller type of gearbox, however, manual operation is often used. When this system is employed a main friction clutch between the engine and gearbox is not essential, and a fluid coupling can be used to take up the slip when starting from rest. The provision of independent friction clutches for each set of change speed gears eliminates the drag torque difficulty.

A *constant mesh gearbox* with internal teeth dog clutches for engaging the gears is extensively used by the Hunslet Engine Co. Ltd for locomotives developing up to 500 H.P. A multiple-plate, self-ventilating friction clutch designed to withstand the severe conditions encountered in locomotive work, is mounted on the engine flywheel or at the input end of the gearbox. Shafts and loosely mounted gears are carried on ball or roller bearings, and bevel reverse gear is incorporated. To simplify gear changing, a pre-selective mechanism operated by compressed air is used. After the required gear has been selected, engagement occurs automatically when the clutch is released. The speed of the input shaft is reduced to the synchronizing point by means of a shaft brake actuated by the clutch operating mechanism. The construction and operation of a six-speed gearbox used for a 500 H.P. locomotive is clearly shown in Plates 8A, 8B and 8C, page 48.

A system of *hydraulic transmission* has recently been introduced by this firm which includes the basic elements

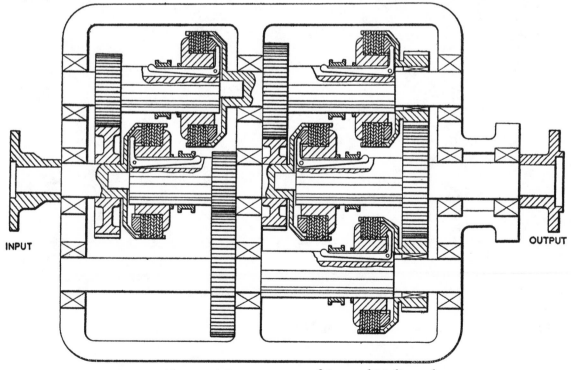

INPUT

OUTPUT

FIG. 24. Diagrammatic arrangement of six-speed Media gearbox.

of the mechanical transmission system already described. The engine drives a hydraulic torque converter from which the drive is transmitted through a multiple-plate friction clutch to a constant mesh gearbox incorporating two speeds and reverse. In common with other types of hydraulic transmission, gear changing takes place automatically at the appropriate rail speed.

Another constant mesh system is that originally introduced by the Swiss Locomotive and Machine Works, known as the S.L.M. transmission, in which each set of change speed gears is provided with an independent friction clutch. The clutches are contained inside the gears which are loosely mounted on the output shaft, and are operated by oil under pressure fed through passages cut in the shaft. Fluid couplings are not used with this type of gearbox, but the clutches are released automatically when the torque is reversed, and thus prevent stalling of the engine. Gearboxes of this type, which have been applied to both locomotives and railcars, can be easily adapted for multiple-unit working.

For railcar work, constant mesh gearboxes have been developed giving up to six or eight speeds, in which two or more friction clutches are in simultaneous engagement. The *Saurer eight-speed gearbox* incorporates four pairs of gears in constant mesh, and six multiple-plate friction clutches. The input and output gears are permanently connected to their respective shafts, but the remaining gears are coupled to the shafts by means of friction clutches in various combinations to give the speeds required. The clutches are operated by oil under pressure fed through passages cut in the shafts, and for each speed three clutches are in simultaneous engagement, thereby reducing wear of the friction surfaces under slipping conditions. The admission of oil to the

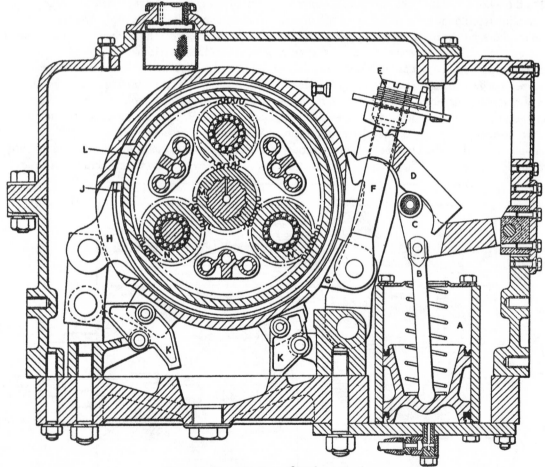

FIG. 25. Cross-section of Wilson gearbox.

(A) air cylinder
(B) operating strut
(C) cam plate
(D) thrust pad
(E) automatic adjuster group
(F) brake pull rod
(G) external brake band
(H) internal brake band
(J) brake lining
(K) brake centralizer
(L) annulus and brake drum
(M) sun wheel
(N) planet wheels

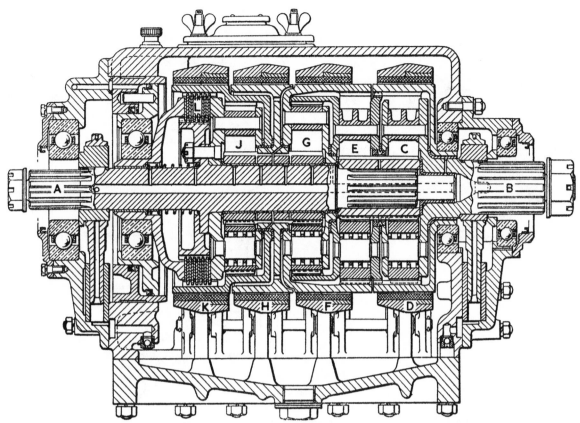

FIG. 26A. Longitudinal section of a Wilson five-speed gearbox.

(A) input shaft	(D) 1st-gear brake	(G) 3rd-gear train	(K) overdrive brake
(B) output shaft	(E) 2nd-gear train	(H) 3rd-gear brake	(L) top-gear clutch
(C) 1st-gear train	(F) 2nd-gear brake	(J) over-drive train	

clutch operating cylinders is controlled by electro-pneumatic valves.

Zahnradfabric manufacture constant mesh gearboxes of the layshaft type for use in railcars, in which engagement is effected by means of electro-magnetic clutches. More recently this firm has introduced the Media gearbox in which gear-changing clutches of the multiple-plate type are used. Engagement of these clutches, which are normally in the release position, is effected by pressing the plates together through the medium of pivoted levers, operated by means of sleeves mounted on the shafts.

The *six-speed gearbox* (Fig. 24) contains a drive shaft and two layshafts with five friction clutches. Two clutches are engaged for each speed. These clutches are not intended to transmit power under slipping conditions such as occur at starting, and for this purpose a main friction clutch or fluid coupling is employed. The clutches are operated by grooved cams cut in a shaft which is turned by a small electric motor.

A *constant mesh transmission for diesel trains*, developed over a number of years by Ganz, incorporates a multiple-plate friction clutch mounted on the engine flywheel, in conjunction with multiple-plate clutches inside the gearbox for engaging the change speed gears. The main friction clutch is used only when starting from rest, and is operated by compressed air. Compressed air is also used for operating the change speed clutches, two of which are in simultaneous engagement for each gear except top, when only one is used. The control system includes a device which automatically reduces the engine speed to obtain synchronization, when the driver operates the gear controller to engage a higher gear.

GEARS: EPICYCLIC

The *Wilson gearbox* (Fig. 25) is probably the best known application of epicyclic change speed gearing to loco-motives and railcars. The first-gear train which gives a reduction of about 4 to 1 consists of a pinion driven by the engine, termed the sun wheel, which meshes with three pinions surrounding it, termed planets. These are free to rotate on pins carried on a circular plate fixed to the gearbox output shaft. The planets also mesh with an

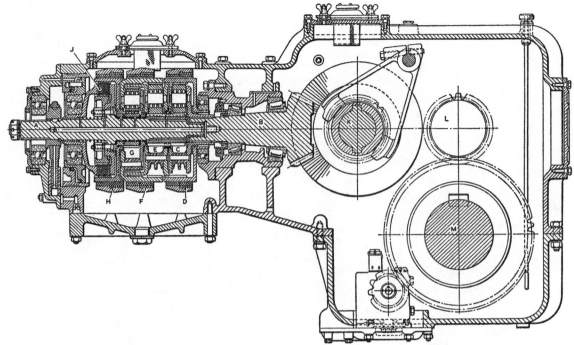

FIG. 26B. Longitudinal section of a Wilson four-speed gearbox and final drive.

(A) input shaft
(B) combined output shaft from change-speed unit and input bevel pinion to final drive
(C) 1st-gear train
(D) 1st-gear brake
(E) 2nd-gear train
(F) 2nd-gear brake
(G) 3rd-gear train
(H) 3rd-gear brake
(J) top-gear clutch
(K) top-shaft final drive unit
(L) intermediate shaft
(M) jack shaft

internally toothed gearwheel termed the annulus, which is held stationary by means of a band brake acting on its periphery (Figs. 26A & B).

To obtain the intermediate gears the band brake is released, and the annulus rotated at varying speeds. This has the effect of reducing the gear ratio. For top gear, a friction clutch is used which causes the complete gear assembly to rotate giving direct drive. To rotate the annulus, additional epicyclic gear trains are used which are brought into operation by means of band brakes.

The band brakes used in the Wilson gearbox are analogous to the friction clutches for engaging the gears used in constant mesh gearboxes. They are not suitable for transmitting the maximum torque under slip conditions at starting, and a fluid coupling is always used for this purpose.

Gear changing is generally effected by means of compressed air cylinders, one for operating each brake; compressed air is also used for operating the friction clutch which gives direct drive. An important feature of the Wilson gearbox is the arrangement for automatically taking up the wear in the brake bands, and ensuring that these are always in correct adjustment. The arrangement of brake bands is such that when the bands are tightened,

there are no unbalanced forces acting on the drum or its supports. Forced lubrication by means of a pump is used throughout. For multiple-unit working, electro-pneumatic valves are used for admitting compressed air to the brake-operating cylinders.

An epicyclic change speed gear which maintains the tractive effort during the gear-change interval has been developed by the Fluid Drive Engineering Co. Ltd, known as the *dual fluid drive transmission* (Fig. 27). Three gear ratios are provided which are brought into operation as required, through the medium of two fluid couplings and a synchro-clutch. The epicyclic gear train employed is basically similar to the one already described, and the first gear which gives a reduction of about 3 to 1 is obtained in the same manner, the drive being transmitted through the sun pinion and planets to the planet carrier which is connected to the output shaft.

The annulus forms part of a sleeve surrounding the output shaft, and the sleeve carries a synchro-clutch which can move axially on helical splines. When first and second gear are engaged the annulus tends to rotate backwards, but is held stationary by the action of the clutch, which engages with internal gear teeth forming part of the casing. A second annulus is provided which transmits the drive in second gear. The impellers of the two fluid couplings are connected together, and driven by the input shaft. The runner of the first- or low-gear coupling is connected to the sun pinion, and the runner of the second- or

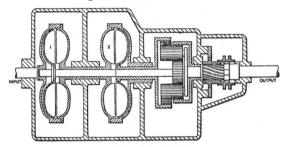

FIG. 27. Diagram of dual fluid drive transmission.

high-gear coupling is connected to the second-gear annulus by means of a sleeve shaft. The planet carrier carries two independent sets of planets which mesh with each other. One set also meshes with the sun pinion and fixable annulus, and the other set meshes with the second-gear annulus.

To obtain first gear, the low-gear coupling is filled and the drive is transmitted through the sun pinion to the planet carrier and output shaft. Second gear is obtained by emptying the low-gear coupling and filling the high-gear coupling. The drive is then transmitted through the second-gear annulus, both sets of planets, to the planet carrier and output shaft, giving a reduction of 1·8 to 1. In top gear both couplings are filled, and the sun pinion, second-gear annulus, and planet assembly rotate as a unit. The fixed annulus is freed by the action of the synchro-clutch, since the tendency to rotate backwards ceases, and the output shaft rotates at almost the same speed as the engine, the slight loss of speed being due to coupling slip.

The fluid couplings are provided with quick-emptying valves, and filling and emptying are carried out by means of a centrifugal pump driven from the input shaft. There is no loss of tractive effort when changing gear. Gear changing is carried out by means of compressed air which operates valves controlling the flow of fluid to the couplings. A locking sleeve operated by compressed air is also provided. This locks the synchro-clutch when first gear is used, and prevents it freeing itself when the engine is used as a brake, and the torque is reversed.

The complete assembly, comprising the two fluid couplings, change speed gear, filling pump and control valves are housed in a casing with a sump which contains the working fluid when the couplings are empty (Plate 16B, page 74).

GEARS: SYNCHRO-MESH

When dog clutches are used, smooth and shock-free engagement depends on synchronizing the speeds of the

two members between which a connection is to be made. One method of achieving this is by means of a clutch brake which in its simplest form consists of a friction disc which makes contact with the clutch runner, and reduces its speed when this is withdrawn to its fullest extent. Some degree of skill is required to change gear when this method of synchronization is used, but the principle of synchronizing the speeds of the two members by means of frictional contact before making a positive connection is embodied in the different systems of synchro-mesh in use.

The system developed by Mylius, which is used both for locomotives and railcars, consists of a main friction clutch and constant mesh gearbox of the layshaft type. The gears are engaged by gear-tooth-type dog clutches, and each dog clutch is provided with a cone-shaped synchronizing member.

Compressed air operation is used in conjunction with manual or compressed-air-operated pre-selection. The *pre-selection process* consists in setting the operating mechanism to operate the correct selector forks when compressed air is admitted to the operating cylinder. The first movement of the operating piston releases the main clutch. Further movement of the piston causes the selector forks to engage the appropriate synchronizing cones. Having reached the full extent of its travel the piston starts to return. The effect of this is to release the synchronizing cones and engage the dog teeth, which since the speeds have been synchronized, slide easily into engagement without shock. The final movement of the piston re-engages the main clutch. When in the case of multiple unit operation, remote control is used, electro-pneumatic valves are used for operating the clutch and pre-selective mechanism.

Another method of synchronization is incorporated in the *S.S.S. Powerflow transmission*, which maintains the tractive effort during the interval of gear changing (Fig. 28). Three speeds are provided, and for transmitting the drive in top gear a friction clutch is used, which is engaged momentarily when changing from first to second gear. This causes a reduction in engine speed, thereby synchronizing the speeds of the engaging members, and also transmitting torque to the driving wheels during the gear-change interval.

The third- or top-gear pinion is connected to the clutch runner by means of a sleeve which surrounds the input shaft, and meshes with two gears fixed to the output shaft and layshaft respectively. First and second gears are engaged by means of a special type of dog clutch termed a synchro-clutch mounted on the layshaft which can slide axially on helical splines, and will move to the right or to the left depending on whether the torque is applied through the dog teeth or through the splines. The synchro-clutch has external teeth and is provided with two sets of pawls facing in opposite directions.

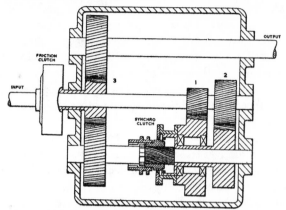

FIG. 28. Diagrammatic arrangement of S.S.S. Power-flow transmission.

When a change is made from first to second gear momentary engagement of the friction clutch reverses the torque and causes the synchro-clutch to move towards the right, so that the pawls overrun the internal clutch teeth of the first and second gears. As soon as the speeds of the synchro-clutch and second gear synchronize, the pawls engage the clutch teeth of the second gear, whereupon the two members engage smoothly and without shock. The synchro-clutch is then firmly locked to the layshaft by means of the locking ring which slides axially.

Gear changing is carried out by moving the locking ring, movement of which in one direction causes the compressed-air-operated friction clutch to engage momentarily. When the friction clutch is engaged to give top gear, the torque exerted on the layshaft causes the synchro-clutch to move over to the right as far as it will go. With this type of transmission a scoop-controlled fluid coupling is always used. It is manufactured in varying sizes designed to transmit from 25 to 500 H.P. and is used only for locomotives. To increase its sphere of operation extra gearing is often provided, so that the three speeds can be made available for both a low and a high speed range.

PROPULSION BY THE FELL SYSTEM
(see page 86)

The Fell System of locomotive propulsion embodies pressure-charged diesel engines in which the power output remains approximately constant throughout the speed range. This is made possible by using independently driven blowers, which supply combustion air at a gradually decreasing pressure as the engine speed increases. The mean effective pressure and the torque, therefore,

diminish with increasing speed in a manner similar to that which occurs in steam locomotives.

Four separate engines, each provided with scoop-controlled fluid couplings, are employed for traction purposes. Each pair of engines drives through a primary differential, and the combined output of the four engines is transmitted through a secondary differential, from which the power is transmitted to the driving wheels through final reduction and reverse gears. The purpose of the differentials is to enable the engines to work together when operating at different speeds, and developing different torques (Fig. 29).

By means of the scoop-controlled fluid couplings, the engines can be connected or disconnected as required, and synchro-clutches are provided which prevent the primary shafts rotating backwards when no torque is being transmitted. Fig. 30 shows tractive effort and power curves when working at maximum output. At starting, power

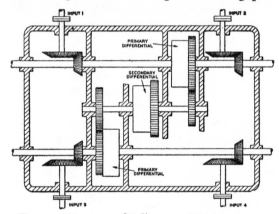

FIG. 29. Diagram of Fell transmission system.

is transmitted to the driving wheels by one pair of engines. In order to increase the starting tractive effort, the gearing which transmits the drive to the secondary shaft is arranged to give a somewhat greater speed reduction than that provided for the second pair of engines. These commence driving when a speed of about 17 m.p.h. has been attained. The flat portions of the T.E. curve represent the periods in which the fluid couplings are slipping, and for the second pair of engines, slipping terminates when a speed of 31 m.p.h. has been reached. Thereafter, all four engines transmit their maximum power output to the wheels. Since the Fell system does not incorporate change speed gearing as the term is generally understood, the maximum speed of the engines is only attained when the locomotive is travelling at maximum speed.

It will be appreciated that since the two pairs of engines commence driving at different rail speeds, each pair must be rotating at different speeds and developing different torques at the same time. These differences are adjusted by the secondary differential which ensures that, neglec-

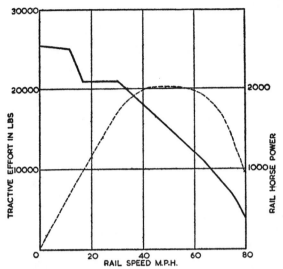

FIG. 30. Performance curves – Fell transmission.

ting transmission losses, the power transmitted to the wheels is equal to the total power developed by the number of engines driving. When the fluid couplings are slipping, a considerable proportion of the power developed will be dissipated as heat.

The locomotive is controlled by a master regulator, which regulates the speed of the engines. Special provision is made to ensure that the engines in each group develop equal power outputs. Four starting levers are provided for engaging or disengaging the fluid couplings. The spur reverse gear can only be operated when the locomotive is stationary. For driving the blowers and other auxiliaries, two separate engines are used, thus permitting the entire output of the four main engines to be used for traction purposes.

When in service, the 120-ton, 2–D–2 type Fell locomotive owned by British Railways, developed a maximum output of 1,900 d.b.h.p. at a speed of 44 m.p.h., and a maximum drawbar pull at starting of 29,400 lb. (Plate 10A, page 68).

Part VI. Diesel locomotives

DEVELOPMENT

(*See* History of Diesel in North America, page 110)

The first successful application of the diesel engine to railway traction was in Sweden in 1913 (*see* Diesel Railcars – Development, page 90).

Diesel locomotives were first successfully used in Russia. As a result of the enlightened thinking of Lenin, Professor G. V. Lomonossoff was, in 1922, granted £100,000 with no "strings" for the construction of three diesel locomotives. Ultimately, designs for four 1,200 H.P. locomotives were developed in collaboration with the German Reichsbahn. Each locomotive had a different transmission system – electric, mechanical, pneumatic and hydraulic. Only the diesel-electric and diesel-mechanical machines were successful. Construction of the diesel-pneumatic locomotive was delayed and then proved impracticable while the diesel-hydraulic was never completed, as experience with several smaller machines in the interim had been a complete failure.

Between 1923 and 1925, transmissions using compressed steam and exhaust gas were tried and discarded. In view of the recent large-scale adoption of the Mekydro and similar transmissions (*see* page 58) it is of interest that the Swiss Locomotive Company experimented with a hydro-mechanical transmission (Schneider) but found it impracticable.

In those early days, it was the diesel-electric locomotive which came to the forefront, and one of the most notable

examples in Europe was the Fiat narrow-gauge B–B locomotive built in 1924, for the Mediterranean Railway of Italy. The diesel engine in this locomotive used the two-stroke cycle. At this time also, a low-powered locomotive having an Atlas diesel was introduced on the Tunisian Railways.

Diesel motive power for railway locomotives was not established until the early nineteen thirties. In the preceding five years, however, the railways of many countries were experimenting with diesels built in Britain, Germany, and some other European countries. South America, India, Ceylon and Japan were early users of this form of motive power, while the Russians steadily increased their usage of fairly high-powered locomotives for use in the vast areas of poor water supplies.

It is interesting that at this time the Germans, who achieved such brilliant success with their diesel railcars and trains (culminating in the "Flying Hamburger" of 1932), had little success in the field of diesel locomotives. This was probably due to their adherence to hydraulic or mechanical transmissions, both of which were, at that stage, inferior to electric transmission for anything other than low horsepower engines.

BOGIE AND ARTICULATED LOCOMOTIVES

The smaller type of bogie or articulated locomotive is mainly used for working over lightly constructed narrow-guage railways. A typical locomotive of this class weighing twenty-five tons, intended for working on a 24-in. gauge sugar plantation railway has been constructed by the Hunslet Engine Co. Ltd. It is provided with two outside plate frame bogies which carry the drawgear, each having four wheels. The Hunslet gearbox, giving four speeds in either direction, is placed in the centre of the underframe and drives the two outer bogie axles, which carry bevel reduction gear boxes, by means of cardan shafts. Coupling rods are used to couple the wheels of each bogie.

An articulated locomotive with Voith hydraulic transmission is manufactured by the Arn Jung Lokomotivfabrik. This consists virtually of two locomotives, each with its own engine and transmission, permanently coupled together, and carrying a common cab structure in the centre, each unit being provided with four wheels. The cab is carried on a resilient three-point suspension system, and is unaffected by the movement of one unit relative to the other. Locomotives of this type are built with outputs up to 440 B.H.P. The smaller locomotives are provided with chain drive or cardan shaft drive to the axles, and the larger ones with jack shafts and side rods (Plates 9A, 9B and 9C, page 67).

Diesel-electric locomotives intended for operating over routes where severe track curvature exists, may have the drawgear mounted on the bogies, which are coupled together by means of a hinge joint. This permits both vertical and radial movement of one bogie relative to the other. The underframe and superstructure is thus relieved from traction and braking stresses. This arrangement reduces the flange wear on curves but gives inferior riding at the higher speeds. Larger locomotives of this type are provided with guiding bogies or radial trucks in addition.

The 1,500 H.P. 2–Co–Co–2 type diesel-electric locomotives, constructed by the English Electric Co. Ltd for the New Zealand Railways, and capable of operating at speeds up to 60 m.p.h., employ a modified arrangement. The main bogies carry the drawgear, and are coupled together by means of a spring-controlled guiding device which helps to reduce flange wear. Traction and braking stresses, however, are transmitted through the bogie pivots to the underframe (Plate 14C, page 72).

The two 1,750 H.P. and the one 2,000 H.P. 1–Co–Co–1 type locomotives Nos. 10201–2 and 10203 (Plate 14A, page 72) operating on British Railways carry the drawgear on the main bogies but these are not coupled together in any way. One of the latest locomotives of this type is shown in Plate 14B, page 72.

In the majority of cases, locomotives carried on two non-articulated bogies are preferred. The space between the bogies can then be used to accommodate the fuel tank, etc., and the riding of the locomotive is improved. This type of construction lends itself well to electric transmission, and is often used for diesel-electric freight and shunting locomotives. Locomotives provided with two four-wheeled bogies, and having all the axles driven, are widely used for all classes of work (Plates 10B and 10C, page 68). Diesel-hydraulic locomotives of this type incorporating two engines and two sets of transmission, are constructed with outputs up to 2,200 H.P. (Plates 15A, 15B, 15C and 15D, page 73). Twin engine and generator sets are often used for diesel-electric locomotives also. To keep the axle loading within prescribed limits it may be necessary to provide non-driving axles, and in this case six-wheeled bogies are used in which the drive is transmitted to the outer axles only. Both diesel-electric and diesel-hydraulic locomotives are constructed with six-wheeled bogies having all the axles driven.

Moderately powered locomotives are usually provided with auxiliaries belt-driven from the engine, but the larger locomotives including those with hydraulic transmission employ electrically-driven auxiliaries.

Superstructures may be of the body type with cabs at both ends (U.S.A. "cab units"), or of the hooded type with a single cab (U.S.A. "hood units"), (Plate 13B, page 71). With the former type of construction, nose ends are often provided in front of the cabs (Plates 11A and 11C, page 69). These help to reduce the effect of "sleeper flicker" on the driver, and are generally considered to improve the appearance of the locomotive. In Germany, a single-turret cab is widely used for all types of diesel-hydraulic locomotive. The radiator may be built into the roof structure, where together with the fan, fan motor and ducting, it forms a complete assembly removable as a unit. Other types of superstructures are shown in Plates 11B, 12B, 13A and 13C, pages 69, 70 and 71.

RIGID FRAME LOCOMOTIVES

Chain drive. The simplest and cheapest method of transmitting the drive to the wheels, by means of roller chains, is used for locomotives with mechanical, hydraulic or electrical transmission ranging from 10 up to 400 H.P. A substantial gear reduction can be obtained by adopting different sizes for the chain sprockets mounted on the gearbox output shaft and driving axle, thereby eliminating the massive and relatively expensive final reduction gears, shafts and bearings required for side-rod driven locomotives. So that the correct chain tensions can be maintained, a *chain tensioning device* must be incorporated. This generally consists of an arrangement whereby the position of the axles can be altered relative to the frame.

Norwegian State Railways

PLATE II. Norwegian State Railway: 400 H.P. diesel railcar.

In four-wheeled locomotives, the gearbox output shaft is sometimes located about midway between the axles with driving chains to each axle. It may also be located outside the wheelbase driving the leading or trailing axles. When the latter arrangement is used, chains and sprockets must also be employed for coupling the axles together.

The 23-ton diesel-mechanical shunters (switchers), constructed by F.C.Hibberd & Co., incorporate a drive of the latter type. This arrangement enables the engine, which carries a fluid coupling to be resiliently mounted in the frame, the drive being transmitted through a long propeller shaft to a four-speed Wilson epicycle gearbox, which forms a unit construction with the air-operated bevel reverse gear. A triplex roller chain is used for transmitting the drive from the gearbox output shaft to the rear axle, and also for coupling the leading and trailing axles.

The frame is constructed from steel plate and rolled steel sections, and is without hornguides. The axles are located by means of adjustable radius rods which connect the axle-boxes to the frame, and are so arranged that the distance between the centres of the sprockets remains approximately constant when spring deflections occur. The laminated bearing springs are rigidly clamped to the axle-boxes, the ends being free to slide on wearing plates fixed to the frame. This type of drive can also be applied to locomotives having more than two driving axles.

Other manufacturers employ outside plate frames with adjustable hornguides for tensioning the driving chains. Locomotives embodying this type of construction, with both hydraulic and electric transmission are used on the S.N.C.F. A 32-ton diesel-electric locomotive of this type, built by De Dietrich, incorporates a single frame-mounted traction motor in conjunction with a reduction gearbox giving two alternative gear ratios, from which the drive is transmitted to both axles by roller chains. The alternating gear ratios enable the locomotive to be employed on shunting work with a maximum speed of 12·5 m.p.h., or for operating on the main line at speeds up to 37 m.p.h. (Plate 9D, page 67).

Individual axle drive. Rigid frame diesel-electric locomotives are sometimes built in which each of the driving axles is provided with a separate traction motor. A disadvantage of this arrangement is that individual wheel slip can occur, and to counteract this, coupling rods are occasionally fitted. On account of the increase in constructional and maintenance costs, however, this procedure is rarely adopted except in the case of four- and six-wheeled coupled shunting locomotives.

During the period immediately prior to World War II, some large locomotives were built in which advantage was taken of this form of construction to incorporate traction motors rigidly fixed to the frames with quill and cup drive to the wheels. In France four high-speed, 2,200 H.P. locomotives of the 2–Co–2 type were built, designed to work in pairs. Two of these were equipped with Sulzer 12-cylinder twin-bank engines. Two locomotives of the 2–Do–2 type provided with similar engines and also designed to work as one unit, were constructed by Brown Boveri for service in Roumania. In recent years, however, bogie-type locomotives have been generally preferred.

Rigid frame diesel-electric locomotives with axle-hung traction motors, and provided with leading and trailing radial trucks or bogies are suitable for operating on narrow-gauge railways where restricted axle loadings are in force. By distributing the weight over a number of axles high-power outputs can be obtained under these conditions. The locomotives of this type constructed by Frich for the Thailand State Railways have already been noted. Recent examples are the 2–Do–2 type passenger locomotives constructed by Metropolitan-Vickers for the West Australian Government Railways (Plate 14D, page 72).

Shaft drive. This type of drive is used by several European builders for small four-wheeled locomotives, provided with mechanical or hydraulic transmission. The locomotives are generally constructed with outside frames, incorporating hornguides and spring gear of the orthodox pattern, each axle being driven. The drive is transmitted through cardan shafts to axle-mounted gearboxes, anchored to the frames, in which it is turned through 90° by means of bevel reduction gears. With this arrangement spur reversing gear can be used.

The Swedish State Railways employ a number of small 28-ton locomotives of this type, for mixed service on branch lines, many of which are provided with Krupp hydraulic transmission. In Russia also, four-wheeled locomotives with shaft drive are employed on shunting duties. These locomotives are provided with engines developing from 200 up to 400 H.P. and Voith hydraulic transmission. They were constructed at Jenbach in Austria.

Side-rod drive. The majority of rigid-frame diesel locomotives are provided with side-rod drive, and the construction of the frames and running gear resembles that used for conventional steam locomotives. This method of transmitting the power to the wheels is more suitable when increased power outputs and intensive service are required, than that provided by roller chains. It is also relatively cheap in comparison with shaft drive, and the constructional technique involved is well understood by the various locomotive builders.

Diesel-mechanical and diesel-hydraulic locomotives usually incorporate a *jack shaft* carried in the main frames, from which the drive is transmitted to the driving wheels by means of *connecting rods*; occasionally, however, the gearbox is mounted on one of the driving axles and flexibly connected to the frames. Almost invariably the engine forms a separate unit, and may be either rigidly bolted to the frames or resiliently mounted, in order to insulate it from traction and buffing stresses.

A *propeller shaft* is generally used to transmit the drive to the gearbox. This may be provided with resilient couplings which not only permit a small amount of angular misalignment, but also provide torsional flexibility and possess valuable shock-absorbing characteristics. When the engine output and gearbox input are arranged permanently out of line, a *telescopic propeller shaft* with universal joints must be used. Sometimes there is insufficient space between the engine and gearbox to permit the use of a propeller shaft, and to meet such cases special types of resilient couplings are available which allow for slight angular misalignment.

The transmission, including the reverse and final reduction gears, is usually housed in a single gear case attached rigidly to the frames by means of fitted bolts. This may be of cast iron or cast steel or fabricated by welding. To facilitate dismantling and assembling, the gear case is split through the shaft centre lines, the sections being firmly bolted together, and located relative to each other by means of dowel pins.

The transmission developed by the Drewry Car Company for diesel-mechanical locomotives of this type, incorporates a Wilson gearbox with cardan-shaft drive from the engine, together with a separate reverse and final reduction gearbox. Both of these units are carried in a sub-frame, one end of which is pivoted to the main frames by means of pins fitted with resilient bushes. The final reduction gearbox is mounted on the jack shaft which is carried in bearings fixed to the frames, and the sub-frame acts as a torque arm for taking the torque reaction. This arrangement helps to insulate the gearbox from traction and buffing stresses.

For locomotives of moderate power, *belt-driven auxiliaries* are usually preferred. These include the radiator fan, air compressor, etc., generally driven from pulleys mounted on the forward extension of the engine crankshaft. An exception occurs in the case of the 637-H.P. diesel-hydraulic locomotives constructed by the Hunslet Engine Company, where an auxiliary diesel engine developing 75 H.P. is employed for this purpose.

The majority of locomotives of this type are provided with coupled wheels only, and the entire weight is available for adhesion. To improve the flexibility without sacrificing adhesive weight, a number of eight-coupled diesel-hydraulic locomotives have been constructed by M.A.K. and other makers, in which the principle of the *Beugniot bogie,* originally developed many years ago for steam locomotives has been revived. The wheels are arranged in two groups with the jack shaft placed approximately midway between. The axle-boxes for each axle are connected by a casing, and are provided with about half an inch of side movement in either direction. Each group is provided with a horizontal lever, pivoted in the centre to a frame stretcher, and with the ends bearing on the axle-box casings in such a manner that the end of the lever can move radially about the frame pivot when the axle-boxes move sideways. This arrangement gives much better riding and curving than would be obtained if the axle-boxes were provided with uncontrolled side movement, and enables a locomotive of this type to operate satisfactorily at speeds up to 55 m.p.h. (Plate 16A, page 74).

Cases sometimes occur where rigid frame locomotives with side-rod drive are provided with guiding axles or bogies, a notable example was the 2–D–2 type *Fell locomotive* which ran on British Railways (Plate 10A, page 68 and *also* page 80). The plate frames were placed outside the wheels, with four-wheeled bogies at both ends. The gearbox was completely spring borne, and the drive was transmitted through hollow quills surrounding the two inner driving axles, which were flexibly connected to the spokes of the driving wheels. This arrangement allowed for the rise and fall of the locomotive on the springs, and by providing torsional flexibility protected the transmission from shocks. The frames were rigidly stayed together at both ends by box girders placed above the bogie centres, which also served as fuel tanks. Tubular cross stays, employed as air ducts to the main engines, were used at other points. Like many other modern locomotives, the driving axles were provided with roller bearings.

Rigid frame locomotives with electric transmission may be equipped with a single frame-mounted traction motor driving a jack shaft through reduction gearing, from which the drive is transmitted to the wheels by means of inclined connecting rods coupled to triangulated side rods. This type of drive was used for the original series of diesel-electric shunting locomotives with inside frames, built by the L.M.S. (Plate 16C, page 74). Generally, however, axle-hung traction motors are employed, one or two motors being used in accordance with requirements. To accommodate the normal type of traction motor with spur reduction gearing, the axle-boxes, frames and spring gear must be placed outside the wheels (Plate 16D, page 74). If, however, the motor is placed with its axis parallel to the longitudinal centre line of the locomotive, and provided with bevel drive, inside frames may be used for standard and broad gauge locomotives. This type of drive in conjunction with spur reduction gearing is used for diesel-electric shunting locomotives built by the Yorkshire Engine Company and by Ruston, Hornsby.

It is often found more convenient in the case of locomotives with electric transmission to drive the auxiliaries, with the exception of the radiator fan, from a pulley mounted on an extension of the armature shaft of the main generator. In addition to the air compressor or exhauster, these include the auxiliary generator or exciter, and blower for forced ventilation of the traction motors.

The majority of rigid-frame diesel locomotives intended for shunting and mixed traffic working are provided with *superstructures of the hooded type*, the cab being placed approximately in the centre or at one end. The conventional type of cab is provided with doors at the sides, but there is an increasing tendency to adopt the modern American pattern with doors opening on to the platform at the front and back. Sometimes access to the cab is from a wide platform at the rear of the locomotive.

It is desirable to give the driver as wide a field of vision as possible, but whatever is achieved in this direction depends ultimately on the limiting height and width imposed by the structural gauge. Some European builders, such as M.A.K., have adopted a turret-shaped cab projecting above the top of the hood and giving a practically unobstructed view in all directions. The controls are generally arranged so that they can be operated from both sides of the cab, and sometimes the driver is provided with two control desks facing in opposite directions.

Locomotives with outputs up to about 300 H.P. are usually fitted with *axial flow radiators* placed in front of the engine, the fan being driven by belts from a pulley mounted on the forward extension of the engine crankshaft. In larger locomotives two radiator banks are used, mounted opposite each other in the sides of the engine housing. The fan is placed vertically in the centre and is driven by shafts and bevel gears from the engine, air being drawn through both radiator banks and discharged through the top of the engine housing.

STRUCTURAL DATA: BOGIES

Locomotives intended to operate at comparatively low speeds are often provided with *fixed centre bogies* of simple and robust construction. Bogies of this type adapted to carry the drawgear, and coupled together by means of a hinge joint which transmits the traction and buffing forces, are used for articulated locomotives. The centre pivots which carry the weight of the superstructure are sometimes made hemispherical so that the bogies can adapt themselves to inequalities in the track without causing excessive deflection of the bearing springs. Side bearers with a small clearance are provided, to check rolling. These may also incorporate lifting lugs for lifting the bogies complete with the superstructure, thus eliminating the need for bogie centre pins (Plate 21A, page 97).

Improved riding at higher speeds is obtained by using bogies with *sprung bolsters*, suspended from the bogie frames by means of inclined hangers, so that the bogie can move both radially and laterally relative to the underframe. The axle-boxes are often connected by equalizer bars, to which the load is applied through nests of coil springs. The equalizers are usually of the swan-neck pattern, bearing on top of the axle-boxes, but straight bars are also used (as in British Railways locomotives Nos. 10000 and 10001), and straight bars slung beneath the axle-boxes are being used to an increasing extent. In bogies of this type, elliptical bolster springs are generally employed to damp out the vertical oscillations (Plate 21B, page 97).

A simplified version of this type of bogie is now being used which incorporates an unsprung bolster carried on inclined swing links, with hydraulic shock absorbers to dampen the action of the coil bearing springs. This type of bogie is used for the 1,200 H.P. Co–Co diesel-electric locomotives built by Metropolitan-Vickers for Coras Iompair Eireann (Plate 12B, page 70). To give softer riding, sprung bolster bogies are sometimes constructed with widely spaced coil bolster springs, and inclined hangers placed outside the frames.

Six-wheeled bogies may be provided with twin bolsters, joined together by a member which passes above the centre axle, and carries the bogie centre pivot. The weight of the superstructure is transmitted through bearers moving in radial guideways, to the corners of the bolster directly above the bolster springs. The centre pivot is used to transmit traction and braking forces only.

A modified design is used by Henschel for the 1,900 H.P. A1A–A1A diesel-electric locomotives supplied to the Egyptian Railways (Plate 20B, page 96). The weight of the superstructure is transmitted to the centre of the H-shaped bolster which is supported on four nests of coil springs, located at the corners, and resting in pockets integral with the bogie frame. No swing links are used, but the coil springs permit a small amount of lateral movement to take place.

With the development of hydraulic transmission, bogies without either bolsters or centre pivots are being increasingly used. This arrangement enables the transmission unit located in the main frames to project downwards into the centre of the bogie, with short cardan shafts driving both axles. The bogie is anchored to the main frames by means of a simple linkage, the pins of which are fitted with resilient bushes. The linkage controls the movement of the bogie, and transmits the traction and braking forces. The weight of the superstructure is carried by spring-loaded side bearers (Plate 21C, page 97).

The bogie frame is required to withstand indeterminate vertical and lateral forces caused by irregularities in the track; it must also be rigid enough in the horizontal plane to resist the tendency to distortion which occurs when the vehicle is running through a curve. Bogie frames may be built up from steel plate and rolled steel sections, by riveting or welding, or may be one-piece steel castings. When an all-welded frame is used, it is important to remove the stresses set up during the welding process by subsequent annealing. It is not always convenient to anneal a complete frame, and some makers prefer to manufacture the sole-bars and cross stretchers as sub-assemblies, and rivet them together. To increase the stiffness in the horizontal plane, sole-bars are often fabricated in the form of a box section. Cast steel frames are made of hollow box section throughout.

Roller-bearing axle-boxes of different types are now almost universal, but in France, *Athermos axle-boxes* with plain bearings and forced lubrication are used. *Hornguides,* instead of being in the form of separate castings, are usually made integral with the sole-bars, and together with the axle-boxes, are provided with manganese steel liners on the wearing surfaces, thus enabling high mileages to be achieved before renewal is necessary.

There is a noticeable tendency to eliminate wearing surfaces completely, and many bogies are constructed without conventional hornguides. An arrangement widely used in Switzerland and other European countries, incorporates wing-type axle-boxes and coil bearing springs. The axle-boxes are located by means of cylindrical guides containing oil, which also function as hydraulic shock absorbers, placed inside the bearing springs. In the system used by Alsthom the axle-box is connected to the frame by two short links located at diagonally opposite corners. These are provided with resilient bushes of considerable length, designed to transmit to the frame the lateral forces set up at the point of contact of the wheel flange with the rail, together with the traction and braking forces. The weight is transmitted to the axle-boxes by means of laminated or coil bearing springs.

The conventional type of bogie with a sprung bolster and provided with a bogie centre pin, makes use of a hemispherical centre pivot, and side bearers provided

with a small clearance. Flat centre pivots of large diameter, without centre pins, and giving a high degree of stability, are commonly used with cast steel bogies. These are provided with renewable wearing plates on the bottom and round the sides.

Alsthom employ a bogie centre pivot consisting of a rubber cone located in suitable housings attached to the bogie and underframe. Pivoting and other movements of the bogie relative to the underframe cause deformation of the rubber, and all wearing surfaces are eliminated. The weight is transmitted to the bogie through the centre pivot in conjunction with resilient side bearers.

A modified arrangement is used for the bogies of the 1,800 H.P. Co-Co type diesel-electric locomotives built for the French National Railways (Plate 13C, page 71). Each bogie is provided with two pivots located on the longitudinal centre line, midway between the axles. Each pivot is virtually a short strut provided with rubber cones top and bottom, pointing towards each other, and when the locomotive traverses a curve the pivots tilt in opposite directions. The tilting action is controlled by means of spring-loaded links, one at either side of each pivot, connecting it to the main frame. Four side bearers are also provided.

STRUCTURAL DATA: FRAMES AND SUPERSTRUCTURE

Rigid-frame diesel locomotives have longitudinal frame members made of steel plate, braced together to form a rigid structure, designed to withstand the buffing shocks and other stresses to which it is subjected. Details such as axle-boxes, hornguides, spring gear and brake rigging are generally similar to those used for steam locomotives.

In *bogie locomotives* the underframe forms a girder supported at the bogie pivots, and must possess sufficient stiffness to carry the imposed loads without excessive deflection. The buffing and drawgear is normally carried on the underframe, which besides carrying the vertical loads due to the weight of the equipment, must also be capable of withstanding buffing shocks, etc.

Such underframes are generally fabricated by welding from steel plate and rolled sections, the principal load-carrying members being the two main longitudinals which are frequently made of deep beam section. These are rigidly braced together at the bogie centres and other points, the attachment to the buffer beams depending on the type of buffing and drawgear used, together with its height above the rails. The outer edges of the superstructure are supported by longitudinals of lighter section, although sometimes the arrangement is reversed and the deep-section longitudinals are placed at the outside. This makes heavier cross-bracing necessary. The top of the

No.	Railway	Builder	Type	Total weight tons	Adhesive weight tons	Maximum tractive effort lb.	Driving wheels diam./in.	Max. speed m.p.h.	Engine output B.H.P./r.p.m.	Maker	Transmission	Maker	Final Drive	Rail Gauge
1	Industrial	Hunslet	B-B	25½	25½	16,150	24	13	230/1200	Gardner	Mech.	Hunslet	Gear & side rod	2' 0"
2	German Federal	Arn Jung	B+B	29	29	22,000	—	21·8	2(145/2000)*	Deutz	Hydr.	Voith	Chain	Metre
3	New Zealand	Eng. Elec.	2-Co-Co-2	102	69	38,500	37	60	1500/850	Eng. Elec.	Elec.	Eng. Elec.	Gear	3' 6"
4	B.R.	B.R.	1-Co-Co-1	133	109½	50,000	43	90	2000/850	Eng. Elec.	Elec.	Eng. Elec.	Gear	4' 8½"
5	German Federal	Krauss-Maffei	B-B	75	75	56,800	—	87	2(1000/1500)	Daimler	Hydr.	Maybach	Gear & Shaft	4' 8½"
6	Industrial	Hibberd	B	23	23	10,650	37½	13·2	117/1200	Dorman	Mech. Wilson	S.C.G.	Chain	4' 8½"
7	S.N.C.F.	De Dietrich	B	32	32	15,400	—	37	150/1500	Renault	Elec.	—	Chain	4' 8½"
8	West Australian	Beyer Peacock	2-D-2	77½	41	28,000	36	55	1105/625	Crossley	Elec.	Metro-Vick	Gear	3' 6"
9	Swedish State	Nohab	B	28	28	20,000	38½	37	295/950	M.A.N.	Hydr.	Krupp	Gear & Shaft	4' 8½"
10	Paita Piura	Hunslet	D	55	55	35,400	48	33	500/1375	Paxman	Mech.	Hunslet	Jack shaft & side rod	4' 8½"
11	German Federal	M.A.K.	D	54	54	37,500	49	50	600/750	M.A.K.	Hydr.	Voith	Jack shaft & side rod	4' 8½"
12	B.R.	B.R.	2-D-2	120	75	33,500	48	78	4(500/1500)*	Paxman	Mech. Fell	B.R.	Gear & side rod	4' 8½"
13	Industrial	Ruston Hornsby	B	28	28	17,000	38½	17·5	155/1250	Ruston Hornsby	Elec.	B.T.H.	Gear & side rod	4' 8½"
14	Coras Iompair Eireann	Birm C & W Co.	A1A-A1A	75	57	41,800	37½	75	960/710	Sulzer	Elec.	Metro-Vick	Gear	5' 3"
15	Egyptian	Henschel	A1A-A1A	132	88	48,500	42	65	2(950/835)	G.M.	Elec.	E.M.D.	Gear	4' 8½"
16	S.N.C.F.	Alsthom	Co-Co	110	110	59,000	41½	81	1800/1500	M.G.O.	Elec.	Alsthom	Gear	4' 8½"

* The total output is available for traction purposes, power for auxiliaries being supplied by a separate engine.

underframe is often covered by a continuous steel deck plate, which prevents the leakage of oil and the ingress of dust. This plate is generally too thin to contribute much to the strength; sometimes the fuel tank when mounted between the bogies is used to stiffen the underframe.

When the body type of superstructure is used and weight saving is a matter of importance, the framework carrying the body is often made integral with the underframe. With this type of construction, the outer longitudinals and body side frames become the main load-carrying members. The side frames are stiffened by means of diagonal bracing and rigidly fixed to the outer longitudinals or sole-bars. Cross-bracing at roof level can only be used to a limited extent, since a large portion of

the roof must be made removable for placing the engine and transmission in position. In the most advanced designs hollow girders and tubes are used for the underframe and superstructure, instead of rolled steel sections.

BRAKES, TYPES OF
(*see* Braking Systems, page 103).

Clasp brakes are generally applied to all wheels, operated by brake cylinders mounted on the bogie. Sometimes one brake cylinder is used for each wheel. Automatic slack adjusters are often incorporated.

Disc brakes are used to a limited extent for locomotives with mechanical or hydraulic transmission, but cannot normally be applied when electric transmission is used.

Part VII. Diesel railcars and diesel trains

DEVELOPMENT

Petrol- (gasoline-) driven railcars employing both mechanical and electrical transmissions were introduced experimentally by several British railway companies during the years prior to World War I. It was in Sweden, however, that diesel-electric railcars were first employed successfully in regular service, for in 1913 an eight-wheeled non-bogie railcar was completed with accommodation for fifty-one passengers, and an empty weight of twenty-nine tons. A 6-cylinder diesel engine developing 75 B.H.P. at 550 r.p.m. was coupled to a D.C. generator which supplied current for two axle-hung traction motors mounted on the inner axles. Later, similar railcars were built provided with engines developing up to 250 B.H.P. and capable of hauling three or four trailers.

The first diesel-electric multi-car train to operate in Great Britain commenced working on the L.M.S. between Blackpool and Lytham in 1928. It consisted of four cars of converted electric stock and weighed 144 tons empty. A Beardmore engine developing 500 B.H.P. at 900 r.p.m. coupled to a D.C. generator was mounted in a special compartment of the leading coach, one of the bogies of which was equipped with two traction motors. Driving compartments were provided at both ends of the train.

A diesel-electric railcar built by Armstrong Whitworth, designed for operating in multiple unit, worked for a period on the L.N.E.R. in 1932. It weighed forty-three tons empty and was equipped with an Armstrong-Sulzer 6-cylinder engine developing 250 B.H.P. at 775 r.p.m. coupled to a single-bearing generator. Two traction motors were mounted on the leading bogie. One of these cars, equipped with a buffet, provided a non-

stop service between Euston and Birmingham during the 1933 British Industries Fair.

In 1934, diesel railcar workings ranging from fast inter-city trains to branch line and parcels service, were inaugurated by the G.W.R. and gradually extended up to the outbreak of war. The railcars employed for these services were developed by A.E.C., and incorporated engines and transmissions similar to those used for London buses. Each car was provided with two engines and two sets of transmission. Later examples were designed to work in multiple-unit (Plate 17B, page 93).

A bogie railcar with hydraulic transmission commenced working in Ireland on the lines operated by the Northern Counties Committee of the L.M.S. in 1933. Two under-floor-mounted Leyland engines drove the inner bogie axles through Lysholm-Smith hydraulic torque converters. Similar engines and transmissions were used for the experimental, three-coach articulated train built by the L.M.S. in 1938. This weighed seventy-five tons empty and was provided with six engines, each driving separate axles, so that only two axles were not driven. Each engine developed 125 B.H.P. at 2,200 r.p.m.

It was in Europe, however, where the most outstanding developments took place. In 1932 the first German high-speed diesel-electric train known as the "Flying Hamburger" was placed in service. This consisted of a two-coach articulated set with engines and generators mounted on the leading and trailing bogies. The centre bogie carried two force-ventilated, axle-hung traction motors. The weight empty was seventy-six tons, and the two Maybach 12-cylinder engines gave a combined output of 820 B.H.P. at 1,400 r.p.m.

In 1936, three-coach articulated sets were introduced,

one of which attained a speed of 127 m.p.h. The two Maybach engines were provided with exhaust turbo-pressure chargers, which enabled the combined output to be increased to 1,200 B.H.P. at 1,400 r.p.m. Two of the three coach sets were provided with Voith hydraulic transmissions, consisting of three torque converters operating in sequence. Engines and transmissions were mounted on the bogies, and the empty weight was 114 tons compared with 124 tons for those sets provided with electric transmission.

High-powered diesel-electric trains were also introduced in France, Belgium, Holland and Denmark, and railcars with mechanical transmission were developed in France.

BOGIE AND ARTICULATED RAILCARS

To increase carrying capacity and to improve riding characteristics, the majority of railcars are provided with bogies. A large number of railcars of this type operating in Great Britain are equipped with engines and transmissions resembling those used for heavy road vehicles. Each car normally hauls one trailer, and twin engines are used with a total output of 300 B.H.P. This arrangement permits a high degree of flexibility in the formation of diesel trains, and the number of power cars and trailers can be varied to meet requirements. Railcars are also constructed with outputs up to 1,000 H.P. and capable of hauling several trailers. When the higher powered engines are used it is generally necessary to divide the drive between two or more axles.

The twin-engined cars built by A.E.C. Ltd for the G.W.R. were equipped with two 6-cylinder vertical engines mounted on opposite sides of the underframe and driving separate bogies (Plate 17B, page 93). The drive from each engine was transmitted through a fluid coupling and Wilson four-speed gearbox to a bevel reverse unit mounted on the end of the inner bogie axle. The radiators were carried on the sides of the underframe ahead of the engines, and the cooling fans were mounted on the propeller shaft which transmitted the drive to the gearbox. This type of drive makes the engine and transmission very accessible without encroaching on the space inside the body, but in some cases is difficult to accommodate within the confines of the loading gauge. A large number of railcars operating on the G.N.R. (Ireland) incorporate this type of drive.

Underfloor-mounted horizontal-type engines are used in a large number of twin-engined cars (Plate 18B, page 94), built for British Railways. Many of these cars are equipped with engines and transmissions resembling those used for A.E.C. four-wheeled railcars (q.v.), but some are equipped with Lysholm-Smith hydraulic torque converters. Recently, cars have been put to work

employing the same layout but provided with engines of increased power. These include cars equipped with Rolls-Royce 8-cylinder horizontal engines and twin disc hydraulic torque converters.

Horizontal underfloor-mounted engines with mechanical or hydraulic transmission are also favoured in Italy, where a large number of railcars are in service with engine outputs up to 480 H.P. The cars with the highest power outputs classified ALn 990, are equipped with Saurer S.B.D.-type 12-cylinder horizontal twin-bank engines, and weigh 48–49 tons empty (Plate 18A, page 94). Two types of transmission are used; cars constructed by O.M. are provided with Lysholm-Smith hydraulic torque converters, forming a unit construction with the engine; cars constructed by Fiat have five-speed synchromesh gearboxes. In both cases the final drive is transmitted through a propeller shaft to a gearbox located in the centre of the driving bogie, from which the drive is conveyed through short cardan shafts to axle-mounted drive units.

The cars are equipped for working in multiple unit with similar cars or trailers in accordance with requirements. Fiat cars of this type supplied to Spain operate in pairs with one trailer, and are capable of maintaining a speed of 75 m.p.h. up a gradient of 1 in 250 (0·4 per cent). The cars which have been constructed by Breda for operating Trans-Europe Express (T.E.E.) trains are of the same general type but with numerous modifications and refinements, and incorporate Wilson five-speed automatic gearboxes (Plate 19A, page 95). Other types of railcar in use incorporate twin engines and transmissions.

In France a number of twin car sets consisting of one power car and one trailer are in service. A number of these cars have all axles driven and are provided with two 300 H.P. engines mounted on the underframe above the bogies. Each engine forms a unit construction with the change speed and reverse gearbox, from which the drive is transmitted to the axles by cardan shafts. A more powerful type of car is also used equipped with a single engine of the 12-cylinder V-type developing 825 B.H.P., and with Mekydro hydraulic transmission. The engine together with the transmission which incorporates the reversing gear are mounted on the underframe. The transmission output shaft is located above the bogie centre with short cardan shafts driving both axles. A modified version of this car is used for T.E.E. trains.

Bogie-mounted engines and transmissions have been widely used in Germany since the first high-speed diesel-electric trains commenced working in 1932. Since the end of World War II a number of railcars have been built, for working in conjunction with trailers, and equipped with single engines developing 1,000

B.H.P. Both Voith and Mekydro hydraulic transmissions are used. The engine and transmission unit are resiliently mounted in the bogie, with short cardan shafts driving both axles. The radiators and fans are built into the roof structure of the car, and the fans are electrically driven. The driving compartment is in front of the engine (Fig. 31).

The new T.E.E. trains built for the German Federal Railways embody some important modifications. The engine together with the hydraulic transmission is mounted in the underframe above the driving bogie, which is without a centre pivot, the transmission casing projects downwards into the centre of the bogie, with short cardan shafts driving both axles. The cars are fully streamlined with nose-ends over the driving bogie, the driving compartment being located behind the engine. Each power car is provided with a diesel-electric generating set for supplying power for lighting and air-operated disc brakes, and magnetic rail brakes (Plate 19B, page 95).

When electric transmission is used, the engine and generator are normally mounted on the underframe of the car. An exception occurs in the case of the 400 H.P. diesel-electric articulated railcars built in Holland by N.V. Allan, where underfloor-mounted generating sets are used (Plate 18c, page 94). Each car is provided with two A.E.C. 6-cylinder pressure-charged horizontal engines with cardan shaft drive to the generators. The two traction motors are mounted on each outer bogie with cardan shaft drive to the axles.

An experimental railcar operating on British Railways is equipped with an underfloor-mounted Paxman 6-cylinder horizontal engine developing 450 B.H.P., direct coupled to an electric generator.

The six-car diesel-electric trains working on the Southern Region of British Railways, between London and Hastings, include power cars at both ends of the train provided with compartments which house the engine, generator and ancillary equipment. The motor bogies, which are standard with those used for Southern Region electric stock, are located at the inner end of each power car (Plate 17D, page 93).

High-speed diesel-electric trains are now under construction for British Railways. These will operate as six- or eight-car sets with two power cars per train. Each power car will be provided with a North British/M.A.N. engine developing 1,000 H.P. at 1,500 r.p.m., coupled to a G.E.C. electric generator. The fully spring-borne traction motors will be distributed throughout the train, each power car supplying current to four traction motors. A separate power plant for air conditioning, lighting, etc., will also be provided.

Seven-car diesel-electric trains of advanced design have been built by De Dietrich for high-speed service on the Moroccan Railways. Each of the two power cars carries a 16-cylinder V-type engine developing 1,000 B.H.P. under site conditions, coupled to an electric-generator. The inner bogie of each power car carries two traction motors. Two auxiliary diesel generating sets of 180 H.P. supply power for air conditioning, etc. On a trial run of 313 miles between Paris and Strasbourg, an average speed of 73 m.p.h. was maintained.

The Dutch-Swiss T.E.E. trains are provided with one power car only for operating in conjunction

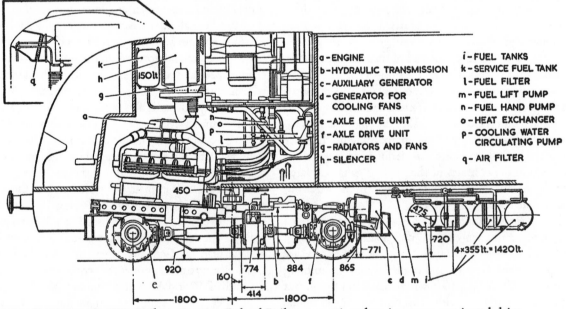

a - ENGINE
b - HYDRAULIC TRANSMISSION
c - AUXILIARY GENERATOR
d - GENERATOR FOR COOLING FANS
e - AXLE DRIVE UNIT
f - AXLE DRIVE UNIT
g - RADIATORS AND FANS
h - SILENCER
i - FUEL TANKS
k - SERVICE FUEL TANK
l - FUEL FILTER
m - FUEL LIFT PUMP
n - FUEL HAND PUMP
o - HEAT EXCHANGER
p - COOLING WATER CIRCULATING PUMP
q - AIR FILTER

FIG. 31. 1,000 H.P. Railcar, German Federal Railways, section showing power unit and drive.

P. Ransome-Wallis

PLATE 17B. Great Western (now British Railways) railcar in express parcels service.

British Railways

PLATE 17D. British Railways diesel-electric trains sets for the London–Hastings service.

P. Ransome-Wallis

PLATE 17A. S.N.C.F. diesel-electric shunting locomotive with booster unit which has two traction motors supplied with power from the generator on the locomotive ("cow and calf").

P. Ransome-Wallis

PLATE 17C. Turkish State Railways: the Bogaziçi Express –Haydarpasa to Ankara. The train is composed of two three–unit railcar sets; each set is powered by two 550 H.P. M.A.N. engines with Voith hydraulic transmission.

93

D. Wickham and Co.

PLATE 18B. British Railways 300 H.P. twin-car M.U. train with two underfloor engines and Wilson gearbox.

Birmingham Railway Carriage and Wagon Co. Ltd

PLATE 18D. New Zealand Government Railways (3 ft 6 in. gauge) articulated two-car set with two 210 H.P. underfloor engines and Wilson-Drewry mechanical transmission.

Fiat

PLATE 18A. Italian State Railways 480 H.P. car-type ALn 990 with Fiat underfloor engine and mechanical transmission.

N. V. Allan and Co.

PLATE 18C. Netherlands Railways articulated two-car set with two 200 H.P. underfloor engines and electric transmission.

PLATE 18. Diesel Railcars

Italian State Railways

PLATE 19A. Italian State Railways two-unit train powered by two Breda 12-cylinder "flat-twin" underfloor engines, each of 490 H.P. The drive is through a Vulcan clutch and a Wilson 5-speed automatic gearbox.

Deutsche Bundesbahn

PLATE 19B. German Federal Railways seven-car set, powered by two M.A.N. 1,100 H.P. engines and with Voith hydraulic transmission. A separate engine provides power for auxiliaries.

PLATE 19. High speed railcars for Trans-Europ Express Services

D. Wickham and Co.

PLATE 20C. Bogie railcar for British Railways showing Wickham integral body-frame construction employing square solid-drawn steel tubing.

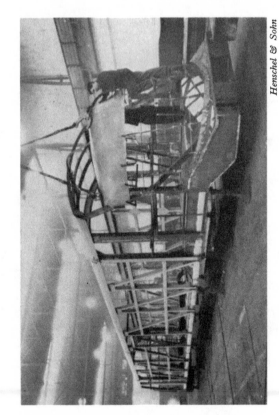

Deutsche Bundesbahn

PLATE 20A. German Federal Railways driving bogie for diesel-hydraulic train with 1,000 H.P. Daimler engine and Mekydro transmission.

Henschel & Sohn

PLATE 20B. Integral body-frame construction for 1,900 H.P. diesel-electric locomotive for Egyptian State Railways.

Swiss Locomotive and Machine Works

PLATE 21A. Driving bogie for Co–Co diesel-electric locomotive. This is a fixed-centre bogie. The weight of the superstructure is carried on resilient side bearers. Cylindrical axlebox guides contain oil and function as hydraulic dampers, used in conjunction with underslung laminated springs.

The Clayton Equipment Co. Ltd

PLATE 21B. Driving bogie for British Railways Bo–Bo diesel-electric locomotive with underslung equalizing beams, coil bearing springs and elliptical bolster springs. The brake cylinders incorporate slack adjusters.

Deutsche Bundesbahn

PLATE 21C. Driving bogie for German Federal Railways 2,200 H.P. B–B diesel-hydraulic locomotive. British Railways (Western Region) B–B locomotives are similar. Bogies have no centre pivot.

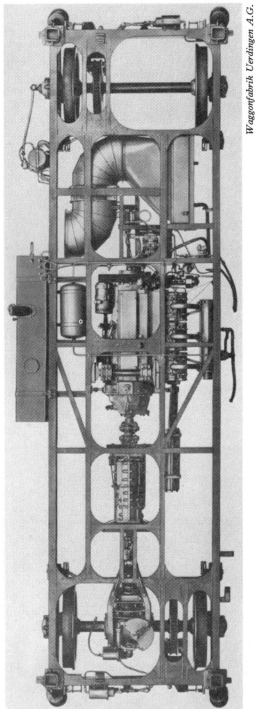

PLATE 22A. Plan view of frames and layout of engine, transmission, etc., of single-engined diesel mechanical four-wheeled railcar, underfloor engine.

Waggonfabrik Uerdingen A.G.

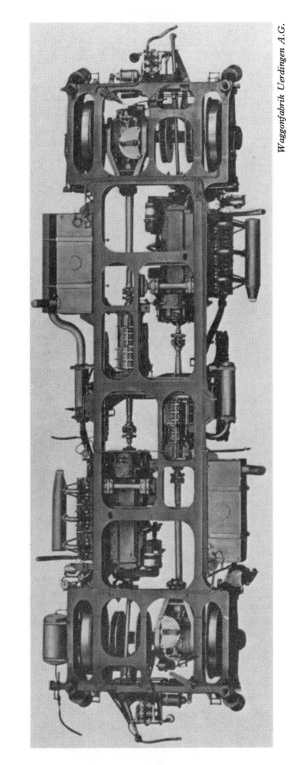

PLATE 22B. Plan view of double-engined four-wheeled diesel mechanical railcar.

Waggonfabrik Uerdingen A.G.

Wagonfabrik Uerdingen A.G.

PLATE 23A. Underframe, suspension and magnetic track brakes of twin-engined diesel mechanical four-wheeled railcar.

P. Ransome-Wallis

PLATE 23C. Single-ended railcar for the 3 ft gauge County Donegal Railway of Ireland, with Walker-Gardner 102 H.P. engine, driving a four-coupled leading bogie (truck) through mechanical transmission.

Wagonfabrik Uerdingen A.G.

PLATE 23B. Single-engined four-wheeled railcar and trailer.

Alco Products Inc.

PLATE 24A. The first commercially successful diesel-electric locomotive in the USA – C.R. of N.J. No. 1000 (*right*) was built in 1925. It is of 300 H.P. but the power unit is of almost the same weight as that of the 1957 Alco road-switcher (*left*) which develops 1,600 H.P. No. 1000 was withdrawn in 1957 and is preserved in the B. and O. Museum at Baltimore, Md.

(B)

Baltimore and Ohio Railroad

PLATE 24B. B. and O. R.R. 300 H.P. "box-cab" switcher was built in 1925, and was still reported at work in 1957.

with three trailers. This is virtually a 2,000 H.P. A1A-A1A type diesel-electric locomotive with twin generating sets together with an additional generating set which supples power to the auxiliaries. The traction motors are force ventilated and fully spring borne. Driving compartments are provided at both ends of the train.

STRUCTURAL DATA

Body and underframe. The performance of railcars and diesel trains is largely governed by the power/weight ratio, the car and equipment therefore should be made as light as possible consistent with strength and durability. Integral construction of the body and underframe is widely adopted, full use being made of the roof to give the complete structure the form of a box girder.

In vehicles constructed to suit the British loading gauge, the side pillars must be bent inwards to meet the sole-bars and are therefore less well adapted for carrying vertical loads than the straight pillars which can be used in many European designs. Specially shaped sections are used in the construction of the body framework, in conjunction with rolled steel sections for the underframe. The body panels also may be designed to contribute to the strength of the structure, and are sometimes stiffened by means of pressed-in horizontal grooves.

Tubular sections of different shapes are being increasingly used in the construction of the body and underframe. This type of construction using square solid-drawn tubes has been employed by D.Wickham & Co. Ltd. for many years (Plate 20C, page 96). By using corrugated steel flooring, the resistance to buffing loads can be greatly increased. Both these features are used in the latest diesel railcars and trailers built for British Railways. The railcars are 57 ft long over head stocks and weigh $27\frac{1}{2}$ tons. Cars for operating T.E.E. services, built in Germany, have steel underframes, with body framing and panelling of aluminium alloy.

Bogies. Railcar bogies of all types should combine good riding qualities with a light and durable construction. Driving bogies must be designed to take the torque reactions, and transmit the traction forces to the body of the vehicle. They are also required to accommodate part of the transmission, including in some cases the engine, and to meet these varying requirements many different types of bogie have been evolved (Plate 20A, page 96).

Railcars with mechanical or hydraulic transmission operating on British Railways are provided with *driving bogies* closely resembling those used for conventional coaching stock. The frames are fabricated by riveting or welding from rolled steel sections and plate, with sole-bars of deep channel section. Laminated bearing springs and rubber auxiliary springs are used, and the swing bolster is carried on nests of coil springs. *Hydraulic dampers* are usually fitted to control the side movement. The axle-mounted final drive and reverse unit is fitted with a *torque arm*, rubber springs being used at the point of anchorage to take the torque reaction. The clasp brakes are usually operated by vacuum cylinders hung from the bogie frame. Bogies of this type weigh about $5\frac{1}{2}$ tons, complete with drive unit.

In many modern designs, the centre pivot is used for locating purposes, and for transmitting the traction and braking forces only. The body of the vehicle is supported on side bearers located directly above the bolster springs, and by spacing these as widely as possible the riding can be much improved. This feature is incorporated in a type of bogie widely used in several European countries, in which the bolster is carried on inverted laminated springs arranged longitudinally and suspended from the outside of the sole-bars. In those cases where the bolster is provided with side movement, *inclined swing links* are used for this purpose. A light and simple type of bolster can thus be incorporated, which can, if necessary, be conveniently arranged to clear transmission units located in the centre of the bogie.

In another arrangement used for diesel-electric railcars constructed by N.V.Allan & Co., four torsion bars of rectangular section arranged longitudinally are used instead of conventional bolster springs. These are supported from the sole-bars, and are secured at the headstock ends. The bolster is hung from short levers fixed to the free ends of the torsion bars. *Torsion bar springing* is extensively used in Switzerland.

The driving bogies used for the T.E.E. trains operated by the French National Railways are without horn-guides and have sole-bars and transoms of box section placed between the wheels. The axle-boxes are of the wing type with coil bearing springs, and are located by flexible steel plates connecting the top of the axle-box to the frame. The weight of the body is transmitted to the bogie by two widely spaced coil springs resting on beams suspended on swing links from the ends of the transoms which project through the sole-bars. The bogie is provided with a centre pivot in the same plane as the axle centres. It is connected to the transoms by two horizontal links one on either side, arranged longitudinally, and provided with resilient bushes. Locating the centre pivot in this position is conducive to good riding. Both axles are driven, and the bogie weighs $7\frac{1}{2}$ tons, complete with drive units.

FOUR-WHEELED RAILCARS

Four-wheeled railcars can be constructed with very low tare weights, and are therefore economical to

RAILCARS AND DIESEL TRAINS

No.	Railway	Builder	Type	M indicates a power car. T indicates a trailer. Formations shown are representative but can be varied	Seating Capacity	Tare weight tons	Total H.P.	Max. speed m.p.h.	Engine output B.H.P./r.p.m.	Maker	No. of engines	Mounting	Transmission	Maker
17	G.W.R.	G.W.R.	4-wh. bogie	M–T	—	65	210	70	105/1650	A.E.C.	2	Side	Wilson	A.E.C.
18	G.N.R. Board Ireland	G.N.R.	4-wh. bogie	M.T.T.M.T.M.	246	200	900	80	150/1800	B.U.T.	6	Side	Wilson	B.U.T.
19	B.R.	Gloucester	4-wh. bogie	M.T.	118	55	300	70	150/1800	B.U.T.	2	Underfloor	Wilson	B.U.T.
20	B.R.	B.R.	4-wh. bogie	M.T.	200	72	476	70	238/1800	Rolls-Royce	2	Underfloor	Hydraulic	Rolls-Royce
21	Italian State	O.M.	4-wh. bogie	M.T.	—	77	480	81	480/1400	O.M.	1	Underfloor	Hydraulic	O.M.
22	Spanish National	Fiat	4-wh. bogie	M.T.M.	174	153	960	75	480/1400	O.M.	2	Underfloor	Mech.	Fiat
23	T.E.E.	Breda	4-wh. bogie	M.M.	90	103	960*	87	480/1500	Breda	2	Underfloor	Wilson	Breda
24	T.E.E.	De Dietrich	4-wh. bogie	M.T.	81	84	825	75	852/1500	M.G.O.	1	Frame	Hydraulic	Maybach
25	German Federal	M.A.N.	4-wh. bogie	M.T.T.	260	110	1000	75	1000/1400	Maybach	1	Bogie	Hydraulic	Maybach
26	T.E.E.	German Various	4-wh. bogie	M.T.T.T.T.M.	105	226	2200*	87	1100/1500	Maybach	2	Frame	Hydraulic	Maybach
27	Netherland	Allan	Artic.	M.M.	133	85	400	75	200/1800	A.E.C.	2	Underfloor	Electric	Smit
28	B.R.	B.R.	4-wh. bogie	M.T.T.T.T.M	288	227	1000	75	500/850	Eng. Elec.	2	Car frame	Electric	Eng. Electric
29	Moroccan	De Dietrich	4-wh. bogie	M.T.T.T.T.T.M.	240	275	2000*	87	1000/1500	M.G.O.	2	Car frame	Electric	Alsthom
30	T.E.E.	Werkspoor	6-wh. bogie	M.T.T.T.	114	221	2000*	87	1000/1400	Werkspoor	2	Car frame	Electric	Brown, Boveri
31	B.R.	A.E.C.	4-wh.	T.–M.–T.	139	40	125	45	125/1800	A.E.C.	1	Underfloor	Mech. Wilson	A.E.C.
32	German Federal	Uerdingen	4-wh.	M–T	104	21	110	50	110/1800	Bussing	1	Underfloor	Mech.	Zahnradfabric
33	German Federal	Uerdingen	4-wh.	M–T	180	41	300	46	150/1900	Bussing	2	Underfloor	Mech.	Zahnradfabric
34	German Private	M.A.N.	4-wh.	M–T	136	35	300	43	150/2000	M.A.V.	2	Underfloor	Hydro-Media	Zahnradfabric

* The total output is available for traction purposes, power for auxiliaries being supplied by a separate engine.

operate. They do not normally give the same standard of riding as that provided by bogie vehicles, but are suitable for working on branch lines where only moderate speeds are required. To obtain the maximum amount of floor space an *underfloor-mounted horizontal engine* is used, in conjunction with mechanical or hydro-mechanical transmission. The cars are often equipped for multiple-unit operation and designed to haul lightweight trailers. When they are required to operate in conjunction with normal rolling stock, a heavier and more robust type of construction must be adopted.

In Great Britain three-car lightweight train sets built by A.E.C., composed of four-wheeled vehicles with a power car in the centre, are used to a limited extent. Each car is 37 ft 6 in. long over headstocks with a wheel base of 22 ft, the maximum speed is 45 m.p.h. The drive from the 6-cylinder horizontal engine is transmitted through a fluid coupling, and short propeller shaft, incorporating a free wheel, to a Wilson four-speed gearbox, thence through a propeller shaft to a bevel reverse unit mounted on one of the axles. The radiator is mounted beneath the car in close proximity to the engine, from which the fan drive is taken. *Electro-pneumatic control gear* is used with driving compartments at both ends of the train. The underframe is of riveted construction, consisting of rolled steel sections joined together by cleats and gussets. The axle-boxes which work in cast steel hornguides bolted to the underframe, are provided with laminated bearing springs.

Four-wheeled railcars are used extensively for branch line service in several European countries including Germany, where some notable vehicles have been produced. The cars, manufactured by Waggonfabrik Uerdingen, have underframes of welded construction, and the body is carried on four coil springs, suspended from the underframe. This arrangement gives much-improved riding. No hornguides are used, and the axle-boxes are clamped to underslung laminated springs which transmit the traction and braking forces. The

wheel base of the earlier railcars was 14 ft $9\frac{1}{2}$ in., but in later vehicles this has been increased to 19 ft 8 in., and flange lubricators fitted. They are designed to operate at speeds up to 56 m.p.h. Power is provided by a 6-cylinder horizontal engine, and the drive is transmitted through a multiplate friction clutch, to a six-speed gearbox with electro-magnetic gear-changing clutches; thence through a propeller shaft to an axle-mounted reverse box. Air-operated *disc brakes* are mounted on the axles, and *magnetic track brakes*, suspended from the underframe, are also fitted, for use in emergencies when running on unfenced lines (Plates 7B, 22A and 23A, pages 47, 98 and 99).

To obtain improved performance when hauling a trailer over steeply graded routes, a modified design has been produced, in which two engines and two sets of transmission are used, driving separate axles (Plates 22B and 23B, pages 98 and 99). The radiators are mounted on the car body beneath the platforms at both ends, the fans being driven by cardan shafts from the engines. The cars are equipped for operating in multiple unit.

A further development has been initiated by M.A.N. who have produced a car mounted on two *Bissel trucks*, thus making it possible to incorporate a wheelbase of 29 ft 6 in. together with a body 49 ft 6 in. in length. The pivots which connect the Bissel trucks to the underframe are provided with resilient bushes. An adjustable spring-centring device in combination with a hydraulic damper is used to prevent hunting when running on straight track. The weight of the superstructure is transmitted to each truck by means of twelve rubber balls rolling in radial guideways. The roller-bearing axle-boxes are of the wing type with coil bearing springs and are connected to the truck frame by means of torque arms fitted with resilient bushes. Hydraulic shock absorbers are also fitted. Each axle is driven by a 6-cylinder horizontal engine forming a unit construction with a *Hydromedia transmission* from which the drive is transmitted through a propeller shaft to an axle-mounted reverse unit.

Part VIII. Diesel locomotives and railcars: other equipment and testing

BRAKING SYSTEMS

Diesel locomotives may be equipped with the vacuum or straight air brake, and are generally provided with equipment for operating continuously braked trains. Since they are not normally required to work in conjunction with other rolling stock, railcars and trailers are frequently provided with special types of quick-release braking equipment. The straight air brake is usually pre-

ferred for locomotives, where the small and compact type of brake cylinders employed can be conveniently accommodated on the bogies. Recent improvements in the vacuum brake, however, have resulted in the introduction of a smaller type of brake cylinder, more suitable for use in such cases.

When the speed of heavy trains descending long gradients is controlled by the brakes, the wear of brake shoes and wheel treads will be considerable. It is an advantage, therefore, in these circumstances to employ some form of dynamic braking, so that the use of the train brakes can be reduced to a minimum.

Dynamic braking. The simplest form of dynamic braking consists in using the *engine compression* to retard the train. This can only be used when the transmission is capable of transmitting torque from the driving wheels to the engine, and is possible with the different types of mechanical transmission incorporating both friction clutches and fluid couplings. As an example of what can be achieved by this method, a train weighing 285 tons hauled by a locomotive weighing 55 tons on a down grade of 1 in 50 (2 per cent) had the speed reduced from 4·5 to 2·6 m.p.h. in a distance of 33 yards, when the throttle was closed with first gear engaged.

To give increased braking power a development of this system is used in which a butterfly valve is fitted in the engine exhaust pipe. When the valve is closed there is a rapid build up of pressure in the exhaust manifold, which can only be relieved when the pressure is sufficient to force the exhaust valves open during the suction stroke and allow air to escape via the inlet valves. The retarding effect may be varied by manipulating the butterfly valve. When the valve is closed, brake power ranging from 60 to 80 per cent of the engine output is available.

The *hydraulic torque converter* also possesses valuable braking characteristics. When the output speed is greater than the input speed, the output torque becomes a negative or braking torque, increasing as the square of the output speed. If, therefore, the engine is stopped when the train is in motion, the retarding effect will be considerable, and can be usefully employed to control the speed of the train when travelling at low speeds. By running the engine at speeds ranging from idling up to the maximum, the speed of the train can be controlled when travelling at higher speeds.

When three torque converters operating in sequence are employed, the braking effect of each converter can be used when the working circuit is filled. The characteristics of the second converter are such that the braking effect can be made available at the upper end of the speed range, but the third converter has little useful braking effect. Useful braking efforts are available down to 20 per cent of the maximum rail speed. During the braking period the engine may be stopped or rotating at low speed, and to ensure adequate cooling of the working fluid under these conditions, the cooling fan should be independently driven.

Rheostatic braking is another form of dynamic braking which can be used in diesel-electric locomotives. By altering the connections, the traction motors can be made to function as generators, when the train is running downhill with the power supply to the motors switched off. The current thus generated is dissipated in resistance grids which are often built into the roof structure, and cooled by means of electrically driven fans. The resistance may be varied, and the speed of the train controlled through the medium of the power controller which is provided with a series of braking notches, but sometimes a separate controller is used. Sometimes the extra weight of this equipment may prohibit its use.

Straight-air brakes. When compressed-air brakes are used, the compressor may be either belt-driven from the engine or driven by an independent electric motor. The locomotive may be provided with the straight-air brake together with equipment for operating vacuum-braked trains, and in this case a comparatively small compressor will be able to meet requirements. Alternatively, the locomotive may be provided with the straight-air brake together with equipment for working trains fitted with the automatic air brake. Sometimes equipment for working trains fitted with both types of brake is provided. The locomotive straight-air brake may also be provided with equipment enabling it to function as an automatic brake, when working trains thus fitted.

Vacuum brakes. The vacuum is created and maintained by means of vacuum exhausters of the reciprocating or rotary type. These may be driven by belt or shaft from the engine, or may be direct coupled to an electric motor. With the former arrangement, the capacity of the exhauster will be lowest when the train is at a standstill with the engine idling, and unless oversize exhausters are employed, brake release will be excessively slow. If the exhausters are motor driven, however, they may be speeded up under these conditions to secure a quick release.

The driver's brake valve is provided with electrical contacts and when a brake application is made by admitting air to the train pipe, communication between the train pipe and exhausters is cut off by means of electro-magnetic valves. A quick brake release can be obtained by means of a trigger-operated switch attached to the driver's handle, which speeds up the motor used for driving the exhauster. Although developed primarily for diesel-electric locomotives, this arrangement is suitable for locomotives with other types of transmission, particularly if they are intended for working in multiple-unit.

For use on locomotives where space is restricted, Gresham & Craven have developed special vacuum brake equipment, which augments the brake power available, thus making it possible to reduce the size of the brake cylinders. The driver's brake valve controls two pipes, the train pipe which communicates directly with the brake cylinders, and an additional pipe which communicates with the exhauster, and is also in permanent communication with the vacuum reservoirs on the locomotive. When the driver's brake valve is in the "off" position, it establishes communication between the exhauster and the train pipe, a relief valve being incorporated which limits the vacuum throughout the system to twenty inches.

In the "on" position of the driver's brake valve air enters the train pipe which is cut off from communication with the exhauster. If a full brake application is made and the vacuum in the train pipe drops to ten inches an automatic valve comes into operation which isolates the relief valve and enables the vacuum in the vacuum reservoirs to be increased to twenty-six or twenty-seven inches, thus increasing the braking force by about 50 per cent. The train pipe is provided with hose connections at both ends of the locomotive for operating vacuum-braked trains provided with the normal type of vacuum-brake equipment. The smaller type of brake cylinder used in conjunction with this system may be placed either vertically or horizontally.

Vacuum-braked railcars with mechanical transmission are usually provided with exhausters driven by belt or shaft from the engine or some convenient point in the transmission. To overcome the disadvantages associated with this type of drive, Gresham & Craven have developed a two-pipe quick-release vacuum-brake system utilizing automatic isolating valves of the type mentioned above. In addition to the train pipe, a release pipe is provided which communicates with the exhausters, and can be connected to the train pipe through the driver's brake valve. The release pipe is connected to a series of high-vacuum reservoirs fitted to each railcar and trailer, the flexible connections between the coaches being designed to prevent coupling to the train pipe in error.

When the train is running and the exhauster capacity is at a maximum, a working vacuum of twenty-one inches is created in the train pipe, and on both sides of the brake cylinder pistons. As soon as this figure is reached, the train pipe is automatically isolated from the exhauster which continues to exhaust the release pipe and high-vacuum reservoirs until a vacuum of twenty-seven to twenty-eight inches is attained. A brake application is made in the normal manner by admitting air to the train pipe. When the driver's brake handle is moved to the release position the air inlet valve closes and communication is restored between the train pipe and release pipe. The high vacuum existing in the release pipe and high-vacuum reservoirs enables a quick release to be achieved

although the engines may be idling and the exhausters operating at their minimum capacity.

In practice it has been found advantageous to prevent the vacuum in the high-vacuum reservoirs from being completely destroyed due to repeated brake applications or some other cause, since, if this is allowed to occur, a considerable period of time may be required to re-create the original vacuum. Automatic isolating valves are used, therefore, to isolate the high-vacuum reservoirs when the vacuum in the release pipe falls to eighteen or nineteen inches.

EXHAUST-CONDITIONING AND FLAME-PROOFING

Exhaust-conditioning. To enable diesel locomotives to operate in mines and closed spaces it is necessary to remove or neutralize irritant and poisonous elements present in the exhaust gases. Exhaust conditioning is carried out by passing the gases through water, which absorbs most of the noxious elements present. Various methods are used to guard against loss by evaporation and to prevent any possibility of water passing back into the engine.

Flame-proofing. When locomotives are required to work in explosive areas such as coal mines, additional precautions must be taken to eliminate any possibility of flames or sparks coming in contact with the surrounding atmosphere, and flame-traps must be provided on both the air inlet and exhaust outlet. These consist of sets of plates made up into grids, and presenting a number of narrow passages through which the gases must pass. All electrical equipment including switchgear must be fully flame-proof.

In the system used by F.C.Hibberd & Co. for 23-ton shunting locomotives, the engine is provided with a water-cooled exhaust manifold. The exhaust gases pass through a flame-trap into a partitioned water tank, where they are made to pass twice through the water, before being exhausted into the atmosphere. A reserve water tank is fitted which automatically tops up the water in the main tank.

If the water in the main tank drops below a certain level, the engine is stopped automatically by means of a device incorporating a fusible element which melts, releasing a spring-loaded cable which operates the engine shut-down. Twin anti-suction valves are fitted to the exhaust pipe between the engine and water tank, which prevent water being drawn back into the engine should a vacuum occur in the exhaust manifold. Twin flame-traps are also fitted to the engine air intake.

FIRE PROTECTION

Oil leakages, together with the high temperatures attained by certain transmission elements during the period of acceleration, increase the risk of fire in diesel-propelled vehicles, and the importance of providing adequate fire protection has been demonstrated on a number of occasions.

In addition to portable fire extinguishers, it is becoming increasingly common to provide a piped fire-extinguishing system which ensures that all parts of the engine and transmission are sprayed in the event of a fire. CO_2 stored in bottles, or some other fire-quenching medium may be used for this purpose.

Unattended vehicles operating in multiple unit present a special problem, and to meet this, an automatic system has been devised. This consists of a device which acts if the temperature becomes dangerously high, causing a switch to close which sets the fire-extinguishing system in operation.

TESTING
(*see* chapter 7 – Diesel Locomotives)

Two kinds of testing are employed for diesel motive power units, firstly the testing of engines, transmissions and ancillary equipment before or during installation, and secondly acceptance tests carried out on the road where the ability of the unit to achieve the specified performance is determined.

Diesel engines are always systematically tested by the makers. These tests are made with a 10 per cent overload, full load and partial loads. Records are kept of the fuel consumption, together with the temperatures of the cooling water and exhaust gases. A certificate recording the results of the tests is supplied with each engine. Sometimes in the case of large orders, the customer may require that one engine selected from the batch should be subjected to exhaustive tests lasting up to 2,000 hours. After these tests have been completed the engine is dismantled and careful measurements are made to determine the amount of wear which has taken place.

When electric transmission is used it is often more convenient to test the engine and generator as a single unit, the generator output being absorbed by means of an adjustable resistor. The governor is tested by opening the circuit at full load, and this test is repeated with lower speed settings. The generator is tested for temperature rise, the insulation and commutation are also tested. Traction motors, control equipment and auxiliaries are all subjected to suitable tests.

Mechanical and hydraulic transmissions may be tested in special rigs to ensure satisfactory operation and to check the temperature rise of the lubricating or transmission oil.

Before leaving the maker's works, the axle loads are recorded, and the unit may be run over a short length of track, where the operation of the brakes and sanding gear can be checked together with the various working clearances.

When the unit is the prototype of a new class, or embodies extensive alterations, a series of road tests are carried out.

TRAIN HEATING

When diesel locomotives are required to haul passenger trains, an oil-fired boiler is usually provided to generate steam for heating the coaches. A feed-water tank with a capacity of 400–600 gallons is also required, together with a fuel-oil tank having a capacity of 90–100 gallons. The weight of the combined equipment is thus quite considerable.

Boilers may be of the vertical multi-tubular type generating steam at a pressure of 50–80 p.s.i. or of the flash type consisting of a single tube through which water is forced by an electrically-driven pump. In both cases an electrically-driven fan is used to accelerate combustion. Once the boiler has been started up, its operation becomes completely automatic.

The weight of this equipment and the space required, make it unsuitable for use in railcars and diesel trains. In this case radiators may be provided through which part of the engine cooling water is pumped. Another system incorporates oil-burning air heaters with electrically-driven fans, placed under each car. The heated air is conveyed through a system of ducts and discharged into the interior of the car at suitable points.

Initials of manufacturers mentioned

M.A.N. = Maschinenfabric Augsburg-Nürnburg A.G. (Germany).

M.G.O = manufactured by Soc Alsacienne Constructions Mécaniques (France).

B.U.T. = British United Traction.

B.R. = British Railways.

M.A.K. = Maschinenbau Kiel A.G. (Germany).

A.M.N = Les Ateliers Metallurgiques Nivelles (Belgium)

S.E.M = Societé d' Electricité et de Mécanique (Belgium).

C.E.M. = Cie Electro-Mécanique (France).

A.F.L. = Soc. des Ateliers et Forges de la Loire (France).

S.L.M. = Swiss Locomotive and Machine Works.

O.M. = Officine Meccaniche (Italy).

S.C.G. = Self Changing Gears, Ltd. (Britain).

S.S.S. = Synchro-Self-Shifting (Powerflow Transmission). (Britain.)

CHAPTER 2

Diesel Traction in North America

by DAVID P. MORGAN

Part I. The conquest of diesel traction in North America

The word "diesel" as used in this chapter refers to diesel-electric traction unless specifically noted otherwise.

U.S. statistics used in this section refer only to Class One roads, a revenue grouping which includes those lines which operate 95 per cent of U.S. rail mileage and handle 99 per cent of its traffic.

In this chapter note that a ton is a short (American) not a long ton. (*See* Appendix II.)

DIESELIZATION IN NORTH AMERICA

The diesel-electric locomotive, easily the most fundamental change in North American railroad practice since the adoption of the air brake, has made good its threat to monopolize motive power in the Western Hemisphere. Not only has it killed steam locomotive development in its prime, but it is rapidly eliminating the original locomotion of railroading altogether. Electrification, a cause the diesel might have been expected to advance, has actually retracted in the face of the newcomer. And no form of competition – turbine or atomic, produced or on paper – has yet been able to match the diesel's economic supremacy.

The first commercially practical diesel-electric locomotive – a 60-ton, 300-h.p. switcher – appeared in 1925, but as late as 1946 the Class One roads' 37,551 steam engines handled 88·15 per cent of freight gross ton-miles, 78·24 per cent of passenger car-miles, and 69·17 per cent of yard locomotive-hours. So rapid was the post-war motive power revolution, however, that during 1956 the diesel was responsible for 88·49 per cent of freight movement, 91·02 per cent of the passenger, and 93·24 per cent of yard work. By 1 July, 1958, steam locomotive ownership had fallen to 1,737 engines, while the number of diesel units had risen from 4,441 in 1946 to 27,615 units, owned or leased by Class One railroads. There were also 567 electric locomotives and 28 gas turbine-electric locomotives in service in 1958.

Canada's two transcontinental systems are both well beyond the half-way mark toward total dieselization, even though the changeover there is almost wholly a post-war occurrence. Indeed, the privately owned and operated Canadian Pacific took delivery of its last steam locomotive, a 2–10–4, in 1949, yet it is committed to complete dieselization by 1961. The government-owned Canadian National is making similar progress. Most of the subsidiaries of these systems as well as the independent railroads of the Dominion are now fully dieselized. In Mexico, finances have not permitted as rapid a retirement of steam as that experienced elsewhere on the North American continent, but the National Railways of Mexico nevertheless expects diesel to handle half its traffic by 1960.

As 1957 began, the remnants of the steam fleet that had powered U.S. railroading through its crisis in World War II was concentrated on just ten railroads: four in the east, one in the south, and five in the west. Of these, only one major carrier – Norfolk and Western – still depended upon steam to move the bulk of its traffic. N. & W.'s reluctance to dieselize stems from the fact that 75 per cent of its tonnage is coal. Yet even this hold-out is now well on the way to complete dieselization, and it concedes that an expensive experiment with a steam-turbine-electric locomotive could not keep it on a coal-burning economy. Elsewhere, steam is largely employed in peak traffic periods by roads with seasonal traffic fluctuations which do not want to buy diesels for part-time service.

The diesel's conquest in the U.S. has not been limited to the rout of steam. Electrification, never really extensive in North America, has lost ground too. In the U.S., Class One railroads operated 2,840 electrified route-miles in 1944, but only 2,063 remain today. Those electrified zones, brought about because of municipal anti-smoke ordinances, or long tunnels, turned out to be expensive interruptions of long diesel runs when steam power vanished. Since the war, third-rail has been removed from New York Central's 8,290-ft tunnel under the Detroit River between Detroit, Michigan, and Windsor, Ontario; diesels have also eclipsed America's first steam-road electrification – 3·63 miles of Baltimore & Ohio through nine tunnels in Baltimore, Maryland, electrified in 1895. De-electrification of Cleveland Union Terminal permits New York Central passenger diesels to make the run between Harmon, New York, and Chicago, Illinois, without change. The catenary has also been removed from the longest tunnel in the U.S., Great Northern's 7·79-mile Cascade Tunnel. On the Milwaukee Road, diesels are being operated in multiple with electric locomotives under one control.

Diesels now work into electrified zones on the Pennsylvania, whose extensive electrified mileage (671 route-miles, 2,248 track-miles) is the largest in North America in point of traffic handled. The diesel has cancelled projected extensions of P.R.R. electrification and has caused the road to re-evaluate its present electric operations. Meantime, New Haven has purchased a fleet of sixty "diesel-electric-electric" locomotives which can operate as diesels or as straight electrics and can make the switchover at speed. These passenger units can operate from Boston, Massachusetts, into New York City without the former engine change necessary at New Haven, Connecticut, and they are expected eventually to reduce N.H. electrification to the area immediately adjacent to New York, which does not permit diesels on Manhattan Island.

Any number of locomotive experiments have been inspired by the diesel in an effort to curb its advance, but to no avail. The list includes four steam turbine-electrics of three different designs, a direct-drive steam turbine, reciprocating steam locomotives with automatic firing and water injection controls, rectifier electrics, and coal and oil-fired gas turbines. An experimental 4,500 H.P. oil-fired gas turbine locomotive tested on Union Pacific resulted in installation of twenty-five production models, and, later, orders for at least thirty 8,500 H.P. turbines of similar but larger design. Other roads have yet to show interest in supplementing diesel power in this manner. Three railroads—New Haven, Pennsylvania, and Virginian—operate new rectifier electric locomotives, but in each case such units were purchased to supplant older engines rather than to extend electrification.

Thus the diesel has virtually wiped out the steam locomotive in the U.S. and will do likewise in both Canada and Mexico. In America diesels now handle more than 90 per cent of all railroad operations. By 1960 this figure will have risen to 98 per cent and more if certain ageing electrified systems cannot compete economically. It will be less only in the unlikely event of a remarkable advance by a form of motive power as yet unexplored.

EASE OF FINANCING

The diesel locomotive remains the best possible collateral on which a railroad could borrow money. More than three billion dollars is invested in the approximately 26,000 diesel units that power Class One roads in the U.S., and the overwhelming majority of these diesels have been purchased on the instalment plan. In a recent example, Illinois Central purchased 70 road-switcher units of 1,750 H.P. each, for approximately $12,150,000 or $175,000 apiece. The railroad itself paid 25 per cent of the total, the balance being borrowed and paid off in twice-annual instalments over a period of fifteen years. In the event of a failure to complete payment, the lender is entitled to seize the locomotives and sell them at auction. Such a financial arrangement is attractive to both lender and borrower because:

(i) the diesels are capable of effecting operating savings immediately that will more than cover their payments.

(ii) the diesels, being movable assets and adaptable to any other railroad, make excellent security.

(iii) since motive power is essential to operation, the borrower can be expected to go to great effort in times of tight money to protect the equipment from seizure.

Indeed, in the history of such equipment-trust financing in the U.S., there have been only two important failures: once when the closing of the Florida real estate boom in the 1920's forced the Florida East Coast to dispose of a number of 4–8–2 steam locomotives for which it could not pay, and in 1957, when New York, Ontario & Western ceased operation by Court order because of bankruptcy, and its diesel units were sold at auction to other carriers.

INDICES OF DIESEL EFFICIENCY

Any itemization of the achievements of the diesel locomotive in North America must necessarily be headed by the fact that, technologically speaking, the diesel is primarily responsible for the continued solvency of the last major network of privately owned and operated railways in the world. Diesels are currently estimated to save U.S. roads 800 million dollars a year. These same Class One roads averaged an annual net income of only 537 million dollars for the years 1946–50, managed 927 million in 1955, only a net of 874 million in 1956. The obvious relationship of diesel performance to such profit as was earned is unmistakable, and constitutes the major cause for the diesel's rapid acceptance.

Virtually all efficiency indices confirm the diesel's enormous impact upon American railroading. Perhaps the prime evidence lies in gross ton-miles per freight train-hour, an index that combines both tonnage and speed. In largely steam 1946, that index was 37,057; in nominally diesel 1956, it reached the all-time high of 57,102. The record of 1956 was 2·3 times the average gross ton-miles per freight train-hour of 1929 and more than

3·75 times the average of all the 1920's. Again, in 1946, the average U.S. freight train carried 1,086 net tons of lading, operated at an over-all average speed of 16 m.p.h.; in 1956, it carried 1,420 net tons, moved at an average speed of 18·6 m.p.h. One more index: in 1946, the average American freight locomotive in the "serviceable not stored" category operated 115·9 miles per day; in 1956 it operated 149·1 miles per day.

OPERATING ADVANTAGES

Statistics, by themselves, cannot do justice to the diesel. It circumvented the steam locomotive on many counts, one of the most important being that it made the maximum-horsepower locomotive available to all railroads, rather than to just the fortunate few. When the practical road freight diesel entered mass production, U.S. steam locomotive builders were turning out 6,000 H.P. 4–8–4's and articulateds, which steam engines were theoretically capable of matching the most powerful diesels available. (In point of fact, rated steam locomotive horsepower output has proven an exaggeration when compared with the diesel; on the average 1 diesel H.P. has replaced 2·2 steam H.P. in the U.S., a ratio not wholly predicated on the diesel's greater availability.) Such steam power, moreover, was restricted to the main lines of those roads whose traffic density and finances permitted generous axle loadings and width and height clearances. The 6,000 H.P., 4-unit diesel, however, spreads its not inconsiderable weight over sixteen or more axles, all of them powered; and the locomotive is not only flexible to the degree of independent units operated in multiple, but each unit is mounted on a pair of swivel trucks, or bogies.

Clearances required are moderate. For example, Santa Fé 5,000-class 2–10–4 steam freight locomotive with 74-in. drivers, 30 × 32-in. cylinders, and 310 p.s.i., is 11 ft wide, 16 ft high, possesses a driver axle loading of 76,000 lb. and a rigid wheelbase of 50 ft 2 in.; the 1,500 H.P. diesel freight unit (four of which would create a 6,000 H.P. locomotive quite equivalent to the 2–10–4) is, by contrast, 10 ft 8 in. wide, 15 ft high, possesses a driver axle loading of only 57,600 pounds and a wheel-base of just 39 ft. To quote a yardmaster: "a diesel can go anywhere a boxcar can", whereas a locomotive of such proportions as the aforementioned 2–10–4 is severely restricted in its operating orbits. Virtually every Class One railroad in the U.S. operates 6,000 H.P. diesel locomotives today, even though ten or fifteen years ago only a few used articulated steam engines and many owned nothing larger than 2–8–2's or modestly proportioned 4–8–4's.

Unlike a steam engine of equivalent power output, the 6,000 H.P. diesel is essentially four independent units or locomotives operated by a single crew under multiple-unit control, hence it is not inflexibly restricted to that size. Assuming each unit is equipped with road operating controls, the 6,000 H.P. locomotive can quickly be translated into two, three, or even four separate locomotives of smaller power, to meet varying traffic demands. A single unit can be withdrawn for maintenance and replaced with another, without placing the entire locomotive in the shop.

Any number of other variables combine to make the diesel a far more attractive tool for the operating department than a steam engine. As an example, the manufacturer usually offers a choice of, say, eight different gear ratios for the axle-mounted traction motors with maximum speeds ranging from 55 m.p.h. to 102 m.p.h. and, for a 1,500 H.P. unit, continuous tractive effort ratings ranging from 52,400 lb. to 25,500 lb. Thus the same basic locomotive may be operated initially as a heavy-duty, slow-speed freight engine, then subsequently adapted by a change in gear ratio to a high-speed passenger machine. An intermediate gear ratio provides a mixed-traffic or dual-purpose unit, at home in freight or passenger work and even in yard switching duty.

Maximum availability is another element in the diesel's success. Aside from inspection, fueling and watering (functions which may usually be carried out during the normal crew change period or lunch hour), the diesel is available for traffic twenty-four hours a day, able to run any conceivable distance without change. The longest diesel locomotive run in North America occurs on Canadian National, which runs its streamlined *Super Continental* between Montreal, Quebec, and Vancouver, British Columbia – 2,930 miles – without change of power. Even here, the limiting factor is the route and not the motive power. This is equivalent to more than seven one-way journeys between Chicago and the Twin Cities, or between London and Edinburgh. Capacity for service such as this explains why it is quite common for U.S. diesel freight units to run up between 8,000 and 12,000 miles a month and for passenger diesels to operate between 18,000 and 30,000 miles a month.

Those diesels which are assigned to services of inherently low mileage frequently turn in interesting statistics from a flexibility viewpoint. In August 1953, a Western Maryland 1,600 H.P. diesel road-switcher unit ran up 4,905 miles in the following services:

	miles
Passenger trains	3,306
Road freight trains	368
Travelling switcher	474
Yard service	504
Running light or "deadheading"	131
Helper duty	122

The average American diesel locomotive consumes 1·6 gallons of fuel oil per mile, and there are numerous ways of expressing what that means. For example, it is estimated that on a railroad a teaspoonful of diesel oil will move one ton over a mile. In 1956, diesels moved 5,299 gross ton-miles for each $1 of fuel oil, whereas coal-fired steam locomotives moved 3,087 gross ton-miles per $1 of fuel; electrics moved 2,959 gross ton-miles per $1 of fuel, and oil-fired steam 1,957. It is obvious, therefore, why U.S. railroads purchased 108·1 million tons of locomotive coal in 1946 but only 12·9 million tons in 1956. It also explains how these roads could spend 553·1 million dollars for fuel in 1946 but only 476·9 million in 1956, a year of more ton-miles and in spite of a decade of supply cost inflation.

These figures, and many more, amply prove that the diesel itself is the overwhelming cause for its widespread and rapid adoption in North America, but contributing factors must needs be studied for a full understanding of this motive power revolution.

STEAM POWER DEVELOPMENT REACHES FINALITY

A cause for the diesel's quick acceptance in North America was that the reciprocating steam locomotive was reaching practical limits of size, power and efficiency – even on those roads with maximum axle loadings and clearances.

In over-all size American steam locomotive manufacturing had attained the Union Pacific "Big Boy" 4–8–8–4, an articulated machine which, with tender, extends 133 ft over coupler faces, stands 16 ft 2½ in. high, and weighs 1,189,500 lb. (535½ long tons). (Plate 73B, page 333.)

To improve steam distribution at speed the Franklin Railway Supply Co. had developed two types of poppet valves, Type A with oscillating cams, and Type B with continuous-contour rotating cams. Both gears were applied to a considerable number of rebuilt and new engines, ranging in wheel arrangement from 4–6–2's to 4–8–4's and rigid-frame 4–4–4–4's.

Several roads, notably Norfolk & Western, had gone to great lengths to increase locomotive availability by modernization and rearrangement of engine terminals.

Boiler pressures of 300 and even 310 p.s.i. were common, as was the all-welded boiler.

Pennsylvania put non-articulated duplex-drive locomotives into production in an effort to utilize fully the boiler output, and to lessen the load on main rods.

Norfolk & Western experimented with a pair of automatic switchers, 4–8–0's capable of supplying their own fuel and water without human attention.

Following extensive research, New York Central developed scoops and tenders capable of picking up water from track pans at 80 m.p.h. These and other developments, often quite spectacular in their relationship to conventional steam power, could not resolve the drawbacks of a fuel efficiency of 6–8 per cent; the delays and costs inherent in boiler washouts and maintenance; the lack of constant torque power application; the handling difficulties inherent with a hard fuel; ash disposal problems; fluctuation in power output because of weather; the damaging effect of high axle loadings and masses of reciprocating metal upon the track structure; and so forth.

Part II. History of dieselization in North America

(*See also* Chapter 1, Development, Parts I (page 27), VI (page 81), and VII (page 90)

1906–23. SELF-PROPELLED RAILCARS

The concept of internal combustion power providing railway locomotion through an electric transmission came to fruition in America in 1906 when the General Electric Co. produced a self-propelled gas-electric coach or railcar for the Delaware & Hudson. The combination baggage-coach car weighed 98,000 lb., extended 68 ft 7 in. over buffers, had an arch roof, rounded ends, a centre drop-door entrance for passengers, and seated ninety-one passengers. Rated at 200 H.P., a gasoline engine powered a generator which fed 600-volt direct current to two traction motors mounted on the forward truck beneath the engine compartment. D. & H. No. 2000 was a refinement of earlier direct-drive gas-engine motor cars in which the railroads were interested to reduce

the costs of steam-train local service over thinly trafficked branch lines. Eventually, more than 700 gas-electric cars (or "doodle-bugs" as they were called) entered U.S. service, most of them being constructed during the 1920's. These cars, some of which could haul light non-powered trailers, enjoyed great favour as a means of reducing costs, but their limitations included a constant danger of fire, inadequate power output, and engine unreliability. Attempts at independent gas-electric locomotives were invariably frustrated by these same limitations.

In 1923, Canadian National purchased from Messrs Beardmore in Britain several diesel-electric railcars and one, No. 15820, operated from Montreal to Vancouver in 67 hours 55 minutes, averaging almost 44 m.p.h. (*see* Chapter 1, Part I, Engines, Development, page 27).

1923. THE FIRST DIESEL-ELECTRIC LOCOMOTIVE

In 1923 three American manufacturers pooled their resources to construct the New World's original diesel-electric locomotive (albeit termed an "oil-electric" at the time of its inception). The unit was a 300 H.P., 60-ton B–B type (two swivel 4-wheel trucks, all axles powered) assembled by the American Locomotive Co. and containing an Ingersoll-Rand Co. (IR) diesel, driving electrical equipment made by General Electric. An inline, vertical type, 6-cylinder IR diesel produced 300 H.P. at 600 r.p.m. and drove a 600-volt D.C. generator which supplied four 60 H.P. D.C. traction motors directly geared to the axles. The control system, following principles established by Herman Lemp of General Electric in a 1914 patent, eliminated manual control of the generator field; that is, the throttle governed the diesel engine itself, with the generator and traction motors automatically adjusting through the control system, to meet the flexible demands of acceleration and grade.

After a preliminary four-month test in the plant yard of the Ingersoll-Rand factory at Phillippsburg, New Jersey, the little locomotive was tested in the West Side Yards of New York Central in New York City, and there it quickly proved what railroading could expect of this new form of motive power. Inasmuch as the unit carried fuel sufficient for forty-eight hours of continuous duty, it could be worked round the clock with time out only for routine inspection and crew change. (Crews, incidentally, were taught to operate the diesel in fifteen minutes, the essential controls being throttle, reverse and brake, far fewer than on a steam engine.) There were no delays chargeable to the diesel for taking water or for the disposal of ashes. Moreover, the diesel was virtually smokeless in operation, a considerable asset in New York City. In switching service on N.Y.C. the unit burned 4·4 gallons of oil per locomotive-hour for an over-all operating cost of barely more than 16 cents per mile. (In that year, 1924, diesel oil sold for 5 cents per gallon.)

A telling point in the diesel's favour was that it could produce as much work on a tank car of fuel oil as the steam engine it replaced (probably a small 0–6–0) could produce on twelve cars of coal. The major handicap aside from the questionable reliability of any brand-new machine was that the diesel was priced at about $100,000, or double the cost of even a large steam switcher.

This experimental diesel switcher demonstrated its talents on at least fourteen railroads or industrial plant yards, and it did an occasional stint of main-line local freight duty. Once, at least, it pulled passengers (a two-car "extra" over the rails of the Reading). Eventually, its mission completed, the pioneer was dismantled.

1925. FIRST "COMMERCIALLY PRODUCED" DIESEL

Jersey Central Lines No. 1000, now preserved in the Transportation Museum of the Baltimore & Ohio Railroad at Baltimore, Maryland, is the locomotive most Americans think of as the "granddaddy" of U.S. dieselization. Actually, No. 1000, a 300 H.P. 60-tonner basically similar to the true pioneer of 1923 and built in 1925 by the same three manufacturers, was the first commercially-produced diesel built in America and the first actually sold to a U.S. railroad (Plate 24A, page 100). In appearance, No. 1000 was nothing more or less than a box-on-a-flat-car. Inside was an operator's compartment at each end with a diesel engine and main generator located in the centre; beneath were two four-wheel trucks (bogies) with traction motors driving on each axle. No. 1000 was 32 ft 8 in. long, 9 ft 4 in. wide. In testimony to its builders' faith in the experiment of 1923, the unit was one of four they built "for stock" without advance order. No. 1000 was sold to Jersey Central on 20 October, 1925, for service at its Bronx Terminal in New York City and where it served until retirement in 1957.

Sister units were sold to Baltimore & Ohio and Lehigh Valley in 1925 and 1926 respectively, and in the latter year Long Island took delivery of a 600 H.P., 100-ton model that had two diesel engines and four 200 H.P. traction motors. This last-named unit is sometimes acclaimed the world's first "over-the-road" diesel-electric inasmuch as Long Island operated it in freight and passenger service as well as in the yard. Essentially, however, the unit, LIRR No. 401, was a low-speed yard unit, and should be so regarded. It was retired in 1951. The Baltimore & Ohio unit is still in its original form and is still (1957) at work in New York City (Plate 24B, page 100).

1925–36. EARLY DIESEL SWITCHER PRODUCTION AND ACCEPTANCE

Despite the events of the mid-1920's, the diesel locomotive was in no sense "accepted" by the railroad industry in North America. During the period 1925–36 only 190 diesel switchers were produced for U.S. service, and almost all of these units were for special assignments where objections to smoke, or a fire hazard, prohibited operation of the more favoured steam engine. Essentially, the diesel's engines were too heavy (1 H.P. for every 60 lb. of weight) and slow (600 r.p.m.) to produce the "snap" of acceleration required in many heavy-duty switching situations. Their cost, approximately double that of a steam locomotive of equivalent horsepower, stymied widespread usage. Nevertheless, these early box-cab units were the practical fulfilment of the diesel-electric principle and their availability and fuel economy virtues were undeniable (Plate 25A, page 117). The ceiling on horsepower was raised in 1925 when the nation's oldest locomotive builder, Baldwin, constructed an experimental 1,000 H.P. diesel and followed it with another in 1929.

1928. THE FIRST ROAD DIESEL LOCOMOTIVE

One of the first prophets of total dieselization and an internationally famous railroad man, U.S.-born Sir Henry W. Thornton, was responsible for the first true non-articulated road diesel-electric locomotive in North America. In 1928 the Canadian Locomotive Co. of Kingston, Ontario, delivered Thornton's Canadian National No. 9000, a 2,660 H.P. machine which weighed 334 tons. Essentially, No. 9000 was two distinct units operated by one crew under multiple-unit (M.U.) control. Each unit possessed a box-cab car body with a control compartment at one end, mounted over a four-wheel lead truck, four driving axles, and a two-wheel idler axle at the rear. Each unit contained a William Beardmore V-12 diesel engine rated at 1,330 H.P. at 800 r.p.m. and this drove a Canadian-Westinghouse generator which fed current to four traction motors geared to 51 in. drivers. No. 9000 was a big, bold step toward complete railroad operation by what Thornton referred to as "self-propelled electric locomotives", and it was adaptable to either passenger or freight service by a change of gear ratios in the traction motor connections to the axles. It was considered approximately equivalent to a 6000-series C.N.R. 4–8–2 steam engine of that period. As designed, the diesel was geared to haul a 2,800-ton passenger train at 40 m.p.h. on the level, or, with a gear change, a freight train of 3,700 tons at 35 m.p.h. on the level. In actual service, No. 9000 did work the $663\frac{1}{4}$ ton *International Limited*'s second section over the 334 miles from Montreal to Toronto in 7 hours 40 minutes despite thirteen intermediate stops, and delays of between two and thirteen minutes at station halts. At one point the train was operated at 73 m.p.h. Compared with the 4–8–2, No. 9000 would accelerate its train up to a mile a minute in half the time and perform equivalent service at one-third the fuel cost.

Unfortunately, the depression prevented allocation of funds for further development of the locomotive. The units were renumbered 9000 and 9001, operated separately, then stored. No. 9001 was retired in 1939 but her sister unit was rebuilt in 1941 with a new engine and armour plating to haul a defence train on the West Coast in British Columbia during World War II. After the conflict, the armour was removed and No. 9000 ran in passenger service for fifteen months between Quebec City and Edmundston, New Brunswick, and was then scrapped.

Somewhat similar road diesel locomotives were built in 1928 by American Locomotive and General Electric for service on the Putnam Division of New York Central, a 145-ton, 750 H.P. freight unit and a 165-ton, 900 H.P. passenger unit. Neither unit proved to be much more than an unsuccessful pioneer and both were eventually retired.

1934. ENTER THE STREAMLINERS

Despite the pioneering efforts of the nineteen-twenties, the key element of dieselization, a high-speed, lightweight engine, did not appear until the age of streamlining dawned. Specifically, the diesel came of age in 1934 with the construction of the Burlington's *Pioneer Zephyr*, a three-unit articulated streamliner incorporating a light-weight 600 H.P. diesel in its power car (Plate 25B, page 117).

Prior to the *Zephyr* there had been many groping efforts toward much the same goal, most of them by the Electro-Motive Co., which acted as designer and contractor for 500 gas-electric railcars during the years 1922–30. In 1929, for example, Chicago Great Western rebuilt three old knife-nosed McKeen railcars into the *Blue Bird* for service between Minneapolis, Minnesota, and Rochester, Illinois. The old gas engines and mechanical transmissions with which each of the cars had been equipped were removed, and in the lead car E.M.C. installed a 300 H.P., 6-cylinder gasoline engine with electrical transmission. Inasmuch as this train provided parlour and sleeper space, it anticipated the streamliner era remarkably. In 1929–30 the Rock Island rebuilt some gas-electrics into all-power units or locomotives, each with a pair of 400 H.P. gas engines driving through an electrical transmission. In 1932 E.M.C. powered, and the Pullman Car Manufacturing Co built, a bigger, different railcar for Santa Fé. Articulated over a common centre truck, the first unit of No. M-190 contained a 900 H.P. distillate engine, which burned a low-grade gasoline but used spark-plug ignition. Electric transmission was used, and the rear unit held train-heating equipment and baggage space. This equipment, which could pull trailing coaches, introduced a new peak in railcar power output and the principle of articulation.

Streamlining itself was formally born on 12 February 1934, when Pullman built and delivered (and E.M.C. powered) Union Pacific's articulated, three-car M-10000 (later, the *City of Salina*). This 85-ton, $204\frac{1}{2}$ ft train was powered by a 12-cylinder, V-type, distillate engine through an electric transmission. It set forth on a 12,625-mile coast-to-coast tour during which more than a million people inspected it, and, on test M-10000 managed 111 m.p.h. This pioneer aluminium train operated until

1942, when it was cut up for its valuable metal in the war effort.

Meantime, the Burlington, whose stainless-steel *Zephyr* was on order with the Budd Co., decided to chance the installation of a true diesel-electric powerplant. The automobile giant, General Motors, had purchased both Electro-Motive and its engine supplier, the Winton Engine Company, in 1930; in the early 1930's G.M.'s efforts were concentrated on a diesel that would free its new subsidiaries from the limitations of both gasoline and distillate engines without necessitating the weight of contemporary diesels, which was then about 80 lb. per H.P. G.M.'s research culminated in the 201A design, a lightweight powerplant that weighed only 20 lb. per H.P., thanks to new alloy steels, welded construction, and two-cycle design. Two such engines were installed for power generating purposes in G.M.'s exhibit at the Century of Progress Exposition in Chicago in 1933, and over the maker's reluctance a third such engine, an 8-cylinder in-line, 600 H.P. job, was ordered for the *Zephyr*.

The test of the new engine came within days of the train's delivery on 18 April, 1934. On 26 May, Burlington dispatched its *Pioneer Zephyr* on an historic dawn-to-dusk run from Denver, Colorado, right to the stage of the Exposition in Chicago. The run was made without a stop, and the 1,015 miles were covered at an average speed of 77·6 m.p.h. to set an unprecedented world's record for speed and distance in train operation. It may be effectively argued that on that notable day the diesel engine truly came of age for railroad operation. It proved that it could absorb sustained, arduous service, that it was freed from its former cumbersomeness, that it had earned a place, a big place, in railroad mechanical practice.

Across the next two years U.S. railroads installed a number of articulated diesel-powered streamliners to meet an overwhelming public demand for the modern high-speed passenger service introduced by U.P.'s M-10000. Between them, Burlington and U.P. installed eight such trains, each longer and more powerful (thanks to multiple engines) than the last. By 1936 U.P.'s *City of Denver* boasted 2,400 H.P. and Burlington's *Denver Zephyr* had 3,000 H.P. The latter train, in answer to the complaint that the *Pioneer Zephyr*'s speed run had been made "downhill" from 5,280 ft above sea level to lake-level Chicago, averaged 83·69 m.p.h. from Chicago *to* Denver on 23 October, 1936, and once hit 116 m.p.h.

At least two drawbacks appeared. First, the locomotives were inseparably linked to their trains and could not be used with conventional equipment or easily cut out for maintenance. Second, the diesel was being handi-capped in its bid to dethrone steam power; at that time, existing power as large as a 4–6–4 was being rebuilt and often streamlined for luxury passenger service, and plans were being laid for brand-new, streamlined, high-speed 4–8–4's.

1935. NON-ARTICULATED ROAD DIESEL PASSENGER UNITS

Practically all the effort to develop the obvious, non-articulated, diesel-electric road passenger locomotives fell into the eager hands of G.M.'s Electro-Motive Co. Among the Big Three of steam power builders, Baldwin did not enter even the diesel switcher field seriously until 1939; Lima stuck exclusively to steam until after World War II; and American Locomotive (or Alco) restricted its efforts to yard diesels and the power for Gulf, Mobile & Northern's articulated *Rebel* streamliners.

What E.M.C. set out to create was essentially a "platform" for the 201A engine that could be coupled to any railroad equipment instead of being restricted to lightweight articulated streamliners. The prototypes, Nos. 511 and 512, appeared in 1935, powered by 201A engines and built under contract by General Electric's Erie, Pa. works, with G.E. generators and carbodies. Each unit weighed 240,000 lb., rode on a B–B wheel arrangement (two swivel, four-wheel trucks, all four axles powered), and mounted two 12-cylinder, V-type diesels, each of 900 H.P. Coupled in multiple-unit control, the demonstrator thus produced 3,600 H.P., which E.M.C. felt was comparable to the average U.S. steam passenger locomotive, in practice if not on paper. Box-cab carbodies, similar to the earliest diesel yard units of 1925, were used. An example of performance occurred when 511–512 hauled a train of ten standard-weight cars weighing 757 tons from Jersey City, New Jersey, to St. Louis, Mo., 1,114 miles, without change, waited two hours, then proceeded from St. Louis to Chicago, 284 miles, with another passenger train weighing 808 tons. As a result of such operation, E.M.C. figured that the 3,600 H.P. locomotive could haul twelve to fourteen cars on steam schedules for 2·7 gallons of fuel per mile at a cost of 40–60 per cent less than that of a steam engine. Repairs were estimated to be correspondingly less.

The first such locomotive sold to a U.S. road (and the first non-articulated road passenger diesel since the C.N.R. and N.Y.C. experiments of 1928) was an 1,800 H.P. unit. This was placed in service on the Baltimore & Ohio in 1935, to be followed in the same year by two units operated as a 3,600 H.P. team by Santa Fé. The latter units suffered a serious fire because of a leaking fuel line, but in the following year they inaugurated the once-a-week *Super Chief* Pullman service with conventional sleepers between Chicago and Los Angeles on the unprecedented schedule of 39¾ hours. There was no engine change in the 2,227-mile run.

Thus far into dieselization, Electro-Motive had not achieved a basic tenet of future locomotive construction policy: standardization. Its engines and design concept had been largely incorporated into articulated streamliners whose carbuilders merely set aside a lead "power unit" for propulsion machinery. Elsewhere, demonstrators 511 and 512, as well as B. & O.'s 1,800 H.P. unit, had been actually assembled by General Electric, and the St. Louis Car Co. had produced the carbody for the Santa Fé tandem; G.E. had supplied electrical equipment for all these units. The cost-bulging horrors of customization continued as railroad after railroad attempted to make its streamliners as distinctive as its steam engines.

In 1937 Electro-Motive (by then a full-fledged division of General Motors) had its own plant in La Grange, Illinois, and began selling a semi-standardized line of 1,800 H.P. passenger diesel units known as E-1's. This locomotive was a considerable advance in both design and dieselization technique. In over-all shape, the carbody featured a high-mounted cab behind a rounded, slanted nose. This feature was designed to protect the engine crew from the danger of hitting highway vehicles at grade crossings. Such accidents took a toll of life from the crews of earlier "shovel-nose" articulated streamliners; moreover, the crews, being near the rail, developed a fixation on rails and crossties at high speed, causing them to miss signals. In wheel arrangement, the E-1 pioneered the now-standard A1A–A1A passenger design (a pair of six-wheel swivel trucks with a total of four traction motors and centre idler axles) and mounted a pair of V-12 engines of 900 H.P. each. E-1's were sold in two versions: cab and booster units. The latter units had blind ends and were intended for M.U. operation with cab units containing controls; the intent was to supplement total horsepower at slightly less cost than for a cab unit. Both B. & O. and Santa Fé purchased E-1's for a long-haul passenger duty.

In 1938 E.M.D. refined this basic model with the 2,000 H.P. E-6, a locomotive notable for being the world's first true mass-production road diesel. It incorporated E.M.D.'s own generators and traction motors; each unit mounted two 567-series 1,000 H.P. V-12 diesels designed for mass production. It made possible, by three-unit operation, 6,000 H.P. diesel locomotives. First sold to Seaboard Air Line for service between Washington, D.C., and Miami, Florida, on the all-Pullman, winter-season tourist train, the *Orange Blossom*

Special, the E-6 was operated as a 6,000 H.P. locomotive with a cab unit at each end and a booster unit in the centre (or in A–B–A fashion). It quickly proved itself, and a rival railroad, which had purchased large 4–8–4 steam engines when S.A.L. bought diesels, quickly reversed itself and ordered E-6 diesels. The E-6 was subsequently sold in large numbers to other roads without compromise of its standardized design. Indeed, for the first instance in any volume, railroads began purchasing diesels instead of 4–6–4 or 4–8–4 steam power for heavy-duty passenger work. Produced until halted by wartime restriction on passenger power, the E-6 was largely responsible for the fact that, by 1944, diesel-electric power was handling 8 per cent of U.S. passenger car-miles, more than all of the straight-electric locomotives of the nation.

In 1940 competition entered the U.S. road diesel locomotive field when American Locomotive and General Electric joined forces to produce a 2,000 H.P. A1A–A1A unit powered by a pair of inline 6-cylinder engines originally employed in yard diesels. Eventually six roads purchased sixty-seven such units. New Haven operated them in both freight and passenger service, marking the first occasion of mixed-traffic, or dual-purpose, usage of road diesels. Otherwise, these units were not particularly successful and today most are retired, stored or have been re-engined.

1936. THE YARDS "GO DIESEL"

Until 1935 the diesel yard locomotive, ten years old, had failed to break out of "special assignment" service (i.e. operations where steam was prohibited because of smoke nuisance or fire hazard). In 1935, twenty Class One roads were operating just eighty-seven diesel yard units in the U.S. Suddenly, though, the cumulative experience obtained with these pioneers, plus the economic advantages of mass-produced standardized units, began to have their impact. In 1936, 52 yard units were ordered; 93 in 1937; 110 in 1938; 176 in 1939; 255 in 1940; and 583 in 1941.

The Alco-G.E. team, which had built virtually all of the yard units prior to 1936, continued to hold a front rank in this field. Electro-Motive began large-scale switcher production in 1936 when it moved to La Grange, Illinois, and issued 600 H.P. and 900 H.P. (1,000 H.P. by 1938) models which incorporated its light-weight, two-cycle engines, first the 201A, then the 567 series. Of interest is the fact that E.M.D.'s refusal to customize its models brought about a significant reduction in their cost. By building twenty-five or fifty switchers in blocks, then selling them "off the shelf" on a first-come, first-served basis, E.M.D. managed to reduce the price of a 600 H.P. unit from about $84,500 to about $70,000 in 1936 and even chopped that to a minimum of $59,750 by 1940.

Baldwin Locomotive did not seriously interest itself in diesels until 1939. Through that year it had built only seven such units of all kinds, including experimentals. In 1940, however, Baldwin began producing 660 H.P. and 1,000 H.P. models equivalent to the stock units of the other two big builders. General Electric, on its own, started a good many short lines toward dieselization with smaller 44-ton, 380 H.P. centre-cab B–B units.

1939. THE ROAD FREIGHT DIESEL APPEARS

Until 1939 no manufacturer made a serious attempt to construct a diesel road freight locomotive. For one thing, the scepticism of the railroad industry toward internal combustion power for road service was not broken down until the advent of the streamliner. Even then, many carriers had to be convinced that the diesel could do more than pull lightweight, articulated trains, that it could perform without velvet-glove attention, and that it was worth more than $100 per horsepower compared with approximately $35 per horsepower for steam. Again, until 1939 the largest commercially produced engine available for locomotive usage was 1,000 H.P., hardly enough to power a heavy-duty freight unit without resorting to the complexity of multiple engines and generators. Finally, the sudden rising demand for both yard and passenger units after 1935 practically absorbed the limited diesel production in the U.S.

In 1939, Electro-Motive had the proper engine, 567-series, 16-cylinder, V-type, two-cycle diesel rated at 1,350 H.P., and the locomotive in which to install it. Demonstrator freighter No. 103 was a 5,400 H.P. locomotive composed of four B–B units of 1,350 H.P. each (two cab or control units with two booster units in the centre, operated in A–B–B–A formation). Each unit measured approximately 50 ft and weighed 110 tons, all of the weight being on driving axles and thus available for adhesion. Although in this original FT-type locomotive the cab and booster units were permanently coupled, No. 103 could be separated in the centre and operated as two distinct 2,700 H.P. locomotives. There were any number of unusual statistics and facts about No. 103. It weighed 900,000 lb. (401¾ long tons) and at that time far more than the biggest steam locomotive ever built. That weight was, however, spread across thirty-two axles, all of them driving, so that power was produced without the locomotive being handicapped by excessive axle loadings. It was 193 ft long (compared with about 83 ft for the Union Pacific's 4–8–8–4, less tender) but flexible to the extent of being four units

mounted on eight flexible trucks. It produced a starting tractive effort of 220,000 lb., approximately twice that of the largest articulated steam locomotive in normal U.S. service. It carried 4,800 gallons of fuel (compared with 6,100 gallons in tenders of the biggest oil-burning steam power), sufficient for 500 miles of operation without refuelling.

On paper, No. 103 looked significant and in practice it justified its promise. For over a year the locomotive ran tests on twenty Class One railroads in thirty-five States – over 83,764 miles in temperatures from —40° F. to + 110° F., and at altitudes from sea level to 10,200 feet. A few examples will suffice to show what No. 103 did in comparison with existing U.S. steam freight power:

On Denver & Rio Grande's Western crossing of the Rocky Mountains via the 6·21-mile Moffat Tunnel, a line which involves fifty miles of continuous 2 per cent (1 in 50) grade, No. 103 covered 127 miles in 5 hours 20 minutes with 1,780 adjusted tons as compared with 6 hours 30 minutes for a 2–8–8–2 with 1,800 adjusted tons.

Over the sawtooth profile of Erie between Marion, Ohio, and Meadville, Pasadena, the diesel covered 212 miles over a series of 1 per cent (1 in 100) grades with from 5,000 to 5,700 tons in the same time that a large 2–8–4 required to pull from 2,800 to 3,100 tons.

On the 2·55 per cent (approximately 1 in 40) grade up the west slope of California's Tehachapi Pass, No. 103 covered twenty-five miles with 1,800 tons in 1 hour 31 minutes compared with 1,100 tons in 2 hours 15 minutes for a Santa Fé 2–10–2 and 1,350 tons in 2 hours 15 minutes for a Southern Pacific articulated, cab-forward 4–8–8–2.

Although steam-heat boilers permitted No. 103 to work passenger trains, it was essentially a freight locomotive, geared for a maximum speed of 75 m.p.h. Nevertheless, a spectacular performance ensued in the mountainous 240 miles between Livingston and Missoula, Montana, a Northern Pacific mainline district involving grades of as much as 2·2 per cent. Train No. 1, the *North Coast Limited*, served as the index of the test. Normally, a 4–6–6–4 articulated steam locomotive handled this train westbound in 6½ hours, with no helpers on 10 cars, one helper on 12 cars, two helpers on more than 12 cars. The diesel took through 17 cars weighing 1,316 tons in 6 hours 19 minutes with no helpers.

On a nationwide basis, No. 103 completed its more than 83,000 miles of testing without a single delay being charged to the locomotive (i.e. 100 per cent availability). Because of its weight distribution, the diesel operated over track restricted to steam engines of less than 75 per cent its pulling power. Two of its units rated at 2,700 H.P. were found to be equal to 2–8–2's and 2–10–2's commonly rated at between 3,000 and 3,500 H.P. The locomotive was operated in both freight and passenger service and as either one 5,400 H.P. engine or two 2,700 H.P. engines without adjustment of any kind. And on a nationwide basis the diesel burned 0·067 cents of fuel oil per million gross ton miles as compared with 0·137 cents per M.G.T.M. for the steam power it encountered.

Such a phenomenal test had immediate reaction as thirteen of the twenty railroads which tested No. 103 ordered similar FT-type diesel freight power. Practically all such power was purchased for mountain work. Santa Fé installed the first diesel freighters in revenue service in 1940, although demonstrator No. 103 was eventually sold to the Southern Railway, resulting in some conflict of claims for "firsts" in freight dieselization!

1941. INTRODUCTION OF ROAD-SWITCHERS

An important step in diesel design was taken in 1941 when Alco-G.E. began advertising a stock-model 1,000 H.P. road-switcher. The unit, essentially a switcher with an extended frame, featured an offset cab. Additional length permitted the mounting of a steam generator in the "short end" nose of the unit as well as water tanks and enlarged fuel capacity between the trucks. The result was a locomotive that could work in the yard or perform on the road with local freight or passenger trains. Electro-Motive sold a similar but non-mass-production 1,000 H.P. unit to Great Northern. There was no particular demand for such power then or immediately after the war, apparently because railroading could not rid itself of the concept of a different engine for each job. In addition, many U.S. roads feared the operation of road-switchers in multiple lest the unions seek a crew for each individual unit on grounds that it contained operating controls. Fortunately, this management-labour situation did not develop.

1941-45. EFFECT OF WORLD WAR II ON DIESELIZATION

It is difficult to determine whether World War II retarded or accelerated railroad dieselization in the U.S. Once America formally entered the war in December 1941, the War Production Board decided what roads would receive how many locomotives of what type from which builder. The construction of purely passenger power was banned for the duration, although dual-purpose 4–8–4 steam engines and a few mixed-traffic New Haven Alco-G.E. diesels were allowed. Elsewhere, the production of road diesels was allocated entirely to their biggest builder, Electro-Motive. Alco-G.E. and Baldwin were restricted in diesel construction to yard units of 1,000 H.P. or less. This move, obvious and efficient, nevertheless cut five years of invaluable design and operating experience in road diesels off the records of Alco-G.E. and Baldwin and increased E.M.D.'s know-how and prestige. The production of road freight units was severely restricted and allocated exclusively to roads with considerable experience or need for this type of power.

Otherwise, the unprecedented demands of war upon locomotives gave the diesel an excellent opportunity to demonstrate its great power and high availability. Harold L. Hamilton, longtime chief of Electro-Motive, summed it up this way: "During the war effort, we had an opportunity to put both types of motive power, steam and diesel, to a full test. Whatever the steam engine could do she was given her chance to do, and whatever the diesel could do it received the chance to do. When it was all over we drew conclusions and marked up the winner, and it was obvious that the logical investment in motive power from that point on was going to be the diesel locomotive." Which is a simplification, but none the less accurate. One statistic proves it. In 1943, Class One roads installed 438 new steam locomotives in service (the highest number since 782 in 1930). Thereafter, particularly after wartime restrictions ended to permit freedom of purchase, steam declined: 326 new steam engines were installed in 1944; 115 in 1945; 86 in 1946; 69 in 1947; 86 in 1948; 57 in 1949; and 12 in 1950.

1946. THE DIESEL AT WAR'S END

When World War II was concluded the diesel locomotive was no longer a novelty or an experiment, but a proven piece of motive power. Every basic unit required for total dieselization, yard, road-switcher, freight and passenger, had been available, and in at least limited use, when war came. After the war only factory expansion and minor model modification was required to begin the job of replacing steam power. It is true that a few roads, particularly those with large coal traffic movements, continued to reject the diesel in favour of modern or experimental steam power. It is also true that at least one major eastern road, New York Central, felt the need for a detailed comparative test between passenger and freight diesels and late-model 4–8–2's and 4–8–4's (*see* Chapter 6, IV, Steam v. diesel tests on the N.Y.C.). But for the majority of Class One railroads the question of steam versus diesel was settled during 1941-45. The only question remaining was how soon one could obtain delivery on diesel power and arrange the necessary financing.

PLATE 25A. Early box-cab 1,800 H.P. units built for test and demonstration purposes in 1935. These non-articulated units were scrapped after tests were completed.

C. B. and Q.R.R.

PLATE 25B. C.B. and Q.R.R. "Pioneer Zephyr" the first of the world-famous "Burlington Zephyr" articulated streamline trains, using an Electro-Motive lightweight 2-cycle diesel engine, and a body of Budd stainless steel construction.

Reading Company

PLATE 26A. Reading R.R. 1,600 H.P. road and yard switching locomotive. Fitted with M.U. controls so that two or more units may be operated together.

Santa Fé Railway

PLATE 26B. A.T. and S.F.R.R. (Santa Fé) 1,500 H.P. road and yard-switching locomotives.

Fairbanks, Morse and Co.

PLATE 27A. Fairbanks, Morse: Two 2,400 H.P. road-switching hood units. These locomotives were used for demonstration purposes.

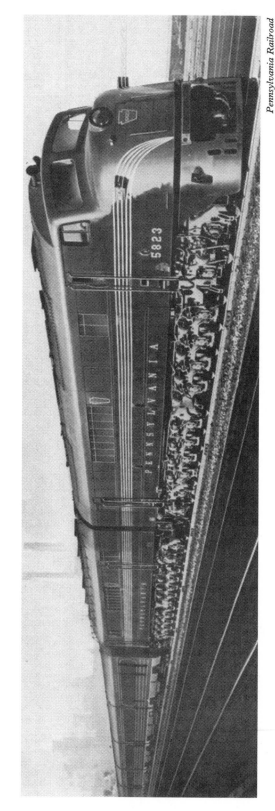

Pennsylvania Railroad

PLATE 27B. Pennsylvania R.R. two 3,000 H.P. 2–D+D–2 passenger locomotives working in tandem. This two-unit combination was unsuccessful and showed up badly against more flexible 2,000 H.P. diesel units coupled in multiple, if necessary, to achieve equal total locomotive horsepower.

Union Pacific Railroad

PLATE 28A. Union Pacific R.R. A1A–A1A booster unit of 2,000 H.P. Controls are provided only for hostler (light engine) operation in engine terminal or similar areas. Normally a booster unit is operated in multiple-unit control from one or more cab units.

Electro-Motive Division, General Motors

PLATE 28B. Three Electro-Motive 1,500 H.P. units in A–B–A formation with multiple-unit control. A booster unit is in the centre with a cab unit at either end.

Association of American Railroads

PLATE 29. Ivy City Engine Terminal, Washington, D.C., showing servicing facilities in the background, and round house, with centrally-placed turntable in the foreground. (*See* Chapter 8, page 445.) A Baltimore and Ohio two-unit (cab and booster) diesel-electric locomotive is on the turntable. Ivy City is operated by the Washington Terminal Company and deals with locomotives of several railroads including electric locomotives of the Pennsylvania R.R.

Santa Fé Railway

PLATE 30A. A.T. and S.F.R.R. (Santa Fé) diesel shops at Barstow, California, which handle major overhauls as well as the routine servicing of diesel locomotives.

Western Maryland Railroad

PLATE 30B. Western Maryland Railway, two-unit diesel locomotive passing through a mechanical washing plant at one of the company's engine terminals.

Chicago, Rock Island and Pacific Railroad

PLATE 31B. Chicago, Rock Island and Pacific diesel shop at Silvis, Illinois. Work is carried out on two levels. The excellent illumination at the lower level will be noted.

PLATE 31A. A.T. and S.F.R.R. (Santa Fé) diesel shop at Argentine, Kansas, showing "three-level" arrangement of platforms and pit.

Santa Fé Railway

(A)

Alco Products, Inc.

PLATE 32A. A 1,500 H.P. Alco road-switcher being wheeled at the builders' works.

Budd Company

PLATE 32B. New York Central System: Budd Rail Diesel Car (R.D.C.). This is an R.D.C.–3 (mail express passenger combination car) powered by two 275 H.P. 6-cylinder diesels driving through torque converters.

Part III. The diesel locomotive

THE BASIC DIESEL UNIT

Since its introduction to America, the diesel-electric locomotive has ranged in size from 60-ton, 300 H.P., B–B yard units to 297½-ton, 3,000 H.P. 2–D+D–2 road units. However, the typical diesel falls into a medium-horsepower bracket for reasons of operating flexibility. It is a B–B type ordinarily powered by one 16-cylinder, V-type engine producing, according to builder and date of manufacture, 1,350, 1,500, 1,600, 1,750 or 1,800 H.P. The outer shell, or carbody, may be of the streamlined type (i.e. cab) or of the hood type (i.e. road-switcher) but mechanical components remain virtually identical.

The frame consists of a single steel plate with sills and other support members welded to it.

On the frame the diesel is mounted, its shaft directly coupled to a generator-alternator producing direct current for traction, and alternating current for such auxiliaries as cooling fans, motor blowers and lights. Also coupled to the engine shaft is a three-cylinder compressor to supply the air-brake system.

The main generator converts the mechanical energy and distributes it to traction motors mounted between the wheels and geared to each axle.

Over-all, a typical 1,500 H.P. cab-type freight unit is 10 ft 8 in. wide; 15 ft high; and approximately 50 ft long. It weighs 115 tons in working order, all of that weight being on the drivers and available for adhesion.

Supplies include: 16 cu. ft of sand, 230 gallons of cooling water, 200 gallons of lubricating oil, and 1,200 gallons of fuel oil.

It is available in eight gear ratios with maximum speeds ranging from 55 to 102 m.p.h. and continuous tractive effort ratings (maximum output which the unit can develop continuously without overheating the traction motors) ranging from 52,400 lb. to 25,500 lb.

It can negotiate a minimum radius curve of 23° (250·8 ft radius). Wheels are 40 in. in diameter.

Multiple-unit connections and controls as well as common air-brake schedules will permit this unit to be operated with others of its type in locomotives of 7,500 H.P. or more; it can be adapted to operate with units of other manufacture as well.

CARBODY DESIGN

The external outline of the American diesel has been rigidly standardized to a handful of basic designs, the only difference among them being the flat and curved surfaces peculiar to one manufacturer or another. They can be listed by the popular terminology commonly accorded them in the railroad and manufacturing fields:

Cab units. The common definition is a streamlined unit, the outer shell of which covers the control compartment and all frame-mounted mechanical components such as engine, generator, steam generator, etc., and extending the full width and length of the unit. The control compartment, or cab, is mounted high and behind a short, streamlined nose which affords collision protection. Such units are sold as "A" and "B", or cab and booster, units. The B, or booster, unit produces equivalent power but must be operated in multiple since it has no control compartment and has blind ends. Simple controls allow it to be worked around engine terminal and shop areas by a hostler (Plates 28A and 28B, page 120).

Hood units. The essential trait of a hood unit is that the hood or metal covering over engine, generator and other mechanical components makes no attempt at streamlining and conforms closely to the height and width of the equipment it protects. In most instances the hood is not as high as the cab and there is space on either side and at the ends for walkways or running boards. The placement of the cab itself varies. Centre cabs prevail on both the small 44-ton, 380 H.P. short-line diesels as well as on certain heavy-duty C–C transfer diesels in the 2,000–2,500 H.P. range. In both examples, tandem engine-generator sets are used, one mounted on each side of the middle cab. The overwhelming majority of yard units are single-engine, however, and have a cab placed at the end of the unit. In yard service the hood unit's exceptional visibility is a tremendous advantage and, when steam was competitive, was quite a feather in the newcomer's cap.

Road-switchers, occasionally called general or all-purpose locomotives, range in horsepower from 1,000 to 2,400 and usually possess offset cab location. In the unit's "short end" the owner can mount a steam generator for optional passenger duty, or a toilet; the "long end" of the hood covers the engine, generator and other mechanical components. The builder will arrange controls so that the unit may usually be operated "short-end first" or vice versa, and dual controls mounted on both sides of the cab are to be found on a few road-switchers (Plates 26A and 26B, page 118).

Upon occasion, *B or booster-type hood units* have been constructed for multiple-unit operation with regulation units. Railroaders refer to yard unit combinations of this type as a "cow-and-calf" or a "herd" in the event of one A and two B units. Booster, or B-type, road-

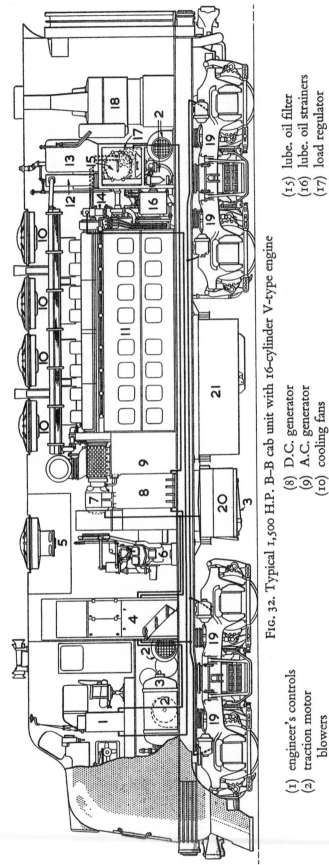

FIG. 32. Typical 1,500 H.P. B–B cab unit with 16-cylinder V-type engine

(1) engineer's controls
(2) traction motor
      blowers
(3) air reservoirs
(4) electrical cabinet
(5) dynamic brake hatch
(6) air compressor
(7) aux. generator

(8) D.C. generator
(9) A.C. generator
(10) cooling fans
(11) diesel engine
(12) lube. oil cooler
(13) cooling water tank
(14) elec. hydr. governor

(15) lube. oil filter
(16) lube. oil strainers
(17) load regulator
(18) steam generator
(19) traction motors
(20) battery boxes
(21) fuel tank

switcher units have been constructed upon occasion, notably for Union Pacific.

Box-cab units. Now dormant in North American construction, the box-cab unit which contained an operating or control compartment at each end and resembled an electric locomotive, typified virtually all pre-1930 diesel development. Both yard and road units followed this pattern. Box-cab design vanished because of the influence of streamlining, the excessive cost of duplicating control positions, and the lack of collision protection.

OPTIONAL EQUIPMENT

All manufacturers offer the customer an extensive list of standard modifications at extra cost to adapt the diesel unit to specialized service conditions. For crew comfort the builder (or the road) can add a third cab seat for the brakeman (on freight trains), rear-view side mirrors, a toilet, water cooler and so forth.

For optional passenger duty the standard unit may be equipped with oil-fired, automatic operation *steam generators* which operate at more than 300 p.s.i. and evaporate from 1,600 to 4,500 pounds of water per hour; in this event extra water tanks are added with capacities from 800 to 1,750 gallons. This feature is necessary in North America since virtually all passenger equipment is steam heated from the locomotive and there are steam-jet air-conditioning systems in use on certain cars as well.

Dynamic braking (see Chapter 1: Braking Systems, page 103) which in effect reverses the traction motors and utilizes their energy as a braking force, may also be added to any stock road-unit at extra cost. It is necessary to install, in addition to the electrical controls involved, a dynamic brake hatch with fan and register grids to dissipate the energy. (In straight-electric operation, regenerative braking returns power to the trolley line.) Dynamic braking effectiveness increases with the speed of the locomotive and is limited by the heating of the motors and the ability to dissipate that heat through the hatch in the roof.

A number of minor modifications are usually available, such as fire extinguishers, jumper cables through the nose of streamlined cab units (for M.U. operation), concrete ballast for adhesive purposes, additional water or fuel capacity, winterization for operation in unusually low temperature zones such as Canada, and different warning horn, bell, and headlight assemblies.

RUNNING A DIESEL

In a typical diesel unit the operating controls are remarkably similar in function, even in location, to their counterparts in steam locomotive cabs. Indeed, manufacturers long ago made a definite point of simplifying locomotive controls so that engine crews could make the transition from steam to diesel power with a minimum

of difficulty and without the necessity of extensive technical knowledge.

The throttle on a typical unit controls an electrohydraulic or electropneumatic governor on the engine, and accelerates it from an idling speed of, for example, 275 r.p.m. to a maximum speed of 800 r.p.m. Each notch on the throttle increases engine speed by approximately 75 r.p.m. in this instance. Incidentally, a *cutout button* permits the engine to be opened up independently of the generator. This is necessary to increase the speed of the air compressor when the locomotive is standing prior to departure and pumping up the train line of the automatic air-brake system. Otherwise, as the speed of the engine increases, its mechanical energy turns the main generator and converts that power into electrical energy for distribution to the traction motors.

A definite, if automatic, sequence known as *transition* (and manually controlled by the engineer on early-model road diesel locomotives) takes place to protect electrical components from damage and adapt electric current production to road conditions. High torque is required to start the train, so the traction motors are connected in *series parallel* (each motor receives full current but only half the voltage of the generator). Once under way, acceleration replaces torque as the key consideration, so the motors are connected in *parallel* (full voltage across the motor terminals). Subsequently, two more changes – *shunt* and *parallel shunt* – take place as resistors and switches indicate the lessened demand of the motors. Most units have both short-time and continuous ratings on traction motors. If these are not observed (meaning a condition whereby the motors cannot use all the current being supplied them at continuous low speed), the heat created gets beyond the control of the motor insulation and motor blowers and softens the copper and solder. Either the motor fails, or its service life is materially decreased. The ideal motor, of course, is one which will slip before it burns out, and practical progress has been made toward this goal.

In addition to the throttle, the engineer of a typical diesel unit has within his reach the following controls:

A reverse lever
Dynamic brake lever
Automatic and independent air-brake handles (for control of the train and engine respectively)
Sander (which is sometimes a pushdown feature incorporated in the automatic brake handle)
Deadman's pedal (which must be depressed while the locomotive is running or the brakes will apply)
Bell valve
Light switches
Horn cords

Ammeter or *Load indicator*
Wheel-slip warning light
Air-brake gauges
Cab heater
Windshield wiper switches

Compared with the backhead of a steam locomotive, a diesel cab is simplicity itself and reflective of the automotive production of the diesel's biggest builder. It is more of a dashboard than a cab.

WHEEL ARRANGEMENTS

The wheel classification or notation used in respect of diesel locomotives in North America is essentially that which is set out in detail in the appendix (*see page 505*).

It is relevant to repeat here certain essential features of this notation:

(1) An idler or non-powered axle is indicated by a cumulative numeral which both defines it and indicates the number of such idlers.

(2) Powered axles and/or trucks are indicated by letters. "A" means single powered axle; "B" means a four-wheel truck, all axles powered; "C" means a six-wheel truck, all axles powered; and so forth.

(3) A dash or a hyphen (–) means a separation between swivel trucks.

(4) A plus mark (+) means articulation joints between trucks.

(5) A number placed outside a wheel arrangement grouped in parentheses indicates the number of such units coupled together in multiple. For example, a typical 6,000 H.P. freight diesel would be recorded as a 4(B–B) locomotive.

The basic units which effected North American dieselization may be itemized as follows:

B–B. Approximately 80 per cent of all the diesel units on the continent fall into this category of the four-motor locomotive with a pair of four-wheel swivel trucks. Yard units of B–B type range from 44-ton, 300 H.P. models for short lines up to 125-ton, 1,200 H.P. locomotives for heavy-duty, Class One road switching. The majority of freight units, whether of the cab or hood design, are likewise B–B's, ranging in horsepower from 1,350 to 2,000, and a considerable number of cab-type B–B units have been produced for mixed-traffic or exclusively passenger duty since the modern four-wheel truck permits safe operating speeds of up to 102 m.p.h.

C–C. In post-war years a modest number of diesels, mainly hood-type units, have been constructed with six-wheel trucks, all axles powered. Indeed, most manufacturers of hood units in the range between 1,500 and 1,800 H.P. offer six-wheel trucks as an optional extra. Engine size remains unchanged but tractive effort at low speed is increased. There is argument as to whether this extra power is worth the added maintenance of two more traction motors per unit and there are stout proponents on both sides of the question. The C–C wheel arrangement is virtually mandatory on 2,250, 2,400 and 2,500 H.P. units to spread the weight over more axles and take full advantage of the high engine horsepower. C–C units are especially favoured for heavy transfer work between freight yards, as pushers and helpers in mountainous territory, on heavily graded lines, and in other low-speed work.

A1A–A1A. Virtually all of North America's exclusively passenger-type cab units, ranging from 1,800 to 2,400 H.P., have been built upon this chassis (Plate 28A, page 120). A centre idler axle on each truck increases its wheelbase from 9 to 14 ft (comparing E.M.D. E and F models) and thereby insures safe riding qualities at speeds in excess of 100 m.p.h. Occasionally, such six-wheel trucks are mounted beneath road-switcher hood units for operation over branch lines which cannot take the axle loadings of either B–B or C–C units.

There have really been only two notable deviations from these orthodox wheel arrangements, at least since World War II.

B–A1A. In 1950 Fairbanks, Morse introduced a series of 1,600, 2,000 and 2,400 H.P. "Consolidation Line" cab units for both passenger and freight service which incorporated an A1A truck at the rear (optional on the 1,600 H.P. model) for weight distribution purposes. More recently, this wheel arrangement appeared for the same cause on Electro-Motive's FL0 "diesel-electric-electric" units on the New Haven.

2–D+D–2. In an effort to concentrate as much horsepower as possible into a single unit, Baldwin constructed a number of diesels of this arrangement for the Pennsylvania and for the National Railways of Mexico; the Seaboard subsequently purchased the same model. The huge machine was built on a cast-steel frame and mounted two inline 8-cylinder 1,500 H.P. engines. It was not considered a success (Plate 27B, page 119).

Part IV. Diesel locomotive builders

America's original diesel locomotives were custom design experimentals resulting from the collaboration of at least three parties:

(i) A manufacturer of diesel engines.

(ii) A manufacturer of electrical components such as generators and traction motors.

(iii) A locomotive builder.

The natural result was a division of responsibility and authority, and an extension of design-thinking accumulated in such foreign fields as stationary diesel power plants, straight-electric or steam locomotives. There was little standardization as it exists today and no mass production.

The dominating force in North American diesel construction was, and is, General Motors, whose Electro-Motive Division built the original factory exclusively devoted to the construction of diesel motive power. This firm enforced standardization, established mass production, and instituted both a parts exchange programme and, later, complete unit remanufacture and upgrading. Approximately seven out of each ten diesel units in North America are of E.M.D. manufacture.

The following tabulation covers major diesel locomotive builders and their Canadian affiliates:

ALCO PRODUCTS INC., SCHENECTADY, N.Y.

This manufacturer was formerly the American Locomotive Co. and was the largest producer of steam locomotives in the U.S. A 1901 amalgamation of eight builders, Alco traces its history back to a 15-ton steam engine delivered in 1849. Less than a century later it constructed its 75,000th locomotive – a diesel – in 1946. Alco constructed the heaviest steam locomotives in existence for Union Pacific in 1941 and 1944. It collaborated with General Electric on straight-electric units as early as 1904. In addition to diesels, Alco is responsible today for the development of a coal-burning gas turbine-electric with funds supplied by an association of coal companies and major coal-hauling railroads.

Diesel locomotive history. Alco's first diesel venture was a 60-ton, 300 H.P. yard unit with an Ingersoll-Rand engine and a G.E. electrical transmission. Completed in 1923 and tested extensively in 1924, this diesel was followed by a number of similar models for revenue service. In 1929 Alco acquired the McIntosh & Seymour Corpn., makers of marine diesels, and by 1931 was able to produce a 600 H.P. diesel switcher for New Haven using a newly developed M. & S. inline 6-cylinder, four-cycle engine. By 1935, diesel switchers were 20 per cent of Alco's total locomotive production, and work was authorized on a high-horsepower, V-type diesel engine for road-unit installation. In 1940 Alco and G.E. (which sold diesels under the joint label of Alco-G.E.) entered the road field with a 2,000 H.P. passenger unit powered by a pair of M. & S. 6-cylinder engines. Some sixty-seven units of this type had been sold when war came and Alco was limited by government authority to 1,000 H.P. yard units in the diesel field.

Design work on V-type engines, begun in earnest in 1944, permitted a full-fledged Alco-G.E. entry into road diesel locomotive production following the war. Alco reconverted a part of its Schenectady plant to a mass-production line for engines and locomotives at a cost exceeding 20 million dollars, and steam engine production was formally halted by June 1948. Two basic engines powered the new line: a 1,500 H.P. V-12 for road-switchers and cab-type freight units, (Plate 32A, page 124) and a 2,000 H.P. V-16 for passenger units. Both power plants were 4-cycle, turbo-supercharged engines, model 244. Gulf, Mobile & Ohio purchased the first 1,500 H.P. unit in 1945; and in 1946 Santa Fé bought 2,000 H.P. passenger units (one of which was Alco's 75,000th locomotive). Subsequently, these engines were boosted to 1,600 and 2,250 H.P. ratings, then discontinued in 1956 in favour of model modification 251, rated at 1,800 H.P., and 2,400 H.P. for the V-12 and V-16 engines, respectively. The Montreal Locomotive Works has produced equivalent models since the war.

Sales record. Alco has consistently been second only to E.M.D. diesel locomotive builders. In 1948, when the demand for diesels ran 40 per cent beyond production facilities in the U.S., Alco sold 40 per cent of domestic diesels, but in typical years its cut of the market has ranged from 10 to 15 per cent. Its major contributions to diesel design have been:

(i) An early (1940) recognition of the usefulness of road-switcher units and steadfast development of such all-purpose units.

(ii) Its development of engines powerful enough to underwrite 2,000–2,400 H.P. units without resorting to multiple engines.

Alco, as indicated by its post-war change of name, is no longer dependent upon the railroad field as it was prior to the war. Locomotives constitute less than half the company's gross (compared with 72 per cent as late as 1949) because of diversification of sales into such fields as oil-drilling rigs, atomic energy development, pipe fabrication, and forgings.

Current production. Inasmuch as domestic demand for cab-type units has virtually vanished, Alco's production is almost exclusively hood locomotives. It actually builds:

The DL-430, a 900 H.P., B–B switcher incorporating a redesigned 251-model, 6-cylinder inline engine.

The DL-701, an 1,800 H.P., B–B road-switcher powered by a V-12.

The DL-702, a C–C or six-motor version of the DL-701.

The DL-600, a 2,400 H.P., C–C road switcher with a V-16 engine.

For the export market, Alco produces the 1,800 H.P.

cab-unit, multi-gauge *World Locomotive*. All models use G.E. generators, traction motors and other electrical components.

BALDWIN-LIMA-HAMILTON CORPN, PHILADELPHIA, PA.

(Plant location: Eddystone, Pa.) – the result of a merger between two of America's "big three" of steam locomotive production – is now completely out of the diesel-electric locomotive business. Its major component, the former Baldwin Locomotive Works, began building steam locomotives in 1832 and was still turning them out for export as late as 1956. It has produced reciprocating, geared steam-turbine, steam-turbine-electric, and, in co-operation with Westinghouse Electric, electric and gas-turbine-electric locomotives. Its failure to establish itself in the diesel locomotive market may be attributed to a late entry and lack of standardized designs and a really modern, V-type engine. Lima-Hamilton Corpn, with which Baldwin merged in 1951, was itself a merger of 1947. Lima Locomotive Works built its first steam engine, a gear-type Shay, in 1879, became America's number three locomotive builder, and continued until post-war days to be the most enthusiastic exponent of steam power. A merger with a diesel engine manufacturer in 1947 turned Lima to manufacture of 1,000 and 2,500 H.P. yard and transfer diesel hood units, but only 172 such locomotives had been built by 1951 when merger with Baldwin cancelled further locomotive production at the Lima, Ohio, plant.

Diesel locomotive history. Baldwin's first diesel effort was early and unusually bold. In 1925 it built an A1A+ A1A locomotive powered by a Knudsen-type 12-cylinder, 2-cycle, 1,000 H.P. engine of the inverted V-type with two crankshafts. Built along electric locomotive principles of that era, No. 58501 weighed 275,500 lb., of which 180,000 lb. rested on driving wheels. The unit was tested in road freight service on the Reading and other roads but Baldwin considered it an experimental and not for sale. In 1929 Baldwin built a 1,000 H.P. B+B diesel of box-cab design powered by a Krupp 4-cycle, 6-cylinder, inline engine. The articulated trucks took the drawbar pull rather than the frame or carbody. The unit, No. 61000 (works number) worked on six roads and in open-pit mining service, hauling trains ranging from 2,000 to 4,500 tons. In 1936 Baldwin drew nearer commercial production with construction of a conventional-design yard unit, No. 62000 which was sold to Santa Fé. It contained a 4-cycle, 6-cylinder, inline, 660 H.P. engine of the De La Vergne Engine Co., a firm with forty years diesel experience which Baldwin had acquired. In 1937 Baldwin delivered three 900 H.P. yard units to New

Orleans Public Belt. These units, which had 8-cylinder De La Vergne engines, were operated in 1,800 H.P. multiple across the Huey Long Bridge over the Mississippi River, over grades of 1·25 per cent (1 in 80) extending 2·4 miles. In 1939 Baldwin built a 660 H.P. switcher for the Reading and thereafter embarked upon a standard line of 660 and 1,000 H.P. yard units. In 1940 Baldwin decided to enter the road diesel field "in the not too distant future" but that venture was postponed until 1945 by the war. The works' initial road unit entry was a 2,000 H.P. A1A–A1A passenger unit of conventional streamlined exterior, built and demonstrated on various roads in 1945. Thereafter, Baldwin – a minority builder – embarked upon a construction programme that ignored standardization to solicit customers who were not satisfied with the basic patterns of Alco or E.M.D. In all cases engines used were 6- and 8-cylinder, 4-cycle, inline De La Vergne design, either normally aspirated or supercharged and attaining maximum power ratings of 1,200 H.P. for the 6-, and 1,600 H.P. for the 8-cylinder unit. Development of a light-weight, high-speed, V-type engine was begun but it never reached the production stage. In addition to matching the standard yard, road-switcher and road units of other builders, Baldwin produced double-ended passenger cab units for commuter service on Jersey Central, two-engine, centre-cab, 2,400 H.P. transfer units, and two-engine, 3,000 H.P., 2–D+D–2, cab-type freight or passenger locomotives (the most powerful single-unit diesels ever built in North America).

In recent years Baldwin limited its production to hood units, left the diesel-electric business entirely in 1956, partially as a result of the same decision by its traditional supplier of electrical components, Westinghouse Electric. The company is well diversified, with production in non-railroad fields dating back to 1929, but its only current locomotive concern is diesel-hydraulic units.

Sales record. Never better than number three in diesel sales (behind E.M.D. and Alco), Baldwin slipped into last place before the stiffened competition of Fairbanks, Morse. Its yard and road-switcher units are to be found on the rosters of most Class One U.S. roads, but only one company – Pennsylvania, an old steam customer – ever purchased its cab-type road units to any appreciable extent. Export sales were similarly restricted. After good sales during the peak market years immediately after the war, Baldwin's share of the market slipped until 1956 – when the firm did not receive a single locomotive order.

Current production. In 1957 Baldwin delivered three light-weight, 1,000 H.P., B–2 type passenger units, each powered by a Maybach 4-cycle, V–12 diesel and *Mekydro Drive* (see Transmissions – Hydraulic, page 52). All units power the low-slung, articulated trains of the *Train X* type. One unit pulls New York Central's *Xplorer* and the others are coupled to either end of New Haven's *Daniel Webster*. The N.H. units are equipped with auxiliary electric motors which drive through the mechanical transmission in third-rail D.C. electrified territory in New York City. Baldwin has been so impressed with their performance that it is engineering an all-purpose, 1,800 H.P. diesel-hydraulic unit for freight service (*see* Chapter 9, Part VI, page 476).

ELECTRO-MOTIVE DIVISION OF GENERAL MOTORS CORPN

(Home plant: La Grange, Illinois; Canadian affiliate: General Motors Diesel Ltd., London, Ontario) is the world's largest diesel locomotive builder, yet its corporate life extends back only to 1922. An idea of its growth may be gauged by the fact that E.M.D. had produced 3,238 diesel locomotive units by the summer of 1947, and more than 18,000 by 1956. The original Electro-Motive Co. of Cleveland, Ohio, was not a builder at all. It designed and solicited orders for gas-electric passenger railcars, then subcontracted the work to manufacturers of engine and electrical equipment. The carbody construction and final assembly was delegated, in turn, to a passenger carbuilder. Some 500 railcars were sold by E.M.C. between 1924 and the advent of the streamliner.

In 1930 General Motors purchased E.M.C.'s power-plant supplier, Winton Engine, then E.M.C. itself. However, Electro-Motive did not get a "home" of its own until completion of its La Grange, Illinois, plant in 1936, and its dependence upon other manufacturers did not end until 1938 when E.M.D. began production of all major components – diesels, generators, and traction motors – at La Grange.

E.M.D.'s modern history may be said to have begun at La Grange in 1938 when the mass-produced 567-series engine was first installed in standardized E-6 type, 2,000 H.P., A1A–A1A passenger units of wholly E.M.D. manufacture. Production of diesel freighters rounded out a catalogue of stock power begun in 1936 with standardized yard units and this established Electro-Motive as a major builder. It built 218 units in 1940 – almost one a working day – and managed 500 road freight units in 1944 to ease the critical shortage of motive power in the U.S. At the peak of the post-war locomotive market boom, E.M.D. turned out as many as ten units per working day at La Grange and in plant number two located in Cleveland, Ohio. The second plant has since been turned over to another G.M. division, but expansion at La Grange permits construction of approximately six units per day when sales warrant.

Diesel locomotive history. The first Electro-Motive diesel application was the 600 H.P. power plant installed in the Burlington *Pioneer Zephyr* of 1934. E.M.D. subsequently powered several other articulated streamliners and contracted with General Electric's Erie, Pa. Works for a pair of non-articulated, 1,800 H.P., B–B demonstrator road units, delivered in 1935. E.M.D. sold a similar unit, built at G.E., to Baltimore & Ohio in 1935 and two, built at the St Louis Car Co., to Santa Fé.

The first diesel unit built at the new E.M.D. plant in La Grange, Illinois, a 100-ton, 600 H.P., B–B switcher, rolled off the erecting floor in May 1936. This and other yard units of 600 and 1,000 H.P., together with the original streamlined passenger 1,800 H.P. units, were powered by 2-cycle engines of Winton manufacture. In 1938 the 567-series engine went into production at La Grange and, with modification, it remains the heart of any E.M.D. diesel. Originally produced in three sizes, the 567-series engine is now produced as follows:

    800 H.P. V-6;
    900 H.P. V-8;
    1,200 H.P. V-12;
    1,750 H.P. V-16;
    2,400 H.P. V-16.

The basic Electro-Motive line of locomotives has now been established with these prefixes:

E	for passenger locomotives;
F	for freight locomotives;
SW	for switchers;
FP	for dual-purpose freight or passenger road cab units;
GP	for general purpose road-switcher B–B units;
SD	for special duty C–C hood units;
TR	for transfer "cow and calf" multiple-unit and permanently-coupled combinations of stock yard units in control-and-booster unit groupings.

With rare exceptions, Electro-Motive has ignored specialized markets in locomotive applications in favour of mass-producing standardized designs of widespread application. To date E.M.D. has made no effort to invade the short-line market for units of less than 600 H.P. and only recently has it produced a very high-horsepower C–C hood unit such as F.-M.'s *Train Master* and Alco's CL-600. However, its rigid standardization has been relaxed to the extent of permitting, for example, a 1,200 H.P. switcher with dynamic braking (a feature ordinarily available on units of 1,750 or more horsepower) and road-switcher booster units minus control cabs.

Today domestic locomotive production is supplemented at E.M.D. by a research and development centre, which has produced an experimental 800 H.P. switcher with torque-converter drive instead of an electrical transmission. It also undertakes remanufacture and upgrading of older E.M.D.'s – those units with a million miles or more service – and re-engining of other manufacturers' units at railroad request. E.M.D. has developed and constructed such railroad equipment as piggyback cars for truck trailers, mechanical refrigerator cars, and the lightweight *Aerotrains*. It has also entered other fields of manufacture such as oil drilling rigs, mobile power-generating plants, and other apparatus incorporating locomotive components.

Sales record. There has never been a serious question of Electro-Motive's diesel locomotive production leadership in the U.S. since the La Grange, Illinois, plant opened in 1936. E.M.D. has always been supported by the financial and technological resources of its parent, and its loyalties have never been divided between diesel and steam or electric locomotives. Most roads own more E.M.D. power than all other makes combined and some, such as Illinois Central, own no other type of diesel. Out of more than one hundred Class One roads, less than twenty own no E.M.D. power. In 1950 alone, a boom year in dieselization in the U.S., E.M.D. built 2,200 diesel units and wrote 65 per cent of all diesel locomotive orders signed by Class One roads. In lean years the plant has enlarged its share of the orders.

Current production. In common with the other builders, Electro-Motive now does its largest business in hood units, notably two 1,750 H.P. models: the GP-9 B–B and the SD-9 C–C. However, it still offers and occasionally constructs a full catalogue of cab-type road units:

The 2,400 H.P. E-9 A1A–A1A passenger unit.

The 1,750 H.P. F-9 and FP-9 B–B's for freight or passenger (an FP-9 has an extended frame to accommodate extra steam-generator water tanks).

A 1,200 H.P. B-2 lightweight, low-slung passenger unit for use with *Aerotrain* and other articulated lightweight streamliners.

E.M.D. also sells a 115-ton, 900 H.P. B–B type and a 125-ton, 1,200 H.P. B–B type, both for yard duty.

For export, the G.M. division manufactures two multi-gauge hood-type units as well as special adaptations of domestic prototypes. In 1957 E.M.D. began production of the FL-9, a 1,750 H.P. B–A1A cab unit for operation as a diesel or as a straight electric in third-rail D.C. zones. In 1958 E.M.D. introduced the SD–24, a C–C fitted with a supercharged 2,400 H.P. V–16 engine.

FAIRBANKS, MORSE & CO.

(Home plant: Beloit, Wisconsin; Canadian affiliate: Canadian Locomotive Co., Kingston, Ontario) has long been a railroad supplier (pumps, scales, etc.) but its motive power experience dates from 1939; 750 H.P., 5-cylinder F.-M. opposed-piston diesels were installed in five motor trains built for the Southern Railway by the St. Louis Car Co.

The F.-M. "OP" 2-cycle diesel, developed in the early 1930's, might have entered locomotive service earlier, but the U.S. Navy took the entire production for submarine use between 1941 and 1944. The first locomotive so powered was Milwaukee Road No. 1802, a 1,000 H.P. B–B yard unit containing a 6-cylinder O.P. engine. Since then F.-M. has secured for itself a secure place in the diesel locomotive field by production of a variety of yard and road units.

Diesel locomotive history. Following the acceptance of its original 1,000 H.P. yard unit in 1944, Fairbanks, Morse quickly added 1,500 and 2,000 H.P. hood-type road-switchers and transfer units to its catalogue. Production soon absorbed the limited manufacturing area available at the Beloit Works, so construction of F.-M.'s first cab-type road units – 2,000 H.P. A1A–A1A units powered by a single 10-cylinder O.P. engine – was contracted to General Electric's Erie, Pennsylvania, Works, as had been the case with Electro-Motive's initial road unit production. G.E. eventually delivered 111 cab and booster units of this type. Known as the "Erie-builts", they marked only the second occasion in U.S. diesel history that A1A–A1A units with idler axles have been offered for freight as well as passenger service.

In 1950 F.-M. moved all production to Beloit and introduced the Consolidation Line cab-type road units of 1,600, 2,000 and 2,400 H.P. sizes. C-Liners, powered by 8-, 10- and 12-cylinder O.P. engines, had a 56-ft 6-in. length regardless of horsepower rating, but a B–A1A (instead of B–B) wheel arrangement on the 2,000 and 2,400 H.P. models for weight distribution. The most powerful unit set a record of $42\frac{1}{2}$ H.P. per foot of length. Offered as freight or passenger units, C-Liners were well received but unfortunately arrived on the verge of the big shift in buying from cab to hood units. Since then, F.-M. has concentrated on hood units, culminating in the 2,400 H.P. C–C *Train Master* unit of 1953 (Plate 27A, page 119).

Sales record. Youngest of all diesel locomotive builders, Fairbanks, Morse had nevertheless surpassed Baldwin-Lima-Hamilton in sales before the latter left the diesel-electric market. Indeed, upon occasion F.-M. has enjoyed a sales volume comparable with that of Alco. The firm has pioneered high-horsepower units and has been responsible for a number of chassis and carbody innovations, exclusive of propulsion machinery.

Current production. F.-M.'s line includes a 1,200 H.P. B–B switcher; 1,600 H.P. B–B and C–C road-switchers; and a 2,400 H.P. C–C road-switcher, the *Train Master*.

PLATE III. Canadian Pacific Railways: transcontinental express "The Canadian", in charge of a three unit diesel-electric locomotive, of 5,100 H.P. The two cab units are separated by a booster unit in the middle of the trio.

All are powered by O.P. engines and General Electric is the major supplier of generators and traction motors. In 1957 F.-M. offered a 1,200 H.P. lightweight passenger unit for service on the new articulated *X*- and *Talgo*-type trains; it can be built to operate as a straight-electric in third-rail D.C. zones. Unusually, its one 1,720 H.P. O.P. engine provides power for such train needs as heat and air-conditioning, with 1,200 H.P. remaining for traction purposes. (In other equivalent units separate diesel-alternator sets do this job.) This requires that the engine be operated at a constant speed because of the alternator. Such units – termed *Speed Merchants* – have been built for Boston & Maine and New Haven.

GENERAL ELECTRIC CORPORATION

(Locomotive plant: Erie, Pennsylvania) has been a household word in the locomotive business since the nineteenth century, both as a builder and as a supplier of electrical components. Its Erie Works have constructed virtually every type of power, except steam, ranging from 1½-ton, 5 H.P., battery-powered industrial mine locomotives to 360-ton, 5,000 H.P. electrics. Today G.E. is the only builder of straight electric and gas turbine-electric locomotives in the U.S.; it also supplies generators, traction motors, and other electrical gear to both Alco and Fairbanks, Morse.

Diesel locomotive history. General Electric's diesel locomotive experience (excluding road units it assembled under contract for both E.M.D. and F.-M.) dates back into the early 1930's and concerns mostly centre-cab, twin-engine B–B switchers for industrial use or export. Mass production of standardized models began in 1940 when G.E. offered a centre-cab B–B unit of 44 tons powered by two 190 H.P. diesels. These units were sold to twenty railroads within a year of their introduction, proving especially popular with short lines whose rail or bridges could not support the other builders' yard units. By 1947, more than 160 such units were operated by more than fifty roads. In 1945 G.E. introduced a larger, single-engine unit of 70-ton, 600 H.P. size, also B–B in wheel arrangement.

Sales record. G.E.'s own locomotive production for domestic operation (as opposed to its role of supplier of components) has never been large when compared with the other builders, but it has filled an important market niche. G.E. has been the builder for the short lines and industrial concerns, also a speciality builder of custom diesels for unusual locations which exclude production models.

Current production. Today G.E. is prepared to build diesels of stock pattern in hood-type carbodies of practically any size from 25 tons or less to 80 tons, from 150 H.P. to approximately 1,000 H.P., twin or single engines, 4-wheel rigid-frame or B–B in wheel arrangement. Most units sold for commercial U.S. railroad service are either 45 or 70 tons. In addition, G.E. is making a considerable bid for the export market with B–B and C–C models of road-switchers rated at 1,320 and 1,980 H.P. and powered by Cooper-Bessemer 4-cycle, V-type, 8- and 12-cylinder engines. Domestic units of smaller size are powered by diesels of Cooper-Bessemer, Caterpillar, or Cummins manufacture since G.E. does not build its own engines.

OTHER MANUFACTURERS

The Atlas Car & Manufacturing Co., the Davenport Locomotive Works, the Vulcan Iron Works, H.K. Porter Co. and Whitcomb Locomotive (a Baldwin subsidiary) have all constructed diesels, usually for industrial use, but all are out of the locomotive market today.

Part V. The operating of diesel locomotives

THE "BUILDING BLOCK" PRINCIPLE

Because North American dieselization did not begin with total abolishment of the steam locomotive in mind, many of the lessons acquired in finally achieving that objective were not followed at first. Diesels of multi-unit make-up were considered as locomotives instead of individual units, often numbered 100 A, B, C and D, and sometimes permanently coupled between cab and booster units. Reluctance to break up the locomotive into units meant that 6,000 H.P. diesels were held for assignments requiring that much power, and that an entire locomotive was out of service when only one unit required repair. The limitations of this steam locomotive concept of operation quickly came to light.

Today an efficient diesel operation is predicated upon the fact that all units of comparable horsepower (1,350 to 1,800 H.P.) and gear ratio have M.U. connections and common air-brake schedules, regardless of whether the unit is of cab or hood carbody type or even produced by the same manufacturer. This permits adoption of the "building block" principle, so named because the units can be coupled at short notice to meet specific demands of tonnage and schedules, and just as quickly cut out for repairs. It is not uncommon to see a diesel "locomotive" made up of two or three units of different ratings, models, makes and ages. Even streamlined cab units are equipped with jumper cables on their noses so that they may be run in multiple with other cab units on a nose-to-nose basis. Such flexibility reduces the total number of units needed to complete dieselization in a dramatic fashion (Plate 28B, page 120).

DIESELIZATION METHODS

North American dieselization began in the yard. In 1941, for example, the diesel performed more than 12 per cent of yard work but only 7·75 per cent of passenger service and just 0·22 per cent of freight operation in the U.S. Historically, the yard unit was the first standardized, mass-production diesel available to the industry, and there were innumerable locations which could take advantage of the diesel's 100 per cent availability for twenty-four-hour-a-day service. By World War II virtually all Class One roads had ceased buying steam switchers except the major coal haulers.

In road service the diesel made its first inroads in specialized assignments, e.g. articulated streamliners. Almost all pre-war passenger units were assigned to streamlined trains and the balance were purchased to meet peculiar operating conditions. Louisville & Nashville, for example, desired steam engines, preferably 4–8–4's, for operation between Cincinnati, Ohio, and New Orleans, Lamont, without change, but it settled for 4,000 H.P., two-unit diesels because the axle load of the 4–8–4 was too heavy for trestle viaducts at the south end of the line, and because such an engine was too long to go into the company's erecting shop.

There were similar examples in the first phases of dieselization. Santa Fé, the first purchaser of E.M.D.'s 5,400 H.P., four-unit FT's, concentrated them between Barstow, California, and Winslow, Arizona, a 459-mile district through desert country so dry that water for steam locomotives had to be imported by tank car. In World War II the great increase in train movements thus tied up tank cars urgently needed elsewhere for oil traffic. Moreover, the heavily graded territory required many helper locomotives which the diesels eliminated. So useful were the diesels that they were utilized 91·5 per cent of the time.

Until V.J. day, however, there was scant talk of total or 100 per cent dieselization. Many felt the diesel was too expensive a tool for such marginal assignments as commutation trains, local freights, small yards and other duties which do not require round-the-clock utilization. Others maintained that almost-new steam power should be fully depreciated. Exceptions to the rule were the roads which frankly planned total dieselization on a time-table, start-to-stop basis. The 541-mile Monon, to cite an instance, owned eighty-four steam locomotives and only four diesel yard units at the end of 1945, but had completely converted to diesel power by early 1949. Monon did so with just fifty-seven diesel units, or 79,600 H.P.,

costing 7·5 million dollars. By so doing Monon avoided the expense of duplicating servicing and maintenance facilities – one set for steam, another for diesels. Larger roads could not, or did not, follow Monon's example of swiftness because diesel manufacturing facilities could not keep pace with the immediate post-war demand for such power, or because management was still sceptical of totally eliminating steam.

As a rule, diesels were first placed in those services which would earn the greatest return on their investment – in other words, high-mileage, high utilization service. Steam was accordingly left with such runs as extra freight trains, branch-line assignments, 8- and 16-hour yard tricks, helper movements, and the like. Not infrequently, the newest, largest steam power in North America was the first to be scrapped because:

(i) diesels had taken over the peak-horsepower demands for which this big steam power had been built;

(ii) older steam power was usually of more simple design, hence easier to maintain for low-utilization services;

(iii) axle loadings prohibited downgrading of big steam power to branch-line services.

Pennsylvania 4–6–2's built during and after World War I were running a decade after the road had scrapped its Class T-1 streamlined, duplex-drive, poppet valve 4–4–4–4 engines delivered in 1945.

Occasionally, a road dieselized a complete division or region at one stroke (Union Pacific did so between Los Angeles, California, and Salt Lake City, Utah) but the usual practice was to allow the diesel system-wide freedom of operation, finally dieselizing individual districts when it became apparent that water plugs and coaling docks were being maintained for just a handful of steam locomotives.

Those railroads which experience violent seasonal fluctuations in passenger traffic or carloadings were the most difficult to dieselize efficiently. The common reaction was to restrict steam to a single division (which saved system-wide maintenance of steam facilities), operating it on that district during peak season only. In one instance Bangor & Aroostock purchased road-switchers to complete dieselization under a reciprocal-use pact with Pennsylvania. In the winter potato-rush season B. & A. operates the units; in the summer Pennsylvania uses them to work ore docks on the Great Lakes which are frozen and closed to navigation in the winter. In another example, Burlington dieselized a thirty-eight-mile commuter zone out of Chicago by tapping a pool of main-line passenger units, which lay over in Chicago between long-distance runs. Not a single diesel in this service has been exclusively assigned to, or purchased for, the inherently low-mileage commutation service. Expectedly, practically all of the roads which have yet to attain total dieselization in the U.S. are subject to seasonal traffic fluctuation or feel a reluctance to part with steam because of a dependence upon coal traffic.

THE ROAD-SWITCHER

If North American railroads had their dieselization to do all over again, the streamlined or cab-type unit would be purchased only for high-speed, de luxe passenger service. Elsewhere, the road-switcher would handle all assignments except exclusively yard jobs.

As compared with cab-type units, the road-switcher affords:

(i) visibility necessary for terminal and local freight-train operations;

(ii) operation at will in yard or road service (including passenger duties if the unit is equipped with a steam generator);

(iii) use as either a lead or control unit or as a booster;

(iv) greater accessibility for inspection, servicing and repair because of its hood-type carbody construction;

(v) operation in either direction without the need for turntables or wyes.

The road-switcher, once it was available in horsepower ratings sufficient for optional mainline duty, achieved a sudden success. In the 1,500–1,600 H.P. bracket at which most diesels are rated for road service, U.S. roads ordered 1,374 units in 1948, of which only 200 were hood type; in 1952 two-thirds of the 1,267 units of this rating ordered were hood type. Indeed, the largest diesels produced today in America – 2,400 H.P. C–C's – are available only as road-switchers. The largest road to dieselize totally with hood units was Delaware & Hudson, which replaced approximately 350 steam locomotives of nine different wheel arrangements with 169 Alco-G.E. diesel units comprising 51 1,000 H.P. B–B switchers and 118 1,500 or 1,600 H.P. B–B road-switchers. In recent times, D. & H. has even found that it purchased too many diesels and has accordingly leased or made plans to sell its older units. In another example, New York Central planned dieselization of one district with fifty cab-type and switcher units, only to find that thirty-eight road-switchers could do the job alone.

ROSTERING OF DIESELS

Many roads operating a high number of fixed schedules in either freight or passenger service assign diesels to definite cycles instead of running them in "first in, first out" pools. For example, Burlington operates 6,000 H.P. diesel freighters on a 6,200-mile, fifteen-day cycle that schedules the locomotive out of Chicago to Denver,

Colorado, thence south to Texas and north through Colorado to Montana, thence home to Chicago. Only fuel and similar supplies and routine inspection are given the locomotive en route.

Western Maryland based a pool of 1,600 H.P. road-switchers at Cumberland, Maryland, to dieselize the coal traffic of a division which operated six days a week and included three helper districts. Since the coal demanded no specific schedule, it was moved to increase diesel utilization. So well executed was the assignment sheet that not only were the units revolved about from road to helper duty to equalize unit mileage but from Saturday night until Monday morning all units were in Cumberland, available for any other twenty-four-hour assignment out of there, or for inspection and maintenance.

TECHNOLOGICAL DEVELOPMENTS IN PHYSICAL PLANT

Introduction of road diesel power, notably in freight service, quickly made apparent how tailored the existing railroad plant was to the steam locomotive. Diesel-length trains could not be accommodated in sidings, passing tracks, or receiving and departure yards designed round steam locomotive performance. For example, in road tests Louisville & Nashville discovered that a 1,500 H.P. diesel freight unit possessed 82 per cent of the tonnage rating of a heavy 2–8–2, the road's standard steam freight engine. Thus, although steam-powered trains were once limited to from thirty to fifty cars per engine, it is common practice today to dispatch freights of 120–150 cars behind multi-unit diesels.

Diesels, by pulling more freight cars per train, have tended to reduce the number of trains operated; this has, in turn, permitted a considerable reduction of multiple-track mileage. Pennsylvania's picture outlines the manner in which diesels have reduced the need for so much track space. At war's end the Pennsylvania Railroad operated 100 million train-miles with more than 4,000 steam engines and 245,000 freight cars; by 1955 the system's 2,009 diesel units plus about 600 active steam and electric locomotives and a freight car fleet of 190,000 cars ran just 70 million freight train-miles.

Two major technological developments have helped to exploit the diesel:

First, centralized traffic control which has permitted more effective utilization of single track reduced from double because of dieselization.

Second, hump-retarder classification yards with automatic retardation and switching. These installations have permitted classification of longer trains in as short a time as the older flat yards once classified the shorter trains of the steam era.

Part VI. Diesel locomotive maintenance and rebuilding

MAINTENANCE FACILITIES
(Plates 29, 30A, 30B, 31A, 31B, pages 121, 122, 123)

Class One U.S. railroads own approximately 550 diesel locomotive repair shops to maintain the more than 27,000 diesel units in service at the beginning of 1958. Most of these plants once served the steam locomotive and have been converted from steam to diesel use to avoid the high cost of new buildings. However, there are several major shops which are new and have been designed from the ground up, for the exclusive task of diesel maintenance.

Conversion of existing facilities, particularly round-houses, has had its drawbacks. The average roundhouse stall can accommodate only two or three diesel units and necessitates the uncoupling of longer multi-unit loco-motives, if only for routine inspection. Also, that part of the stall nearest the turntable (and occupied by the tender in steam practice) is cramped in terms of space, since the tracks are laid out on a fan basis extending out-ward from the turntable. The ideal diesel maintenance running track is a tri-level arrangement with:

(i) carbody floor-level platforms for access to the cab, diesel engine and generator;

(ii) a rail- or truck-level area for inspection and repair of trucks, traction motors, fuel and water tanks, etc.;

(iii) a between-the-rails drop pit for inspection of running gear;

(iv) a drop table to permit removal of a truck without the necessity of lifting the locomotive itself.

Such an arrangement involves expense which many roads have felt should be placed in a new building rather than an old one. In a roundhouse occupied by both steam and diesel power, the diesel section must be walled off to escape the water, grease, steam, soot, smoke and other dirt agents associated with steam.

Major repair or overhaul shops for diesels – those which permit removal of the engine and generator from the carbody and complete dissembly and repair of all components – are far fewer in number for diesels than they were for steam as well as being different in layout and equipment. The diesel's freedom from clearance or axle loading restrictions permits a centrally located shop to overhaul all the units of even a major system. More-over, by its very nature the diesel is more economical of shop space. A diesel engine due for overhaul can be pulled from the carbody, and replaced with a rebuilt engine in a matter of hours, thus releasing the unit to the operating department during the time its components are being overhauled.

Steam shops have proved acceptable for diesel usage (as opposed to the limitations of the roundhouse as a light repair site) because of the generous space allowances and the existence of heavy-duty overhead cranes. An instance is Burlington's West Burlington, Iowa, Shops which handle all heavy overhaul work on more than 500 diesel units powering 8,858 route-miles of railroad. Power returned to West Burlington is scheduled in on two-year cycles for road units (each 300,000 miles for freight units; each 400,000 miles for passenger units) and every ten years for switchers, and it is overhauled accord-ing to its over-all cumulative mileage.

Until the end of World War II West Burlington remained primarily a steam shop and had constructed 4–8–4's as late as 1940. Conversion meant erecting a wall across the main or erecting shop to separate it from machine tools, a new flooring, movement of some 400 machines, purchase of new shop tools (but not exceeding 20 per cent of the steam tool investment) and re-educa-tion of existing staff. The nature of the work performed before and after the conversion may be gauged by the fact that the number of electricians rose from three to fifty, and blacksmiths declined in strength from one hundred to three. So efficient is the revised setup that a four-unit, 6,000 H.P. freighter, which has run 900,000 miles, can have all its components pulled and overhauled, be cleaned and repainted, be reassembled and tested, between Monday morning and Friday afternoon of the same week.

MAINTENANCE PROCEDURES

The majority of North American railroads follow some form of progressive maintenance on their diesel locomotives. That is, the unit is scheduled into the shop on established dates determined by its mileage, fuel con-sumption, or time on the road. Work proceeds according to the needs of each individual component part. For example, traction motors have brushes and leads checked at 2,500 miles; are inspected and cleaned each 30,000 miles; have load meters checked for proper calibration each 100,000 miles; and are overhauled each 200,000 miles. This is the average builder's recommendation. In actual practice, certain roads have run traction motors more than 600,000 miles without incident before overhaul. In any event, a traction motor re-mains with the locomotive unit only until it is pulled for maintenance; after overhaul, the motor is placed "in stock" and mounted on a similar unit when needed.

Diesel engines are treated in similar fashion. Engine air filters are usually pulled and cleaned each 2,500 miles, lubricating oil filters each 10,000 miles, engine main bearings each 100,000 miles. The engine itself is removed from the carbody for complete dissembly and overhaul every million miles in passenger service, every 750,000 miles in freight service and once every six years in switching duty.

REBUILDING OR UPGRADING?

There is considerable discussion in the United States as to whether a diesel unit reaching the major overhaul stage – approximately a zone of 1,000,000 to 1,750,000 miles for a freight unit and accumulated over ten to fifteen years – should be rebuilt "in kind" or upgraded. This is what might be called the second stage of the diesel unit's overhaul life. At the first stage, perhaps six years after its construction, components such as engine and main generator can be removed from the carbody and replaced without detaining the unit from service for more than a few hours. At the second stage, however, reworking of running gear, electrical cables, carbody and so forth, necessitate removal from service of the unit as well as of its components.

Rebuilding "in kind" means that a 1,350 H.P. cab unit, for example, is returned to service at its original power rating and in its original design, at a cost of from $35,000 to $60,000. Upgrading, which is usually done by the original manufacturer, means salvaging certain components, remanufacturing them, and installing them in a new carbody with other new parts. Electro-Motive upgrades an FT 1,350 H.P. unit into a 1,750 H.P. GP-9 road-switcher (with a new locomotive warranty) for approximately $140,000. Rebuilding in kind might cost half that sum and a brand new GP-9 would cost approximately $175,000. However, the upgraded unit – which uses thirteen major parts from the original unit including crankshaft, auxiliary generator, truck assemblies, traction motors, etc. – can pull $33\frac{1}{3}$ per cent more tonnage than the FT model and return 12·4 per cent on its rebuilding investment. An increasing number of roads are returning their oldest units to Electro-Motive and Alco for this upgrading procedure. Other builders such as Fairbanks, Morse have thus far limited their upgrading programme to components such as engines, rather than the complete unit, and in a few instances railroads have undertaken upgrading in their own shops, with and without the manufacturer's assistance. Other roads prefer to rebuild "in kind" because they either doubt the realism of projected upgrading savings or have a substantial investment in company shop facilities to protect.

Regardless, upgrading establishes a circumstance apparently unique among locomotives at large, i.e. the oldest diesel units are as powerful as the newest.

Part VII. By-products of dieselization

DEMONSTRATIONS

It is possible for a railroad to assay the operating potential of a diesel locomotive while it is still in blueprint form, and there are frequent examples of this in North American dieselization. For example, New York Central had fifteen 2,000 H.P. Fairbanks, Morse C-line units on order in 1950 before the original prototype left the builder. However, the familiar pattern for either dieselization or model introduction has been a national tour of builder-owned units. In this way the railroad has an opportunity to evaluate the diesels under actual operating conditions against existing motive power and with no obligation to buy, and, if the units are of a new design, the builder also is able to stage a "shakedown" test under controlled conditions and correct any troubles that develop in subsequent production models. Such demonstrations were the exception to the rule prior to the diesel age, because of the custom design of the steam locomotive and the sharply varied limitations on its operating range because of clearance and axle-loading restrictions.

THE EXPORT MARKET

U.S. and Canadian locomotive builders are in an enviable position to claim a major share of the world diesel market because of their extensive manufacturing facilities backed by an unparalleled experience in diesel locomotive design, construction and operation. The U.S. market has been largely reduced to replacement or upgrading of existing units and manufacture of parts, thus it is imperative that the builders look elsewhere to keep their plants intact. Locomotive components, notably diesel engines, have been adapted to, and sold for, such non-railroad purposes as mobile power-generating plants and oil-drilling rigs. However, there is a practical market limit for engines rated at 1,750 H.P. or more. Thus the attractiveness of world dieselization. From a North American builder viewpoint, the world market has been variously estimated as equivalent to the U.S. field (27,000 units' worth between 1925 and 1958) or infinitely bigger. During 1956, at least, the export field was good enough for Electro-Motive to produce one diesel unit for overseas use per working day.

Initially, North American builders introduced modified versions of domestic models into the export market, but the generous axle loadings and clearances of U.S. roads tended to restrict the adoption of such units elsewhere, especially in countries using less than the standard North American track gauge of 4 ft 8½ in. General Electric adhered to the continental custom of building custom-design diesel to fit specialized overseas operating conditions, but this approach denied the virtues of standardization and mass-production, which played such an important role in the swift dieselization of North America. Finally, U.S. builders encountered tariff problems and a dollar shortage in the export trade.

Today three builders are making a major effort to dominate the foreign market: Alco, Electro-Motive, and General Electric. All three have developed standardized export designs which combine standardized components with adaptability to many gauges. For example, Alco and its subsidiary Montreal Locomotive Works offer an 1,800 H.P. "World Locomotive" cab unit adaptable to any gauge from 3 ft 3⅜ in. to 5 ft 6 in. and available as a B–B and A1A–A1A, or a C–C in wheel arrangement. Both Alco and E.M.D. have licensed various foreign manufacturers to produce their export models. Electro-Motive, for example, has such associates in Australia, Belgium, Germany and Sweden.

INFLUENCE OF THE DIESEL ON
OTHER TYPES OF MOTIVE POWER

The reciprocating steam locomotive reached the peak of its development and performance under the impetus of competition from the diesel. Introduction of diesel power in North America, with its fresh interpretation of such indices of efficiency as utilization, availability and continuous runs, was directly responsible for such steam achievements as a New York Central 4–8–4 piling up 30,000 miles a month; Santa Fé 4–8–4's operating more than 1,700 miles without change; and Norfolk & Western engine turnarounds in periods of less than sixty minutes.

The strong influence of diesel carbody design upon the straight-electric locomotive may be gauged by the contrast between new electrics delivered to the Virginian in 1948 and 1957. The earlier units were streamlined, 500-ton, 8,000 H.P. giants consisting of two permanently coupled units with a B–B–B–B+B–B–B–B wheel arrangement; the latest units, rectifiers, are hood-type, 197-ton, 3,300 H.P. units of C–C wheel arrangement, equipped to operate singly or in multiple with each other. Thanks to diesels, the power has become far more flexible in unit size and horsepower as well as more accessible for maintenance. Significantly, the 1948 units

were purchased when Virginian was otherwise all steam and the 1957 were ordered after the road disposed of its steam power in favour of diesel road-switchers.

NON-LOCOMOTIVE USES FOR THE DIESEL ENGINE

Rapid dieselization introduced the diesel engine itself to non-locomotive service in railroading, if only for the reason that steam fuel, water, ash disposal and repair facilities were no longer available. For example, the wrecking (breakdown) crane turned out to be an ideal application for diesel conversion because in an emergency there is no delay necessary to build up steam nor is it ever necessary to "cut and run" for water. Other equipment including pile drivers has been similarly converted.

The steam generator – that necessary item for diesels operated in passenger service – has been utilized to replace ordinary boilers in pile drivers and as a source of power for roundhouses, store rooms, etc.

A number of steam rotary snow ploughs have been converted to diesel-electric power under a system whereby locomotive-type traction motors mounted in the rotary itself, turn the blades and are supplied with current from the main generator of the pusher diesel locomotive. In a Southern Pacific conversion of this type four traction motors develop 1,250 H.P. and spin the twenty 11-ft blades of the rotary at speeds up to 150 r.p.m. S.P. thinks four such rotaries can replace six steam ones because of the freedom from having to halt operations and return to water plugs and fuelling stations. And each summer the traction motors can be removed and returned to locomotive duty.

RAIL DIESEL CARS

The gas-electric (petrol-electric) passenger railcar, dating back to the first decade of the century and extremely popular in North America in the 1920's, died out of favour in the 1930's because of:

(i) the space and weight of its electrical transmission;

(ii) limitations on the size of gasoline (petrol) or distillate engines;

(iii) the fire hazard of an inflammable fuel.

There remained, however, a need for self-propelled equipment to handle runs which could not be economically operated with locomotive-hauled trains. Several attempts with diesel-electric railcars, dating from 1925, proved too cumbersome from a power-to-weight ratio standpoint.

In 1949 the Budd Co., carbuilders of the *Pioneer Zephyr* and specialists in all-stainless-steel passenger equipment, introduced the R.D.C. (rail diesel car), an 85-ft car powered by a pair of General Motors inline, 6-cylinder, 275 H.P. (later 300 H.P.) diesel engines driving through torque converters to the inside axles of four-wheel trucks. These engines and their transmissions were mounted entirely under the carbody, leaving the entire floor space between vestibules available for revenue loading. A distinctive, hump-backed dome on the roof housed radiators, cooling fans, engine air intake, exhaust pipes and bell. The car had a control compartment in both vestibules, could be operated in multiple with other R.D.C.'s under a single control, featured disc brakes, and provided its own heating and air conditioning.

The following details were relevant to the original R.D.C.:

Power-to-weight ratio: 8·68 H.P. per ton.
Initial acceleration: 1·4 m.p.h. per second.
Centre of gravity: 52·6 in.
Maximum speed: 85 m.p.h.
Weight, unladen: 63 tons.
Seating capacity: 83.

Later models of greater horsepower increased these performance figures. R.D.C.'s became available in five versions:

R.D.C.-1 all-coach;
R.D.C.-2 coach-baggage;
R.D.C.-3 coach-baggage-mail (Plate 32B, page 124);
R.D.C.-4 baggage-mail (73 ft 10 in. long because of a heavier floor loading);
R.D.C.-9 an all-coach but single-engine car without controls to supplement other R.D.C.'s in M.U. operation.

Since 1949 Budd has sold (to January 1957) 329 R.D.C.'s to twenty-nine U.S. and foreign roads for runs ranging from 7 to 1,051 miles. In early 1958 a pair of R.D.C.'s on Western Pacific passed the million-mile mark each. Most of these cars have been operated in local or commutation service but there are exceptions which point to R.D.C. long-distance expresses. Baltimore & Ohio operates a three-car R.D.C. train with dining facilities each way daily between Philadelphia and Pittsburgh, Pennsylvania, 428·7 miles. These R.D.C. trains cut $1\frac{1}{2}$ hours off previous locomotive-powered train schedules, raised revenue 13 per cent, cut costs 47 per cent. Also significant is the fact that in 1956 a Canadian Pacific R.D.C. schedule covering 22·2 miles in 19 minutes at an average speed of 70·1 m.p.h. was the only train in the British Empire operated at more than 70 m.p.h. on a start-to-stop basis.

Part VIII. The future

THE DIESEL OF TOMORROW

Based upon evidence now at hand, the typical diesel built for North American service during the 1960's will be an all-purpose or road-switcher hood unit of B–B or C–C wheel arrangement, powered by a single V-type diesel engine rated at between 1,800 and 2,500 H.P. It will operate satisfactorily on extremely low grades of fuel oil and it will be scheduled progressively greater mileages between scheduled overhauls of its components. In the foreseeable future, such a unit will continue to employ an electrical transmission because of its flexibility and stamina.

There has been limited experimentation in North America with non-electric transmission. Electro-Motive has turned out yard units with torque-converter transmission. Baldwin-Lima-Hamilton sponsored the construction of three lightweight 1,000 H.P. passenger units with German-built Mekydro torque-converter mechanical drives. Imported M.A.K. diesel-hydraulic switchers are now at work in Canada and Cuba. If Baldwin can make good its promise of an all-purpose 1,800 H.P. hood unit with torque-converter mechanical drive, the effect upon dieselization will be interesting indeed. However, it is doubtful if generators and traction motors will lose their dominance within the next decade.

COMPETITION FROM OTHER FORMS OF MOTIVE POWER

There is as yet no practical evidence that the diesel locomotive faces technological displacement in North America in the foreseeable future. Indeed, the diesel's rivals seemed to be in much more of an aggressive position a few years ago than is the case today. We may consider these rivals as follows:

Electrification. Many people believe that rising oil costs will inexorably force the electrification of main lines of high traffic density, and they point toward European use of commercial power and the development of rectifier locomotives, as barometers of the future. However, the question of who will supply the capital for electrification is anything but resolved. Indeed, as 1958 began, large Eastern roads were having difficulty obtaining funds for the purchase of freight cars on equipment trusts. Unless some form of government subsidy develops, it now seems unlikely that electrification will emerge from what has been called its "financial imprisonment".

Atomic energy enthusiasts have argued that an atomic locomotive could be operating by 1960 and that by 1970 such units could be competitive in over-all costs with diesels in high-speed, high-horsepower service. Yet in common with electrification, the high initial cost of an atomic locomotive appears to be an impossible barricade short of government subsidy.

Gas turbine power is the prime mover that offers the diesel its most immediate, most practical competition. Union Pacific is now handling about 10 per cent of its gross ton-miles with a fleet of twenty-five 4,500 H.P., oil-fired gas turbine-electric freight locomotives built by General Electric. The same builder is now delivering thirty 8,500 H.P. two-unit gas turbine-electrics to replace U.P.'s 4–8–8–4 steam locomotives (*see* chapter 5, page 324, example no. 24). As a rule of thumb, the promise of these units was that they burned twice as much oil as a diesel but that fuel cost half as much because of its lower grade. Being thus competitive in fuel costs, the turbine would pull ahead of the diesel because of reduced maintenance due to its freedom from a reciprocating engine. However, the diesel has been successfully adapted to lower grades of oil and it is a mute, unanswered question as to what cost differential, if any, exists in the turbine's favour. Again, because of its excessive idling speed and consequently its high fuel consumption when not working at maximum power output, the turbine is rendered inept for anything but long-haul, high-power operating conditions. Work on a coal-burning gas turbine power plant, suitable for application in existing diesel carbodies, has been progressed by Alco under the sponsorship of coal companies and coal-hauling railroads, and it was hoped that an actual locomotive could be started by 1957. Such a unit might succeed on a fuel-cost basis where its oil-fired sister cannot (*see* chapter 10, page 487).

PREDICTION

The North American reaction to diesel locomotives and their future was suitably summed up in a 1955 statement by Fred G. Gurley, when he was president of the Santa Fé. He said: "We could not have carried on without it. Its efficiency and economy – particularly its flexibility – are hard to beat. I don't think we'll be putting it out to pasture in the next few years."

CHAPTER 3

Electric Motive Power

by F. J. G. HAUT

Part I. Development of electric traction

1835–70. EARLY ATTEMPTS TO USE ELECTRICITY FOR RAILWAY TRACTION

According to existing records the first attempts to drive rail vehicles by electric power were made by Strattingh and Becker in Groningen, Netherlands, in 1835, who tried to drive a two-axled vehicle by battery power. Also in 1835, Thomas Davenport tried a similar experiment in the U.S.A. The first electric locomotive which ever worked was run in Scotland in 1842 by R. Davidson on the Glasgow-Edinburgh line. It is supposed to have weighed seven tons and it ran on two axles. It hauled a load of about six tons at a speed of 4 m.p.h. Each axle carried a wooden cylinder on which were fixed three iron bars, parallel to the axle. Four electro-magnetic units were arranged in pairs on each side of the cylinders and current was produced by a battery. The electro-magnets attracted the bars on the cylinder, then alternately the current was cut off and on, and rotation was produced.

1870–95. EARLY LOCOMOTIVES IN THE UNITED STATES AND EUROPE

1879. Werner von Siemens' locomotive. The first major success was due to Werner von Siemens, who showed in 1879 a two-axle locomotive at an exhibition in Berlin. It hauled a small train on a 300-yard long railway in the exhibition grounds. Voltage used was 150 and the output was approximately 3 H.P. Current was brought to the vehicle by a third rail between the running rails; the transmission was by gear wheels and the motor was started by a liquid rheostat. The locomotive is today preserved in the Technical Museum in Munich.

1884. The work of René Thury. Another successful attempt was made by a Swiss engineer, René Thury (1860–1938), who built in 1884 an experimental rack railway in Territet, a suburb of Montreux. The purpose was to connect a mountain hotel, about 1,000 ft up the mountain slope, with Montreux, and the project was quite satisfactory. The locomotive was a two-axle vehicle running on a gradient of 1 in 33. It could take four persons. It is reported that the motor was used on

descending as a generator for braking purposes. The inventor, René Thury, was an important engineer. He went to the U.S.A. where he worked for T.A.Edison. He was responsible for numerous railway designs and invented many types of electric machinery, especially multi-pole machines, and the series coupling of electric motors. He later worked for many years for Dick, Kerr & Co., of Preston. (Today part of the English Electric-Co. Ltd.)

1885. Van Depoele and F.J.Sprague. In 1885 Van Depoele first introduced a roller running under the overhead wire, and pressing upwards against it for the purpose of current collection – the beginning of our trolley system. The first installation was on a tram-line in Toronto; F.J.Sprague three years later fixed the roller to a wooden, spring-loaded rod. F.J.Sprague is well known as an electrical engineer and inventor, especially for the multiple-unit system of combining electrically-driven vehicles in such a manner that they can be operated by a single driver from any one driver's cab.

1883–8. L.Daft, S.D.Field and T.A.Edison. The first main line locomotive for a standard gauge line was built by the American Leo Daft, in 1883, for the Saratoga & Mount McGregor Railroad. It was called "Ampere" and incorporated field control for speed variation, although this was effected by parallel and series connections of the field windings. It also had an electro-magnetic brake (Plate 33A, page 161). This 12 H.P. locomotive pulled 10 tons at speeds of from 6 to 9 m.p.h. Current came from a central rail. In 1885 the same inventor built an improved model for the New York Elevated Railroad. It was called "Benjamin Franklin", weighed 10 tons, and was 4·35 m.(14 ft 3 in.) long; it took its 250 v. current supply from a central rail. There were two 48-in. driving wheels and two 33-in. trailing wheels. In 1888 the locomotive was re-equipped with four driving wheels and a 125 H.P. motor and could then haul an eight-car train at 10 m.p.h. The "Benjamin Franklin" used friction drive from the armature shaft to wheels on the axle to transmit power.

T.A.Edison interested himself in electric railways and built electric test-lines between 1880 and 1884 at Menlo Park. One of his locomotives was "The Judge" developed together with his partner Stephen D. Field in 1883 (Plate 33B, page 161). In the same year, two electric railway lines were opened in the British Isles, one being the Giant's Causeway Electric Tramway in Ulster and the other Magnus Volk's line in Brighton.

Following these early experiments with electric traction on rails came attempts to electrify minor railway lines.

1890–1910. THE FIRST MAIN LINE ELECTRIFICATIONS

1890–1901. London underground railways. In 1890 the City & South London Railway was opened as the first electric railway in England and the first electric underground railway in the world. It ran from King William Street to Stockwell and its sixteen locomotives were supplied by Messrs Mather & Platt and Siemens Bros. (Plate 33C, page 161). Each weighed 20,700 lb., with two 50 H.P. motors of the so-called gearless type, in which the armatures were mounted on the axles. The locomotives were 14 ft long, had 27-in. wheels and could run up to 25 m.p.h. They hauled 40-ton trains with a speed of approximately 14 m.p.h. The original line was 3·5 miles long; current used was 500 v. D.C. supplied from a conductor rail.

In 1901 the Central London Railway was opened from Shepherd's Bush to the Bank, a distance of 6·5 miles. The line was first operated by two-axle locomotives, having gearless motors supplied by the General Electric Co., but the excessive unsprung weight caused difficulties and the motors were replaced by geared ones, built by B.T.H. In 1903 locomotive traction was abandoned in favour of motor-coach traffic. The locomotive had 3 ft 6 in. driving wheels, a bogie wheelbase of 5 ft 8 in. and a distance between bogie centres of 14 ft 8 in. The locomotive weighed about 45 tons and the tractive effort was about 13,500 lb. supplied by two motors.

1894–5. Baltimore & Ohio Railroad. The Baltimore & Ohio Railroad crossed the town of Baltimore on sections consisting of overhead and underground lines, totalling seven miles. The town council requested that by 1894–5 steam locomotives should be replaced by electric ones and the then unique problem of electrifying a busy main line across a city was solved by the General Electric Co., of U.S.A. The line used 675 v. D.C. and a rigid overhead conductor line in the form of a Z-section on which glided a kind of pantograph. This surrounded and gripped the conductor rail with two Z-pieces, thus transmitting current to the four-axle 1,440 H.P. locomotives, which weighed 96 tons. They had four gearless

motors of 360 H.P. each, transmitting power to the wheel-spokes by means of rubber blocks. The locomotives could start and haul a 1,870-ton train. Other particulars were: Total lengths 19 ft 6½ in., wheelbase per bogie 6 ft 10 in. Wheel diameter 62 in. (Plate 33D, page 161).

1899. Burgdorf–Thun Railway. It was obvious that three-phase A.C. would have considerable advantages for electric railways, these being especially the absence of commutators, lower weight and current regeneration. The first 3-phase railway using electric locomotives was the line from Burgdorf to Thun in Switzerland. It used current at 750 v., was twenty-eight miles long, and had gradients of up to 1 in 40. It was completed in 1899 by the firm of Brown, Boveri. Motor coaches were used for passenger services and locomotives of the 0–4–0 or –B– type for goods trains. The latter had to haul loads of up to 100 tons with speeds of up to 20 m.p.h. The motors drove the four wheels by means of gear wheels, jack shafts and rods, closely following steam locomotive design (Plate 34A, page 162).

Main particulars of these locomotives were as follows:

Wheel arrangement:	0–4–0 or –B–.
Wheelbase:	3,140 mm. (10 ft 3¾ in.).
Length over buffers:	7,800 mm. (25 ft 7¼ in.).
Number of motors:	2.
Total output:	300 H.P.
Gear wheel ratio:	1 : 1·88 or 1 : 3·75 for slow speeds.
Wheel diameter:	1,230 mm. (4 ft 0½ in.).
Total weight:	29·6 tonnes (29 tons).

1901–3. The Zossen–Marienfelde tests. Between 1901 and 1903 the firms of Siemens and A.E.G. tried out on a five miles long railway line near Berlin two test cars and one test locomotive, the test-cars reaching the then phenomenal speed of 135 m.p.h. The experiments were conducted by Walter Rathenau and Dr. Reichel. Current used was 10,000 v., 3-phase A.C. of 50 cycles, and was transformed on the vehicles down to 1,150 v. (Siemens) or 435 v. (A.E.G.). The tests were very important as they proved the possibility of high-speed services with electrically-driven vehicles. After substantial alterations to the track and overhead lines, initial difficulties were overcome and the vehicles were quite satisfactory (Plate 34B, page 162).

1902–10. The Simplon and other Alpine electrifications. Up till now only minor railway lines had used electric current. The major test of using electric power to operate heavy goods trains and fast regular passenger trains was to come when the Simplon and other Alpine lines with their long tunnels were electrified. The work was mainly carried out by the firm of Ganz & Co., of Budapest, and Brown, Boveri, of Baden, between 1902–10;

15-cycle, 3,000 v., 3-phase A.C. was used, and a considerable number of three- and four-coupled locomotives developed. The locomotives had an output of 900–1,000 H.P. and a weight of about 60–65 tons. Later, a large number of similar machines, three-, four- and five-coupled, were built, all following similar principles (Plate 34C, page 162).

1903. H.T. direct current locomotive for St George de Commiers–Le Mure. While the Baltimore & Ohio Railroad proved that a D.C. railway on a larger scale was possible it was the firm of Secheron of Geneva which built the first high-tension electric main line. This was the line from St George de Commiers to Le Mure (Isère) of the French State Railways, and in 1903 the first five locomotives were delivered. The two-wire system supplied 2,400 v. D.C. to the 500 H.P. locomotives with wheel arrangement Bo–Bo and which weighed 50 tonnes (49¾ tons). The electrical equipment consisted of four pantographs, resistances and four nose-suspended motors operating at 1,200 v. The railway line is 31 km. (19¼ miles) long and has gradients of up to 1 in 38; trains of up to 110 tonnes (107¾ tons) weight could be hauled (Plate 35A, page 163).

1905. Seebach–Wettingen electrification. It had by now been proved that 3-phase A.C. current was better than the early D.C. installations. Three-phase current has, of course, many disadvantages, and the choice of single-phase A.C. requiring one contact wire only was an obvious one. The great difficulty encountered was to produce a large motor with safe commutation. The earliest electric locomotive using 15-cycle, single-phase A.C. (at 15,000 v.) was first an A.C.-D.C. converter locomotive later transformed into a single-phase A.C. locomotive. In 1901 the firm of Oerlikon suggested to the Swiss Federal Railways that they would be prepared to electrify, at their own expense, a fourteen mile long line from Seebach to Wettingen to prove the advantages of their traction system, as developed by E.Huber-Stockar. The line was completed in 1905, three locomotives being used. After the line was run very successfully for about 1½ years the Swiss Federal Railways decided against purchasing the line. Thus the electrified installations were dismantled, but the value of the experiments remained. Other lines decided to electrify their mountain sections, especially the Loetschberg Railway (Plate 34D, page 162).

The first locomotive constructed for the Seebach–Wettingen line was, as mentioned, originally designed as a converter locomotive; the single-phase current was transformed on the locomotive down to 700 v. and then fed into a converter set with a direct current generator of 400 kW.–600 v.–1,000 r.p.m. There were two D.C.

shunt-wound traction motors of 200 H.P. each. In the meantime, however, a satisfactory single-phase series commutator motor for low-frequency current had been built, and the locomotive was rebuilt into an ordinary single phase type. It now had 250 H.P. motors with running speeds up to 38 m.p.h. It weighed 42 tons.

1907–18. NOTEWORTHY ELECTRIFICATION SCHEMES IN AMERICA

1907. The first electric locomotives for the New York, New Haven & Hartford Railroad. This was the first main line railway in the United States to operate, over a considerable distance, heavy passenger traffic, both local and high-speed, with electric locomotives. The electrification of the line as far as Stamford, Conn., thirty-three miles from New York, was effected at the same time as the Grand Central Station and its approaches were electrified. With increasing traffic, the latter electrification had become essential on account of the smoke nuisance in the long tunnel approach under Park Avenue, New York. The terminal electrification was operated with D.C. at 600 v. In the case of the New Haven Line, however, it was advantageous to operate the entire passenger traffic electrically to Stamford and later to New Haven, seventy-two miles from New York. Here it was decided to use 11,000 v. single-phase A.C. of 15 cycles collected from an overhead catenary. Electric services were started in 1907 and operated with thirty-five Baldwin-Westinghouse locomotives which at that time represented many innovations in locomotive design and worked from the start with considerable success. There was no precedent for electrifications of this kind and size and a considerable amount of pioneer work had to be done.

The first New Haven electric locomotives were of the Bo–Bo double-bogie type, weighing 100 tons each. They were designed to handle local trains of 200 tons weight, making frequent stops, at an average speed of 26 m.p.h.; and express trains of 250 tons at higher speeds, the maximum running speed being about 60 m.p.h. The nominal capacity of each locomotive was 1,000 H.P. When handling heavier trains, two locomotives could be coupled together and operated as a unit from the leading locomotive. In actual service, however, the locomotives proved themselves capable of considerably exceeding their rating. The locomotives were each equipped with four gearless motors, so designed that the entire motor weight could be spring-borne. Instead of the motor armature being carried directly on the axle, it was mounted on a quill surrounding the axle, and with sufficient annular clearance to permit the required amount of vertical play. The torque developed by the armature was transmitted to the driving-wheels through specially designed helical springs. In this way shocks were cushioned, and the amount of non-spring-borne weight was reduced to a minimum. The bogie-frames were positioned outside the wheels, and followed steam locomotive practice in design. The equipment included transformers for reducing the voltage from 11,000 to 500 as required by the motors. The transformers and motors were force-ventilated. The transformers were, of course, not in use when operating on direct current. While these locomotives were fully capable of handling the trains which they were designed to haul, serious difficulties developed in their riding qualities. This was remedied by adding a two-wheeled running truck at each end, and equalizing the trucks with the adjacent group of driving wheels.

During the first few years of electrified service on the New Haven, only passenger traffic was hauled by electric locomotives. Thirty-six Baldwin-Westinghouse goods locomotives were placed in service in 1912 and 1913, which were similar to the passenger locomotives already described, there being four driving axles, grouped in two independent bogies with a two-wheeled running truck at each end. The weight was 219,450 lb., with 42,000 lb. on each axle. These locomotives developed a maximum tractive force at starting speeds of 40,000 lb., and a tractive force of 18,600 lb. at a speed of 27·5 m.p.h., on a one-hour rating. In 1912, Baldwin-Westinghouse supplied sixteen shunting locomotives for service in the New York terminal yards. These locomotives are of the Bo–Bo type and the entire weight of 80 tons is carried on the driving axles. They develop a maximum tractive force of 40,000 lb. A later type of passenger locomotive was introduced in 1919, after an exhaustive study of the service requirements. Seventeen locomotives were built of the 1–C–1+1–C–1 type, the two sets of frames being joined together.

As in the earlier locomotives, the body extends the whole length of the locomotive, but in this case the body underframe is spring-borne on the main frame, and is not used to transmit any of the tractive forces, thus giving good riding qualities; it is held in proper alignment by means of the two centre pins.

Turning to the electrical equipment, there is one twin-armature motor for each driving axle, or a total of six motors per locomotive. The twin-armature motor weighs less and occupies less space than a single-armature motor of equivalent capacity. The drive is through a quill of improved design, and the entire motor weight is spring supported.

These locomotives were designed to haul trains

between the Grand Central Station and New Haven, seventy-two miles, in ninety-nine minutes; representing an average speed of 44 m.p.h. With four intermediate stops, the run can be made in 115 minutes with the same weight of train. The locomotives are also used between New Haven and the Pennsylvania Station in New York, operating over the Hell Gate Bridge route.

1916. The locomotives of the Chicago, Milwaukee & St Paul Railroad. Among the early D. C. electrifications were some of the main lines of the Chicago, Milwaukee & St Paul Railroad. A total of 647 miles was electrified by 1918 and is operating under the severe weather conditions of the Rocky and Bitter Root Mountains. D.C. is used at 3,000 v. The equipment for the original electrification was manufactured by the General Electric Co., of America, who supplied forty-two locomotives for freight and passenger service and four shunting locomotives. Of this original equipment, the freight and passenger locomotives were practically the same and differed from each other only in the gear ratio between motors and driving axles (Plate 37c, page 165). Further locomotives of a different design were developed for the Cascade electrification. The locomotives were of the bi-polar gearless type, with motor armatures mounted directly on the driving axles. They followed the principle of the gearless locomotives used on the New York Central Railroad, which proved very satisfactory.

The Chicago, Milwaukee & St Paul locomotives weighed 265 tons with 229 tons on drivers. They had fourteen axles, twelve of which were driving and two running axles. The weight of the armature and wheels was the only dead weight on the track and this was approximately 9,500 lb. per axle. The total weight on drivers (458,000 lb.) was 86 per cent of the weight of the locomotive and, distributed among twelve axles, resulted in a weight of 38,166 lb. per axle.

The locomotive was designed for handling trains up to 1,000 tons trailing load on gradients of 1 in 50 at 25 m.p.h. This performance required 56,500 lb. tractive effort which is equivalent to a coefficient of adhesion of 12·3 per cent of the weight upon the driving axles. For continuous operation, the locomotive was designed to operate at 42,000 lb. tractive effort at a speed of 27 m.p.h.

The motor was bi-polar, two fields being supported upon the bogie springs with full freedom for vertical play of the armature between the pole faces. For maximum speed operation, the twelve motors were connected three in series with 1,000 v. per commutator. Control connections were also provided for operating four, six or twelve motors in series. Additional speed variation was obtained by tapping the motor fields in all combinations. Cooling air for each pair of motors was supplied by a small motor-driven blower.

The 3,000 v. contactors and grid resistors were mounted in the streamlined cab at each end of the locomotive. In one of these cabs there was also located the 3,000 v. D.C. air compressor and storage battery. In the other was located a small motor-generator set and the high-speed circuit breaker. The driving cabs contained the master controller, indicating instruments, and a small air compressor, operated from the battery circuit with sufficient capacity for raising the pantograph when first putting the locomotive in operation. Near the controller were the usual air-brake handles for standard braking equipment.

Another set of locomotives was supplied by the firms of Baldwin-Westinghouse with wheel arrangement 2–C–1+1–C–2. The H.P. ratings are 4,680 for one hour and 3,396 continuously, corresponding to tractive forces of 72,600 and 49,000 lb. respectively, at a speed of approximately 23·3 m.p.h. for one hour and 26 m.p.h. continuously. With 33 per cent adhesion the starting tractive force is 126,000 lb.

The two main frames of the 2–C–1 parts are connected by a central link, which transmits the traction and buffing stresses. Each of the two main driving wheelbases has a three-point equalization system with its central point of support at the fulcrum of an equalizing lever which connects the driving springs with the four-wheeled truck at the end of the locomotive.

Main particulars of the G.E.C. locomotives are:

Length inside knuckles	76 ft 0 in.
Total wheelbase	67 ft 0 in.
Rigid wheelbase	13 ft 11 in.
Diameter driving wheels	44 in.
Diameter running wheels	36 in.
Weight electrical equipment	235,000 lb.
Weight mechanical equipment	295,000 lb.
Weight complete locomotive	530,000 lb.
Weight on drivers	458,000 lb.
Weight on guiding axle	36,000 lb.
Weight on each driving axle	38,166 lb.
Number of motors	12
One-hour rating	3,240 H.P.
Continuous rating	2,760 H.P.
Tractive effort – one-hour rating	46,000 lb.
Tractive effort – continuous rating	42,000 lb.
Tractive effort – 2 per cent gradient with 960-ton train	56,500 lb.
Coefficient of adhesion ruling grade	12·3 per cent
Starting tractive effort – 24 per cent coefficient of adhesion	115,000 lb.

1902–22. RAILWAY ELECTRIFICATION AND ELECTRIC LOCOMOTIVES IN EUROPE

The Austrian Alpine Railway. One of the early single-phase electrifications was that of the so-called Austrian Alpine Railway from St Poelten to Maria-Zell. The line has very heavy freight and passenger traffic, is 91·3 km. (56·6 miles) long, and rises to 3,000 ft above sea level; it has severe curves, and gradients up to 1 in 40. The country is very mountainous and there are no fewer than 155 bridges or viaducts and seventeen tunnels, one tunnel being 892·5 m. (976 yd) long. The line is built with the unusual rail gauge of 760 mm. (2 ft 6 in.); it was completed in 1907 and first worked with steam; this proved unsatisfactory because, owing to the many tunnels and curves, double-heading was undesirable. Electrification appeared to be a suitable solution. Plenty of water power exists in the surrounding country and it was decided to build two special power stations to supply power to the railway and also power and lighting to neighbouring towns. The current is transmitted to the locomotives by overhead catenary line with a line pressure of 6,500 v. single-phase A.C. of 25 cycles.

Eighteen locomotives were ordered between 1909 and 1912. They weighed 48 tonnes (47 tons) and had the wheel arrangement C–C; they were bogie locomotives, and were specially designed for negotiating sharp curves. Each bogie had three coupled axles.

The locomotive would pull a 100-ton train at a speed of 25 m.p.h. on a 2·5 per cent (1 in 40) gradient. The drawbar pull on this gradient is about 5 tons. The maximum tractive effort at the rim of the drivers was 10 tons. The motor was placed at one end of the bogie, transmitting its power through gearing on to an intermediate shaft which was coupled by means of a crank to the wheels. Each locomotive had two self-ventilated motors with an output of 300 H.P. at a speed of 700 r.p.m.; motor voltage was 220 and the frequency 25 cycles per second.

The Lancashire & Yorkshire Railway. In 1904 the then existing Lancashire & Yorkshire Railway decided to electrify the important suburban lines from Liverpool to Southport and two years later the line from Liverpool to Aintree. The Liverpool–Southport line is 18½ miles long, with fourteen intermediate stations and the whole work was entrusted to one of the forerunners of the English Electric Co., Dick Kerr & Co. Ltd; the rolling stock was made at the railway company's works at Horwich and at Newton Heath. Train services, as compared with steam, were doubled, and the running times reduced to forty minutes (Liverpool–Southport).

The railway stock was made up of standard four-coach trains (two first- and two third-class), the latter being at the outer ends and equipped with two motor bogies, each axle being motored with one nose-suspended 150 H.P. motor; in all there were, therefore, eight motors per train. D.C. at 630 v. reached the motors from the third rail, through a cast-steel shoe attached to a beam on each side of the motor bogie. The vehicles were 60 ft long and 10 ft wide, and had 8 ft wheelbase bogies, distance between bogie centres being 40 ft 6 in. The motor cars were divided into two main compartments, with a luggage and driving compartment. The capacity of the coaches was 270 passengers for the four-coach train.

New stock was bought in 1926, when the Metropolitan-Vickers Electrical Co. supplied forty-eight motors to equip eleven motor coaches, with some additional spares. Each of these motors developed 265 H.P. at 530 v. It is of interest to note that at about the same time a further eighteen motors installed in the London suburban stock of the L.M.S. were designed to be suitable for alternative operation in the Liverpool and Southport stock. Each motor developed 280 H.P. at 580 v. (full field) or 265 H.P. at 530 v., and was fitted with two field tappings, the weaker being employed only when the motors were used on the Liverpool and Southport route.

The Loetschberg Railway. The opening of the Loetschberg line resulted in providing one of the most important railway links in Europe, connecting Paris via Berne to Italy. Considerable difficulties were encountered in building the line, especially in boring the 14·6 Km. (9 miles) long tunnel. The line starts in Frutigen and ends in Brigue, where it connects with the Simplon line, the total distance being 75 Km.

From the start in 1913 single phase A.C. electric working was proposed, the heavy gradients and long tunnels allowing of no other traction method. The maximum gradient is 1 in 38. Already by 1908 the level section from Spiez to Frutigen was electrified and electric traction was tested while the tunnels and the rest of the line were being built.

The railway administration ordered two 2,000 H.P. locomotives of wheel arrangement C–C and also three motor coaches.

The locomotives were ordered from Messrs Oerlikon and had two 1,000 H.P. motors. They proved so satisfactory that the consulting engineers to the Loetschberg Railway ordered twelve more, which had still greater capacity. The design, however, did not correspond entirely with that of the first engine. Some of these locomotives were built in the works of Brown, Boveri & Co., to the designs of the Oerlikon Co. The new locomotives were of type 1–E–1, and had a capacity of 2,500 H.P., for speeds up to 50 Km/h. (31 m.p.h.), and developed a

tractive force of 18,000 Kg. (39,600 lb.). The maximum speed was 75 Km/h. (46·6 m.p.h.). (Plate 37A, page 165.)

The two 1,000 H.P. motors were mounted on two independent six-wheel bogies, all wheels of which were coupled. The total weight of the engine was 90 tonnes (88¼ tons).

After many tests, it was shown that the locomotive could develop 2,000 H.P. continuously at 42 Km/h. (26 m.p.h.), with a drawbar pull of 12,700 Kg. (27,940 lb.).

The locomotives have two transformers, one for each motor. In order to control the voltage of these transformers, therefore, two controllers combined together are now employed, and the contact fingers of these are connected to the tappings of the transformers, so that the secondary voltages can be changed. The controller is of the revolving drum type. With these two controllers it is possible to connect both motors in series or parallel, or the transformers can be dealt with in a similar manner. It is also possible to control the two motors by series and parallel connections from a single transformer and controller.

The engines were required to haul a load of 310 tonnes (304 tons) up a gradient of 1 in 38 at a speed of 42 Km/h. (26 m.p.h.) and to haul 500 tonnes (492 tons) up a gradient of 1 in 64 at the same speed.

The design of these locomotives marked an important step in the development of electric railway motive power. Their dimensions and main characteristics are:

Wheel arrangement	1–E–1
Length over buffers	50 ft
Wheelbase	38 ft 6 in.
Diameter of driving wheels	4 ft 4½ in.
Gear ratio	1 : 3·25
Maximum speed	70 Km/h. (43·5 m.p.h.)
Maximum T.E.	28,650 lb.
Weight of mechanical part	46·7 tonnes (46 tons)
Weight of electrical part	44·7 tonnes (44 tons)
Weight of total	91·4 tonnes (90 tons)

The London, Brighton & South Coast Railway. The first of the railway companies serving South London which electrified some of its lines was the London, Brighton & South Coast Railway. The first stretch to be converted, mainly to compete with the London tramways system, was from Peckham Rye to Battersea Park, and later extended to London Bridge and Victoria. A well-known engineer, Philip Dawson (later Sir Philip Dawson), took charge, and the whole scheme was opened in 1909.

High-tension, single-phase A.C. of 6,600 v. and overhead supply lines were chosen. The overhead conductor line was hung from compound catenary spans carried by portal structures spanning the rails. Double and single brackets with central mast were also used. Some of these rather clumsy structures can still be seen as signal bridges, for example at Clapham Common (Plate 36A, page 164).

The South London trains consisted of open compartment type coaches with side gangways, but there were no connections through the train. A unit comprised one motor coach, two trailers and a control trailer. There were four 150 H.P. motors per motor coach, each being supplied at 600 v. through the motor coach from the 6,600 v. line. Current collection was by twin overhead bow collectors. The results were satisfactory and in 1911 the electrification scheme was extended from Battersea Park Junction via Clapham Junction, Balham, Streatham Hill and Crystal Palace to Selhurst (where substantial repair and storage sheds and works were built), and later extended to Coulsdon.

The Metropolitan Railway. This railway acquired two lots of ten engines each. The first set was supplied in 1904 by Metropolitan-Cammell and Westinghouse. The second delivery was in 1906 by the same makers with B.T.H. electrical equipment. The first one was of the "electric iron" type with central cab, the second had a box superstructure. The total of twenty locomotives were rebuilt in 1922 by Metropolitan-Vickers, when they received their present appearance (Plate 37B, page 165).

There are two 2-axle bogies, each axle carrying a 300 H.P. motor. Wheel diameter is 43·5 in. and the total wheelbase is 29 ft 6 in. The total length is 39 ft 6 in. Control is of the electro-magnetic contactor type; and current is collected by four shoes. In addition there is a power rail to connect shoes distributed through the train to span wide gaps in the conductor rail. Two or more locomotives can be coupled together for multiple working, and driven from the leading cab.

Acceleration is automatic and the controller is provided with slow-speed notches for shunting and marshalling work. Like all Metropolitan Line stock, the locomotive is fitted with devices which cut off current and apply brakes should the driver overrun a stop signal. These trip cocks are worked from lineside contacts, automatically positioned in accordance with the state of the signals. The weight of the average train hauled is about 180 tons, and speeds up to 65 m.p.h. can be reached. The locomotives weigh 56 tons in working order. Until the oubreak of World War II, these were the only main-line electric locomotives in the British Isles which were in full use.

The Midi Railway locomotive trials. In 1902-8 the Midi Railway of France electrified the branch line from Villefranche-de Conflet to Bourg-Madame (Pyrénées Orientales), 55 Km. (34 miles) long with metre gauge, and later, in 1908, the main line from Perpignan to Villefranche on its section Ille–Villefranche. The electrification

was carried out as an experiment to see whether the system proposed was suitable for use on the Transpyrenean lines which were soon to be opened. Current used was 12,000 v. single-phase A.C. of $16\frac{2}{3}$ cycles. The locomotives were of special interest as the same wheel arrangement (1–C–1) was ordered in six different units from the leading manufacturers in France and abroad. These trials became, in fact, something of an Electric Rainhill! (Plate 36B, page 164.)

The locomotives were required to haul:

308 tonnes (303 tons) up a gradient of 22 in 1,000 at 40 Km/h. (24 m.p.h.);

440 tonnes ($431\frac{1}{2}$ tons) up a gradient of 17 in 1,000 at 60 Km/h. (36 m.p.h.).

The adhesion weight was limited to 54 tonnes (53 tons) with a total weight of about 80 tonnes ($78\frac{1}{2}$ tons).

Six 1–C–1 type locomotives were entered for the trials.

(i) No. 3001 built by General Electric–Thomson Houston. Two single motors, centrally placed, drove the wheels through jack shafts and side rods.

(ii) No. 3101 built by A.E.G.-Berlin, had identical rod drive, but the motors were placed at the outer ends of the engine room.

(iii) No. 3201 built by Westinghouse, had series-compensated motors, with gear wheels and jack shafts, but the drive was of the triangular rod type so familiar on the Italian 3-phase lines at the present day.

(iv) No. 3301, built by Brown, Boveri, also had a triangular rod drive, but had Deri induction propulsion motors.

(v) No. 3401, built by Messrs Jeumont, had individual axle drive.

(vi) No. 3501, built by Schneider, had ordinary rod drive similar in all respects to that of Nos. 3001 and 3101.

The most interesting locomotive was undoubtedly the fifth, No. 3401, submitted by Messrs Jeumont; it was the only one to have individual axle drive. It had three 500 H.P. series-compensated motors, each driving an uncoupled axle.

The motors could be used as repulsion induction motors for braking runs, being permanently coupled in a series of three. Each drive was different – three different joints being used to enable comparisons to be made.

The trials proved No. 3101 to be the most powerful, the Jeumont engine No. 3401 to be the most versatile, and No. 3001 the easiest to maintain.

The Jeumont engine had good commutation and proved that single-phase current regeneration was possible – up till then a very debatable point.

The outcome was an order for eight locomotives of 95 tonnes (93 tons) weight and with wheel arrangement 2–C–2 from Westinghouse and Jeumont, which type became ultimately the well-known 2–Co–2 locomotive with three vertical motors.

Finally, some of the main particulars of the test locomotives:

Number		3001	3201	3401	3101	3301
Wheel arrangement				All 1–C–1		
Maker		TH.H.-G.E.C.	West.	Jeumont	A.E.G.-West.	C.E.M.-B.B.C.
Total weight	tonnes	88	81	85·3	85	84
	tons	86¼	79¼	83¾	83¾	82¼
Length		13·75 m.	11·37 m.	14·27 m.	13·14 m.	13·14 m.
		45 ft 1¼ in.	37 ft 4 in.	46 ft 9½ in.	43 ft 1¼ in.	43 ft 1¼ in.
Wheelbase		9·6 m.	8·8 m.	11·6 m.	—	9·2 m.
		31 ft 5½ in.	36 ft 1 in.	38 ft 1 in.	—	30 ft 1¾ in.
Diameter of driving wheels		1,310 mm.	1,200 mm.	1,400 mm.	1,310 mm.	1,600 mm.
		4 ft 3½ in.	3 ft 11¼ in.	4 ft 7 in.	4 ft 3½ in.	5 ft 3 in.
Number of motors		2	2	3	2	3
H.P. per motor (continuous)		600	600	500	800	750

The Silesian Mountain Railways. The line from Breslau to Hirschberg, Lauban, Koenigszelt and to Goerlitz on the main line from Berlin to Breslau was electrified with 15,000 v., 16⅔ cycles, single-phase A.C. Work started before World War I but was only completed in 1922. Not only was the line a difficult one, there was heavy suburban and summer traffic as well as very heavy freight traffic. Average distance between stations was 5 Km. (3 miles), there were gradients rising at 1 in 50, and curves of 180 m. (196 yards) radius were frequent. The total length ultimately electrified was 274 Km. (164½ miles). The electrification programme, started with great hopes, remained isolated and, during World War II, suffered severe damage; the whole installation was taken to Russia and is now used as a freight railway in Siberia.

The electrification is, however, interesting and of great importance to the student of electric locomotives, by reason of its numerous electric locomotives, some of which were of very unusual design. Among the locomotives were the following:

27	B–B	11	1–C–1
13	B–B–B	23	2–B–B–1
10	C–C	12	2–D–1
30	C–C	1	1–B–B–1

They ranged from the 2–D–1 weighing 108 tonnes (105¾ tons) with a single 3,000 H.P. motor (weight 22 tonnes (21½ tons), and 3·5 m. (11 ft 6 in.) housing diameter) to a number of goods locomotives with the unusual wheel arrangement B–B–B (recently revived in

New Zealand and Italy), with an output of only 1,200 H.P. for 103 tonnes (101 tons) weight and a maximum speed of only 50 Km/h. (31 m.p.h.).

The 2–D–1 locomotives were built between 1918–23, by Bergmann of Breslau, and were intended to be express mountain locomotives and to fulfil the following requirements:

To haul 400 tonnes (394 tons) trains on gradients of 1 in 50 (2 per cent) at a speed of 50 Km/h. (31 m.p.h.) and on the level 500 tonnes (492 tons) trains at 90 Km/h. (54 m.p.h.). This required about 3,000 H.P. motor output; at that time a difficult problem. The solution was a single motor with rod drive over double jack shafts and driving rods. The wheel diameter was 1,250 mm. (4 ft 1 in.) giving, at 90 Km/h. (56 m.p.h.), 380 r.p.m., considered high for such heavy motion components. A heavy plate frame with bolted-on motor supports and other parts were clearly developed from steam locomotive practice. These locomotives, although they saw many years of service, were very costly to maintain, owing to breakages of motion components. They were the guinea-pigs whereon were learned the lessons of giving up rod drive in favour of motor power divided between individually driven axles.

Another type was the B–B–B already mentioned (Plate 36D, page 164), built by Siemens in thirteen units between 1915 and 1921. They had three motors, each of which drove the two coupled axles of each unit through gear wheels, jack shafts and fly-cranks with outside frames. Each unit had its own transformer. Again maintenance costs were enormous.

Further particulars of the Silesian locomotives are given in the following table:

Lawrence S. Williams, Upper Darby, Pennsylvania

PLATE IV. Pennsylvania–Reading Railroad: Budd R.D.C.–I railcar in operation on the Pennsylvania–Reading Seashore lines.

LOCOMOTIVES FOR THE SILESIAN RAILWAYS

No.	Wheel Arrangement	Type	Year built	Number built	Tare weight tonnes	Length mm. (ft in.)	D.W. diam. mm. (ft in.)	No. of motors	Output per 1 hour H.P.	Max. speed Km/h. (m.p.h.)
1	2–D–1	EP.235/46	1918–23	12	104/108	14,400/14,800 (47' 2"/48' 6")	1,250 (4' 1")	1	3,000	90 (56)
2	2–B–B–1	EP.209/12 and EP.233/44	1918–23	23	115 and 101	16,500 (54' 0") 13,160 (43' 1")	1,700 (5' 7") 1,250 (4' 1")	2 1	1,400 3,000	90 (56)
3	1–B–B–1	EG.509/10	1910	1	94·9	15,750 (51' 7")	1,270 (4' 2")	2	1,600	75 (46·5)
4	B–B–B	EG.538/50	1915–21	13	103	17,210 (56' 5")	1,300 (4' 3")	3	1,200	50 (31)
5	C–C	EG.551/52	1920–21	10	96	15,934 (52' 2")	1,250 (4' 1") k	4	1,400	50 (31)
6	C–C	EG.	1924–25	30	110·4	16,500 (54' 0")	1,200 (3' 11")	4	2,550	55 (34)

The Simplon Railway. Between 1902 and 1920, a number of railway lines across the Alps, all of international importance, were electrified, among them the Simplon line. Following the success of other minor electrifications such as the Berlin–Zossen experiments and the electrification of the Valtellina line in Italy, the 3-phase A.C. system was decided upon. In 1905, the Simplon Tunnel, 23 Km. (14·29 miles) long, presented a severe bottleneck; from the start electric services were considered, as steam services would have presented almost insurmountable ventilating problems. The first locomotives, with 1–C–1 wheel arrangement, were supplied by Brown, Boveri, voltage used being 3,000; the current was 3-phase A.C. of 15 cycles frequency. The tractive effort of these locomotives (6,000 Kg. (13,200 lb.) at 35 Km/h. (21·7 m.p.h.)) was quite remarkable, although there were only two effective running speeds according to the number of poles used (8 or 16). In 1908, two further locomotives with wheel arrangement –D– followed. Here, the two inner axles were driven from the motors by a quadrangular driving rod and the outer axles connected by coupling rods with side-play. These locomotives form the basis of most of the 3-phase locomotives designed later. The locomotives had driving wheels 1,250 mm. (4 ft 1 in.) diameter, the total length being 11·65 m. (38 ft 2¼ in.). Speeds were 26 and 52 Km/h. (16·1 and 32·3 m.p.h.), but by using 8 and 16 poles instead of only 6 and 12, speeds of 35 and 70 Km/h. (21·7 and 43·5 m.p.h.) could be obtained. The tractive effort was 11·5 tonnes (11¼ tons) and the output 1,700 H.P. The final design was a 1–D–1 locomotive which followed the above-mentioned principles (Plate 35B, page 163).

In 1930, the Simplon line was converted to normal Swiss single-phase A.C. Three-phase traction was taken up by Italy, which electrified about 400 Km. (248 miles) of lines within a few years, such as the Valtellina line, the Giovi line and the line from Torino to Modane. The name of a young Hungarian engineer, Kando, and an Italian, Bianchi, are outstanding among the leading engineers.

1919–39. WORLD DEVELOPMENT OF ELECTRIC LOCOMOTIVES AND MOTOR COACHES

Gotthard Line locomotives. From the inception of the Gotthard Railway in 1882, every effort was made to make the Basle–Gotthard–Chiasso route a major European rail link across the Alps. The electrification of the mountain section in 1921 contributed enormously to the efficiency of the line. The first electric Gotthard express locomotive was of series Be 4/6, which was built for a one-hour rating of 2,040 H.P. at 52 Km/h. (32·3 m.p.h.) and a maximum speed of 75 Km/h. (46·6 m.p.h.). This 106·5-ton locomotive was capable of hauling a trailing load of 230 tonnes (226¼ tons) up the steepest gradient, 1 in 38·5, at a speed of 60 Km/h (37·5 m.p.h.); this represented a considerable advance on the performance of steam locomotives. Later, locomotives of the series Ce 6/8 (Plate 38B, page 166) were brought into service for handling freight traffic; with a one-hour rating of 2,460 H.P. and weighing 131 tonnes (129 tons). These locomotives were capable of hauling a trailing load of 520 tonnes (511¾ tons) up the maximum gradient at 35 Km/h. (21·7 m.p.h.). Ultimately, however, neither type could compete with the increasing loads and speeds. The consequent need to use pilot locomotives on the Gotthard and Monte Ceneri sections proved very uneconomical and wasted much time.

To avoid double heading, two twin locomotives, series Ae 8/14 (Plates 47B and 55B, pages 193 and 219) were put into service in 1932 and a third locomotive of similar design in 1939. On the main gradient these locomotives, weighing 246 tonnes (242 tons), hauled a trailing load of 770 tonnes (757¾ tons) at 65 Km/h. (40·4 m.p.h.). The available output of these twin locomotives could only be utilized when they were employed for hauling freight trains. In service with express passenger trains they proved uneconomical because the high output was needed only for comparatively short distances on the main gradients. On the easier sections of the route the locomotives were unnecessarily heavy for the trailing-load hauled.

It was therefore decided to split the double locomotives into two independent units and in 1945 locomotives of the series Ae 4/6 appeared which could be employed individually or coupled in pairs, with multiple unit control. These locomotives proved to be well suited for the existing requirements, since, in 1949, not more than approximately 3·6 per cent of the Gotthard express trains weighed more than 600 tonnes (590½ tons) while the average trailing load of such trains was approximately 380 tonnes (374 tons) (Plate 46A, page 192).

Great Indian Peninsular Railway – British-built locomotives. In 1925, Messrs Metropolitan-Vickers Electrical Co. supplied a number of locomotives to the then existing Great Indian Peninsular Railway. They are typical of the design of the time. The freight locomotives supplied are of C+C type (Plate 38C, page 166), weighing 120 tons with an output of 2,600 H.P. There are two independent bogies with the body supported between them. Each bogie has two 650 H.P. motors which drive through twin helical gears on to a common jack shaft, the wheels being coupled in sets of three. The body rests on the main frame, the latter connecting the two driving bogies to which it is itself connected by spherical pivots. It transmits all tractive forces. The inner ends of the bogies are also coupled. There are two driving cabs. The lower hoods at both ends of the locomotive contain control apparatus. Current is collected from a 1,500 v. overhead catenary by two double pantographs and, as the motors are each wound for full-line voltage, double-series parallel control is employed. Electro-pneumatic equipment is installed and is arranged to provide nine running positions, three in each combination, together with regenerative braking, this latter being effective at speeds from 8 to 35 m.p.h. Excitation for the traction motors during braking is provided by a small axle-driven generator.

Three different types of passenger locomotive were ordered from different makers; all with wheel arrangement 2–Co–1 and 2–Co–2. They served to investigate carefully the relative advantages of the various types of power transmission, etc. Typical of these is the one supplied by Metropolitan-Vickers Electrical Co. Ltd. The locomotive has six 360 H.P., 750 v. driving motors, mounted in pairs above the three driving axles, each pair driving through an intermediate gear to a hollow gear wheel surrounding the axle, but carried in journals mounted on the locomotive frame. Each gear wheel flexibly drives its particular axle by a floating connection which allows the axle universal movement relative to the gear wheel. The form of drive employed not only keeps the main motors above the flood levels prevalent around Bombay during the monsoon period, but also gives a relatively high centre of gravity which was then considered essential for good riding qualities at high speeds.

The body of the locomotive has a driver's cab at each end, the cabs being connected by a central gangway. Adjoining one of the driving cabs is a compartment containing auxiliary machinery such as vacuum pumps, air reservoirs, brake apparatus, and blowers for main motors. A centre compartment contains the various cam groups, etc., a compartment at the other end holding the resistances and unit switches. The apparatus is mounted on frames placed on either side of the central gangway. All live parts are protected by interlocking doors to prevent access while current is on. The two pantographs, which may be operated from either cab, are mounted on a shield plate insulated from the pantographs and also from the main roof.

The control apparatus is similar to that on the freight locomotives with the exception that regenerative braking is not provided for, as a freight locomotive is also coupled to every passenger train when ascending or descending the Ghats. The motor combinations are arranged to give one-third speed with all six motors in series, two-thirds speed with two circuits of three motors in series, and full speed with three circuits of two motors in series, all with full field. In addition, a field tapping can be used with any of the three combinations, giving a total of six running speeds without resistance in circuit.

Main particulars are as follows:

Voltage		1,500 v. D.C.
Motors		6 × 360 H.P.
Weight:		
	mechanical part	62 tons
	electrical part	38 tons
	total	100 tons
Adhesive weight		60 tons

Gauge	5 ft 6 in.
Length over buffers	53 ft 6 in.
Total wheelbase	39 ft 0 in.
Rigid wheelbase	7 ft 6 in.
Driving wheel diameter	5 ft 3 in.
Hourly rating	21,500 lb. at 37 m.p.h.
Continuous rating	17,500 lb. at 39 m.p.h.
Maximum tractive effort	33,600 lb.
Maximum speed	85 m.p.h.

Italian Railways – standard 3,000 v. D.C. locomotives.
Electrification of the Italian Railways main lines goes back to 1912 when it was first decided to use 3-phase A.C., following the experiments on the Giovi and Valtellina railways. Italy indeed was the only country to expand and remain faithful to the 3-phase system; the experiments during 1925–34 with 3,000 v. D.C., however, decided the State Railways to use only this power for future electrifications. From the beginning, it was decided to reduce the number of locomotive types to an absolute minimum and to let the various manufacturing firms work to standardized designs developed by the Locomotive Development Bureau. Since then five types have appeared; Series E 424, 326, 428, 626 and, later, 636; in addition several types of motor-coaches have been developed.

Standardization of mechanical parts has been far advanced: there are four types of main axle only, and one type of four-wheel bogie; three types of axle-box are used.

Locomotives of types E 424, 626, 636 are driven through nose-suspended motors while types E 326 and E 428 have fully-sprung motors using the Bianchi or Negri Quill Drive. With this design the tractive force is transmitted from the hollow axle to the wheels by leaf springs. Suitable ball-ends on the wheel end of the springs allow movements of the hollow axle relative to the main axle. All locomotives have a central machinery compartment with cabs both ends; these are connected by a gangway. The central compartment contains all H.T. equipment and is suitably interlocked with the pantographs so that it can only be entered if the current is switched off.

Standardization of the electrical equipment has also been achieved to a high degree. One type of pantograph only is used. All resistance units are equal. One single motor type has been designed. Its one-hour rating is 350 kW at 700 r.p.m. and its continuous rating 315 kW. at 730 r.p.m. The weight is 3,500 Kg (7,700 lb.).

The first locomotives of the E 626 class were built in 1927, the last one in 1938. The first of the E 428 in 1934, the last in 1940. To complete the survey, the units of the series E 636 and E 424 were built in 1940 and 1943

respectively; it has been decided to build further units of the two latter types.

The Bo–Bo locomotive series E 424 is typical of modern Italian designs. The mechanical part follows the principles laid down in the general standardization scheme described. The drive is by Negri quill-drive via a single reduction gear with an idle intermediate gear wheel.

There are four single-armature, 4-pole type motors of a total rating of 1,410 kW. (continuous) or 1,590 kW. (one hour). The armature voltage is 1,500. The motors are ventilated from two blowers, each driven by a 3,000 v., 10 kW. motor. The contactor gear is of the electro-pneumatic type; the motors can be arranged in series or in parallel.

There are twelve running speeds (of which ten are with reduced fields obtained by tappings).

Other main particulars are:

Type of service: Passenger	gear ratio 16:65
Fast freight	gear ratio 21:65
Weight: Total	72 tonnes (70½ tons)
Mechanical part	44 tonnes (43 tons)
Electrical part	28 tonnes (27½ tons)
Overall length	15,500 mm. (50 ft 9 in.)
Wheelbase: Total	10,500 mm. (34 ft 5 in.)
Rigid	3,150 mm. (10 ft 8 in.)
Wheel diameter	1,250 mm. (4 ft 1 in.)
Maximum tractive effort	13,200 Kg. (29,040 lb.)
Maximum safe speed	140 Km/h. (87 m.p.h.)
Auxiliaries: Two 2-cylinder, 3,000 v., 10 kW., 35 cu.ft/min. compressors, regenerative braking by compound field excitation. Two 4·5 kW. generators charging 24 v. battery for lighting.	

Two further important designs of standardized electric locomotives, classes E 626 and E 636, were built between 1927 and 1943 in no less than 529 units. Both locomotives have the unusual wheel arrangement Bo–Bo–Bo.

The series E 626 was the first of the new standard D.C. locomotives to be developed and met the following requirements: the locomotive was to exert a maximum tractive effort of 18,000 Kg. (39,600 lb.) at 40 Km/h. (25 m.p.h.), or approximately 2,000 kW. per one hour; maximum axle-load was limited to 16 tonnes (15½ tons) and maximum speed to 90 Km/h. (56 m.p.h.). The locomotives run on three 2-axle bogies, which are standard also for E 424. The bogies are coupled together so as to allow for side movements only, eliminating vertical movements of one bogie against the other. They can run through curves of 90 m. (98·4 yards) radius, and run quite smoothly at speeds up to 110 Km/h. (68 m.p.h.). Each outer bogie has its own frame, the centre bogie carrying the main frame and the single body rigidly, weight being transmitted from the main frame to the (outer) bogie frames. The outer bogies move round central pivots with a minimum side-play of ± 100 mm. (3·9 in.). Strong return springs and hydraulic shock-absorbers allow quick return movements. Draw and buffing gear is carried on the bogie frames and such loads are not transmitted to the body. The bogies carry, on spring-loaded cross-members, the six nose-suspended motors, which can be removed easily through the floor of the body when necessary. The motors drive the wheels through single reduction gearing, two different gear ratios being provided for passenger and freight service. The electrical equipment is situated under the locomotive ends in front of the two cabs, and in a central machine compartment. The electrical and braking equipment is practically identical with that used for series E 424.

The main particulars of series E 626 are:

Total weight	93 tonnes (91 tons)
Total length	14,950 mm. (48 ft 11 in.)
Total wheelbase	11,500 mm. (37 ft 9 in.)
Rigid wheelbase	2,450 mm. (8 ft 0 in.)
Wheel diameter	1,250 mm. (4 ft 1 in.)
Traction motors	6, single-armature, 4-pole
Gear ratios	24:73 and 29:68
Tractive effort (Maximum continuous)	11,000 Kg. (24,200 lb.)

The series E 636 (Plate 41A, page 187) was supplied first in 1939, and while the principles of the design remained the same as in the case of E 626, the details were considerably altered, following experience with the experimental types. These have again the three four-wheel bogies, and the body consists of two coupled halves, each half resting on a central pivot of an outer bogie, the weight being transmitted to the centre bogie by special levers and springs, the whole forming a statically controlled unit.

The buffing and tractive gear is carried on the body. The series E 636 was originally designed for freight and slower passenger services. A later series of fifteen is, however, intended for speeds up to 120 Km/h. (74·5 m.p.h.).

Following are some of the main details of these locomotives:

Total weight	101 tonnes (99 tons)
Overall length	18,250 mm. (59 ft 9 in.)
Total wheelbase	13,900 mm. (45 ft 6 in.)
Wheel diameter	2,250 mm. (4 ft 1 in.)
Motors	6, single-armature, 4-pole
Motor voltage	1,500 v.
Maximum tractive effort	19,600 Kg. (43,120 lb.)
Maximum safe speed	140 Km/h. (87 m.p.h.)

North Eastern Railway express passenger locomotive (Plate 38A, page 166). In 1922, the then Chief Mechanical Engineer of the North Eastern Railway, Sir Vincent Raven, built a high-speed 1,800 H.P. electric locomotive, No. 13, with wheel arrangement 2–Co–2, intended for experimental work on the proposed main line electrification between Newcastle and York. The project never materialized and No. 13 only saw service on the Shildon–Newport line. The electrical equipment was supplied by Messrs Metropolitan-Vickers Electrical Co. Ltd, and the mechanical part was made at the railway company's works at Darlington. The locomotive had individual axle drive, with two motors driving each axle.

The six motors were of the twin-armature type, transmitting power through quill drives. Each motor developed 300 H.P. and the six together could exert a total tractive effort of 15,900 lb. at 43 m.p.h. at the one-hour rating, or 9,480 lb. at 51·5 m.p.h. continuously. The control system was of the electro-pneumatic type. The six motors could all be connected in series, or in two or three parallel groups, and were force ventilated. Twelve speeds were available for any particular tractive effort. Driving wheels were 6 ft 8 in. in diameter, of unusual design. From the boss radiated a small number of half-spokes, each of which divided into a fan of three before joining with the rim of the wheel.

The locomotive weighed 102 tons (17 tons axle load), and was 53 ft 6 in. long. The starting tractive effort was 16,000 lb. and it was required to start a train of 450 tons on a gradient of 1 in 78 and to haul the same train on level sections at 65 m.p.h. Maximum speed was 90 m.p.h. The locomotive fulfilled all these conditions and was thus far ahead of its time; it was never properly used and was finally stored, and scrapped.

Pennsylvania Railway electrification and its locomotives. The Pennsylvania Railroad, operating in the area between the Atlantic seaboard, the Great Lakes, and the Mississippi River, is the largest private railroad system in the world. Chief main-line mileage operated is about 11,000, and the mileage of all tracks is over 25,000 (at the end of 1934). The main line electrification project, first announced in 1928, involved an expenditure of more than $100,000,000 (at prices ruling at that time), 60 per cent of the total being for catenary, transmission line, and sub-station construction, and the remainder for rolling stock.

For more than fifty years, the Pennsylvania Railroad has been using electric traction on portions of its extensive system. Electric operation of the Long Island Railroad was begun as early as 1905. The New York Terminal Division, with the Pennsylvania Station, began electric operation in 1910. In 1915, a suburban service on the main line from Philadelphia to Paoli with multiple-unit trains, was inaugurated. Later, the line to Germantown and Chestnut Hill was electrified in 1918, the branch from Allen Lane to Whitemarsh in 1924, the main line to Wilmington, and the branch to West Chester in 1928, and the branch to Norristown in 1930.

Electrification of the main line between New York and Washington had been under consideration for a number of years. Through main-line electric services were started between New York and Philadelphia in January 1933, and extended to Wilmington the following month. The remaining work, including the electrification of the lines from Wilmington to Washington, was continued rapidly during 1934, and on 10 February, 1935, the north- and south-bound *Congressional Limited* trains, between New York and Washington, began electric operation between those two cities. By 15 March, all passenger trains between the two cities had been taken over, either by electric locomotives or multiple-unit motor-coach trains. The handling of freight was taken over in May 1935, and built up as rapidly as new electric rolling stock could be delivered. The entire electrification embraces 1,405 track miles and 364 route miles.

For handling passenger and freight equipment over the New York–Washington line, including the suburban service at Philadelphia and other points, the Pennsylvania Railroad Co. acquired 100 passenger locomotives, 64 freight locomotives, 28 switchers, and 388 motor passenger cars, with which are operated 43 trailer coaches. Two types of passenger locomotives, known as Types P-5A and GG-1, were built for long-distance passenger services. In addition to these two standardized designs, the passenger equipment includes five Type P-5 and P-5A locomotives, eight O-1 locomotives, and one experimental R-1 locomotive.

The Type P-5a streamlined locomotive is a modification of the original box-cab design, adapting these locomotives to high-speed operation and at the same time locating the driving compartment at the centre instead of at the ends of the locomotive. This type of locomotive weighs, complete in running order, 394,000 lb., with 229,000 lb. on the three driving axles. The chassis consists of a single main frame casting with oil, water, and air tanks cast integral. The driving wheels are carried under the central part of this main frame, with two-axle running bogies at each end. The motors, flexible drive, spring rigging, draught gear, brake rigging, etc., are assembled together with the chassis. The transformer, control apparatus, and other equipment are then assembled on the deck, and the cab, which is constructed of aluminium alloy, completes the assembly. With this design, the motor-man's compartment is located at the central part of the locomotive, together with indicating lights, master controller, brake

valves, and cab signals. From the centre or operating compartments, the cab slopes in streamline, forming hoods which house the remaining installations. An oil-fired train-heating boiler is located in the No. 1 end hood of the locomotive. A passageway through the central part of the hoods permits access to the equipment and to the end doors of the locomotive.

Three single-phase, 25-cycle, twin-armature motors mounted in the main frame provide a total of 3,750 H.P. (at continuous rating) for this locomotive. The three main axles are driven through quill and flexible cup drive by the 1,250 H.P. motors. The wheel centres are of cast steel and have eight spokes with pads so located as to receive the torque of the motors through spring cups mounted on contact arms carried on the gear centre on one end of the quill. The motor has a three-point support on the main locomotive frame and, in turn, supports the quill in bearings on the under-side of the motor frame. Roller bearings are used on both the driving and guiding axles. The weight of the locomotive is distributed to the wheels by an equalizing system, consisting of main springs and equalizers, providing a stable system in lateral as well as longitudinal planes. Protection in case of wheel slipping is provided by a slip relay. When the speed differential between drivers becomes greater than 8 m.p.h., the relay operates, lighting an indicating light in the engineer's compartment and sounding an alarm buzzer. If the warning is not heeded, and the differential becomes greater than 20 m.p.h., all motor switches are automatically opened.

Separate single-phase induction-motor blower units are provided for the traction motors, and for the main transformer.

Type GG-1 locomotives were chosen to handle the main-line passenger trains. The side doors and windows of the body are recessed, all corners are rounded and smoothed to minimize air resistance, and marker-light projections are carefully streamlined. The two front cab windows at each end of the locomotive are arranged to provide maximum visibility of the track from the driver's and assistant's positions.

The locomotive weighs a total of 460,000 lb. with 300,000 lb. on driving wheels. The overall length is 79 ft 6 in. The running gear consists of two main cast-steel trucks, each with three driving axles, and two-axle guiding trucks at each end. The six driving axles are each driven by twin motors developing a continuous capacity for the locomotive of 4,260 H.P., at speeds from 63 to 90 m.p.h.

The two bogies are coupled at the centre of the locomotive by an articulated coupling. The cab is supported on centre plates located between the inside guiding axle and the outside main driver. One of the centre plates is built to slide longitudinally to provide for the change in distance between the centre pins, due to the action of the articulated joint on curves. Side bearings located between the two inside main driving axles on each truck provide additional support. Each locomotive carries an oil-fired train-heating boiler of the vertical tubular type having an evaporating capacity of 4,500 lb. of water per hour at 200 lb. pressure (Plate 39, page 167).

Other locomotive designs comprise the Types O-1, which preceded the Type P-5; these are similar in construction to the P-5A, except that they have two instead of three driving axles, with two instead of three twin motors. The locomotives have a weight of approximately 150,000 lb. on drivers and have a continuous rating of 2,500 H.P.

A total of fifty-nine of the box-cab Type P-5A locomotives which handled the initial passenger service on the main line are being used for freight services. Except for the box-type construction of the cab, and the different location of the apparatus, these locomotives are similar to the streamlined units already described. Two or more units can be operated in multiple. The freight equipment also includes two Type L-6 and one Type L-6A locomotives – the type originally planned for freight work. This type of locomotive consists of a 1–D–1 wheel arrangement, with each main axle driven by a single-armature 625 H.P., axle-mounted motor.

Also used are twenty-eight three-axle shunting (switching) locomotives, known as Class B-1, weighing approximately 75 tons each. These locomotives are designed for low-speed service, and each of the three axles is equipped with a single-armature motor. The hourly rating of the locomotive is 730 H.P., and the maximum operating speed is 25 m.p.h.

The twin-motor design used most generally for the locomotives consists of two armatures carried in a single fabricated frame with the driving pinions meshing in a single solid gear carried on a quill. Spiders, with six sets of spring cups each, provide a flexible driving connection between the gear and the wheel. The type GG-1 locomotive uses the flexible drive for both wheels on each axle. while the P-5 and O-1 types drive through the wheel on one side only. A three-point suspension is used for these motors, with provision for adjustment to maintain the quills central with the axle. A fabricated gear case, split horizontally in two parts through the gear centre, covers the gear and pinions.

Ventilating air is taken from the cab into the motors over the commutators, and exhausted through frame openings at the pinion ends, the stator and its armatures being ventilated by multiple paths through the motor.

South African Railways' mixed traffic locomotives. For use on the electrified section of the Natal Railway between Pietermaritzburg and Glencoe, Messrs Metropolitan-Vickers supplied ninety-five electric locomotives up to 1927. Rapidly increasing freight traffic, together with the exceptionally difficult traffic conditions, led to the adoption of the electric locomotive as the only practicable solution in place of steam traction which had reached the limit of its capacity.

The line is 175 miles long, the altitude varying from approximately 2,210 ft to 4,980 ft, including several long inclines, one of which has a gradient of 1 in 50 for fourteen miles. As both passenger and goods traffic are handled over numerous small curves on this narrow-gauge line, a short wheelbase was necessary, and it was decided to employ relatively short electric locomotives arranged for multiple-unit control (Plate 40A, page 168). Three locomotives coupled together controlled by one driver and his assistant, handle trains of 1,500 tons (the previous maximum with steam traction being 1,000 tons) while single electric locomotives can be employed for passenger trains.

The locomotives have the wheel arrangement Bo–Bo with four 300 H.P., 3,000 v., axle-hung motors. The high-tension compartment is in the centre between two groups of auxiliary machinery, all accessible from a side corridor interconnecting the driving cabs at each end. Regenerative braking is provided.

Southern Railway scheme – motor coaches and trailers. The Southern Railway – now the Southern Region of British Railways – operates the largest surburban electrification in the world, there being 720 route-miles and 1,796 track-miles of railway using direct current at 660 v. on the third-rail system.

The former Southern Railway was created by the amalgamation in 1923 of the London & South-Western Railway, or Western Section, the South-Eastern & Chatham Railway, or Eastern Section, and the London, Brighton & South Coast Railway, or Central Section. Before the amalgamation, two of these three railways had undertaken electrification of a number of their suburban services round London.

Electrification of the Western Section, using 600 v. D.C., started in 1915, with a service from Waterloo to Wimbledon (via East Putney); it was extended to Claygate (via Earlsfield), Hampton Court, Shepperton, and the Hounslow loop in 1916, and to Guildford and Dorking North in 1925, totalling 91 route-miles, or 254 track-miles. The mileage includes the Waterloo and City line (opened independently in 1898) and lines also used by some Metropolitan District trains.

The Eastern Section electrification, also using 600 v.

D.C., was carried out in three stages, starting in 1925 with the section from Victoria and Holborn Viaduct to Orpington, and the Crystal Palace (High Level) branch, in 1926 the lines from Charing Cross to Orpington, Addiscombe, Hayes and Bromley North were electrified and, finally, the Dartford lines and the Cannon Street connecting line, making a total of 92 route-miles or more than 250 track-miles. The East London line from New Cross and New Cross Gate to Shoreditch was also electrified with 600 v. D.C. in 1913.

Competition due to the electrification of the London County Council Tramways caused the London, Brighton & South Coast Railway (now the Central Section of the Southern Region) to electrify some of its suburban lines on a single-phase A.C. system in 1909 (see Railway electrification in Europe, page 148). When the Southern Railway Group was formed, conversion of these lines to D.C. at 600 v. was decided upon, further lines being electrified at the same time to form a total of 85 route-miles, or 239 track-miles on this section. The D.C. electrification commenced in March 1928, and included the A.C. sections between Victoria and London Bridge, and from Victoria and London Bridge to Sutton and Coulsdon North (via West Norwood and via Thornton Heath), also the following steam lines: London Bridge to Crystal Palace (Low Level) and Norwood Junction (via Forest Hill), and the Tattenham Corner and Caterham branches; Tulse Hill to Streatham Common, and the Epsom Downs branch; Balham to Epsom (via Mitcham Junction), Streatham to Wimbledon, and various small connecting lines.

Southern electric services are operated by multiple unit trains consisting of motor coaches and trailers. Three-phase A.C. at a pressure of 11,000 v. and a periodicity of 25 cycles (replaced by 50-cycle supply since 1948) is converted in substations along the route to 660 v. D.C., in which form it is fed into the conductor rail (third rail) for collection and use by the trains. The current is returned through the running rails which are bonded.

All new electrification in the Southern Region will be at a pressure 750 v. and this voltage will ultimately be used throughout the whole Region. This decision was made largely to try to overcome the inherent disadvantages of the third-rail system in conditions of ice and snow.

The rolling stock consists of motor coaches and trailers which are made up into units composed of two, four, five and six vehicles. Trains of up to twelve vehicles are made up from these units. The stock may be broadly classified into suburban, express, and semi-fast.

Suburban stock. Much of the early suburban stock was converted from steam train coaches, a process which went on as late as 1937. The original units were composed of two motor coaches with a trailer between. These were made up, usually, into six-car trains, but a two-coach unit (one motor coach and one trailer) could be added to make an eight-car formation when required. The converted steam stock inherited from the L. & S.W. electrification is now all scrapped – many vehicles had seen up to sixty years service.

Some of the L.B.S.C. (overhead system) stock, converted to use third rail D.C. is still in service, but the new all-steel stock is replacing these old vehicles. Since 1941, all the suburban three-car units have been converted to four-coach units by the addition of another trailer coach, and all new construction has been of four- or two-car units. The three-car suburban unit is, therefore, now extinct.

A modern four-car suburban unit consists, typically, of a motor coach at each end and two trailers between. The seating accommodation is 386 and the motor coaches have a driving compartment at one end, with a guard's compartment, or brake, immediately behind. The rest of the coach is of the open vestibule, or saloon type and seats eighty-four passengers. Each motor coach has one motor bogie, under the driving compartment, and one trailing bogie. Two nose-suspended, self-ventilated traction motors each of 250 H.P., drive the two axles of the motor bogie through gears having a ratio of 17 : 65. The control gear is installed under the coach frames and is of the electro-pneumatic type. (Earlier stock had the control gear behind the driving compartment.)

Self-capping electro-pneumatic brakes are fitted in addition to Westinghouse air brakes. The trailer coaches usually consist of one saloon-type coach and one compartment coach in each unit.

Two-car suburban units follow much the same pattern having one motor coach and one trailer. The trailer in this case, however, has a driving compartment at one end.

None of the suburban stock has vestibule connections.

Main line express stock. The Southern main line electrification schemes were completed from London:
(1) to Brighton and Worthing in 1933;
(2) to Sevenoaks in 1935;
(3) to Eastborne and Hastings in 1935;
(4) to Portsmouth via Guildford in 1937;
(5) to Portsmouth via Horsham in 1938;
(6) to Gillingham and Maidstone in 1939 and extended;
(7) to Ramsgate and Dover via Faversham in 1959.

The electrification of all the main lines in Kent is due for completion in 1962.

These services called for rolling stock of a different type from that used on the suburban services. More space per passenger, adequate lavatory facilities, buffet restaurant and kitchen cars had to be provided with also a high percentage of first-class seats.

For express train services the trains have vestibule connections throughout, and in the case of the *Brighton Belle* all-Pullman trains were provided made up of five-car units. For other express trains on the Brighton services some six-car units were built. In both these cases the two motor coaches of each unit had the axles of both bogies motored. Four 225 H.P. nose-suspended motors were provided – 900 H.P. for each motor coach, 1,800 H.P. for each five- or six-car unit. The maximum speed of these units is 75 m.p.h.

Express trains on the Portsmouth line are made up of four-car units. The two motor coaches, however, have only one motor bogie each aggregating 1,000 H.P. for each unit. The latest examples of this type of stock built under British Railways regime, have axle-hung, self-ventilated traction motors of 250 H.P. (one hour rating). They have electro-pneumatic control mounted below the frames, whereas in the earlier motor coaches the control gear was housed behind the driving compartment. Braking and electrical equipment is largely standard with that of the latest suburban stock.

This latest stock is the present standard of British Railways and, with a slightly altered gear ratio, will form the express trains for the Kent services to Ramsgate and Dover.

Semi-fast and outer suburban stock. The requirements for this stock fall between those of the suburban stock and the express train stock. Such stock was originally developed by the Southern Railway to meet the varying requirements of each main-line extension. They are made up into two- and four-car units and vary considerably in detail and in equipment. The existing motor coaches are powered by two 275 H.P. nose-suspended motors. The latest stock is similar to the latest suburban stock already described and has an improved rate of acceleration over the older units. The accommodation is both of the compartment and the saloon type. There are adequate lavatory facilities in each coach, but the coaches are not vestibule-connected, and there are no buffet or restaurant cars.

Battery driven baggage cars. Most of the boat trains to Dover and Folkestone will, in future, be operated by multiple-unit express trains. In order to avoid impairing the performance of these trains by making them haul extra dead loads, ten baggage cars are being built each with a motor bogie. These cars are also being equipped with traction batteries so that they will be able to operate on the non-electrified lines at the quayside.

When not required for boat-train service, these vehicles may be used as locomotives for van trains, as they are capable of hauling a trailing load of 100 tons, and are fitted with vacuum brake exhausters.

Swedish iron ore railways and their locomotives. The first main line electrified in Sweden on the single-phase system was the so-called Riksgrans or Frontier Railway, 120 Km. (74½ miles) in length, between Lulea and Riksgransen in Lapland. In 1910, orders were placed with Asea and Siemens-Schuckert, these two firms not only making themselves responsible for the supply of all equipment, but also giving a guarantee as to the running costs over a number of years. The power station at Porjus was begun in 1910 and was completed by the end of 1914, as well as four transformer sub-stations. There are 240 Km. (149 miles) of overhead contact wire. The electric locomotives which comprised fourteen 1,800 H.P. freight engines of type 1–C+C–1 and 2,100 H.P. passenger engines of type B–B+B–B, were delivered in 1914 and 1915. The installation worked so well from the outset and the running costs were so substantially lower than had been guaranteed, that the two firms were soon released from their undertaking in this respect and the Swedish State Railway Board took over the line complete. In 1916–17 the State Railway Board decided to continue the electrification from Kiruna to Lulea and further powers were granted to electrify the Kiruna–Gellivare–Nattavara section. For this section a further four 1,800 H.P. 1–C+C–1 locomotives were necessary and the order for the electrical equipment of these was placed with Asea.

The Kiruna–Gellivare section was put into use at the beginning of 1920 and the Gellivare–Nattavara section during 1921.

In 1920, the Swedish Parliament granted powers for continuing the electrification over the section Nattavara–Boden–Lulea–Svarton. In addition, an order was placed for twelve electric locomotives, including ten 1,200 H.P. freight engines type –D– and two 2,400 H.P. passenger engines of type B–B+B–B. This completed the electrification of the Lapland iron ore lines which have a total length of about 450 Km (279½ miles). In 1923 the electrification of the section of the Norwegian State Railway between Riksgransen and Narvik was completed. Electric trains accordingly can now be run from the Gulf of Bothnia to the Atlantic Ocean.

The freight locomotives (wheel arrangement –D–) have the following main dimensions:

Length over buffer	11,250 mm. (36 ft 10 in.)
Rigid wheelbase	3,450 mm. (11 ft 4 in.)
Total wheelbase	6,350 mm. (21 ft 2 in.)
Diameter of driving wheels	1,350 mm. (4 ft 5 in.)
Number of motors	2
Tractive effort (continuous) at a speed of	
36 Km/h. (21½ m.p.h.)	6,299 Kg. (6·2 tons)

F.J.G. Haut

PLATE 33B. Experimental locomotive *The Judge* built in the USA by Edison and Field in 1883.

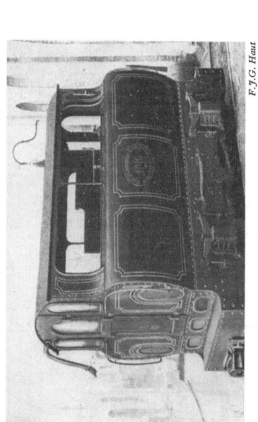

F.J.G. Haut

PLATE 33D. Baltimore and Ohio Railroad: Bo-Bo locomotive, 1,440 H.P. for use in Baltimore area, 1895.

F.J.G. Haut

PLATE 33A. Experimental locomotive *Ampere* built in the USA by Daft in 1883.

F.J.G. Haut

PLATE 33C. –Bo– locomotive for the City and South London Railway; built by Mather and Platt in 1890.

Siemens

PLATE 34B. Bo–Bo locomotive for the Zossen–Marienfelde tests, 10,000 v., 3-phase A.C.

Oerlikon

PLATE 34D. Bo–Bo locomotive for the Seebach–Wettingen Line, the first single-phase A.C. railway, 1901–5.

Brown, Boveri

PLATE 34A. –B– locomotive for the Burgdorf–Thun Railway, the first 3-phase A.C. railway in the world.

F.J.G. Haut

PLATE 34C. 1–C–1 locomotive for the Simplon Line, 3-phase A.C.

Secheron

PLATE 35A. Bo–Bo locomotive for the St George de Commiers–Le Mure Line; the first high tension (2,400 v.) D.C. railway in the world.

Brown, Boveri

PLATE 35B. 1–D–1 rod-driven locomotive for the Simplon Line.

F.J.G. Haut

PLATE 36B. I–C–I test locomotive for the Midi Railway of France, 1902.

F.J.G. Haut

PLATE 36D. Silesian Railways: B–B–B locomotive, 1915–21.

F.J.G. Haut

PLATE 36A. L.B. and S.C. Railway, 6,600 v. single-phase A.C. electrification. Electric train near Clapham Junction, 1909.

F.J.G. Haut

PLATE 36C. Hungarian 50-cycle A.C. test locomotive by Kando.

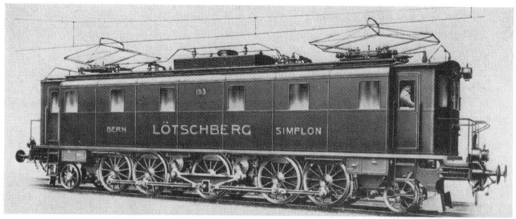

Oerlikon

PLATE 37A. B.L.S. Railway 1–E–1 rod-driven locomotive of 2,500 H.P., using 15,000 v., 16 2/3-cycle single-phase A.C.

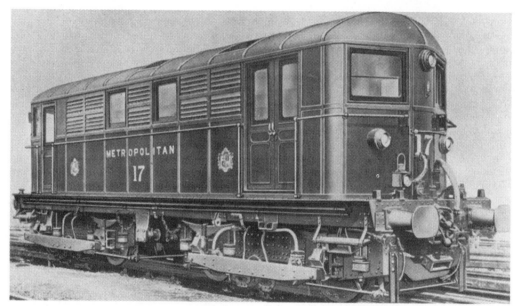

Metropolitan-Vickers Electrical Co. Ltd

PLATE 37B. Metropolitan Railway Bo–Bo locomotive.

C. M. & St P. RR

PLATE 37C. C.M. and St P. Railroad (The Milwaukee Road) 1–Bo–Do+Do–Bo–1 locomotive.

S.L.M. Winterthur

PLATE 38B. Swiss 1–C+C–1 freight locomotive with jack shaft and rod-drive, 1922–24.

Brown, Boveri

PLATE 38D. Sweden–Norway iron ore line 1–C+C–1 using 15,000 v., single-phase A.C.

Metropolitan-Vickers Electrical Co. Ltd

PLATE 38A. North Eastern Railway 2–Co–2 express passenger locomotive No. 13, built 1922.

P. Ransome-Wallis

PLATE 38C. Great Indian Peninsular Railway 1C+C locomotive with jack shaft and rod-drive, 1925. Photographed at Bombay.

Pennsylvania Railroad

PLATE 39. Pennsylvania Railroad 2–Co+Co–2, Class GG–1, 4,620 H.P., electric locomotive working "The Congressional Limited".

PLATE 40A. South African Railways: four 1,200 H.P. Bo–Bo locomotives, Class 1E on test train.

PLATE 40B. South African Railways: 2,000 H.P. Bo–Bo locomotives, Class 5E.

Tractive effort (one hour) at 30·5 Km/h.
(18¼ m.p.h.)   9,652 Kg. (9·5 tons)
Tractive effort, maximum   18,288 Kg. (18 tons)
Motor output (continuous)   880 H.P.
Motor output (one hour)   1,130 H.P.
Gear ratio   3:82
Maximum speed   60 Km/h. (37·5 m.p.h.)
Weight of electrical equipment 26·9 tonnes (26·5 tons)
Weight of mechanical parts   42·7 tonnes (42·1 tons)
Total weight   69·7 tonnes (68·6 tons)

The driving axles are driven by coupling rods from a layshaft carried in bearings fixed in the main frames. Power is transmitted from the driving motors to the layshaft through gearing mounted on each side of the motor. One of the design characteristics is a slotted connecting rod between the inner pairs of wheels. This is necessary as the diameter of the wheels could not be made sufficiently large to use a connecting rod of standard pattern. The layshaft centre accordingly had to be placed 110 mm. (4·3 in.) above the driving wheel centres, and a slotted connecting rod construction used.

Power transmission between the motor shaft and the gear pinion is by a quill drive system with six radially laminated springs.

From the overhead wire the current is collected by two pantographs and brought to the oil switch and from there to the primary winding of the transformer. The door of the oil switch cubicle is provided with a short-circuiting arrangement which earths the high tension system when the door is opened. The transformer is designed for a continuous output of 640 kV.A. and can be arranged with the following ratios: 14,000/120, 280, 480 and 720 v. It is of shell type, oil immersed and arranged for forced air cooling. The traction motors are force-ventilated, single-phase commutator motors totally enclosed and carried directly on the locomotive main frame. The cooling air is obtained from an electrically-driven double centrifugal ventilating fan mounted over the motors. The motors are permanently connected in series. The fields can be reversed by an electromagnetically operated reverser mounted over one of the motors. The locomotive is equipped with a driver's cab at each end.

The passenger locomotives (type B-B+B-B) were specially designed for the passenger traffic between Lulea and Riksgransen and are constructed as two close-coupled units, both halves being identical in their mechanical and electrical details. Each locomotive half is provided with two driving motors, driving through gearing on to a common layshaft from which the power is transmitted to the driving wheels by connecting rods. The arrangement is almost identical to that adopted

for the freight locomotives, with the difference that the centres of the driving wheels and the layshaft are at the same height above the rail level, an arrangement which has made it possible to use connecting rods of ordinary standard pattern in place of the slotted coupling rod used in the freight locomotive.

The principal dimensions are as follows:
Length over buffers   21,400 mm. (70 ft 3 in.)
Rigid wheelbase   3,450 mm. (11 ft 4 in.)
Total wheelbase   16,200 mm. (51 ft 0 in.)
Diameter of driving wheels 1,340 mm. (4 ft 4¾ in.)
Number of motors   4
Tractive effort:
  Continuous rating   4·75 tonnes (4·7 tons) at 78
                            Km/h. (45½ m.p.h.)
  One-hour rating   8·9 tonnes (8·8 tons) at 66
                            Km/h. (41 m.p.h.)
Maximum tractive effort   16·25 tonnes (16 tons)
Motor output (continuous rating)   1,760 H.P.
Motor output (one-hour rating)   2,260 H.P.
Gear ratio   1 : 1·76
Maximum speed   100 Km/h. (62 m.p.h.)
Weight of electrical part   53·4 tonnes (52·6 tons)
Weight of mechanical part 71·7 tonnes (70·6 tons)
Total weight   125·17 tonnes (123·2 tons)

Each locomotive half is arranged on the same principle as the type –D– freight locomotive except that the driving cab at one end is dispensed with.

The freight locomotives (Type 1–C+C–1) (Plate 38D, page 166) were required to fulfil the following requirements:

"Two locomotives shall be able to haul on the Kiruna–Riksgrans Section a train of 1,855 tons, up a gradient of 1 in 100 at a speed of not less than 30 Km/h (18·6 m.p.h.). The maximum speed allowed is 50 Km/h. (31 m.p.h.), and the total running time 3 hours and 20 minutes. One locomotive shall be able to haul on the Kiruna–Riksgrans Section a train of empties of 455 tons tare weight and one idle locomotive, and the speed on a gradient of 1 in 100 shall be 40 Km/h (24·8 m.p.h.). The total running time for the section is scheduled as 2 hours and 56 minutes. The locmotives shall be designed for a maximum speed of 60 Km/h. (37·2 m.p.h.). Two locomotives together shall be able to haul two ore trains to Riksgransen and return with two trains of empties to Kiruna each day. In similar train services the locomotives shall work six days in succession and each locomotive shall be able to cover a gross distance of 90,000 Km. (55,923 miles) per year. The energy consumption shall not exceed 22·6 watt-hours per ton-Km. for ore trains, and 23·9 wat-hours per ton-Km. for empty trains."

The locomotives are of the articulated type with wheel

arrangement 1–C+C–1. In each half locomotive is installed a motor, connected to the driving wheels by cranks, jack shaft, and connecting rods. Both the locomotive halves are identical, each being divided into a driver's cabin with the necessary control apparatus and a machinery compartment containing transformer, motor and switchgear.

A compressed air system in accordance with the New York Air Brake System is used. The locomotives are furnished with both automatic and direct-acting brakes. A 10 H.P. compressor is installed in each half locomotive for supplying the compressed air to the brake system and also for the whistle, sanding, and for the pneumatic control of electrical apparatus.

Main dimensions are as follows:

Total overall length over the buffers	18,600 mm.	(59 ft 1 in.)
Total weight of the locomotive	140·2 tonnes	(138 tons)
Total weight on the drivers	106·6 tonnes	(105 tons)
Weight of mechanical parts	79·2 tonnes	(78 tons)
Weight of electrical parts	60·9 tonnes	(60 tons)
Diameter of driving wheels	1,100 mm.	(3 ft 7 in.)

Part II. An evaluation of the principal electrical systems on railways, and locomotive types employed

After World War I, the Central European and Northern countries decided on single-phase A.C. electrification, with 15,000 v. line tension and 16 2/3 cycles frequency, while the United Kingdom, France and Holland decided to use 1,500 v. D.C., and Spain, Italy, Belgium and Russia, 3,000 v. D.C. The United States remained fairly evenly divided, and the British Empire followed Great Britain with 1,500 v. D.C. As early as 1920, in the Pringle report, 1,500 v. D.C. was recommended and this was confirmed later in the Weir Report. However, in the early nineteen-twenties, a Hungarian engineer, K.Kando, conceived the idea (following earlier American developments) of using single-phase alternating current of 50-cycle frequency on railways. The current supply, which could be taken from the national grid, would replace the established 16 2/3 cycle system which had been adopted following the earlier pioneering work in Switzerland and the U.S.A. First in Hungary and then in Germany, test lines were built, and by 1950, these experimental lines had been operating for between twenty and twenty-five years, and could be termed a complete success. The Hungarian system operates at 16,000 v. and the German at 20,000 v.

Single-phase 50-cycle current may be used in three different ways:

(1) the supply is converted on the locomotives to direct current: a rectifier, with or without grid control, is incorporated in the locomotive;

(2) in phase converter locomotives as in Hungary;

(3) direct use of 50-cycle A.C. current.

Fifty-cycle traction has now passed the experimental stage and appears to be a serious competitor for the three main electrification schemes (1,500 v. D.C., 3,000 v. D.C., and 162/3-cycle, single-phase A.C.).

The initial French 50-cycle, single-phase A.C. schemes have been so successful that the main lines of the Nord and Est Regions of the S.N.C.F. are now in an advanced stage of electrification using this system. All future electrification of British Railways is also to employ 50-cycle traction except in those areas of the Southern Region (e.g. in Kent) where the extension of the existing third-rail D.C. system is the only logical and economical method.

In any historical review of this particular system the ideas of Kando must come first. In the early 'twenties the most successful electric locomotives were the –E– three-phase A.C. locomotives of the Italian State Railways, and following this design, a ten-coupled locomotive was developed (Plate 36c, page 164), which consisted mainly of a plate frame carrying two traction motors with switch gear and a single pantograph, tractive effort being transmitted through triangular coupling rods to the wheels. The interesting part of the locomotive was the electrical equipment which consisted of two synchronous rotary phase-converters, taking 15,000 v. single-phase line current and transforming it into three-phase A.C. The phase-converters fed the traction motor by means of slip rings. The stator of the phase-converter was oil-cooled, the rotor water-cooled. Regulation of speed was by pole-changing of the driving motor, resulting in four effective speed positions, and was carried out by a cascade starter with a water resistance. The locomotive was tested extensively on a special line near Budapest for several years and resulted in the well-known electrification scheme from Budapest to the Austrian frontier for which the Metropolitan-Vickers Electrical Co. Ltd supplied part of the installation. The locomotives used are still at work.

Further developments followed in Germany where it was decided to electrify the Hoellenthal Railway in the

Black Forest. The wholly successful running of the four test locomotives there gave considerable encouragement to French engineers. France was committed, as was England, to electrification with 1,500 v. D.C., as suggested in the Pringle report on railway electrification of 1921 and in the recent report on British Railway Electrification of 1951.

The railway electrified in Germany has a total length of 47·5 miles and is mainly single track. It connects the Rhine Valley with the Black Forest and the beautifully wooded country has a very considerable tourist traffic. The line has heavy gradients of up to 1 in 18, where formerly rack locomotives had to be employed, and curves down to 787 ft radius. The actual electrification comprised 34·5 miles with another 35 miles of track to be electrified later. A single sub-station takes current from the 100,000 v. county grid system and transforms it into 20,000 v., single-phase A.C. Four types of locomotives were developed.

These locomotives, which were of such importance for the future development of 50-cycle traction, consisted of:

(i) No. 244/11 built by Siemens, which transformed the 20,000 v., single-phase, 50-cycle A.C. in the same manner as the normal low-frequency locomotive, except that it used eight 14-pole motors.

(i) No. 244/21, built by Brown, Boveri, transformed the single-phase A.C. by means of a mercury-arc rectifier to D.C.

(iii) The third type, built by A.E.G., and numbered 244/01, was also a single-phase A.C./D.C. rectifier locomotive.

(iv) No. 244/31, built by Krupp-Punga-Schoen, follows the Kando system whereby the live current is transformed to 3-phase A.C. in rotary converters.

At the end of World War II the Hoellenthal Railway came into the French zone of occupation of Germany and French railway engineers became closely acquainted with its working. It had by then been used over ten years and French engineers who took over the running were very impressed by its success. The S.N.C.F. decided in consequence to electrify a 48·5 miles long section on the Geneva–Chamonix line between Aix-les-Bains and La-Roche-sur-Foron on the single-phase, 50-cycle system with 20,000 v. line tension, which has since been changed to 25,000 v. Though the high-frequency, single-phase locomotives are more costly than the D.C. types, the fixed equipment such as sub-stations and overhead structure is very much cheaper. Only one sub-station – at Annecy – was provided. The line is single track, has several tunnels, and grades as steep as 1 in 50. At Aix-les-Bains there is a junction with

the 1,500 v. D.C. Culoz–Chambery–Modane line of the old P.L.M. Railway, where special arrangements were necessary including a neutral section between the D.C. and A.C. catenary systems. A test locomotive was built by Messrs Oerlikon and S.L.M. Winterthur and delivered in 1951. This locomotive, originally CC 6051 and since renumbered CC 20001 (Plate 42B, page 188), has the Co–Co wheel arrangement and is able to work over 1,500 v. D.C. lines, although at reduced power, as well as over the 50-cycle, high-voltage lines, a requirement made to test the operation of dual power units for possible future use.

The success of the La-Roche-sur-Foron – Aix-les-Bains line and its test locomotives decided the S.N.C.F. to electrify the line from Thionville to Valenciennes, via Hirson and Mezières, a distance of 303 Km. (188 miles). This is one of the most important and busy lines in Europe, with very heavy mineral traffic. Its success would have far-reaching consequences on future electrification programmes. Up to then, French electrified railways had all radiated from Paris and were mainly concerned with fast express services to the provincial centres of France, and with frequent suburban services for the Parisian population. The freight traffic on these lines was considered of secondary importance.

The Thionville–Valenciennes line has many important connections; it connects at Apach with the German Federal Railways, while at Metz and Lille it connects Great Britain with Switzerland, Austria and Italy. It moves coal and iron ore to the blast furnaces in Belgium, Luxembourg, the north of France and the Saar, and carries the products of these metallurgical areas back to Paris and Central Europe. The passenger traffic consists in the first instance of slow services for the local population, not more than two trains in each direction being required in the rush hours. In addition there are a number of railcars for local services.

The express trains consist of the important Calais–Basle expresses – a total of six trains in the twenty-four hours. There are also two express services per day from Lille and Nancy.

Freight and mineral traffic, however, is extremely heavy and track occupation on this section is among the highest in Europe. Six types of freight train are dealt with:

(i) Complete iron ore trains of the following types:

(a) Trains of from 1,350 to 1,700 tonnes (1,323–1,666 tons) from Valenciennes to the blast furnaces in the north of France.

(b) Trains to Ecouviez which supply the blast furnaces in Belgium: these trains have a gross weight of from 1,750 to 2,400 tonnes (1,715–2,352 tons).

(c) Short runs for the supply of blast furnaces in the Lorraine and the Saar, again with trains of approximately 1,750 tonnes (1,715 tons) gross.

(ii) Complete coal and coke trains:

(a) Coming from the north and Calais, and running over the whole section; these trains gross approximately 1,350 tonnes (1,323 tons).

(b) Coming from the Sarre or the Lorraine coal mines via Thionville and running as far as Longwy or Mont St Martin, trains of 1,150 tonnes (1,127 tons) gross.

(iii) Complete trains of metallurgical products running over the whole section.

(iv) The empty return trains arising from the before-mentioned traffic.

(v) A substantial number of ordinary freight trains connecting the important shunting yards of the district.

(vi) The express freight services, no less than twenty per day, running mostly over the whole section.

The Thionville–Valenciennes line is steeply graded, the ruling gradient being 1 in 87. The maximum permitted axle load is 20 tonnes (19½ tons).

The 105 electric locomotives originally purchased, replaced 304 steam locomotives. The electric locomotives are of four types, two of which are of the Co–Co wheel arrangement, and two of the Bo–Bo wheel arrangement.

The two types of Co–Co engines have similar tractive effort: 23·2 tonnes (22·7 tons) and 24 tonnes (23½ tons) respectively. They are required to be able to start a train of 2,080 tonnes (2,038½ tons) on a gradient of 1 in 87, and to haul a similar train up a gradient of 1 in 166 at 60 Km/h. (37 m.p.h.) which is the maximum permitted speed for mineral trains.

The Bo–Bo engines again are of two types, having tractive efforts of 13½ tonnes (13·2 tons) and 17½ tonnes (17·1 tons) respectively. The former are intended for freight trains (as opposed to mineral trains) and can start a 1,000 tonnes (980 tons) train up a gradient of 1 in 100. The latter are mixed traffic engines and they are intended for express train duties at speeds up to 110 Km/h. (68½ m.p.h.).

The operating results of the Thionville–Valenciennes electrification have been very successful and since the pilot scheme came into operation in 1954, electrification with industrial A.C. current has been extended right through from Lille to Basle and the line Paris–Amiens–Lille began electric operation on 7 January, 1959. The Est line Paris–Chalons-sur-Marne–Strasbourg is next scheduled for 50-cycle electrification.

Part III. Design and construction of electric motive power

BASIC DESIGN PRINCIPLES

The development of the electric locomotive suffered from the beginning from several disadvantages. Two different designers are involved, the electrical engineer and the mechanical engineer. Until quite recently the two rarely worked together, and designs consequently lacked cohesion. Between 1905 and 1935 it was usual for an electrical manufacturer to supply the electrical equipment, motors, transformers, resistances, and the locomotive works then had the job of putting these into a vehicle. Very often the mechanical part suffered badly, there being insufficient weight and space left for the development of the vehicle section. Designers of one school of thought took over as many as possible of the existing features of steam locomotives; others followed the existing ideas developed for electric tramways. In the first group the locomotive was simply an engine where the steam was replaced by an electric motor, the rotary movement of the motor being transformed by means of jack shafts and coupling rods into a reciprocating movement and then again into a rotating one on the wheels. Usually one or two very large motors were used, first without gears but later with gears, as high-speed motors were developed. The other group of designers followed the tram-car, and used smaller motors, usually one or two per axle. In the first instance the armature was put on the drive shaft (gearless motors), but later, gears were used. This latter development proved the most important, and practically all locomotives today have individually driven axles.

The electric locomotive is divided into two parts, the electrical part and the mechanical part.

THE ELECTRICAL PART

In an electric locomotive, the layout of a single-phase A.C. machine is quite different from that of a D.C. machine. While describing these differences, it is here most convenient to consider the electrical parts of *single-phase A.C. locomotives*. These consist normally of the H.T. circuit and of the traction motor circuit; there are also circuits of the various secondary installations, such as heating, ventilating and lighting equipment, protecting devices, etc.

The *high tension circuit* comprises the pantographs, the main switch and the primary side of the transformer and goes via the wheels to the rails. From the secondary winding of the transformers the H.T. current is taken in a number of steps, thereby varying the tension to the traction motors.

The *step transformer* is the usual method of varying the speed of the single-phase A.C. locomotive. From the transformer the current goes to the contactor gear and the traction motors.

The main parts of a single-phase A.C. locomotive are:
  Current collectors.
  Main circuit breaker.
  Traction motors.
  Transformers.
  Regulating equipment, and
  Resistances.

Current collectors. The electric current is supplied to the train either by third rail or overhead lines. In the case of third-rail supply, the current, which is always D.C., is usually collected by shoes; the shoes are normally carried at the side of each power bogie, and in the case of motor coach trains all are coupled together so that the current supply is not interrupted when passing gaps in the current rail. Shoes are either overrunning or underrunning, the first relying on gravity for contact pressure, which is between 15 and 60 lb. The shoes, made from cast iron, and fixed on cast-steel brackets, are attached to the shoe beam, which is made from teak or other high-quality wood.

For higher voltages, overhead catenary lines are used and current is collected by pantographs. The pantograph has the advantage of collecting equally well in both directions, and also of taking from the overhead line an uninterrupted flow of current. The normal types are:
  (i) raised by compressed air; lowered by gravity or springs;
  (ii) raised by springs; lowered by compressed air.

Pantographs are mounted on a steel base fixed, with insulators, on to the roof of the locomotive. The whole structure is made from steel tube, carrying aluminium castings, which in turn hold the two collecting pans. The collecting pans carry copper or carbon wearing strips having a life of from 30,000 to 80,000 miles. Various other forms originating from the collecting bow were fitted to older rolling stock. By means of rocker shafts, which are operated by air cylinders, the pantograph can be moved up and down. The pans have an independent spring movement to adjust themselves to small variations of the overhead line. The pantograph shown has a working range of 6 ft and a contact pressure of between 15 and 35 lb. (Plate 53C, page 217).

Main circuit breaker. The main circuit breaker controls the H.T. current and is used as a protection against overloads or short-circuits in the traction motors. The main circuit breaker is simply a switch with several fixed contacts and contact-fingers for their operating gear. The whole unit is placed in a container filled with compressed air or oil, which is used as a coolant as well as an insulating material. Another design has the circuit breaker in air, with arrangements for an air-blast to extinguish the arc at the correct moment. The main circuit breaker is remotely controlled by direct current or compressed air, from the driver's cab, but may be operated by hand in case of failures.

Traction motors. Traction motors are divided mainly into two types:
  (i) the series commutator motor, as used for D.C. and, with some modification, also for single-phase A.C.;
  (ii) the 3-phase A.C. induction motor, as used on the 3-phase system and on phase converters.

The series motor is the most suitable and is generally used for railway traction purposes, as the momentum exerted by the armature is proportional to the square of the current. This results in a high tractive effort when starting and a low tractive effort at high speed. Greater tractive effort can be obtained by reduction of the main field flux. To increase the number of running positions or speeds, the operation of tapping or shunting the motor field is carried out, again by using contactors, and including blow-out arrangements at the contacts. This causes an increase in the armature current and torque for any given speed. The traction motor is a far more robust machine than its stationary brother, as it has to withstand the severe flux vibrations and sudden overloads of railway services (Plates 53A and 53B, page 217).

Another problem is ventilation and filtering, as the motor has to be shielded to avoid the dust and grit from the permanent way entering the motor. The main frame of the motor has either a cast or fabricated steel body structure, with bearings at both ends and suitable openings for access to the commutator and brush gear. There are also the lugs, etc., for the transmission gear. Design details of electric motors are outside the scope of this work. Mention must be made, however, of the importance of the ventilating of the motor, which is either totally enclosed and ventilated from the outside, or is of the self-ventilating type in which air, drawn from the outside, is circulated through the motor; the more modern method is that of forced ventilation, the cooling air being supplied by separate machinery and blown through the motor (Plate 56A, page 220).

For single-phase A.C. the series commutator motor requires a number of modifications. Although the A.C. has no effect on the direction of the torque, due to the alternating flux, an electro-magnetic force (E.M.F.) proportional to the frequency is induced in the armature coils and this complicates considerably the problems of commutation. For example, the motor must be designed for a lower voltage and will have a larger number of poles than a similar D.C. motor. The frequency of the current has to be reduced from the industrial one of 50 or 60 cycles to a special railway frequency (16 2/3 cycles in Europe, 25 in the U.S.A.).

The *three-phase A.C. induction motor* has a constant speed characteristic and is unsuitable for modern individual axle drives for this reason. It is very rarely, if ever, used for modern railway traction.

Transformers, regulating equipment and resistances. Transformers are used in A.C. locomotives for the reduction of voltage and for speed regulation by means of a number of taps. The main constructional difficulty in a transformer, apart from size and weight, is the dissipation of heat, and this raises the problem of suitable ventilation and cooling. Today, it is usual to cool them by air or oil and to force-ventilate those transformers which are housed in a special container (Plates 55A and 55B, page 219).

Transformers supply the working voltage for traction motors, heating, lighting, etc.; normal voltages are as follows: 220 v. for the secondary motors, 800–1,000 v. for heating and 60–600 v. for the various speeds of the traction motors.

The Swiss invented some years ago the, so-called, high tension control whereby the current is no longer regulated in the low-tension side of the transformer, but in the high-tension side. The number of coils on the high-tension side varied, and the obtained voltage is transformed in a second transformer (in a fixed ratio) to the required voltage. The resulting advantage achieved is a substantial reduction in size, as the current on the secondary side is usually about 5,000 Amp., and on the primary side only about 300 Amp. The new type of regulating equipment is, therefore, very much lighter and thus considerably cheaper and easier to maintain.

Regulating equipment is employed to operate the steps of the transformer or, in the case of D.C. locomotives, to actuate the resistances. Switches or contactors were developed which operate either electro-magnetically or electro-pneumatically. Rotary transformers or metadynes allow stepless and therefore very smooth regulation. In 1897 Sprague invented the multiple-unit control, improved later by E. Thomson (1898), whereby a number of electric vehicles can be coupled together and operated from any one cab.

Another control development is the dead-man's handle, with which the driver has to press constantly, either by hand or foot, a button on the controller, and the vehicle is automatically braked as soon as he releases the button (e.g. in case of illness). This invention allows extensive application of one-man drive of electric locomotives.

In a single-phase A.C. locomotive, contactor gear is used, as for D.C. locomotives, for regulating speed and tractive effort. Both vary proportionately with the voltage supplied to the traction motors. Normally there are between twenty and forty tappings or steps. Three types of control gear are in use for regulating the low-tension side of the transformer:

(i) Step-switches of the double-throw type have contactors with one fixed and one movable contact; they engage and disengage by spring action. The contacts of this type are on the one side connected with the secondary of the transformer, and on the other side with a choke.

(ii) Another type of control gear is a sliding contact switch tap changer. These contactors are remotely controlled from the cab by three methods, either electrically, electro-pneumatically, or mechano-pneumatically.

(iii) The firm of Sécheron has developed a mechano-pneumatic control gear whereby the controller is turned on the footplate by the driver and, with the help of notches, opens air-valves which engage the contactors. Mistakes are avoided by mechanically interlocking the contactors – the type shown has twenty-four steps (Plate 56B, page 220).

The method employed to obtain speed control of a D.C. motor is by a combination of resistances, grouping of motors and field control, the resistances being cut out progressively as the motor accelerates. This has the effect of limiting the current and also the torque and tractive effort developed on each step. The number of steps depends on the actual service requirements of the locomotive. The useful or economic running speeds are those with no resistance in circuit, and they can thus be used without time limit. Recent designs provide for ventilation of resistances, which largely removes the time limit formerly imposed by overheating when resistances were in circuit. It is not, however, economical to work thus for long periods as current is being dissipated and wasted as heat.

Resistances are either fixed in sub-assemblies in the engine body or, in the case of motor coaches, are housed on the roof or beneath the floor.

THE MECHANICAL PART

The mechanical part consists of:
    The body.
    The driver's cab.
    The drive.
    The frame.
    Running gear.

The body. The outer shape of the electric locomotive follows two main principles; it is either a square box of equal height over the whole locomotive, or it has a central body part with lower sections at each end; these usually contain the auxiliary machinery. Today the first form is more general than the second one; two cabs are in both cases almost universally provided, as these avoid the necessity of turning the locomotive. The body is mainly made from steel or aluminium and carries on the roof the pantographs and brake resistances. It contains windows, repair openings and ventilation louvres, and is rigidly fixed to the frame.

The driver's cab. The driver's cab or footplate is quite differently laid out than for a steam locomotive. It gives a considerable amount of comfort to the crew and affords an unrestricted view of the line. The control elements are carefully laid out and the driver is usually allowed to sit. Most electric locomotives are symmetrical and can run equally well in both directions.

The drive. In the beginning, the transmission of motor power to the wheels followed one of two trends: either it was copied from the steam locomotive, one or two motors driving through gear wheels, jack shafts, rods and axles; or the rotor was directly fixed to the axle shaft. At a later date, this became the nose-suspended, or tram, motor which rested partly on the unsprung axle and partly on the sprung vehicle frame. The first design was not successful. It was expected that the rotating machinery of an electric locomotive would obviate all the disadvantages of the reciprocating steam engine and thereby all its critical movements. However, it was found, while it was possible to balance all moving masses completely, the irregularities in dimensions in the rods, and in the permanent way, gave rise to new disturbances which were even more difficult to overcome than those in the steam locomotive. Fractures of the mechanical parts, such as rods, axles, etc., were caused. These problems, together with the successful development of small series D.C. and single-phase A.C. motors, brought about the end of the rod drive. The nose-suspended motor, however, still exists today, mainly for slower moving locomotives.

In the U.S.A., an individual axle drive was employed, whereby the rotor was put directly on the axle shaft. Later S.H.Short put the stator on a hollow axle which surrounded the driving axle and linked, flexibly, with the wheels. With the growth of locomotives, larger motors were used, and these had to be positioned in the body itself, usually mounted on the frame. They were therefore considerably above the wheels and means had to be found to couple the motors flexibly to the axles. The problem was solved by the introduction of one or

two blind or hollow axles which were rigidly connected by connecting rods to the motors. Suitable extension pieces of the hollow axle allowed movements and then transmitted the motor power to the wheels. The triangular connecting rods system with sliding pieces, which was used on the Simplon line locomotives, was very successful (Plate 35B, page 163).

As soon as it was possible to build smaller motors (one or two for each driven axle) the rod drive, which was copied from the steam locomotive, disappeared, and during the last twenty-five years practically no more rod-driven locomotives have been built.

Today, one or two motors drive each axle by means of flexible couplings. The motors usually sit in the locomotive frame or bogie, rigidly, and the flexible medium takes the form of quills (spring-loaded pots) which drive the wheels or sprung pinions, flexible gear-wheel segments or, lately, flexible steel discs.

In the earliest electric locomotive designs, the electric motor was put with its rotor directly on to the driving axles. Later it was fixed to the simple frame and drove the wheels by coupling rods. The problem appeared in the first instance, simple enough. A rotary prime mover revolved at a certain speed and it was necessary simply to transmit its power to the wheels. If the motor had to revolve at a different speed, intermediate gearing could be introduced between the motor and the wheels.

These simple concepts were unfortunately mistaken, though in the first electric locomotives they were the principles which were followed. Often the electric rotor was fixed directly to the wheel axle and drove the wheel directly. One of the first designs of this type was the Bo–Bo locomotive of the London Underground Railways. The whole weight of the motor was unsprung, and the locomotives proved quite unsatisfactory.

An improvement was brought about by Batchelder, who introduced the gearless type of drive. This design appeared at the beginning of the century in the United States of America. The armature was fixed, spring-loaded, on each driving axle, and the stator rested, also spring-supported, on the frame. Rubber pads transmitted the tractive effort. The unsprung weight was now substantially reduced, but the locomotives only worked well at high speeds as the rotational speed was but a fraction of the running speed of the motor. Although there was no transmission gear, this design disappeared quickly because the wheels had to be kept small, which meant an increase in tractive resistance.

Next came an attempt to use the Batchelder design with a hollow axle (Short), whereby the advantages of direct drive remain, without the heavy load on the driving axle, power transmission to the wheels being carried out by spring-loaded dogs between the wheel spokes. Similar designs in Europe (Ganz and A.E.G.) had, in place of the spring-loaded dogs, universal coupling rods (Valtellina Railway) or spring-loaded coupling rods.

The next development was known as the tram motor or, more correctly, the nose-suspended or axle-hung motor. This motor rests with one side on the driving axle and with the other spring-loaded on the frame; the unsprung weight is thus considerably reduced. Connection between motor and axle is made by a train of gear wheels. For many years this drive was used on tram-cars and then, hesitantly at first, for slow-moving locomotives as it was considered harmful to the track. The combination of the axle-hung motor with flexible couplings was introduced more recently.

It was obvious that the motor which rested in the bogie or between the wheels was limited in capacity; there had to be many more motors with many parts and the idea, therefore, arose to use the space above the wheels to position the motors. In the first instance the motor was placed rigidly in the locomotive frame above the axles and drove the wheels through spring-loaded plungers (Westinghouse, for the New York, New Haven & Hartford Railroad in 1910). The further advantage was that the motor dimensions could be freely developed, unrestricted by the limited space between wheels. It was still a direct drive using slow motors. Very soon the drive was transmitted by coupling-rods, following the ideas of steam locomotive designers. Very large motors began to be used, weighing up to 30 tons, with outputs up to 1,500 H.P. Usually one or two motors were rigidly fixed on to the locomotive frame and drove through cranks and connecting rods to jack shafts or on to blind axles. Often the motor-axle was positioned high in the frame (owing to the large motor diameter) and drove, first of all, a set of intermediate gear trains. The driving axles were interconnected in the usual way by coupling rods (Plate 37A, page 165). Later a sliding coupling was so arranged that the connecting rods could work directly on to the crank pins. In most cases the motor and wheel speed was the same. Subsequently, motors were built which had higher speeds and which were also much lighter in weight; they worked through gears on to small triangular connecting rods (Kando).

Another arrangement used two motors which worked on to the same set of gears driving a jack shaft with gears at both ends, and a flat coupling rod. The rod drive was a faithful copy of steam locomotive practice; the working parts had to be made to close tolerances, and, at the start, worked often quite satisfactorily. But even with careful workmanship, irregularities in the dimensions increased

quickly through wear and resulted in frequent, and often very substantial, fractures of frames and rods.

It became recognized that rod drive for electric locomotives would have to be abandoned. It was not then appreciated that an unsprung wheel follows all the irregularities of the rails lying on the ground, while the motor was usually fixed to the frame and therefore spring-loaded; some form of flexible coupling was necessary.

It was soon discovered that a way would have to be found to do away with the large moving masses employed in rod drives. After World War I, substantial electrification programmes were put into operation in many countries and a variety of electric locomotives was urgently required – both for fast- and slow-moving services.

While, in principle, it was clearly recognized that the rod-driven locomotive was only a temporary solution, a considerable number of electric locomotives were still designed with this power transmission, until it was finally supplanted by the individual axle drive.

The development of the modern individual axle drive is a story of engineering achievement, which has taken about thirty-five years. From the early types – cumbersome, weighty, costly to make and maintain – has come the ability to transmit with a simple steel disc a force of not less than 1,000 H.P. per axle, and able to run up to half a million miles without major examination. Developments of this type of power transmission started with the American Westinghouse quill drive, whereby the gear wheel carries a quill and transmits power through helical springs or spring-loaded plungers to the wheels. All springs are connected at one end with a driven gear wheel (which is again fixed in the locomotive frame) and at the other end fixed to the driving wheel. A development of the Westinghouse design is that due to Sécheron, in the first case the spring-cups sit on the quill; in the second on the wheel. While in the two earlier designs mentioned the springs are in the open and, therefore, exposed to dirt and humidity, the latest types (Brown, Boveri) have the springs enclosed in the gear wheel, which itself is carried by roller bearings mounted on a short-arm quill, and fixed to the motor. Since the quill is fixed and only provides for vertical movements of axle against frame, the hole in the quill is oval and allows the motor to be located nearer to the axle than would be possible with a rotating quill (Plate 54B, page 218). Power is transmitted by springs in the same way as with the earlier Sécheron or Westinghouse drives. There are also a number of deviations from these basic designs, such as combinations with rods and universal links (S.L.M. drive).

PLATE V. South African Railways: "The Blue Train" leaving Johannesburg for Cape Town behind an electric locomotive of type Co–Co, Class 3E. These locomotives, built by Metropolitan Vickers in 1948, use 3,000 v. D.C.

One of the most important transmission types is the Buchli-Drive, developed by Brown, Boveri (Plate 54C, page 218). Although it appears complicated it is still in use and has an excellent record for long life and low maintenance costs. In this type a driven gear wheel is fixed to the frame. Inside this driven gear wheel are two levers which carry gear segments which mesh with one another. The other ends of these levers carry universal joints and are connected to tension bars. These tension bars are in their turn connected at their other ends again with universal joints to pins fixed on the driving wheel. The pins go, with sufficient play, through the motor gear wheel. If, therefore, the wheel moves vertically against the frame, the gear segments also move, and if the driving wheels move axially against the frame, then the universal joints move in their bearings. The driving wheels are, therefore, free to move both horizontally and vertically while transmitting the momentum of the motor. This individual axle drive has been very successful and has been used on approximately 1,500 locomotives. Today, several hundred designs exist and show that the problem has until now not been solved to complete satisfaction.

One of the latest developments of this type of drive is the Brown, Boveri disc type (Plate 46C, page 192). This disc drive has been incorporated even into bogie designs and is suitable for power transmission of up to 1,000 H.P. per axle. The gear wheel is mounted on the axle. The pinion is carried in the gear case on two roller bearings. The gear case itself is carried at one end on the axle by means of roller bearings and the other end is attached to the frame of the bogie. It is, therefore, fixed in the same way as the nose-suspended motor. The relative movements between the centre line of the pinion and the centre line of the motor are taken up by a cardan shaft passing through the hollow motor shaft and which is articulated at each end by flexible steel discs; these dies form the actual flexible link. A substantial number of these units have been built and have given great satisfaction.

Final developments follow mainly Swiss and French experimental work. The French firm of Alsthom developed an individual axle drive employing Silentbloc rubber cushions combined with universal couplings for the transmission of power (Plate 54D, page 218). The design is of great simplicity; about 200 units have so far been built and given quite trouble-free service.

One of the latest Swiss locomotives, the 6,000 H.P. Ae 6/6 (see page 208) developed by Messrs Brown, Boveri, employs a spring-loaded large gear wheel fixed to a hollow stub axle to transmit power through springs to a driving star fixed to the wheel (Plates 45A, 45B and 54A, pages 191 and 218).

The frame may be a single rigid unit, or it may comprise units in the form of bogies. In some designs the bogies support a shorter centre frame, while in others, especially in modern layouts, there are bogies only, the body and frame forming an integral part. In this last case, the bogies are either coupled and transmit the tractive and buffing forces, or they are uncoupled and the forces are transmitted by the body. These principles are in line with developments in coachbuilding in general. The frame is either riveted or of welded construction. The fabricated (welded) design is considerably simpler and cheaper. Rivets tend to loosen and repairs are difficult to effect as the rivets are often in inaccessible places. The disadvantage of welding is that welding techniques require considerable individual skill and care, as a welded frame is over-stressed and is therefore inclined to show cracks. But this type is being more and more used as experience of welding grows. The frame or the bogie frame also carries the buffers, coupling-gear and axle boxes.

Running gear. In an electric locomotive, the running gear consists of wheels, axles and couplings. Identical problems are met with in their design as those encountered in steam locomotive practice. Later electric locomotives had two fixed driving axles and one, two or three axles having some lateral movement. Where there were long wheelbases, some of the driving wheels were often flangeless. Radial wheels or bogies were provided at one or both ends of the locomotive. Radial wheels were, in some designs, combined with the leading driving axle to form a guiding unit or bogie (Java bogie or Zara truck).

Bogie designs have changed little for locomotives with nose-suspended traction motors. The usual designs with spring bolster and swing links have been extensively used but are now outmoded. Heavy and slow-moving locomotives were built without guiding wheels, having two- or three-axle bogies, and thus using all the locomotive weight for adhesion. After it was found that these locomotives ran satisfactorily, some railway administrations, notably those of Switzerland and France, built for express duties all-adhesion locomotives, and these recent designs have been so highly successful that in the last few years most designs are of the Bo–Bo or Co–Co design.

In the new Alsthom design, the body is attached to the bogie by two swing links pivoting on conical rubber seatings in the bolster between the traction motors. Two spring assemblies exert transverse forces to restore the pivot to the vertical. The axle boxes have no guides as generally accepted. They are attached to the bogie frame by short rods mounted in Silentblocs. When subjected to vertical displacement the rods move and the

Silentblocs are twisted and to compensate for the decrease in the horizontal length of the rods the axlebox rotates very slightly. The normal bogie pivot and axlebox guides have thus been dispensed with. The name "swinging pivot" is given to this type of attachment.

Modern bogies are either coupled or uncoupled. New designs often include very careful layouts to avoid weight transfer (unloading of leading axle) especially under starting conditions; special axle load equalizers are being installed; the link between the bogie is so designed as to allow controlled lateral movement but no vertical movements of one bogie against the other.

Part IV. Modern electric locomotives : a survey of current practice

BRITISH RAILWAYS

Bo+Bo and Co–Co locomotives, 1,500 v. *D.C.* Apart from the twenty locomotives of the Metropolitan Railway (*see* page 149) used on the London Transport line from Baker Street via Watford to Rickmansworth, and the Vincent Raven locomotives of the North Eastern Railway (*see* page 156) there were no electric main-line locomotives in Great Britain until 1939. In the late nineteen-thirties, when it was decided to go ahead with the electrification of the route across the Pennines, the former L.N.E.R., under its then Chief Mechanical Engineer, the late Sir Nigel Gresley, laid out a Bo+Bo mixed traffic locomotive, Class EM-1 (Plate 44B, page 190), which forms the basis of the new rolling stock for the Manchester-Sheffield line and which was followed by a Co–Co type, Class EM-2 (Plate 44C, page 190). All traffic was to be hauled by electric locomotives; by so doing the capacity of this very busy main line would be greatly increased.

When World War II broke out in 1939, and work on the scheme was stopped, a considerable amount of preliminary work was still in hand and orders had been placed for most of the equipment required. It was not until the summer of 1947 that the work could be resumed, and the following lines are now electrified, covering 75 miles of route and 330 miles of track:

(i) Manchester (London Road Station) to Sheffield (Rotherwood Sidings).

(ii) Penistone (Barnsley Junction) to Wath via Worsborough.

(iii) The Glossop Branch.

(iv) Oldham, Ashton and Guide Bridge Branch up to and including Ashton Moss Sidings.

(v) Fairfield to Manchester Central and Trafford Park Sidings.

(vi) Ashbury to Ardwick Goods Yard and Midland Junction.

Sixty-five mixed traffic electric locomotives have been supplied, of which fifty-eight have the Bo+Bo wheel arrangement, and the remainder Co–Co. Eight three-coach multiple-unit electric trains have also been put into service.

The Bo+Bo prototype was completed at Doncaster by the former L.N.E.R. and ran trials on the Manchester–Altrincham line in 1941 and 1947. In September 1947, this locomotive was lent to the Netherlands Railways (also electrified with 1,500 v. D.C.), and it ran over 200,000 miles in Holland, hauling such diverse trains as 500-ton international expresses and 2,000-ton mineral trains between Amsterdam and Maastricht. Considerable experience was thus gained in the working of the first locomotive and various refinements were incorporated in the later locomotives introduced in 1950. The Bo+Bo machines give the impression of a very simple and robust locomotive type, fully suited to the hard working conditions to which they are subjected. They have, for example, proved capable of hauling trains of 1,750 tons, and of starting such trains on gradients of 1 in 80. The maximum speed of the Bo+Bo locomotive is fixed at 70 m.p.h.

The mechanical layout is very simple; it has a single body with driver's cabs at both ends, welding technique being used extensively for fabricating the various parts. The body rests on two bogies which are coupled together and carry all traction gear, thereby relieving the body and the bogie pivots of all stresses except weight bearing.

The electrical part is again the usual one on D.C. locomotives; it consists of two pantographs, switchgear, and four 465 H.P. force-ventilated motors. The control system consists of electro-pneumatic resistances and series parallel control, with electro-pneumatic contactors giving ten economic running speeds. The traction motors are axle-hung, and drive through resilient gearing.

The seven Co–Co locomotives (Plate 44C, page 190) were introduced in February 1954. As in the case of the Bo+Bo engines, the parts were built at Gorton Works of British Railways and the electrical equipment supplied by Messrs Metropolitan-Vickers Electrical Co. Ltd, of Manchester. These locomotives are mainly intended for passenger working and are therefore provided with an electric boiler for train heating. They are designed for maximum speeds up to 90 m.p.h. The locomotives are 59 ft long over buffers. Total wheelbase is 46 ft 2 in., and the distance between bogie centres is 30 ft 6 in. Weight in working order is 102 tons. The layout of the cabs is similar to that of the Bo+Bo locomotives and there is a corridor connecting the two cabs along one side, from which access can be obtained to high-tension and resistance compartments as well as to those housing the boiler and machinery. The door giving access to these compartments is mechanically and electrically interlocked by the reverser key on the master controller, so that it can be opened only when current is shut off and the pantographs lowered.

The locomotives are equipped with dead-man's foot pedals and sanding switches. Push buttons are provided on the right-hand side of the driver's desk so that operation of dead-man's device can be delayed should the driver wish to cross the cab.

The bogies have box-type fabricated frames made from 7/16-in. plate and their centres are located on the centre line of the axle-boxes. The bogies are cross-braced by two cast-steel cross-stays in which the double bolster can slide laterally and through which the tractive force is transmitted, via the bolster, to the bogie centre on the body itself. The weight of the body is carried by four spherical bearers sliding sideways on each bolster, which in turn rest on laminated springs supported by swing links from the bogie cross-stays. Equalizing beams resting on the axle-boxes are arranged inside the box frames and suspended from these by tension bolts and coil springs. The locomotives are equipped with Timken roller bearings and horns are lined with manganese steel liners.

The six motors are of the nose-suspended type, all axles being motored. They are supported by a link from the cross-stays fitted with Silentbloc bearings and also located laterally by smaller bearings of similar type, so that the wheel bosses and motor bearings do not come into contact and all movements of the motor are accommodated by the Silentblocs. Gear ratio is 17:64, each axle being driven through single reduction silent gears. Two of the three traction motors in each bogie are permanently connected in series and there are three different combinations so that three speed ranges can be obtained. The motors are independently excited by a motor generator set which also drives the blower for cooling the motors. An identical blower is provided in the other machinery compartment for the same purpose and is driven by a motor generator set supplying current at 50 v. for charging the battery, control gear, lighting and heating.

The control system is operated by contactors from the master controllers in the driving cabs.

Brake equipment consists of a Westinghouse vacuum-controlled straight air-brake with independent locomotive brake. Both systems have automatic action for dead-man's emergency or break-away. The locomotive brake can be operated separately or synchronized with the train brake. Compressed air is supplied by a motor-driven compressor and the vacuum is provided by a Westinghouse motor-driven exhauster. There are four brake cylinders on each bogie and the rigging is operated through Westinghouse slack adjusters. In addition to the air brake system the locomotives are fitted with regenerative braking and should the system fail an immediate equivalent air brake application is made.

Main particulars of the Co-Co type are as follows: The six motors have a continuous rating of 2,298 H.P. at 46 m.p.h., and a one-hour rating of 2,490 H.P. at 44·3 m.p.h.; their maximum starting tractive effort is 45,000 lb. Weight in working order is approximately 102 tons; they are designed for a maximum speed of 90 m.p.h

Co-Co locomotives, 660 v. D.C. An interesting and unusual Co-Co design was developed by the former Southern Railway for three mixed traffic locomotives, Class CC (Plate 44A, page 190). As these locomotives take their current from conductor rails, very unusual design problems arose. To avoid trouble from stalling, it was necessary to provide a continuing tractive effort to bridge the gaps in the conductor rail system. These gaps frequently occur at the end of stations, just where, under starting load, the largest output is required. It was decided, in view of the high tractive effort (of up to 3,000 Amps.) to use a booster-motor generator control with flywheel. The first locomotive was ready in 1943 and, since then, two further locomotives were developed, the third one being completed in 1948. All the three differ considerably the one from the other as experience was gained. Basically, these locomotives are among the most ingenious ever developed.

Bo-Bo locomotives 675 v. D.C. Thirteen locomotives have been built for operating freight and passenger trains on the third-rail electrified lines in Kent. Freight trains to be operated will, in many cases, be loose coupled, and such trains have to pick up vehicles in sidings and yards where third-rail electrification would be impracticable. In such locations, overhead conduction is being used, and the locomotives are therefore equipped with two-pan pantographs as well as current pick-up shoes. The locomotives will also be required to haul fast freight trains for the Continental train-ferry service, and express passenger trains, such as the heavy "Night Ferry".

A true mixed traffic locomotive was therefore required, but one which would be capable of fast speeds. The Bo-Bo locomotives are of 2,500 H.P. (one hour rating) with four bogie-suspended traction motors, driving through

22 : 76 reduction gearing and Brown, Boveri spring drives. A maximum speed of 90 m.p.h. is possible, but the locomotives are capable of hauling loads of up to 900 tons.

The method of booster control used in the earlier Southern Co–Co locomotives has been perpetuated in these locomotives, but there is only one booster motor generator set instead of two. The output of this machine is 945 kW. at 1,750 r.p.m. The motor and generator armature shaft each carry a heavy flywheel to provide kinetic energy to maintain the speed of the set over breaks in supply caused by gaps between the conductor rails.

Like their Co–Co predecessors, these locomotives have many modern and unusual features, and are one of the most outstanding electric locomotive designs in the world.

Future policy. The British Transport Commission has recently announced its future policy towards electrification. Fifty-cycle, single-phase A.C. traction is to become standard instead of the 1,500 v. D.C. system to which Great Britain was hitherto committed.

It is the intention to use the new system on 1,210 of the 1,460 route-miles of British Railways, the electrification of which has already been announced, and on all subsequent electrification schemes other than in the Eastern and Central Sections of the Southern Region, and on London Transport. The lines involved in the A.C. scheme are:

(i) From Euston to Birmingham, Crewe, Liverpool and Manchester.

(ii) King's Cross to Doncaster, Leeds and (possibly) York.

(iii) Liverpool Street to Ipswich, including Clacton, Harwich and Felixstowe branches.

(iv) Suburban electrifications, including:

(*a*) The London, Tilbury and Southend line.

(*b*) Liverpool Street to Enfield and Chingford.

(*c*) Liverpool Street to Hertford and Bishop's Stortford.

(*d*) King's Cross and Moorgate to Hitchin and Letchworth, including the Hertford loop.

(*e*) Glasgow suburban lines.

The Crewe–Manchester section and the Colchester–Clacton–Walton lines will be the pilot schemes and valuable training grounds for the personnel required in large numbers for the completion of the whole scheme.

There are 1,796 miles of track in the Southern Region which are already electrified on the 660 v. D.C. system. To alter this to single-phase A.C. would be very costly, and would seriously delay extension of the electrification in this region; the 250 miles of extensions, including all

the main lines in Kent, will, therefore, be carried out on the existing third-rail system. The voltage on all the Southern Region electrified lines will, however, be raised to 750 v. D.C. for existing and future schemes.

Sir Brian Robertson, Chairman of the Commission, has said that the A.C. system must be regarded as new and as yet untried on a large scale under British conditions, which are in many ways more exacting than those obtaining on any other system in the world. The decision therefore implies a high measure of confidence in the rightness of the technical advice, and a bold acceptance of the risks which are necessarily attached to the adoption of something new. The fundamental designs of electric locomotives and other equipment must be made and proved by trial, at the same time maintaining the programme of electrification to which British Railways are committed.

The 50-cycle A.C. system, although new in the sense that it was only recently that the French Railways made the first large installation, does not necessarily present novel problems to the traction equipment manufacturers. Apparatus, such as transformers and circuit breakers, is already fully developed for industrial purposes, using the standard frequency supply, and can readily be adapted for traction service. One of the trends with the new system is to use D.C. traction motors supplied through a rectifier, and the experience of British manufacturers in the design of such motors is very considerable, as also is the design and manufacture of mercury-arc rectifiers, where world-wide experience has been gained by British industry. Furthermore, development in the use of germanium rectifiers for traction is making good progress in Britain, and the first motor coach equipped with them has recently undergone trials on the British Railways Lancaster–Heysham line with encouraging results (*see* page 226). This type of rectifier may prove a very attractive alternative to the mercury-arc rectifier.

FRENCH STATE RAILWAYS (S.N.C.F.)

Co–Co and Bo–Bo locomotives, 1,500 v. D.C. Locomotives with wheel arrangement B–B have been built in France for about twenty-five years, and there are now more than 600 locomotives of this wheel arrangement in service. They vary of course very widely in their design due to the development of electric locomotives during the last twenty-five years.

Following the general trend of modern electric locomotive design, the French Railways decided that the all-adhesion locomotive would probably be the future motive power for both fast passenger and freight traffic, and they developed the Co–Co, Bo–Bo–Bo and Bo–Bo types, following the success of Series CC 7001, of which

53 were ordered for France, 30 for Holland and 80 for Spain (Plate 46c, page 192).

The Co–Co series, CC 7001 (Colour Plate VI, page 204), is interesting as being a continuation of the ideas of Swiss designers on an all-adhesion high-speed locomotive. Two prototypes were first developed, tested for about three years, and then incorporated in a larger series. The French firm of Alsthom, together with the S.N.C.F., developed a method of body suspension which dispenses entirely with the conventional bolster and pivot. Instead, the body is connected to each bogie by two vertical swing links, which move in opposite directions to permit the bogie to turn relative to the body on curves, or both tilt the same way to allow some lateral displacement of the body from the bogie centre line, Two springs acting transversely return each link to the vertical after it has swung one way or the other. This restoring force is proportional to the amount of movement of the front or rear end of the bogie.

This method of suspension fulfils two opposing requirements, which are difficult to meet with conventional systems. When centrifugal force on a curve causes lateral displacement of the body, both links act together in providing a restraining and restoring effort. Local shocks due to sideways movement of an axle caused by track irregularities, on the other hand, affect only the nearer link, so that there is ample resilience to cushion them. A further advantage is that the arrangement of the bogie frame and the main frame is always parallel and the weight distribution among the axles is constant.

In June 1949 the CC 7001 locomotive was put into regular service after a month of varied and severe tests, the most typical of which are given below:

(i) From Paris to Poitiers, hauling a 1,000-ton (22 coaches) fast train, 335 Km. (208 miles) in 3 hours 9 minutes. This gives an average speed of 106 Km/h. (66 m.p.h.), and was achieved without exceeding 120 Km/h. (75 m.p.h.).

(ii) From Paris to Toulouse, hauling a 950-ton (21 coaches) fast train, exceeding by more than 200 tons the usual weight, on a line of considerable difficulties which includes long 1 per cent gradients and many curves.

(iii) From Paris to Bordeaux, hauling a special light-weight 5-coach train, 579 Km. (360 miles) in 4 hours 26 minutes. An average speed of 138 Km/h. (86 m.p.h.) was maintained between Paris and St Pierre-des-Corps (232 Km., 144 miles) and 131 Km/h. (81 m.p.h.) was the average for the whole journey of 360 miles.

(iv) Starting tests, hauling 1,200-ton freight trains on 0·8 per cent gradients and 950-ton freight trains on 1 per cent gradients, were made successfully.

(v) Speed tests between Paris and Blois, during which a speed of 180 Km/h. (112 m.p.h.) was maintained over 3 Km. (1·8 mile).

The CC 7001 is the first total adhesion locomotive manufactured for speeds higher than 125 Km/h. (78 m.p.h.). Up to then it was customary to take for granted that bogies or pony-trucks were necessary for guiding fast locomotives. As a result, the 2–D–2 locomotives of the S.N.C.F. were designed, in which only 88 tons is available for adhesion out of a total weight of 145 tons.

In 1946, the firm of Alsthom carried out trials on express locomotives with no guiding axles, and in which the whole weight was available for adhesion. Their researches were successful, and the all-adhesion locomotive proved to be cheaper to manufacture and enabled either an increase in the hauled tonnage or a noticeable reduction of the power required. Another object achieved was the reduction of the axle load without an increase in the number of axles. Actually, the greatest part of the stresses acting upon the track is directly proportional to the local loads at each rail-wheel contact and to the division of these loads along the whole length of the locomotive. These stresses may become dangerous as maximum speeds increase, and they increase the maintenance cost of both equipment and track. If, however, when higher speeds are desired, axle loading can be reduced, the wear on the track will be correspondingly less. These results have been obtained by the improvements effected both in the electrical design by increasing the motor power/weight ratio, and in the mechanical design by using electric welding and by adopting tubular construction for the underframe and rigid sides for the body. So satisfactory were the results obtained with the CC 7001 and 7002 locomotives that the S.N.C.F. decided to order CC locomotives for their high-speed motive power. The CC 7101 locomotives are now hauling the fastest trains on the Paris–Lyons line: the 512 Km. (318 miles) are covered in 4 hours 10 minutes, at an average speed of 123 Km/h. (77 m.p.h.), in spite of the Seuil de Bougogne gradients.

Mechanical part. The body and bogie frames are electrically welded either by spot or arc welding. The main parts, such as girders, beams, sills and hood-rounds, are made of closed sections, securing great rigidity for minimum weight. The welding of these various parts ensures, without sliding or permanent distortion, the transmission of stresses and an effective participation of the greatest number of members to the general resistance. The body is of the tubular type, with side walls adding to the rigidity of the whole. The side sheets, stretched between sills and hoods, and jointed to rigid uprights, compel all the body elements to resist together the longitudinal and vertical stresses. The sides have no openings other than "port-holes" which are framed with circular rings. The main body elements, i.e. the floor frame, the two side panels, and the two driving cabs, are separately prefabricated and then joined together by welding. The central part, between the cab walls, has an opening at the top (45 ft 11 in. in length) for placing in position the equipment and auxiliary generators.

One of the main features of these locomotives is that the body rests on the bogies through two oscillating, or swinging, pivots articulated at top and bottom on conical rubber pads. These pivots rest on the cross members fixed between the traction motors. Being laterally connected to the body by means of spring rods, these two pivots are constantly brought back to their vertical position. The support of the body on each truck is completed by four side bearers.

For equal and opposing tilting motions the oscillating pivots enable the necessary rotation between body and trucks, and the maximum tilting motion corresponds to negotiation of curves of 262 ft radius. If the tilting motions are equal and in the same direction, side translation is secured: if the tilting motions are different, there is a combined motion of translation-rotation.

The oscillating pivot system (also known as the Alsthom swinging pivot system) has some very important advantages:

(i) *Manufacture*. The arrangement is simpler and cleaner than it is with the usual method of 3-axle trucks where the single pivot and side bearers are offset owing to the presence of a traction motor at the centre of the trucks.

(ii) *Stability*. It carries out in a simpler way the duty performed by the usual system of the truck bolster. Moreover, the transverse motion possible between body and trucks through the draw-back spring rods, can be adjusted by modifying the initial tension and flexibility of the springs, whereas the draw-back motion of a truck bolster cannot be modified after manufacturing.

(iii) *Adhesion*. The use of two pivots for each truck ensures that the truck frames are parallel with the body frame (except for deformation of the rubber pads). The primary suspension is then wholly balanced. The traction motors being completely suspended and producing no direct vertical thrust upon the axles, the axle load transfer is reduced to the minimum transfer resulting from the difference between the tractive effort at the rail level and the resistive effort at the coupling-hook level.

It is claimed by the makers that the double pivot system has all the advantages of rigid machines, as regards load transfer, as well as the advantages of articulated machines as regards stability.

The motors are completely suspended; they are fixed upon the truck frames and drive each axle through bilateral gearing mounted upon a hollow shaft. Flexible coupling between hollow shaft and axle is provided by a floating ring and connecting rods fitted with Silentblocs. This simple device requires neither lubrication nor maintenance (*see also* Part III, The Drive).

The axle-boxes are connected to the frame by means of rod fitted with Silentblocs. These rods are fixed on the box by two lugs at each end of an oblique diameter, enabling vertical displacements of the axle to occur while keeping its centre line in the vertical plane. This arrangement is held to be more satisfactory than an axle-box guide. Lubrication is no longer necessary. There is no wear, and no clearance between boxes and frame. The transverse flexibility secured by the Silentblocs facilitates negotiation of curves and track defects, by reducing to a large extent stresses upon the rail.

Electrical part. There are six traction motors; they are of the fully suspended type and drive the axles through hollow shafts. Their three connections: series, series/parallel, and parallel are realized as follows:

Series connection – 6 series motors.
Series/parallel – 2 parallel sets of 3 series motors.
Parallel – 3 parallel sets of 2 series motors.
There are 40 notches for full field starting:
Series – 18 notches.
Series/parallel – 12 notches.
Parallel – 10 notches.

When advancing from one notch to another, the ratio of the peak effort to the acceleration effort is lower than 1:1·15. This is considered to be satisfactory, especially having regard to the high adhesion properties of this locomotive.

The series, series/parallel and parallel connections include five balanced steps, with field reduced by shunting the inductive circuit through resistances and inductive shunts. The shunting rates are 24·5, 40·5, 65 and 73 per cent.

Unbalanced intermediate shunting notches, for each connection, are fitted according to Alsthom practice, to facilitate building-up the effort. Transitions are effected by shunting, through three sets of four cam-operated contacts. Starting resistances are disconnected by self-contained electro-pneumatic contacts. The cam-operated reversing switches are controlled by electro-pneumatic servo-motors.

The main particulars of these interesting locomotives are:

Voltage	1,500 v. D.C.
Wheel arrangement	Co–Co
Length over buffers	18,922 mm. (62 ft 1¼ in.)
Total wheelbase	14,140 mm. (46 ft 6 in.)

Rigid wheelbase   4,845 mm. (15 ft 11 in.)
Wheel diameter   1,250 mm. (4 ft 1 in.)
Total weight, all available for adhesion

                                     107·6 tonnes (106 tons)
Weight of mechanical part   64·9 tonnes (63·9 tons)
Weight of electrical part   42·7 tonnes (42·1 tons)

Performance

Light express trains   350 tonnes (343 tons)
                       at 120 Km/h. (75 m.p.h.)
Heavy express trains   750 tonnes (735 tons)
                     at 105 Km/h. (65 m.p.h.)
Freight trains   1,300 tonnes (1,274 tons)
                       at 50 Km/h. (31 m.p.h.)
Express freight trains   680 tonnes (66½ tons)
                     at 80 Km/h. (48 m.p.h.)

Rating and efficiency

One-hour rating at 1,500 v.   5,000 H.P.
Corresponding tractive effort 16,500 Kg. (36,300 lb.)
Corresponding speed   80 Km/h. (49·7 m.p.h.)

Continuous rating at 1,500 v.   4,650 H.P.
Corresponding tractive effort 14,880 Kg. (32,736 lb.)
Corresponding speed   82 Km/h. (50·9 m.p.h.)

Maximum service speed   160 Km/h. (99·3 m.p.h.)
Maximum permissible speed 180 Km/h. (111·8 m.p.h.)
Maximum speed achieved 243 Km/h. (150·9 m.p.h.)

One of these CC locomotives, No. 7121, in February 1954 set up a new world rail speed record of 243 Km/h. (151 m.p.h.). This performance was remarkable in that it was achieved with standard equipment, no alteration having been made to the original design of the locomotive.

The results of these tests encouraged the French National Railways (S.N.C.F.) to carry them further, and a further series of tests was carried out in March 1955, when another Alsthom CC locomotive, No. 7107, reached a speed of 331 Km/h. (206 m.p.h.).

The Bo–Bo series, BB 9000. Two experimental units were ordered based on the Loetschberg design (Series AC 4/4) from Swiss manufacturers and two more from French firms (Nos. BB 9003 and 9004). They were supplied by Le Materiel Electrique de France, and Schneider, from the designs of the French State Railways. BB No. 9004 attained a record speed of 205·6 m.p.h. (Plate 42A, page 188). The main features to be incorporated into these locomotives were as follows:

(i) A highly efficient transmission gear between the motors and wheels, there being the minimum amount of friction and unbalanced disturbing forces; the correct machining of the gear wheels was regarded as of great importance.

(ii) To avoid loss of adhesive power, arising from bad track, uncontrolled oscillation of the locomotive had to be damped out. Bad springing arrangements and weight transference owing to the tractive effort of the motored axles, were to be avoided.

M. Jacquemin, in the *Revue Generale des Chemins de Fer* of June 1952, showed that, in a Bo–Bo locomotive in which the tractive effort (Tr) is transmitted through the frame, with free bogies and uncoupled wheels, the maximum unloading of an axle is given by:

$$Tr = 2z\left(\frac{H-h}{L} \times \frac{h}{l}\right)$$

where:

z = tractive effort at the treads of one pair of wheels;

H = height of the point of application of the tractive resistances on the body (in practice this lies between 1·1 m. (3·61 ft) when starting and 1·35 m. (4·43 ft) at high speed, when the aerodynamical resistance must be considered);

h = height of the points at which the body is moved by the bogies;

L = distance between bogie centre pins;

l = bogie wheel base.

In a contemporary BB locomotive the coefficient of utilization of the adhesion, i.e. the ratio of the load on the axle mostly unloaded ($P-Tr$) to its static load (P) can therefore vary for a tractive effort equal to 20 per cent of the adhesive weight from 0·85 (when $h=H$) to 0·95 (when $h=0$).

If the two pairs of wheels of a given bogie be coupled, it reaches 1 when $h=H$. The best solution then is to have the body moved from a point about 1·2 m. (3 ft 11¼ in.) above the rail and to couple the pairs of wheels. Contrarily, if the bogies are not free from one another in the vertical plane, the coefficient of utilization of the adhesion falls in all cases. The prevention of weight transference is consequently a very important question because, properly solved, it can increase the maximum tractive effort by 10 per cent.

To enable the locomotive to run well at high speeds it was necessary to avoid the creation of play through wear, and use two stages of spring suspension, each with different characteristics, whereby the locomotive is divided into three masses:

(i) The non-suspended parts (wheels and axles).

(ii) The part carried on the primary springing (bogies).

(iii) The part carried on the secondary springing (body).

The three separate masses linked together vertically by the spring gear, in addition are linked together laterally between themselves by devices which allow them to move relatively to one another under damper control. The weight of the non-suspended parts must be kept as low as possible and the spring gear must have a long period of rolling to damp the hunting of the bogie. In practice it was found necessary to accept in the primary suspension of the bogie a period of rolling of low value and, in the secondary springing, a longer period. Reduction of hunting of the bogies is achieved by carrying the weight of the body on side bearings with sliding friction.

It is also important to reduce the weight and radius of gyration of the bogies. While the possibilities of weight reduction are very limited, the radius of gyration can be reduced by moving the motors towards the bogie centres; this necessitates, on the other hand, an intermediate gear wheel for the transmission. Another feature of the design is the bogie spacing, the centres being as far apart as possible. As it was essential to avoid complicated layouts, and have a locomotive with the lowest possible operating and maintenance costs, no action was taken to reduce lateral thrust of the flanges in curves (such as tyre-lubrication or horizontal coupling between bogies). Other requisites were ease of access for lubrication, dismantling, etc.; the installation of switch gear in dustproof containers; light-weight construction, and the use of welding where possible. Furthermore, all necessary steps were to be taken to ensure ease and safety of driving, especially so far as braking arrangements were concerned.

Summarized, the requirements were as follows:

(i) Spring gear of the carriage type on two stages of springs, the second stage between body and bogie being on the pendulum system.

(ii) Fully-sprung motors, arranged in the centre of the bogies.

(iii) Transmission of tractive forces through the body, the transmission of tractive forces by the bogie being as low as possible.

(iv) Coupling of the two pairs of wheels of each bogie.

(v) A long locomotive.

Two groups of coiled springs on equalizers and plain axle-boxes support the frame and body on the bogies. Compound rubber rings are fitted for the purpose of damping out wheel oscillation. The longitudinal connections between axles and bogies are formed by two Silentblocs. Most interesting is the manner in which the motors and transmissions are installed. The two motors per bogie are arranged close to one another and the pinions run in a gear case on the same side of the locomotive. The main gear wheel is mounted on the axle and the intermediate gear wheel is in the same gear case, resulting in a strong, single gear train. An intermediate wheel couples both axles of each bogie. It is interesting to note that this coupling of motors through intermediate gear wheels is a continuation of an earlier 2–Do–2 design, which was completed early in the last war. Transmission

of power from the large gear wheel to the axles is made through a cardan type of drive, the ends of which are fitted with Silentblocs; in addition, the main gear wheel has a resilient frame. The drive also transmits lateral forces between frame and axles.

The connections of the body with the bogie in the three planes (vertical, transverse and longitudinal), normally provided by a single axial pivot, are definitely separated in this case: the haulage in the longitudinal direction is obtained by drawbars passing under the bogie and attached thereto; the vertical bearing and the transverse guiding are provided by a pendulum suspension having the following arrangement: On each side of the bogie the body rests on anti-vibration devices carried on the ends of laminated springs. The buckle of this spring is suspended from inclined links at the end of a load-carrying bolster. This bolster bears on the bogie through side bearings. It is linked transversely to the bogie by a rod and longitudinally by two links fitted with Silentblocs. As an experiment, provision has been made for fitting a friction shock absorber on the pendulum suspension. This arrangement prevents nosing by the friction of the side bearings.

The main particulars of the Bo–Bo express passenger locomotives are:

Output per 4 motors (continuous)	4,450 H.P.
Maximum speed	140 Km/h. (87 m.p.h.)
Total weight	80 tonnes (78½ tons)
Weight of mechanical part	49·4 tonnes (48½ tons)
Weight of electrical part	30·6 tonnes (29½ tons)
Weight of spring-borne part (body)	37 tonnes (36¼ tons)
Weight of spring-borne part (bogies)	2×15=30 tonnes (29·4 tons)
Weight of part not spring borne	4×3·25=13 tonnes (12¾ tons)
Total weight, spring borne	67 tonnes (65½ tons)
Length overall	16,200 mm. (53 ft 2¼ in.)
Distance between bogie centres	9,200 mm. (30 ft 2½ in.)
Total wheelbase	12,400 mm. (40 ft 8½ in.)
Bogie wheelbase	3,200 mm. (10 ft 6 in.)
Wheel diameter	1,250 mm. (4 ft 1 in.)
Gear ratio	1 : 4·2

The Types CC 7000 and BB 9000 show the latest development of high-speed universal D.C. locomotives. They have been highly successful and have resulted in substantial repeat orders being obtained from foreign countries.

The Bo–Bo series, BB 0401 (Plate 41C, page 187), was developed after World War II and over 100 units have been delivered. They are intended for hauling 680 tonnes

Le Mat. El SW

(B)

P. Ransome-Wallis

(A)

PLATE 41A. F.S. (Italy) Bo–Bo–Bo locomotive series E636 for 3,000 v. D.C.

PLATE 41B. S.N.C.F. Bo–Bo–Bo experimenta locomotive of 3,810 H.P.

PLATE 41C. S.N.C.F. Bo–Bo locomotive, 2,400 H.P. series BB-0401.

Alsthom

(C)

Le Mat. El SW

(C)

S.L.M. Winterthur

(B)

Le Mat. El SW

(A)

PLATE 42A. S.N.C.F. Bo–Bo experimental high speed locomotive, No. BB–9004.

PLATE 42B. S.N.C.F. Co–Co locomotive, No. CC–6051, the first locomotive built for the French 25,000 v. single-phase, 50-cycle A.C. electrification.

PLATE 42C. S.N.C.F. 3,102 H.P. Bo–Bo locomotive, No. BB12-018 for 25,000 v. single-phase, 50-cycle current, using ignitron rectifiers,

Virginian Railroad

PLATE 43A. Virginian Railroad Co–Co locomotive of 3,300 H.P. for 11,000 v. single-phase, 25-cycle A.C.

General Electric Co., USA

PLATE 43B. N.Y., N.H. and H. Railroad (The New Haven Railroad) Co–Co locomotive of 4,000 H.P., using rectifiers to convert 11,000 v., single-phase, 25-cycle A.C. to 600 v. D.C. for traction motors.

British Railways

PLATE 44A. British Railways – Southern Region Co–Co locomotive using
660 v. D.C. (third rail conduction).

P. Ransome-Wallis

PLATE 44B. British Railways – Eastern Region Bo+Bo locomotive using
1,500 v. D.C. Manchester–Sheffield electrification.

P. Ransome-Wallis

PLATE 44C. British Railways – Eastern Region Co–Co locomotive using
1,500 v. D.C. Manchester–Sheffield electrification.

Brown, Boveri

PLATE 45A. S.B.B. Co–Co locomotive series Ae 6/6, 6,000 H.P. using 15,000 v. single-phase, 16 2/3-cycle A.C.

Brown, Boveri

PLATE 45B. S.B.B. Motor-bogie for series Ae 6/6 locomotive.

Oerlikon

PLATE 46A. S.B.B. 1–D–1 locomotive series Ae 4/6, 5,700 H.P.

The English Electric Co. Ltd

PLATE 46B. New Zealand Government Railways 3 ft 6 in.-gauge,
Bo–Bo–Bo locomotive of 1,800 H.P.

The English Electric Co. Ltd

PLATE 46C. R.E.N.F.E. (Spain) 5 ft 5 15/16 in.-gauge, Co–Co locomotive of 3,600 H.P. Fitted with Brown, Boveri disc drive.

Brown, Boveri

PLATE 47A. S.B.B. Bo–Bo lightweight locomotive series Ae 4/4.

Oerlikon

PLATE 47B. S.B.B. (1–Bo–1–Bo–1) × 2 heavy freight locomotive, series Ae 8/14, 8,800 H.P. for the Gotthard Line.
(*See* Plate 55B, page 219)

S.L.M. Winterthur

(C)

PLATE 48A. S.B.B. Rack and adhesion locomotive for the Brunig line, series HGe 4/4.

PLATE 48B. Power Bogie of series HGe 4/4.

PLATE 48C. S.B.B. Driver's controls of series HGe 4/4.

S.L.M. Winterthur

(A)

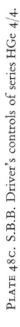

S.L.M. Winterthur

(B)

($666\frac{1}{2}$ tons) trains at a speed of 80 Km/h. (49·7 m.p.h.). The maximum permissible weight is 92 tonnes (89·1 tons) the maximum axle load being 23 tons. The weight can actually be adjusted between four nose-suspended motors driving the four axles and two bogies by means of flexible couplings. The bogies are of welded construction and coupled together; they support the body which is entirely welded and consists of sub-assembled units such as side walls, frame, cross members, cabs and roof. All electrical equipment is built up in sub-assemblies which can easily be removed. A feature of the design is the connection of axle boxes and bogies by single coupling links mounted on silentbloc mountings. The locomotive is capable of the following performance: 1,300 tonnes (1,274 tons) freight trains at 82·5 Km/h. (51 m.p.h.) on the level, or at 53·5 Km/h. (33 m.p.h.) on 1 in 200 gradient; 650 tonnes (637 tons) passenger trains at 115 Km/h. (71·4 m.p.h.) on the level or at 87·5 Km/h. (54 m.p.h.) on 1 in 200 gradient. Main dimensions are:

Total length	12,930 mm. (42 ft 5 in.)
Bogie wheelbase	2,950 mm. (9 ft 9 in.)
Wheel diameter	1,400 mm. (4 ft 7 in.)
Gear ratio	1 : 4·14
Continuous output	2,400 H.P.
Maximum speed	115 Km/h. (71·4 m.p.h.)
Tractive effort	14,935 Kg. (32,857 lb.)

Before the last war, the S.N.C.F. required a heavy locomotive for hauling 1,200 tonnes (1,176 ton) trains at speeds of 45 Km/h. (28 m.p.h.) on the 1 in 100 gradients of the Brive-Montauban section. After careful consideration it was decided to build a Co–Co type – and also a Bo–Bo–Bo, a design not hitherto used in France.

The Co–Co series, CC 6001 has a single body which rests on the two bogies of three axles each, all driven by nose-suspended motors. The bogies carry all coupling and buffing gear and are themselves joined by a special coupling. This is designed to allow lateral and vertical movements of the bogies whilst transmitting all driving forces. The body and bogies are completely arc-welded. The body rests on three points over axles 1 and 2, and 5 and 6, suitably adjusted to allow for side movements and movements on curves; the body is also supported over axles 3 and 4, which two axles are connected by a beam.

The locomotive has a very unusual electrical layout in its automatic multi-notch control gear with fine gradation. The six motors can work in the following combinations:

(i) All in series.
(ii) Two parallel groups of three in series.
(iii) Three parallel groups of two in series.

There are 78 full-field notches and 9 shunt-field notches, making 105 notches in all – of which 13 are running positions selected automatically by the controller. Arrangements are made for hand control by a separate lever. The principle used in this design is the so-called balanced bridge connection, which does not interfere with the tractive effort exerted. The whole installation is based on a servo-motor-driven camshaft-operated contactor gear which contains seventy-nine contactors. Main dimensions are:

Length over buffers	18,600 mm. (61 ft 1 in.)
Total wheelbase	13,500 mm. (44 ft 3 in.)
Bogie wheelbase	4,750 mm. (15 ft 11 in.)
Wheel diameter	1,350 mm. (4 ft 5 in.)
Weight of mechanical part	72 tonnes (70½ tons)
Weight of motors with drive	3 tonnes (2·9 tons)
Weight of control equipment and installations	17 tonnes (16·6 tons)
Total weight	120 tonnes (117·6 tons)
Continuous output	3,600 H.P.
Maximum speed	105 Km/h. (65 m.p.h.)

The Bo–Bo–Bo (Plate 41B, page 187) was developed for comparison with the Co–Co, and it came into service during 1948. The body, which is a single unit, rests on three two-axle bogies, all motored. The body rests on the two outer bogies with a central pivot and two side supporting pieces, the pivot having a side displacement of 20 mm. (¾ in.). The centre bogie does not take weight and has a side displacement of 280 mm. (11 in.). The body is all-welded, the bogie frames are cast in steel and equipped with Athermos axle-boxes. The six traction motors can be arranged in series, in series/parallel (2×3 motors in series) or in parallel (3×2 motors in series). This arrangement can be used during motoring as well as during regeneration, when running downhill. There are two cabs, each containing a master controller which gives the following running positions: 26 notches with the resistances inserted, 1 without resistance and full field, 15 shunt-field positions, and 9 by varying the excitation of the motors. The motors are nose-suspended and transmit their power through flexible gear wheels. The locomotive is equipped with automatic and direct air brakes as well as hand brakes. There are the usual auxiliaries, such as compressors, ventilators, etc. The main particulars are as follows:

Total weight available for adhesion	120 tonnes (117·6 tons)
One-hour rating	3,810 H.P.
Continuous rating	3,450 H.P.
Bogie wheelbase	3,150 mm. (10 ft 0 in.)
Total wheelbase	13,690 mm. (42 ft 7 in.)
Total length over buffers	18,700 mm. (61 ft 3½ in.)

Heavy shunting (switching) locomotives. 1,500 v. D.C. In 1938 the S.N.C.F. took delivery of two unusual shunting locomotives, type CC 1001 of the C–C wheel arrangement. They were originally ordered from Messrs. Oerlikon and Batignolles-Chatillon by the former Paris–Orleans–Midi Railway for work in the hump yards at Vierzon and Juvisy, electrified at 1,500 v. D.C. The two locomotives embody the Ward-Leonard type of control without resistances, so as to avoid overheating of resistances when shunting heavy trains at low speeds. The locomotive has two six-wheeled bogies with outside frames. Each bogie contains two motors driving the two outer axles through twin gears with a ratio of 14 : 100 and the other axle by means of coupling rods connected to fly cranks. The total weight in working order is 90·4 tonnes ($88\frac{1}{2}$ tons). The main frame carries the whole body and rests (with three-point suspension systems) on the two bogies. The locomotive is fitted with automatic Westinghouse air brakes, having a non-automatic regenerative braking system.

The electric equipment consists of the following units: a motor-generator comprising a compound-wound motor and an anti-compound-wound generator, four traction motors, two excitation rheostats, an auxiliary excitation generator and a second auxiliary generator.

Live current is fed to the compound motor which drives the other compound-wound generator. The latter feeds the armature of the traction motors (connected permanently in series); the generator voltage is regulated by means of one of the excitation rheostats.

The excitation generator feeds the field winding of the traction motors. The voltage of the excitation generator is regulated by the second excitation rheostat.

The auxiliary generator excites the field windings of the compound-wound motor, anti-compound generator, and excitation generator.

The locomotive is operated by means of a servo-motor-worked controller. The controller moves the contactors of both rheostats thereby controlling, at the same time, first, the voltage delivered by the other compound-wound generator which feeds the motor armatures, and secondly, the voltage of the current fed by the auxiliary generator to the motor field winding. This combination gives a definite locomotive speed independent of the tractive effort required, which varies according to load and profile of the line. The variation of the excitation of a compound-wound generator results also in a variation of the excitation of the traction motors, thereby giving a change in speed. Speeds required with trains up to 2,000 tons are 6–8 m.p.h. in the yards, and

1–2 m.p.h. on top of the hump. A maximum speed of 30 m.p.h. is permissible. The main motor has an output of 400 kW. (continuous) and 525 kW. (one-hour). Generator and traction motor current is rated at 300 Amp. (continuous) and 385 Amp. (one-hour).

Single-phase, 50-cycle A.C. locomotives: four basic designs (see page 171) (Plate 42C, page 188). There are two Co–Co types and two Bo–Bo types. The lighter locomotives have either A.C. motors or Ignitron rectifiers (see Pennsylvania Locomotives, p. 205), while the heavier locomotives are all of the mixed type, converting the single-phase A.C. either into D.C. or 3-phase A.C. The locomotives were ordered in the latter half of 1953. They are, both in their mechanical and electrical parts, of very unusual design. Dealing first with the transformers, the current at 25,000 volts is received into oil-cooled transformers which are designed in the case of the Ignitron and direct motor types as step transformers, while in the others they are static.

The Co–Co locomotives of the type A.C./D.C. have a converter group with a 4-pole synchronous self-ventilated motor, running at 1,500 r.p.m., and receiving current at 2,700 volts, the motor being rated at 2,900 H.P. The converter group contains two further D.C. generators coupled on the same motor shaft, and each supplying the three traction motors which are force ventilated, and have the following characteristics: 600 v., 550 Amp., 305 kW., with a tractive effort of 3,860 Kg. (8,150 lb.) each, at 1,100 r.p.m.

The Co–Co locomotives of the A.C./3-phase A.C. type require variable frequency 3-phase A.C., and for this purpose they are equipped with two converter groups, one designed as a phase converter group and the other one as a frequency converter group, supplying the six traction motors which are of the asynchrone, squirrel-cage type. These six motors are again force ventilated and have the following characteristics: 930 v., 330 Amp., 440 kW., under 91·3 cycles and 1,055 r.p.m., resulting in a continuous output of 4,000 Kg. (8,818 lb.) each (Colour Plate VII, page 229).

The Bo–Bo locomotives with direct motors are of a very simple design and contain four single-phase motors with series excitation without compensating shunts. The characteristics of these motors are as follows: 235 v., 3,200 Amp., 500 kW. at 681 r.p.m., giving a total output of 3,390 Kg. (7,470 lb.) continuous.

The fourth type, Bo–Bo locomotives with rectifiers, contain eight single anodes supplying four 6-pole D.C. motors, having the following characteristics: 675 v., 950 Amp., 575 kW. at 880 r.p.m., giving an output per motor of 4,400 Kg. (9,700 lb.).

Drive. The Co–Co locomotives of the A.C./D.C.

type have flexible couplings between motors and wheels, while the Co–Co locomotives of the A.C./3-phase A.C. types have nose-suspended motors, again with flexible rubber couplings. In the case of the Bo–Bo machines, the traction motors are housed in the bogies, but their pinions are coupled by an intermediate gear wheel, forming together with the pinions and the motor gear wheels, a complete gear train on each side of the bogie coupled to the other side by a cardan shaft.

Their main particulars are given in the table below.

New Bo–Bo and B–B locomotives for the Paris–Lille 25 kV. A.C. electrification. To work the Paris–Lille line the S.N.C.F. have built four classes of electric locomotives. Two of these are of the same types as those already described, there being twenty units of Series CC 14.000 and forty-three of Series BB 12.000.

The other two classes are new designs, and comprise sixteen Bo–Bo express locomotives of Series BB 16.000 and twenty B–B lightweight locomotives of Series 16.500 for passenger or freight duties. Ultimately, fifty-

Current supply	25,000 v., single-phase A.C., 50 cycles frequency.			
Series:	CC 14.000	CC 14.100	BB 13.000	BB 12.000
Wheel arrangement	Co–Co	Co–Co	Bo–Bo	Bo–Bo
Length over buffers (mm.)	18,890	18,890	15,200	15,200
	(62 ft 0 in.)	(62 ft 0 in.)	(49 ft 11 in.)	(49 ft 11 in.)
Total wheelbase (mm.)	14,140	14,140	11,400	11,400
	(46 ft 5½ in.)	(46 ft 5½ in.)	(37 ft 5 in.)	(37 ft 5 in.)
Rigid or bogie wheelbase (mm.)	4,670	4,670	3,200	3,200
	(15 ft 4 in.)	(15 ft 4 in.)	(10 ft 6 in.)	(10 ft 6 in.)
Wheel diameter (mm.)	1,100	1,100	1,250	1,250
	(3 ft 7 in.)	(3 ft 7 in.)	(4 ft 1 in.)	(4 ft 1 in.)
Weight, total in working order (tonnes)	120	120	80	80
(tons)	117½	117½	76½	76½
Weight, adhesive (tonnes)	120	120	80	80
(tons)	117½	117½	76½	76½
Maximum axle load (tonnes)	20	20	20	20
(tons)	19½	19½	19½	19½
Number of motors	6	6	4	4
Type of motor	4-pole synchronous 2,700 v.	930 v. asynchrone squirrel-cage	Series-excited s.ph. without compensating shunts	Series-excited s.ph. without compensating shunts
Electrical transmission	Transformer and Arno-converters	Transformer and phase and frequency rotary converters	Transformer	Transformer and Ignitrons
Power transmission	Nose-suspension	Nose-suspension	Cardan coupling with Silentblocs	Cardan coupling with Silentblocs
Total output (continuous) (H.P.)	2,490	3,590	2,724	3,120
Maximum tractive effort (Kg.)	42,000	40,000	24,500	24,500
	(92,400 lb.)	(88,000 lb.)	(53,900 lb.)	(53,900 lb.)
Maximum speed (Km/h.)	60	60	105	105
	(37 m.p.h.)	(37 m.p.h.)	(65 m.p.h.)	(65 m.p.h.)
Performance (continuous)	2,490 H.P./ 23,200 Kg. (51,040 lb.) at 27·9 Km/h. (17¼ m.p.h.)	3,590 H.P./ 24,000 Kg. (52,800 lb.) at 39·2 Km/h. (24¼ m.p.h.)	2,724 H.P./ 13,500 Kg. (29,700 lb.) at 53 Km/h. (33 m.p.h.)	3,120 H.P./ 17,600 Kg. (38,720 lb.) at 47 Km/h. (29 m.p.h.)
Builder	Alsthom	Oerlikon and Batignolles	Jeumont and Schneider	Mat.El.S.W. and Schneider

one of Series BB 16.000 and 119 of Series BB 16.500 are to be built for the Nord-Est electrification.

Series BB 16.000. The electrical part of these locomotives is similar to that of Series BB 12.000 but the D.C. motors use current at 920 v. 1,040 Amp., instead of at 675 v. 1,000 Amp. Each motor has a continuous rating of 1,230 H.P. as against 840 H.P. for the BB 12.000. The ten-inch Ignitron rectifiers are of improved design and of increased output. The two pantographs are of the two-pan AM type and are > shaped instead of the conventional lozenge shape. They are specially designed for high-speed running.

The mechanical part is based on that of the highly successful Bo–Bo locomotives of Series BB 9.200 operating on the D.C. electrified lines of the S.N.C.F. There is a driving cab at each end and the superstructure is of the "box-cab" type.

The bogies are without centre-pivot and follow the Alsthom swinging pivot system as fitted to the high-speed D.C. locomotives. The motors are suspended in the centre of each bogie and they are geared for a maximum speed of 160 Km/h. (100 m.p.h.).

Series BB 16.500. These locomotives are probably the most interesting of the present decade. They are of the B–B type and are of variable gear ratio to enable them to be used either for passenger or freight service. Their total weight of 66 tonnes compares with that of 84·5 tonnes for Series BB 16.000.

The electrical part is largely standard with that of other classes of Ignitron rectifier locomotives. The two pantographs are of the AM type, and there is an air-blast circuit-breaker. The transformer comprises a tapped auto-transformer winding with a ratio of 25 : 15 kV., a main power transformer with centre-tapped secondary

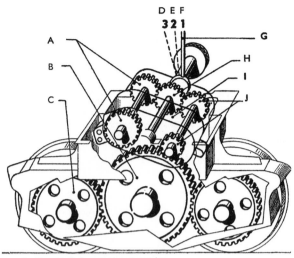

A. Gear train for passenger service.
B. Driving gear wheel for both axles.
C. Gear wheel of axle.
G. Gear lever { D. Passenger working
                E. Neutral position
                F. Freight working
H. Driving pinion of motor
I. Pivotted gear casing which can occupy any of the three positions determined by the gear lever
J. Gear train for freight service

Schematic drawing showing gear arrangements of the Series BB 16.500.

and an auxiliary winding. The high-voltage tap-changer is servo-motor operated and provides forty running notches by the selection of various combinations of the twenty transformer tappings.

The transformer, which is oil-cooled, supplies 2,850 kVA. for traction, 140 kVA. for auxiliaries and 675 kVA. for locomotive and train heating, a total rating of 3,665 kVA.

Two pairs of ten-inch diameter Ignitron rectifiers supply a full-wave-rectified output for each of the two

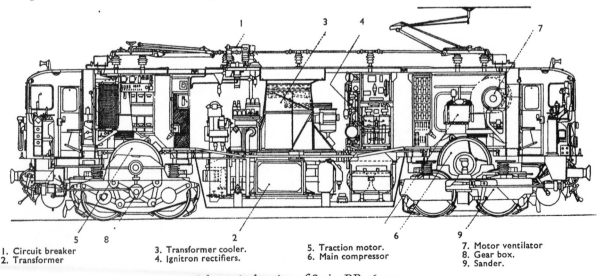

1. Circuit breaker
2. Transformer
3. Transformer cooler.
4. Ignitron rectifiers.
5. Traction motor.
6. Main compressor
7. Motor ventilator
8. Gear box.
9. Sander.

Schematic drawing of Series BB 16.500.

traction motors, which are permanently connected in parallel.

The mechanical part provides much that is unusual in design. Each bogie is powered by a simple D.C. traction motor of 1,750 H.P. (continuous rating). Each motor is spring-borne on the bogie and drives the two axles of its bogie through a change-speed gearbox, the coupling being made by an Alsthom floating ring. The final drive is through a large gear which meshes with pinions, one on each axle, and on the outer side of the carrying wheels. The gearbox enables either of two gear ratios to be selected by the movement of a lever in the side vestibule of the locomotive. Gear change is effected only when the locomotive is at rest, suitable locking devices being provided to prevent misuse. The higher gear ratio, for passenger-working allows a maximum speed of 140 Km/h. (87 m.p.h.). While the low gear engagement provides a maximum speed of 85 Km/h. (52½ m.p.h.)

for freight working. In high gear the maximum tractive effort is nineteen tonnes, in low gear thirty-two tonnes.

In order that the final drive can be made through a single centrally-placed gear wheel driving, and so coupling both axles of a bogie, the two axles have to be placed much closer together than is usual, and the wheelbase of each bogie is only 5 ft 3¼ in. Excellent stability has been obtained by use of the Alsthom swinging pivot system, and the axleboxes are attached to the bogie frames by horizontal links mounted on Silentbloc bushes.

The results which have already been obtained with these locomotives in both passenger and freight service have been such that in this system, the S.N.C.F. have found an ideal mixed-traffic machine, and the same locomotive is now envisaged as working suburban trains, or in multiple unit, the heaviest freight trains.

Principal characteristics of the Series BB 16.000 and BB 16.500

	BB 16.000	BB 16.500	
		Passenger gear	Freight gear
Horse power-continuous rating	4,920 at 85 Km/h.	3,500 at 82 Km/h.	3,500 at 48 Km/h.
Corresponding tractive effort	15·2t.	11t.	18·8t.
Maximum tractive effort	32·2t.	19t.	32t.
Maximum speed	160 Km/h.	140 Km/h.	85 Km/h.
Weight of electrical part	38.9t.	28t.	
Weight of mechanical part	45·6t.	38t.	
Total weight	84·5t.	66t.	
Length over buffers	16,200 mm.	14,400 mm.	
Traction motors	4 bogie suspended motors – (rectified) D.C. (920 v. 1,040 Amp.)	2 motors, bogie suspended (rectified) D.C. (1,100 v. 1,250 Amp.)	

INDIAN RAILWAYS

Co–Co 3,600 H.P. locomotives, 1,500 v. D.C. In 1951, the Indian Railways were supplied with seven Co–Co 3,600 H.P. electric locomotives for operation on the Central Railway. They were built by the English Electric Co. and the Vulcan Foundry. The locomotives are for operation on the Bombay-Igatpuri sections, in all about 200 route-miles. The routes include long 1 in 40 gradients combined with sharp curves. The purpose of these new locomotives is the introduction of greater speeds in the movement of heavy passenger and freight traffic. The locomotive underframe and superstructure are of welded steel structure. The bogies are also all-welded and carry the weight of the locomotive superstructure through a fabricated bolster. The locomotive weight is transferred from the bogie frame to the axle-

boxes through four pairs of coil springs, each pair of springs being carried on one of the four equalizing beams on each bogie. The ends of the equalizer beams rest in slots in the top of the roller-bearing axle-boxes. Manganese-steel liners are fitted to the rubbing surfaces between the bolster and bogie frame, and between the bolster and underframe.

Each bogie contains three traction motors. They are of the axle-hung type and drive the axles through single-reduction resilient spur-gears. The traction motors are force ventilated, each having a continuous rating of 530 H.P. and a one-hour rating of 600 H.P. Special provision has been made in the design of the traction motors to ensure that the locomotive may safely operate on flooded track during the monsoon period.

The controller has nine running notches without

resistances in the motor circuits. These notches correspond to: all six motors in series, two parallel circuits of three motors in series, and three parallel circuits of two motors in series. In each of these three groupings, the motors may be run in full field, intermediate or weak field.

There are fourteen regenerative brake notches which are available in series, series/parallel or parallel motor groupings. The regenerative braking is interlocked with the locomotive air brakes, so that when the locomotive is regenerative the air brakes are held off should the train vacuum brakes be applied, thereby eliminating any danger of locking the locomotive wheels. Should regenerative braking be unobtainable, an emergency brake application is automatically made. In addition, should the driver make an emergency brake application, regeneration automatically ceases.

An unusual design feature of the locomotive is that should there be a failure of any part of the power, control or auxiliary equipment, half of the equipment may be isolated and the locomotive may proceed on the remaining half. Main particulars are:

Gauge	5 ft 6 in.
Length over buffers	68 ft 4¼ in.
Wheelbase of bogie	15 ft 9 in.
Distance between bogie centres	34 ft 10 in.
Wheel diameter (new)	4 ft 0 in.
Minimum curve negotiable	350 ft 0 in.
Weight of locomotive in running order	123 tons
Weight per axle	20·5 tons
Gear ratio	16 : 59

Tractive effort:

Maximum at 25 per cent adhesion	69,000 lb.
At one-hour rating at 28·5 m.p.h.	48,000 lb.
At continuous rating at 30 m.p.h.	39,000 lb.
Operating speed	75 m.p.h.

Compared with the G.I.P. locomotives of 1925, these locomotives show clearly the advances made in locomotive design of the same type.

NETHERLANDS RAILWAYS (N.S.)

Co-Co locomotives, 1,500 v. D.C.: European and American designs. At the end of World War II the Netherlands State Railways were faced with the reconstruction of their almost completely destroyed railway system. It was decided to extend electrification over the whole country by easy stages. Before the war, only motor-coach trains were used – with considerable success. After the war these were to be continued, and to obtain full benefits from electrification, express and freight trains were to be hauled by locomotives. The first order was for ten express passenger locomotives (Nos. 1001–1010)

with wheel arrangement 1–Do–1, following closely the Swiss Ae 4/7, the design being developed by Oerlikon and S.L.M. Winterthur. Further orders followed, fifty Bo–Bo locomotives from Alsthom (Nos. 1101–1150), following closely the French Bo–Bo designs and a further order for Co–Co types, again following the well-known S.N.C.F. experimental locomotives types CC 7000/7100.

It was also decided to order twenty-five Co–Co locomotives of American design manufactured in Holland under Westinghouse licence. This decision, from the start, has led to considerable controversy, as some U.S. electric locomotives have not been very successful in Europe. However, the Dutch locomotives will form a most interesting basis of comparison between the Swiss and French types as well as the British Bo–Bo, which was on loan to Holland for some years and which is now at work on the Pennines electrification.

In design and appearance the new locomotives are most unusual. They have to haul trains of 205 tonnes (201 tons) at 125 Km/h. (77½ m.p.h.) and freight trains of 1,650 tonnes (1,617 tons) at 60 Km/h. (37·2 m.p.h.). Their construction follows closely U.S. practice: cast-steel bogies, which are entirely unconnected; heavy superstructure of all-welded construction; driver's cabs with all modern conveniences placed behind a high nose to give protection to crews against head-on collisions.

Other main particulars are as follows:

Wheel arrangement	Co–Co
Total weight = weight on drivers	105 tons
Length over-all	18,104 mm. (59 ft 5 in.)
Total wheelbase	13,868 mm. (45 ft 6 in.)
Rigid wheelbase	4,724 mm. (15 ft 6 in.)
Wheel diameter	1,100 mm. (3 ft 7 in.)
Maximum operating speed	140 Km/h. (87 m.p.h.)
Minimum curve radius	100 m. (328 ft 1 in.)
Continuous rating of locomotive at full field and 72 Km/h. (44¾ m.p.h.)	2,940 H.P.
at shunted field and 123 Km/H. (76½ m.p.h.)	3,000 H.P.

Tractive effort:

at start	26 tonnes (25½ tons)
at 72 Km/h. (44¾ m.p.h.)	11 tonnes (10¾ tons)
at 123 Km/h. (76½ m.p.h.)	2·8 tonnes (2¾ tons)
Number of traction motors	6
Type	TM 94 force ventilated
Gear ratio	20 : 71

NEW YORK, NEW HAVEN AND HARTFORD R.R. (U.S.A.)

Rectifier locomotives for single-phase A.C. and D.C. In December 1954, this railway purchased ten rectifier locomotives from the General Electric Co. of America. The locomotives are of unusual design on account of the specialized nature of the New Haven passenger locomotive operation into New York City. The railroad's main line is electrified with single-phase A.C. at a frequency of 25 cycles and a pressure of 11,000 v. Current is taken from an overhead wire. This system extends from New Haven, Conn., to Pennsylvania Station in New York, N.Y. From Woodlawn, N.Y., to New York's Grand Central Terminal, over the Park Avenue Viaduct, the line is electrified with a 660 v. D.C. third rail system. Passenger locomotives must, therefore, be designed for operation from these two power sources; they must also carry complete train heating equipment and supplies, yet must not exceed the axle loading limit of 58,000 lb. over the viaduct.

The design (Plate 43B, page 189) incorporates a body of box-type construction with cabs at both ends which are streamlined. It is completely fabricated from sheets and is of all-welded construction. The cab sides are designed as load-bearing girders which support the underframe and equipment. Each end of the underframe is fitted with a pilot, rubber draught gear and tight-lock coupler.

Arrangement of equipment on the roof provides for an A.C. pantograph mounted on a hatch cover at each end. These hatch covers are built as ducts. Ventilating air for the equipment is taken in from the side of these ducts at the roof level. The solid centre portion of the roof carries the accelerating resistors and blowers.

The body is carried on two 3-axle, swing-bolster trucks, all axles being motored. In order to provide room for the motor on the centre axle, the centring plate is located between the first and second axles. Spring-loaded sliding plates on each side of the truck frame between the second and third axles also carry load and provide uniform distribution of weight on the axles. Inside equalizers, supported on top of the roller-bearing journalboxes, carry the cast-steel truck frame on helical springs. The frame, in turn, carries the swing-bolster and centre plate on four swing links, spring plank, and elliptical springs.

The main transformer, having high-voltage bushing on the top, protrudes through the hatch cover to confine the 11,000 v. circuits to the roof. Secondary taps and tap switches are mounted on one end of the tank. The transformer is liquid cooled by means of Pyranol. A motor-driven pump circulates the liquid through the tank and windings to a cooler on the right side of the tank. Ventilating air from the duct in the cab underframe is forced through the cooler.

All of the D.C. control, together with auxiliary and change-over switches, is located in the two control cabinets. The control cabinet structure supports the accelerating resistors and blowers on the roof. This arrangement keeps the length of the leads between the resistors and the switches at a minimum. The resistors are of the ribbon type and are ventilated by means of motor-driven blowers connected across the resistors. With this scheme, the amount of ventilating air is proportional to the voltage across the resistor.

Twelve main circuit rectifiers and four auxiliary rectifiers, together with the firing circuit equipment, are arranged in the two rectifier cabinets. The rectifier cooling system pump, with its controls, is located in the lower part of the blower compartment adjoining the rectifier cabinets. Standard 8-in. rectifier tubes have been modified for this application to provide heavier anode construction and changes in the mercury pool to assure ignition contact under the most adverse conditions of locomotive movement. Cooling water is circulated through the tubes, control valves and air-cooled radiators by means of a motor-driven pump.

Two reactors, located between the two rectifier cabinets, are air-cooled from the main ventilating duct. One is an air-core type reactor with six separate coils for limiting the arc-back current in each of the rectifier circuits. The other is an iron-core reactor used to smooth the current ripple in the rectifier output to the traction motors. This is necessary for proper commutation of the motors. The auxiliary motors also have smoothing reactors for the same purpose.

In the D.C. zone, power is supplied from the third rail at 660 v. It is collected by means of third-rail shoes located on each side of both trucks. Connections are made from fuses through two line breakers and a change-over switch to the accelerating resistors and traction motors. Arrangement of the traction motor circuit provides two motor combinations; two in series, three in parallel for starting and low-speed operation, and full parallel for high-speed operation.

In the A.C. zone, power is supplied to the main transformer primary from the overhead wire at 11,000 v., 25 cycles. From the transformer secondary, power is supplied through tap switches and current limiting reactors to twelve main rectifier tubes arranged in three separate bridge-connected circuits. Power output from the rectifier supplies six traction motors through the change-over switches, the main smoothing reactor, line switches and accelerating resistors. Arrangements of the traction motor circuit provides two motor combinations: three in series, two in parallel on the 1,407 v. top, and two in series, three in parallel on the three higher voltage taps. Voltage variation on the traction motors is obtained by means of the D.C. accelerating resistors in the traction motor circuit in combination with the four transformer taps.

These locomotives are the first built for the New Haven railway with all-adhesion arrangement.

In a recent building programme, an important viaduct, the Park Avenue Viaduct, was modernized and its load-carrying capacity increased. When the rectifier locomotive was proposed with all weight on two widely spaced, three-axle, swing-bolster trucks, it was approved for 26 tons per axle. With six axles this means that the total locomotive weight was limited to 152 tons.

Main particulars of locomotive:

Weight:

Total locomotive fully loaded	152 tons
Per driving axle	26 tons

Dimensions:

Length inside knuckles	68 ft 0 in.
Total wheelbase	52 ft 6 in.
Rigid wheelbase	15 ft 0 in.
Length between centre plates	44 ft 0 in.
Wheel diameter	40 in.

Ratings:

Tractive effort at 25 per cent adhesion	87,000 lb.
Tractive effort continuous	34,100 lb.
Speed at continuous rating	44 m.p.h.
Power at continuous rating	4,000 H.P.
Maximum speed	90 m.p.h.

NEW ZEALAND GOVERNMENT RAILWAYS

British-built Bo–Bo–Bo locomotives, 1,500 v. D.C. Typical of a modern British locomotive layout are the seven 1,800 H.P. electric locomotives (Plate 46B, page 192) supplied by the English Electric Co. Ltd, which were required to increase the services operated with locomotives in the Wellington area and also to operate new services on the 23-mile extension of the existing electrification to Upper Hutt. Current collection is from an overhead contact wire at 1,500 v. D.C. The track gauge is 3 ft 6 in. and the loading gauge is severely restricted by a series of short, single-line tunnels south of the terminal station at Paekakariki, 26 miles distant from Wellington.

The following performances were specified for the new locomotives:

Freight trains – one locomotive to haul 600 tons trailing between Wellington and Upper Hutt at a free running speed of 30 m.p.h., and be capable of starting a 500-ton train on a 1 in 57 up grade and continuing to run.

Passenger trains – one locomotive to haul 400 tons trailing between Wellington and Upper Hutt at a speed of 55 m.p.h. on level tangent track with a maximum safe speed of 60 m.p.h.

The leading particulars are:

Weight	75 tons
Maximum axle load	12½ tons
Length over drawgear	62 ft 0 in.
Wheel diameter	3 ft 0½ in.
Number of traction motors	6
Motor groupings	6, 3 and 2 in series
One-hour rating of traction motor	300 H.P.
Continuous rating of traction motor	250 H.P.
Gear ratio	17:61
Rating (one-hour)	23,300 lb. T.E. at 28 m.p.h.
Rating (continuous)	18,400 lb. T.E. at 30 m.p.h.

The arrangement of body and axles, namely two articulated superstructures carried on three motored bogies, provides great flexibility on the sharply curved New Zealand tracks, and good riding qualities. Each half superstructure is carried on bogie centre bearings and is supported by side rubbing blocks on the centre bogie only. This provides three-point suspension for the half superstructures and leaves the outer bogies free to accommodate themselves to track inequalities without influencing the weight distribution on the rails. Although the side bearer clearances have been adjusted so that the superstructure is controlled by the centre bogie (the outer bogies supporting the superstructure weight on the centre pivots only), provision has been made to enable side bearer clearances to be adjusted so that it would be possible to control the superstructures by the outer bogies and not the inner one. The articulation between the two superstructures is so arranged that one half structure rests on the centre bearer and two side bearers of the second half superstructure, this latter half then being carried on the bogie. The movement of the bogies relative to the superstructure is only approximately one half of that for a completely rigid superstructure, and the side forces on the track and the clearances to structures are considerably reduced. This design also permits the over-all length to be greater than that of a single superstructure locomotive and makes the positioning of the electrical equipment easier.

The body forms a sturdy box formed from riveted members and carrying in front a Janney Yoke draw-gear. At each free end is a driver's cab. Two main resistance compartments are constructed behind each cab and are separated by a central corridor. The main compartments are situated behind the resistance compartments and extend to the articulation in each case. These contain the main equipment frames and auxiliary machines. Each roof carries a pantograph and is removable. Air for ventilating the traction motors is drawn in through oil-wetted filters in the locomotive sides. Hinged access

PLATE VI. S.N.C.F.: south-bound "Mistral" express with an electric locomotive of series CC 7001 using 1,500 v. D.C.

J. Stone and Co., (Deptford) Ltd.

doors are provided to enable the auxiliary machines to be easily withdrawn during overhauls. The flexible connection between the superstructure consists of a canvas vestibule supported on lazy tongs. Flexible electric cables and air hoses are carried across the articulation and are separated from the central gangway by sliding steel covers. The floor through the gangway is a roller-mounted steel fall-plate. Foot-steps and grab handles are secured to the superstructure ends outside the vestibule to allow access to the locomotive roof.

One-piece cast-steel bogies are used, with outside swing links, a combination of semi-elliptic leaf springs and coil springs for bolster springing and the use of a one-piece frame casting with all brake hanger and motor suspension brackets cast integrally with the frame. The frame casting is a box structure, but openings have been left so that the laminated bearing springs over the axle boxes are completely accessible and visible. Experience with bogies of this type on coaches has shown that they will ride extremely well over rough track. Ballast is carried on each bogie to counteract the out-of-balance which would otherwise be caused by the arrangement of brake gear and air-operated automatic slack adjuster.

Air for braking, control and auxiliary operation is provided by two motor-driven compressors, one in each half-superstructure. The locomotive is fitted with both automatic and independent air valves, and brake pipes are provided for use with air-braked stock.

The locomotive is powered by six axle-hung, nose-suspended, force-ventilated traction motors driving the axles through spur gearing. A total of thirty-eight notches is provided. The motor groupings and the corresponding resistance notch positions of the master controller are as follows:

First group, series: six motors in series – fourteen resistance notches.

Second group, series/parallel: two parallel groups of three motors in series – eight resistance notches.

Third group, parallel: three parallel groups of two motors in series – seven resistance notches.

Shunt transition is provided between motor groupings and combined field diversion and field tap control of the traction motor fields gives three free-running notch positions for each motor grouping. These nine notches therefore comprise the series, series/parallel and parallel connections with full field, intermediate and weak field connections. Furthermore, the resistances have a long time-rating in the series connection to enable the locomotives to carry out a certain amount of shunting and pick-up of trains *en route*. The main isolating switch is interlocked with the air supply to the pantographs. The interlock cock is operated by means of the master

controller reverse key and this ensures that when the main isolating switch is opened to isolate and earth the equipment, the pantograph air cylinders are exhausted to atmosphere to allow the pantographs to lower.

The locomotives were designed in co-operation with Robert Stephenson & Hawthorns Ltd, who also supplied the mechanical parts. In these locomotives we see again the simple D.C. locomotive, which is reliable, cheap, easy to maintain and operate.

PENNSYLVANIA RAILROAD (U.S.A.)

Ignitron rectifier locomotives. The Ignitron rectifier is a mercury-arc rectifier which, by means of an igniting device, initiates the arc of each positive half-cycle and permits the arc to cease at each negative half-cycle.

The Pennsylvania Railroad bought, in 1951–2, two freight locomotives from the Westinghouse Electric Corpn, which form an important stage in the application of modern rectifiers on locomotives. During World War II an efficient pumpless single-anode rectifier was developed and an experimental motor coach was equipped with Ignitron-type single-anode rectifiers to study the behaviour of the installation under actual working conditions. The trials were satisfactory and in 1950 the two above-mentioned locomotives were ordered. Both locomotives are two-unit types, rated at 6,000 H.P. One has the wheel arrangement (Bo–Bo–Bo)×2, and the other (Co–Co)×2. Standard D.C. electric motors are fitted. The continuous tractive effort is 132,000 lb. at 17 m.p.h., with respective starting tractive efforts of 187,750 lb. and 182,000 lb. Total weights of the locomotives are 335 and 325 tons respectively, with axle loads of 28 and 27 tons. Current taken from the overhead catenary line is 11,000 v., 25-cycle, single-phase A.C., which is converted by means of Ignitron rectifiers to D.C. A transformer is used, whose secondary windings have a centre tap with accelerating tappings at both sides. There are six 500 H.P. motors in each half, all permanently coupled in parallel and served by two Ignitron units for each half-locomotive. The motors are of the six-pole and inter-pole type. There are thirty-five motoring notches on the controller and fourteen braking notches. Rheostatic braking is employed. Voltage variation is achieved by a combination of transformer tappings and adjustment of the point in the half-cycle at which each Ignitron fires. There are the usual auxiliary systems for ventilation, cooling water circulation (for Ignitron cooling), etc. As the Pennsylvania Railroad is the leading exponent of A.C. electrification in the U.S.A., and generally uses straight A.C. locomotives, the attempted use of converter locomotives is a remarkable development. Recently, the Virginian R.R., which so far has always used phase-converter and motor-generator loco-motives, also ordered twelve locomotives from the General Electric Co. of U.S.A., which will operate under the high-voltage A.C. installation and employ standardized D.C. motors and Ignitron rectifiers (*see* Virginian Railroad, page 221).

SOUTH AFRICAN RAILWAYS

British-built Bo–Bo locomotives, 3,000 v. D.C. In 1925 the Natal section of the South African Railways had almost reached the limit of its capacity under steam operation. It was therefore decided to electrify the main line between Glencoe and Pietermaritzburg, using 3,000 v. overhead system. In later years this was extended in both directions until, in 1937, the electrification was completed between Durban and Volksrust, the total route-mileage being 404, including a branch from Ladysmith to Harrismith. As the electrification proceeded, locomotives of 1,200 H.P. were ordered at intervals. Some of the earlier locomotives now require replacement, and no fewer than 225 locomotives, each of 2,000 H.P., have been ordered in Britain (*see* Colour Plate V, page 177, showing Class 3E).

These locomotives, designated class 5E, are of Bo–Bo type, weigh 83 tons, are capable of a maximum service speed of 60 m.p.h., and will negotiate a minimum curve of 4 chains (264 ft). They will be used over the whole of the electrified system. They are equipped with regenerative braking (Plate 40B, page 168).

The superstructure forms an integral unit with the underframe, and is attached to the bogie by means of a cross transom having downward projections rigidly bolted to a cast-steel bolster. This, in turn, is supported by two semi-spherical grease-lubricated side bearing pads. These bearers form the upper half of manganese-steel-lined sliding surfaces, the lower halves being integral with a cast-steel tie-beam which joins a set of leaf springs at their centres at either side of the bogie. Each spring is fitted with rubber dampers. The bogie frame is carried on the axle-boxes by laminated springs with auxiliary coil springs and rubber dampers. Each wheel has its own brake cylinder and automatic slack adjuster operating a pair of clasp brakes. With the exception of the traction motors and roof-mounted items, all the electrical equipment, comprising control gear, starting and braking resistances, motor generator sets, fans and other auxiliary machines, is installed in the superstructure.

The four force-ventilated, nose-suspended traction motors each develop 500 H.P. at the one-hour rating. Three stages of field weakening are provided by a combination of a field tap and divert resistances. Transmission of power from the motors to the wheels is by single reduction spur-gearing and transmission shocks are reduced by resiliently mounted gearwheels. Most of the H.T. control gear is carried on two frames to which access is normally prevented by interlocking between the compartment doors, main isolating switch and main pantograph interlock cocks. The low-tension control equipment consists of electro-magnetic contactors for compressor, exhauster, battery charging, train line supply and regenerative brake excitation control, together with a low-tension change-over switch and lighting switch.

Mounted on the roof are two pneumatically operated pantographs, a surge absorber, pantograph isolating links and a horn gap. Each pantograph has a double collector pan mounted on its apex joints by a spring mechanism which enables the pan to take up sudden minor irregularities in the overhead contact without causing the main structure of the pantograph to move. The brake system is of the air-vacuum type, with a proportional valve. On moving the driver's vacuum brake valve to the release position, the speed of rotation of the exhauster is increased, ensuring a quick release of the train brakes. Operation of the driver's air-brake valve affects the locomotive brakes only.

On starting, all four motors are connected in series with the starting resistance, which is progressively cut out in steps as the accelerating handle is notched up. On reaching the full field position all this resistance is cut out of circuit. An increase of speed can then be obtained by advancing the accelerating handle to any of the weak field positions. Still higher speeds are obtained by connecting the motors in two parallel groups of two motors in series, by selecting parallel operation on the combined transition and reverse lever and returning the accelerating handle to the first notch. The accelerating handle can now be notched up again to the full field position and, if required, any of the weak field positions are also available. Provision is made so that, after a period of coasting, the driver can return immediately to the parallel operation without first having to notch up in series.

When braking regeneratively, the motor armatures are connected in series or in two parallel groups of two armatures in series, as may be required, whilst for both conditions the motor fields are connected in series and are separately excited from the exciter driven by one of the motor generator sets. The strength of the exciter field can be varied under the driver's control by means of the regenerative braking handle, and thus the degree of braking effort can be determined.

Diameter of wheels	4 ft 0 in.
Bogie wheelbase	11 ft 3 in.
Distance between bogie centres	25 ft 9 in.
Length over couplers	50 ft 10 in.

These locomotives are typical examples of a British design of great simplicity – modern D.C. equipment being used; they compare well, on initial and maintenance costs, with any similar designs throughout the world.

British-built 1–Co + Co–1 locomotives, 3,000 v. D.C. The South African Railways ordered from the North British Locomotive Co. Ltd, and The General Electric Co. of England, forty 1–Co+Co–1 locomotives, 4E class, for 3 ft 6 in. gauge. One of the limiting features as regards loading is the twenty miles of 1 in 66 gradient between Hex River and New Kleinstraat.

The locomotives were required to be capable of starting a 1,070-ton freight train on a 1 in 66 gradient or a 610-ton passenger train on 1 in 50. To meet these requirements, a 3,030 H.P. six-motor locomotive, with three motor combinations and a moderate range of field control in each combination, was provided. At the time the contract was placed, it was intended to construct a new line with a maximum gradient of 1 in 66 between De Doorns and New Kleinstraat. This would have eliminated the considerable stretches of 1 in 40 grade on the existing line. Subsequently it was decided not to proceed with this project and therefore two locomotives will have to be used to take the heaviest trains over this section of the route.

The locomotive has two bogies, each bogie being provided with three driving axles and a running axle. To avoid draw-gear stresses being transmitted to the body, the draw-gear is mounted directly on the bogie headstocks, and the bogies are inter-connected by a link. To avoid this link from having any influence on the movement of the bogies relative to one another, its pins are mounted in spherical bearings. The king-pin assemblies do not carry the weight of the body which is supported by two longitudinal and two transverse bearings on each bogie, all provided with suitable spring loading.

To ensure good riding, control of the bogies is obtained by the frictional effect of the bearers mentioned above, and by special interbogie control of a spring type which tends to keep the bogies moving in phase, but allows independent movement above a predetermined controlling force.

Control of the side movement of the pony truck relative to the bogie frame is such that it provides a constant resistance to such movement and tends to return the truck to the centre. The pony truck and leading driving axle are fully side and cross equalized, while the two other driving axles are side equalized only. Further particulars are as follows:

Gauge	3 ft 6 in.
Length over couplers	71 ft 8 in.
Bogie wheelbase	22 ft 5 in.
Total wheelbase	60 ft 4 in.
Wheel diameter (driving)	4 ft 3 in.
Wheel diameter (running)	2 ft 6 in.
Total weight in working order	155 tons
Adhesive weight	129 tons
Tractive effort (at 25 per cent adhesion)	72,000 lb.

The locomotive body has a driving cab at each end and in the centre is the machinery compartment; the middle portion of the locomotive containing the enclosed high-tension compartment. The cabs contain all the master control equipment for the power, auxiliary and braking equipment, together with certain switches and lighting circuit breakers. The machinery compartments contain the various auxiliary machines and much of the L.T. control equipment associated with them. The H.T. chamber contains all the H.T. power and auxiliary control equipment such as the camshaft changeover switches, main contactors and main H.T. auxiliary circuit breaker. Included in this chamber are the two resistance compartments which house all the various resistances associated with the H.T. circuit. These compartments are self-contained in order to prevent the hot air entering the apparatus section. Entry into the H.T. apparatus compartment is possible only through an interlocked door in the corridor.

Current is collected from the overhead line by one or other of two pantographs, each of which is provided with an isolating switch.

The locomotive performance demands three groupings of the six driving motors both when motoring and regenerating. These combinations are series, series/parallel and parallel, and it is arranged for the motors to be connected in series across the line as one group of six, two groups of three, or three groups of two respectively.

In each motor combination four continuous running steps are provided by field excitation control, namely:

(i) Full field excitation.
(ii) Field diversion.
(iii) Further field diversion.
(iv) Weakest field excitation.

When regenerating, the motor fields are separately excited by means of a variable voltage exciter, the changeover from motoring to regenerating being effected by cam groups which operate certain controls. An inverse compound regenerative characteristic is obtained by feeding the excitation current through a stabilizing resistance which also carries the main current. Thus any increase in the load current will reduce the excitation of the motor fields and ensure stability. The correct regenerative characteristic is determined by the cam groups which vary the arrangement of the stabilizing resistance according to the motor combination. One cam group initiates the main change-over to regenerative connection, while the second governs the changes in the combination of the motors during regeneration. There are no closed transitions from one combination to another. A combination may be selected and, after the correct generated voltage has been obtained, the motors are connected to the line and the starting resistance gradually cut out while the central zero ammeter is observed.

As lightning surges are a constant danger in South Africa, special precautions are taken to safeguard the locomotives. In addition to the conventional double spark gap on the roof, a Ferranti surge absorber, also mounted on the roof, and auxiliary spark gaps with blowouts are associated with both the power and auxiliary circuits. In addition, the insulation of all high-tension equipment, including machine windings, is designed to withstand a pressure test of 10,500 v. A.C.

All the power contactors are of the electro-pneumatic type, so that the master controller handles only small currents at 112 v. for operating the valves.

In order that more than one locomotive can be driven from one driver's cab, the locomotives are provided with control train lines and couplers, sixty train lines being needed for the complete power and auxiliary control.

The 3,000 v. auxiliary equipment comprises the motors driving the exciter and L.T. generator. Each set is provided with automatic resistance starting equipment and may be isolated from the supply in the event of failures. No regenerative braking is possible if the exciter set is isolated, but this set can be used to provide some L.T. supplies in an emergency.

The motor generator set supplies current for battery charging, for the fields of the exciter set, for the two exhauster motors, compressor motor, lighting, control circuits and heaters, its voltage being controlled by a regulator in the generator field.

The battery floats across the L.T. generator and is automatically kept charged. Its normal function is to supply the lighting and control circuits if the L.T. generator is not running.

Each axle-hung traction motor is of the 4-pole, series type arranged for forced ventilation and weighs $5\frac{1}{4}$ tons complete with gears and gear-case.

The lateral forces imposed on the track have been considerably reduced by allowing the motor a limited amount of transverse movement relative to the axle, under the control of rubber springs between the motor and the bogie frame. These springs are pre-stressed so that the motor is normally held in a central position on the axle by the action of opposing springs. Thrust faces are provided on the suspension bearings but act only as stops to prevent excessive axial movement of the motor.

The single spur reduction gearing through which the motors drive the axle and the medium carbon steel pinion are shrunk on to a tapered seating on the armature shaft, no keys being used. The teeth are taper-ground to ensure even distribution of the load over the gear face. The resilient gear wheel comprises a carbon chrome toothed rim mounted on a cast-steel hub, the drive being transmitted from the rim to the hub through Silentbloc rubber units. A large degree of resilience has been obtained by using two Silentblocs in series, one being mounted in the rim and the other in the hub with a steel pin forming the connecting link. The gears are enclosed in a fabricated gear-case supported from the motor frame.

SWISS FEDERAL RAILWAYS (S.B.B.)

Co–Co locomotives, series Ae 6/6, A.C., 15,000 v., 16 2/3 cycles. In 1949, the Swiss Federal Railways decided to introduce a new six (all-driving) axle locomotive of type Co–Co, series Ae 6/6 (Plate 45A, page 191). The first two locomotives, No. 11401/2, were ordered from the Swiss Locomotive & Machine Works (S.L.M.), Wintherthur, and Brown, Boveri & Co. Ltd, Baden, Switzerland. It was possible to utilize as a basis for the design the excellent experience gained with locomotives of Series Ae 4/4 (Plate 47A, page 193) and Re 4/4, which also have all axles driven. The former series is used for hauling trains on level sections, and the latter for duty in the mountainous regions of the Loetschberg Railway.

Further development of rail traffic over the Gotthard line has shown that locomotives having only four driving axles, and an adhesion weight of 80 tonnes ($78\frac{1}{2}$ tons), are are not able to deal single-handed with express and through-freight traffic on this line.

Although design experience of six-axle locomotives was already available, construction details were lacking with respect to the high one-hour rating of 6,000 H.P. foreseen, and an axle load of 20 tonnes (19·6 tons). As a result of the experience gained with Series Ae 4/4 on the Gotthard, the Swiss Federal Railways, in 1954, placed an order for twelve further locomotives of Series Ae 6/6 Particulars are as follows:

Overall length		18,400 mm.	(60 ft $3\frac{1}{2}$ in.)
Total wheelbase		13,000 mm.	(42 ft $5\frac{1}{2}$ in.)
Diameter of driving wheels		1,260 mm.	(4 ft $1\frac{1}{2}$ in.)
Gear ratio			38 : 82
		Continuous	*One-hour*
Rating H.P.		5,400	6,000
Corresponding speed			
	Km/h.	78·5	74·0
Corresponding tractive			
	effort Kg.	18,000	21,200
		(39,600 lb.)	46,640 lb.)

Maximum speed	125 Km/h. (77½ m.p.h.)
Weights:	
mechanical part	65·71 tonnes (64½ tons)
electrical part	53·37 tonnes (52¼ tons)
other equipment	0·52 tonnes (½ ton)
	———
Total weight	119·60 tonnes (117¼ tons)
	———

The service requirements laid down for the construction of the locomotives were as follows:

(i) To haul a trailing load of 600 tonnes (588 tons) up the steepest Gotthard gradient (1 in 38·5) at a speed of 75 Km/h. (46½ m.p.h.). On the remaining sections of the run, with gradients up to 1 in 48, the load is increased to 750 tonnes (735 tons).

(ii) On the Gotthard, employed as a pilot or intermediate locomotive when hauling freight trains, the locomotive must be capable of hauling the same trailing load at speeds between 35 and 75 Km/h. (21¾ and 46½ m.p.h.), and with these loads to be able to start repeatedly on the gradient.

(iii) At the maximum speed of 125 Km/h. (78 m.p.h.) and with a voltage of 15 kV. at the overhead line, the tractive effort shall still be at least 8,000 Kg. (17,600 lb.).

(iv) On the 1 in 38·5 gradient at speeds of 35 to 75 Km/h., the locomotive, with a trailing load of 300 tonnes (294 tons) must be continuously braked by means of regenerative braking. It must be possible to increase the braking power by 20 per cent for five minutes. It must also be admissible to employ electrical braking up to the maximum speed.

(v) It must be possible for fifteen minutes to exceed the one-hour rating by 10 per cent.

The mechanical part. The locomotive comprises two similar, three-axle bogies and a self-supporting locomotive body, which carries the drawing and buffing gear (Plate 45B, page 191). The traction motors are secured to the spring-borne bogie frames. Whereas the centre motor is mounted vertically above the associated axle, the two outer motors are offset somewhat towards the centre of the bogie. The locomotive body is supported at the centre of each bogie.

As all the space above the centre driving axle is fully occupied by the traction motor, it was not feasible to arrange a centre pin in the manner employed for modern two-axle S.L.M. bogies. It was, therefore, necessary to devise an imaginary centre of rotation, and this was achieved by causing the sliding pieces of the locomotive body supports to move in arcs whose centres coincide with the centre of the bogie. As the gliding tracks in question are secured in the horizontal plane to the spring bands of the double laminated springs, the latter, together

with their supporting device, follow every lateral movement of the locomotive body. In order to facilitate angular deflections in the vertical direction as well, the bearing plates of the tracks under discussion are curved. The track is of bronze, while the bearing surfaces and plates are of case-hardened steel. All the supporting elements are enclosed in an oil trough, which is sealed against dust.

Each bogie frame is provided with two transverse beams, one on either side of the centre traction motor, each of which supports a carrier pin. Both pins extend downwards into spherical bearings resting in transoms securely attached to the underframe of the locomotive body and extended beneath the sole bars of the bogie. The carrier pin bearings are capable of lateral movement in the transom; clearance is also allowed in the longitudinal direction, i.e. in the direction of the outer and inner driving axles. This distribution of the longitudinal play ensures not only a natural movement of the bogie about its centre, but enables the bogie to haul and not push, regardless of the direction in which the vehicle is travelling. A single transverse bogie coupling is used. The Brown, Boveri spring drive is employed for all locomotives.

As in all the latest locomotives of the Swiss Federal Railways, the body is a self-supporting, stress-free construction consisting of side walls, underframe and roof; the entire body is of steel plate and welded throughout.

Brakes. In conformity with requirements, the locomotives are provided with automatic as well as direct-acting compressed-air brakes and, in addition, with the so-called shunting brake. This is the first time that a main line locomotive of the Swiss Federal Railways has been provided with brake gear for applying and releasing the brakes in stages.

The electrical part. The experience gained with the electrical equipment of the Ae 4/4 locomotives belonging to the Loetschberg Railway was extensively utilized as a basis for the Ae 6/6 locomotives (Plate 45B, page 191). The motor casing of the single-phase A.C. traction motor has built-on ventilation and inspection openings, as well as fixing lugs; it presents interesting welding innovations which enable the weight of the motors to be kept down to a minimum. The centre motor is mounted vertically in the bogie above the axle; the two outer motors are arranged obliquely above the axles. The motors are installed so that the air-inlet connections are all on the same side, towards the outer wall of the locomotive, with the commutators on the same side as the service gangway inside the locomotive. The rotor is supported at both ends by grease-lubricated roller bearings. The windings are insulated with mica silk and glass tissue. The traction motors are cooled by forced draught; the air, introduced through louvres in the side of the locomotive body, is delivered in each case by a double fan to the three traction motors of each bogie. The oil cooler for the transformer and the supplementary gear belonging to the regenerative brake are also located in the airstream, so that the ventilation plant is fully utilized.

The locomotive has two electro-pneumatically operated current collectors.

A type of airblast circuit-breaker is employed as the main locomotive switch.

For tractive-effort-speed control the Brown, Boveri high-voltage control gear is used. Owing to the fact that the high-voltage control in the motor circuit does not require any additional gear beyond the motor contactors, it is basically feasible to connect the traction motors in parallel. This and the large number of regulating steps, together with the magnetic damping of the effort peaks in the transformer, contribute to the best possible utilization of the adhesive weight of the locomotive for difficult starts; moreover, the undesirable consequences for the traction motors when there is a tendency to slip, are reduced. The tap-changing switch has one preliminary notch and twenty-seven running notches, and is built on to the transformer tank so that the complete unit can be removed from the locomotive for inspection purposes.

The locomotive transformer has a continuous rating of 4,500 kVA. and is built with a radially laminated core.

The motor circuit, per bogie set of traction motors, comprises an electro-pneumatically operated reversing and brake change-over switch with four positions for running and braking in the forward and reverse directions.

The control system consists of a system of mechanically operated position switches, which control the tap changer. This follow-up control is so arranged that when the controller handwheel is turned to a definite step, a contact is closed and the control motor of the tap changer is switched in. As soon as the tap changer reaches the step at which the controller is set, the contact is interrupted and the control motor brought to a standstill. A shaft, driven by the tap changer and connecting the latter with the switchgear in the controller, is employed to achieve this interruption. The same shaft operates a position indicator in the master controller. This device can also be used as a mechanical emergency drive for the tap changer if, for any reason, the control current is not available or if a fault develops in the control unit. In this case a handwheel is placed on the controller in the driver's cab, on the mechanical position-indicating shaft. When the change-over from motorized to manual opera-

tion has been made, by means of a special lever on the tap-changer switch box, regulation can be carried out in the same way, as in the case of normal operation. Thus two completely independent methods are available for controlling the locomotive, ensuring at all times the operational readiness of the vehicle; a factor of major importance for a mountain railway. The master controllers installed in both drivers' desks are fitted with handwheels for the tap-changer control, by means of which the speed is regulated in twenty-seven notches when running and in ten notches when braking electrically; in addition, there are only the position-indicating device for the tap changer and a handle with contact drum for reversing the direction of running, on each master controller. All locomotives are fitted with equipment for the electrical braking of the locomotive. This locomotive can be considered the latest development of single-phase A.C. design, and it is a typical example of the high standard of design reached today.

Heavy shunting (switching) locomotives, series Ee 6/6. The Swiss Federal Railways are using in the hump yards at Zurich and Muttenz (Basle) two six-axle electric shunting locomotives of 90 tonnes (88¼ tons) weight, designed to handle 2,000-ton trains when necessary. In general design these locomotives are simply two of the well-known and old-established three-axle locomotives of the 16300 series used all over the country by S.B.B., the only difference, apart from necessary changes to the cab and controls, is a minor one in the gear ratio, 16·2 from the old 1 : 5·58.

Classified as Ee 6/6, these two locomotives, Nos. 16801 and 16802, have the C–C wheel arrangement. They were built by the Swiss Locomotive & Machine Works (S.L.M.) and have Brown, Boveri motors and Sécheron contactor and control gear. Above the two frame structures, now in the form of bogies, are the underframe with central cab and two casings resting on two spherical pivots and four spring-loaded side bearers.

The motor of each bogie and its force-ventilating blower projects into the adjacent end casing; and these end casings also house the transformer, control gear, compressors, and batteries, the compressor itself being of S.L.M. make and of a capacity great enough to supply the air needed for the Westinghouse straight and automatic brakes.

The wheels are 41 in. diameter and the centre wheel-and-axle assembly in each bogie has ±8 mm. (5/16 in.) side play; special provision is made for relative vertical movement of the two bogies to suit the sudden change in gradient at the hump top and at the bottom on the yard side. The bogie wheelbase is 13 ft 3 in. and total locomotive wheelbase 35 ft 8 in.

Each motor has two one-hour outputs, namely 400 kW. at 920 r.p.m., and 56 kW. at 870 r.p.m.; the one-hour wheel-rim tractive effort is 31,000 lb., continuous effort 25,500 lb., and starting effort 55,000 lb. The control gives a special track speed regulation to suit hump work from zero to 5 Km/h. (0 to 3·1 m.p.h.), but in the higher speed range, up to the maximum of 28 m.p.h., the two motors are coupled together in parallel.

Metre-gauge, rack and adhesion locomotives. (see Chapter 9, Part V, page 472). During the last war the Swiss Federal Railways electrified their only narrow-gauge rack and adhesion railway, the Brunig line, which connects Lucerne with Meiringen and Interlaken, a distance of forty-six miles. Maximum gradient is 1 in 8·5 on the rack part, and 1 in 40 on the adhesion section; the railway is metre gauge. Current used is single-phase A.C. at 16 2/3 cycles and 15,000 v.

The locomotives, Series Fhe 4/6, are termed Luggage Van Motor Coaches to give them the official Swiss name. They contain a central engine room, two luggage compartments and two cabs. Fifteen locomotives were built by Messrs. Oerlikon, Brown, Boveri, Sécheron and the Swiss Locomotive Works. The wheel arrangement is as follows: there are two four-wheel bogies with two motors each, and in addition a centre bogie, also running on four wheels, which contains two cogged wheels driven by two motors. All three bogies are coupled together. The axle load is limited to 12 tonnes (11¾ tons). The six motors have a total continuous output of 880 kW. at 16·77 m.p.h. Maximum speeds are 18·6 Km/h. (11 m.p.h.) for rack working, and 46·6 Km/h. (30 m.p.h.) for adhesion working. The rack bogie contains two cogged-wheels each driven by a motor with double gear transmission.

The cogged-wheels and the large gear transmission wheels are sprung so as to ease entering into rack sections.

The elaborate braking arrangements consist of five independent systems:

(i) Automatic Westinghouse air brakes operating on the adhesion bogies, the adhesion wheels on the rack bogies, and the whole train.

(ii) The rack pinion brake, acting directly on the cogged-wheels of the rack bogies with compressed air.

(iii) A compound air brake working brake-blocks on the wheels of the adhesion bogies.

(iv) An electric resistance brake acting on the adhesion and rack motors; these are connected together with an additional braking exciter, thereby allowing a variation of speed between 5 and 18 m.p.h. The electric brake is the normal service brake when travelling down gradients.

(v) A hand brake operating on the adhesion bogies as well as on the transmission gear axles of the cogged-wheels.

The Swiss Federal Railways ordered in 1954 two further single-phase locomotives with an hourly output of 2,300 H.P. for the Brunig line. These are Series HGe 4/4 (Plates 48A, 48B and 48C, page 194).

The mechanical part consists of the body and two power bogies. The body is of self-supporting all-welded construction divided into two cabs and the central engine room. The two bogies, again of welded construction, are coupled together and transmit all tractive forces. The body rests on the bogies on sliding pieces which, in their turn, sit on leaf springs. The tractive effort is transmitted directly to the body frame so that the support units are independent of the tractive forces.

The four motors are of the nose-suspended type. In addition to the driving axle, the sprung gear-wheels drive through a layshaft a further train of gears which are loosely fixed on to the driving axles and run free during the work on adhesion sections.

The locomotive is equipped with five independent brake systems:

(i) Electric resistance brakes for running downhill on rack sections.

(ii) Automatic air brakes acting on the locomotive and the whole train.

(iii) Direct air brakes on the locomotive only.

(iv) A special disc brake acting on the cogged-wheels.

(v) Hand brakes operated from the cab.

In addition, there is the usual Brown, Boveri type of dead-man's handle, and a centrifugal contactor gear which operates the automatic air brake in case a certain speed is exceeded on the rack section. The electrical equipment is the usual one for A.C. locomotives and consists of a main transformer of 1,250 kVA. continuous output which is oil-cooled and air-operated, main circuit-breaker and four traction motors.

The motor units are of a compact welded construction. The whole electrical layout is much more simple than that of any previous designs as each driving motor is used for both rack and adhesion duties allowing of an extremely simple and satisfactory layout. The locomotive has the usual auxiliaries. Other main particulars are as follows:

Gauge	1,000 mm. (3 ft 3·4 in.)
Total length over buffers	13,130 mm. (43 ft 1¼ in.)
Total wheelbase	9,350 mm. (30 ft 8¼ in.)
Bogie wheelbase	3,150 mm. (10 ft 4 in.)
Driving wheel diameter	1,028 mm. (3 ft 4½ in.)
Gear ratio (adhesion)	1 : 6·505
Gear ratio (rack)	1 : 5·702

Number of driving motors		4
Total weight=adhesive weight	54 tonnes (53·1 tons)	
Maximum starting tractive effort:		
Adhesion	14,000 Kg. (30,800 lb.)	
Rack	28,000 Kg. (61,600 lb.)	
Tractive effort per 1 hour at 31 Km/h.:		
Adhesion	14,000 Kg. (30,800 lb.)	
Rack	19,000 Kg. (41,800 lb.)	
Total output	2,800 H.P.	
Maximum speed:		
Adhesion	50 Km/h. (31 m.p.h.)	
Rack	33 Km/h. (20·5 m.p.h.)	

TURKISH STATE RAILWAYS

Bo–Bo locomotives, single-phase, 50-cycle A.C. In 1953, the Turkish State Railways decided to electrify the suburban line from Istambul to Halkali, along the European side of the Sea of Marmara, the total length of line being 27 Km. (16¾ miles). It was decided to use 25,000 v., single-phase, 50-cycle A.C. and the electrification was entrusted to a French syndicate led by Messrs Alsthom and the firms of Jeumont, Le Materiel Electrique S.W., and Schneider. The order comprised eighteen three-coach trains and three electric locomotives. The locomotives are designed to haul steam-stock passenger trains and freight trains. Train loads up to 700 tonnes (686 tons) are to be hauled and the distance of 27 Km. (16¾ miles) is covered in 20 minutes, with maximum speeds of 90 Km/h. (56 m.p.h.), the line having maximum gradients of 1 in 80.

Other main particulars are as follows:

Wheel arrangement	Bo-Bo
Total weight in working order	77·5 tonnes (76 tons)
Weight, mechanical part	47·04 tonnes (46 tons)
Weight, electrical part	30·46 tonnes (30 tons)
Gauge	1,435 mm. (4 ft 8½ in.)
Total length	16,138 mm. (53 ft 0 in.)
Total wheelbase	11,200 mm. (36 ft 9 in.)
Bogie wheelbase	3,000 mm. (9 ft 10 in.)
Wheel diameter	1,300 mm. (4 ft 3 in.)
Gear ratio	16 : 79
Output of motors:	
Per 1 hour	2,320 H.P. at 62·5 Km/h. (39 m.p.h.)
Continuous	2,200 H.P. at 60·3 Km/h. (37·5 m.p.h.)
Maximum speed	90 Km/h. (56 m.p.h.)
Maximum tractive effort	16,000 Kg. (35,200 lb.)

The mechanical part. The bogies, designed by Alsthom, contain several devices to reduce wear, maintenance and lubrication to the minimum. The bogie frames are fabricated from sheet steel by welding. The pivot seating is of conical shape and of rubber, and permits all movements between body and bogie without any rubbing movement. Side play of ± 10 mm. (25/64 in.) is possible, limited by rubber stops. The axle-boxes are linked to the bogie frame by Silentbloc-ended link members, which allow vertical movements and also small side movements. The primary suspension consists of helical springs only – the Silentbloc mounting of the axle-boxes providing sufficient damping effect. The secondary system consists of the above-mentioned rubber pivot and side equalizers consisting of laminated springs which support a bearer plate registering with a corresponding plate on the body; the rubbing surfaces are lined with manganese steel. The body is a load-bearing structure consisting of main frame, side walls and two roof girders; there are driving cabs at both ends.

The electrical part consists of four Jeumont traction motors; these have 2,320 H.P. total output and are 14-pole machines with interpoles and compensating windings, working at a nominal 300 v. In order to obtain satisfactory commutation over the whole speed range, two types of interpole shunting are used. From 0 to 35 Km/h. (21¾ m.p.h.) the shunts consist of resistances and capacitors, and above 35 Km/h. of resistances alone. These circuit changes are made by a contactor gear arrangement controlled from the speedometer. The capacitors are transformer-coupled to the interpole shunt circuits. The motors transmit power to the wheels by means of the Alsthom transmission system whereby the motors drive a quill shaft through Silentblocs and a floating ring. The control system consists of an Alsthom 400-Amp. air-blast circuit-breaker. Speed control is by means of transformer and tap-changer. The transformer is a twin unit forming a complete sub-assembly and weighing 8·75 tonnes (8½ tons). The first of the two transformers is connected directly to the line and the second (across which the four traction motors are connected in parallel) is fed with a variable voltage through a number of tappings on a winding in series with the main transformer primary. The tap-changer consists of a rotary, oil-immersed selector switch and normal air-brake contactors; it is driven by an electric servo-motor. There are twenty-two main running steps. The transformer is oil-cooled and has a rating of 3,100 kVA. on 25,000 v. at the primary, while the secondary (which is the main traction circuit) has a continuous rating of 1,270 kVA., corresponding to 6,040 Amps. There is also a tertiary winding for heating (1,000 v., 400 kW.) and auxiliary equipment (380 v.). A further control item are the Jeumont reversers, operating electro-pneumatically. There are the usual auxiliaries such as motor-blowers for providing cooling air for the traction motors, Westinghouse compressor, etc. All apparatus is mounted centrally in the body, leaving a wide passageway each side to

P. Ransome-Wallis

PLATE 49A. F.S. (Italy) high-speed D.C. motor-coach train, series ETR-300, at Rome (Termini) before departure for Florence and Milan.

Heemaf

PLATE 49B. Netherlands Railways: high-speed four-car motor-coach train. These trains are painted green with a red band. Trains of similar appearance, but painted blue with a yellow band, are equipped to use either 3,000 v. D.C. or 1,500 v. D.C. They are known as the Benelux trains and operate between Belgium and Holland.

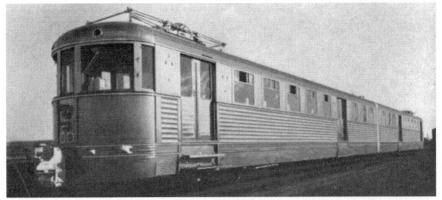

F.J.G. Haut

PLATE 50A. S.N.C.F. stainless steel suburban motor-coach set.

Asea

PLATE 50B. S.J. (Sweden) suburban train set for 15,000 v. single-phase A.C. electrification.

British Railways

PLATE 50C. British Railways multiple-unit stock for 1,500 v. D.C. electrification (Liverpool Street–Shenfield services, Eastern Region).

London Transport Executive

PLATE 51A. London Transport Executive: new tube train stock for the Piccadilly line. These trains are panelled with unpainted aluminium alloy. Each train consists of seven cars.

British Thomson-Houston Co. Ltd

PLATE 51B. London Transport Executive: Metadyne stock for the District Line.

Paris Metro

PLATE 52A. Paris Metro rubber-tyred motor-coach.

Brown, Boveri

PLATE 52B. Rome Metro: twin motor-coach unit.

British Thomson-Houston Co. Ltd

PLATE 52C. Toronto Subway: twin motor-coach unit.

A.C.E.C.

PLATE 53A. Installation of twin motor units into bogie of Bo–Bo locomotive for the Belgian Congo.

Metropolitan-Vickers Electrical Co. Ltd

PLATE 53B. Rotor of a 2,500 H.P. traction motor.

Metropolitan-Vickers Electrical Co. Ltd

PLATE 53C. South African Railways 2,160 H.P. loco-motive showing pantograph raised. Also shown are the lightning arrester and choke coil (in front) and the main fuse (behind).

British Thomson-Houston Co. Ltd

PLATE 53D. B.T.H. Germanium rectifier.

S.L.M. Winterthur

PLATE 54A. B.B.C. spring drive showing "driving star" on right.

Brown, Boveri

PLATE 54B. Brown, Boveri–Buckli drive.

Alsthom

PLATE 54D. Alsthom universal link drive, with silent-blocs and floating ring: as used on S.N.C.F. loco-motives Nos. CC. 7000 and 7100

Metropolitan-Vickers Electrical Co. Ltd

PLATE 54C. Quill-cup drive.

Oerlikon

PLATE 55B. Transformer with tap changer. Swiss series Ae 8/14 locomotive. (*See* Plate 47B, page 193).

Oerlikon

PLATE 55A. Transformer set for Swiss 12,000 H.P. locomotive.

Oerlikon

PLATE 56A. S.B.B. series Ae 6/6 6,000 H.P. locomotive showing traction motors and ventilating equipment, and transformer in the middle.

Secheron

PLATE 56B. B.L.S. series 6/8. Mechano-pneumatic contactor gear.

connect the two cabs and allow easy access to all parts. There are also a number of protecting devices against faulty operation.

The locomotives were tested, before being sent abroad, on the line Valenciennes–Thionville of the S.N.C.F.

VIRGINIAN RAILROAD (U.S.A.)

Rectifier locomotives. The Virginian Railway Co. was one of the last in the United States to electrify its main line, a distance of 134 miles between Roanoke and Mullens. The change was made in 1926, using single-phase A.C. of 11,000 v., 25 cycles. This is a most difficult section of line, with continuous gradients of 1 in 50 and 12° curves. It crosses the Allegheny Mountains and reaches a maximum elevation of 2,526 ft. The amount of traffic is very considerable, and no less than 8,000,000 tons of coal are handled annually at one of the terminal points. Severe load conditions exist because the trains call for power in large increments. At times there may be only one train on the line and at others four or even more requiring power simultaneously. Instantaneous load increments may therefore be of the order of not less than 14,000 kW.

The locomotives originally supplied for the electrification were of the split-phase type with cab-mounted induction motors driving through jack shafts and side rods.

These were supplemented in 1948 by motor-generator locomotives using standard nose-suspended D.C. traction motors. Semi-permanently coupled as two-cab units, these locomotives have a continuous rating of 6,800 H.P. and weigh approximately 450 tons, all on drivers.

In August 1955, twelve rectifier-type locomotives were ordered from the General Electric Co. of U.S.A., and delivered during 1956–7. Although they differed from the motor-generator locomotives built in 1948, the new locomotives have the same ability to operate from a high-voltage A.C. supply line and, at the same time, utilize the standard D.C. traction motor with its advantageous characteristics. The locomotives are of somewhat unusual mechanical design, as they are intended both for haulage and shunting (switching) duties, and in general appearance are very similar to a diesel-electric road-switcher (Plate 43A, page 189).

The entire cab is of fabricated steel construction, electrically welded. The shorter end cab is designated as the No. 1, or front, end since the locomotive will operate with this end leading the majority of the time. Immediately behind the No. 1 end cab is the driver's cab, occupying the full width of the platform. The long-end cab, following the driver's cab, houses the principal equipment. Outside runways lead from the driver's cab to either end on both sides of the platform. Doors opening on to these runways provide easy access to the equipment for maintenance. A spring-raised, air-lowered pantograph trolley is mounted on top of the No. 1 end cab. Canopies extend over the outside runways underneath the pantograph to protect personnel from contact with high voltage. A high-tension cable extends along the roof from the pantograph to the No. 2 end of the locomotive, where a connection is provided to permit multiple-unit operation of two locomotives back to back with power supplied from one pantograph.

The body is carried on two three-axle bogies with a centre bearing and side supports; they have cast-steel frames with rigid bolsters.

The rectifier compartment houses the twelve 12-in Ignitron rectifier tubes with their associated firing circuit apparatus. The compartment consists essentially of two cubicles, each housing six tubes, separated by middle aisles. Access to the rectifiers is obtained by means of outside doors. Four iron-core reactors are located between the rectifier cubicles. These are installed and removed through an overhead hatch. The main smoothing reactor is located directly underneath the rectifier section below the locomotive platform. It is an iron-core, two-coil type, and is used in the motor circuit to suppress the ripple current in the output of the rectifier tubes. It is of the dry type, forced-air cooled, insulated for operating voltages in excess of 2,000, and carries the current of three motor circuits. Its location results in direct inter-connections for the large power circuit cables. Being beneath the locomotive platform, some objectionable magnetic features are eliminated and, at the same time, the centre of gravity of the locomotive is kept low. Also included in the rectifier compartment is a water pump and temperature regulating system required to maintain the temperature of the rectifier tubes within its operating values. A three-way regulating valve ensures that the water temperature is closely controlled at approximately 43° C.

A transverse aisle through the cab separates the rectifier and control compartments and provides access to the rectifier temperature regulating equipment as well as to the central aisle of the control compartment.

Single-phase A.C. power is supplied from the overhead wire at 11,000 v., 25 cycles. This is fed through the pantograph to the transformer. Voltage on the transformer secondary is varied by means of the tap switches, and is fed to twelve Ignitron rectifier tubes arranged three in a group, bridge connected.

Rectified power flows from the tubes through a smoothing reactor to the six traction motors. These are series-wound, axle-hung D.C. motors of the type widely used on diesel-electric locomotives and also on the rectifier locomotives for the New Haven Railroad. For this application the motor is modified by the addition of a ground brush and flash ring. The six motors are connected permanently in three parallel groups, each group consisting of two motors in series. Acceleration and speed control are obtained by tap changing on the transformer secondary and by the use of accelerating resistors in series with the motor armatures. The same resistors are also used for dynamic braking. They are forced-air cooled, the amount of air being proportional to the heat to be dissipated.

Operating requirements were that these locomotives be capable of multiple-unit operation, up to four units in multiple, in either direction; and that they also be capable of performing both road and switching duty. Moreover, they must be able to operate in the same train with the motor-generator locomotives which were put into service by the Virginian Railway in 1948.

To meet these requirements, the locomotive was designed to have a continuous rating of 3,300 H.P., and a maximum operating speed of 65 m.p.h., with a gear ratio of 18 : 74 and 40-in. wheels. The total weight is approximately 174 tons, all on drivers. At 25 per cent adhesion the tractive effort is 98,500 lb. At the continuous rating of the motors, the locomotive develops 79,500 lb. tractive effort at 15·75 m.p.h. The maximum braking effort of 63,000 lb. is available from 25 to 15 m.p.h.

From Clark's Gap eastbound to Roanoke two units handle a 9,000-ton train against a ruling grade of 1 in 172 (0·58 per cent). Westbound, from Roanoke to Mullens, two units handle an empty train of 3,844 tons (approximately 162 cars). On acceptance runs, two units hauled a 3,140-ton train from Mullens to Clark's Gap, and a 10,000-ton train from Clark's Gap to Roanoke.

Main particulars are:

Weight, fully loaded	174 tons
Weight, per driving axle	29 tons
Length inside knuckles	69 ft 6 in.
Total wheelbase	52 ft 9¾ in.
Rigid wheelbase	13 ft 0 in.
Distance between centre plates	45 ft 0 in.
Wheel diameter	40 in.

Part V. The electric motor coach and motor coach train

BASIC CONSIDERATIONS

Railway operation under modern competitive conditions requires the provision of frequent services. On electrified lines, such services can often be provided more economically and efficiently by electric motor coach or unit trains than by trains hauled by locomotives. Furthermore, suburban electrification schemes dealing with commuter traffic are subject to peaks in traffic requirements which can be met by coupling several motor coach trains together up to the limit of existing station platform capacity. During off-peak periods such trains can be divided and units surplus to requirement at those times are sided or garaged at convenient locations. Motor coaches may also be used for light freight and parcels services, and have often enabled small railways or branch lines to be kept economically in existence.

The layout of a modern motor coach follows that of normal passenger carrying vehicles, but with the addition of the necessary electrical equipment. Such equipment is basically the same as that required for electric locomotives, but on a very much smaller scale. Pantographs, main switches, and resistances are usually carried on the roof, the latter sometimes under the floor. Electric motors are mostly of the nose-suspended type. The auxiliary equipment, such as generators and compressors, is often housed in the driver's cab (of which there are either one or two) or under the floor.

In some countries, notably Italy, motor coach trains are designed for high speed luxury travel between large cities. Such expresses run only on the D.C. sections of the Italian electrified lines (Plate 49A, page 213).

A major problem in motor coach design in A.C. traction is the positioning of the transformer so that the space for the payload is not too heavily encroached upon. In modern practice the transformer is usually mounted under the body.

Special attention must also be paid to insulation against noise and vibration so as not to interfere with the comfort of the passengers.

THE GERMANIUM POWER RECTIFIER FOR MOTOR COACHES

Germanium, an element long neglected until its unique properties were recognized, has lately assumed major importance in the fields of communication and power rectification. Among these developments is the use of germanium rectifiers for traction service, developed by the firm of British Thomson-Houston Co. Ltd (B.T.H.), of Rugby (Plate 53D, page 217).

In general railway practice, the power which is supplied to the train may be any of several types: A.C. or D.C., high or medium voltage. With an A.C. supply, either A.C. commutator motors or D.C. motors may be used, the latter in conjunction with rectifiers built into the locomotive or motor coach. The sturdy D.C. series-wound traction motor has ideal characteristics for this service, and has always found favour in Great Britain, although, until comparatively recently, it has generally been used with a D.C. supply, taken from an overhead conductor or a third rail.

The widespread distribution of electric power has rendered practicable the utilization of high-voltage A.C., at standard 50-cycle frequency, for the overhead supply. Compared with D.C. installations, an A.C. supply has many advantages; fewer and simpler sub-stations are required and less copper is needed for the overhead line, which, in turn, permits a lighter construction of the line supports. By using rectifying equipment mounted on the locomotive itself, these advantages can be combined with those of the D.C. series-wound motor, allowing economy in cost of installation without affecting performance or reliability.

A number of locomotives and motor coaches have been equipped with mercury-arc rectifiers on 50-cycle systems, and they have given excellent service. The advantages of the germanium rectifier are, however, self-evident, and with its rapid development in recent years, attention soon turned to traction applications. In December 1955, the British Transport Commission carried out successful trials of an electric train equipped with a germanium rectifier, developed by B.T.H. In these and subsequent tests the equipment gave every satisfaction, and it has been in regular service since April 1956.

Description of the germanium rectifier. Germanium is a silvery-looking metal and has a specific gravity of 5·47. Traces of it are to be found in the earth's crust in many regions. In this country, however, the commercial source is soot or flue dust, in which it occurs in the proportion of between 0·1 and 0·5 per cent.

The germanium is extracted from the flue dust, purified chemically, and converted to germanium dioxide. This is then reduced to the metal by heating to above its melting point in an atmosphere of hydrogen. After reduction, the amount of impurity is only about 0·0001 per cent, but, before the germanium crystals can be used as rectifiers, the percentage of impurity must be decreased, by a special process of high-frequency heat refining, to the value of one ten-millionth of one per cent.

After purification, the germanium crystal is cut into slices, or wafers, each about 0·020 in. thick. The slice is soldered to a metal base, and after careful cleaning, a layer of indium is fused on to the top surface. A top connection is soldered to the indium, and, finally, the rectifier is enclosed in a hermetically-sealed case to exclude moisture.

In operation, rectification takes place inside the germanium slice, on the junction line between the original pure crystal and that portion into which indium has penetrated during the process of fusion. Current can now pass from the indium to the germanium with a voltage drop of only about half a volt, but, in the reverse direction, the application of 500 v. will only produce a current of a few milliamperes, the voltage being concentrated across a barrier only about 0·0004 in. thick. The average voltage gradient is thus more than 100,000 v. per in., equal to the breakdown voltage of some insulators.

The first traction rectifier. The first germanium power rectifier ever to be used in traction service is a B.T.H. equipment, rated at 750 kW., or 500 Amp. at 1,500 v. D.C. It was designed for mounting in an existing motor coach, in which mercury-arc rectifiers had previously been used for power conversion. The new equipment is made up of twelve strings of rectifier cells, each string comprising fifty cells in series. The twelve strings are mounted in six vertical trays, two strings per tray; the trays are on rollers, and can readily be withdrawn for inspection or maintenance.

The complete rectifier is mounted on an angle-iron frame, and housed in a stout sheet-steel cubicle arranged for resilient mounting. The dimensions of the cubicle are 3 ft × 3 ft × 6 ft 6 in. The rectifier cells are cooled by an extractor fan mounted above the trays, the air entry being via grilles at the bottom of the cubicle.

For protection against rectifier-cell failure, a fuse, rated at 100 Amps., is fitted on the input side of each string of cells.

Subsidiary equipment. For satisfactory operation of a germanium rectifier, a certain amount of reactance must be included in the main transformer secondary circuit. Normally, the transformer secondary is designed to possess this reactance, but since the transformer used in this first traction application had been originally designed for use with mercury-arc rectifiers, additional reactance had to be inserted. This takes the form of an external oil-immersed steel-tank reactor, consisting of two legs, each fed from one side of the transformer split-secondary winding. A 5-microfarad capacitor is connected across the output ends of the two legs, and thus across the rectifier equipment, as an additional aid to steady operation. A 1-megohm discharge resistor is connected in parallel with this capacitor.

Two more reactors, acting as smoothing chokes, are arranged in series on a common magnetic core between the two pairs of traction motors. The centre tap of these reactors acts as the common negative, and is connected to the secondary windings of the main transformer.

Tests and trials. The motor coach equipped with the first germanium traction rectifier was tried out on the Lancaster–Morecambe–Heysham line of British Railways in December 1955. With its high efficiency at all loads, its reliability, and its lack of moving parts, this rectifier has proved ideal for traction service. A second series of trials on the same equipment was carried out in London, England, in March 1956, between Fenchurch Street and Bow Junction, the test-train being run at night so that the 1,500 v. D.C. supply to the overhead conductor could be disconnected and replaced by 6,600 v. A.C.

Among the many searching tests included in this second series of trials was a motor short-circuit test. The master controller was so arranged that, in its last-notch position, instead of diverting more of the motor field-current through a resistance, it put a dead-short across the field. The motors flashed over and the sub-station circuit-breaker cleared the fault; on examination after the tests, the rectifier was found to be completely undamaged, proving its ability to stand up to the most arduous conditions and abuse.

The power loss occurring during the reverse part of the cycle can be ignored, since it is negligible compared with the forward loss, which itself is very small. The efficiency of the germanium rectifier can therefore be calculated from the forward-power loss only. For a 750 kW. rectifier equipment, this forward loss is about 2·0 per cent at full load. The cooling fan and other auxiliaries account for a further 0·2 per cent, giving a total rectifier loss of about 2·2 per cent. As the loss in the main transformer is about 1·8 per cent on full load, the over-all efficiency of the equipment is in the region of 96 per cent.

The only losses occurring at no-load are the transformer core loss and the fan loss, and thus the over-all efficiency remains high, down to one-tenth of full load.

Satisfied by the efficiency and excellent performance in service of the equipment here described, British Railways have ordered a further thirty-five germanium rectifier equipments for multiple-unit motor coaches, to be used on the first section of the main-line electrification scheme which will eventually link London and north-west England.

MODERN EQUIPMENT FOR BRITISH RAILWAYS

Multiple-unit stock for 1,500 v. D.C. During the last fifty years traffic from Liverpool Street Station to the East London suburbs has increased continuously with the rising population, and by 1933 the line had quadruple tracks as far as Shenfield, over twenty miles from London. Traffic ultimately became so heavy that electrification of the line was considered essential, the steam service having reached the limit of its capacity. In 1936, electrification was decided upon by the former L.N.E.R. and it was also decided to use 1,500 v. D.C., with overhead catenary lines. The very considerable civil engineering work

necessary was making good progress, when the outbreak of World War II caused a suspension of the scheme. Work was resumed in 1946, and by November 1949 the whole of the suburban traffic was being run electrically. Ninety-two three-coach trains were built each consisting of motor, non-driving trailer, and driving trailer coaches (Plate 50c, page 214).

The coach bodies and underframes are of welded and riveted all-steel construction, the underframe being of rolled-steel section. The bogies are of orthodox single-bolster type with plain axle-box bearings. Electro-pneumatic sanding equipment is provided to all motor coach wheels. The stock is fitted with electro-pneumatic braking and door gear, the latter being actuated (for opening only), from push-buttons in the passenger compartments, under over-riding control by the guard. The three-coach units are normally operated in multiple unit, as nine-coach trains, although three- or six-coach trains may be run as required during off-peak hours.

The frequency and running speed of passenger trains on the Liverpool Street–Shenfield line has been considerably improved since electrification. During peak business hours a maximum of nineteen trains per hour leave Liverpool Street for the Shenfield line compared with a maximum of fourteen with the previous intensive steam service. The overall journey time of an all-stations train between Liverpool Street and Shenfield has been reduced, from 61 minutes with steam operation, to 45 minutes by electric train, a reduction of 26·2 per cent.

Four nose-suspended traction motors are provided on each motor coach driving 43-in. diameter wheels through single-reduction spur gearing. The traction motor is a 4-pole series-wound machine with a one-hour weak field rating of 210 H.P., and a continuous rating of 157 H.P., at 675 v. Each traction motor is self-ventilated by means of a fan mounted on the armature shaft, drawing cooling air through louvred openings in the coach side. The four traction motors are arranged in two groups, each group comprising two motors permanently connected in series.

Each motor coach equipment provides series/parallel control of the two motor groups with automatic acceleration and field tap control. Eleven series and six parallel notches are provided, while automatic acceleration is controlled by a current limit relay. To reduce current peaks with the motor groups in parallel, whilst maintaining the same initial acceleration, the current relay setting is reduced. The sixth parallel notch is the weak field notch and is controlled by a separate current limit relay.

Should either of a pair of traction motors become defective, the power supply to the pair can be cut off by a motor cut-out switch mounted in the reverser case. A control cut-out switch, operated from the motor coach cab, is provided for the isolation of a complete control equipment should a fault occur. The power supply is taken from the overhead line through a single-pan pantograph and main isolating switch to the two motor circuits. Each motor circuit includes an overload relay and two line-breakers, and passes through one group of two motors and associated resistances to a running rail negative return.

Each motor coach control equipment consists of seventeen electro-pneumatic contactors. These contactors, together with those for the motor-generator set, compressor and train heating, also the electro-pneumatic reverser, line-breakers, motor cut-out switches and current limit relays are housed in three main equipment cases mounted longitudinally on one side of the underframe.

Compressed air for braking, sanding, electro-pneumatic control gear and coach-door operation is provided by a two-cylinder reciprocating compressor.

Line voltage	1,500 v. D.C.
Over-all length of three-coach unit over buffers	177 ft 7 in.
Length of motor coach over-all	60 ft 4 in.
Tare weight of motor coach	50·8 tons
Tare weight of three-coach train	104·7 tons
Motor bogie wheelbase	8 ft 6 in.
Wheel diameter	43 in.
Maximum passenger capacity per three-coach unit	176 seated
	220 standing
Motor voltage	750 v.
Gear ratio	71 : 17
Maximum starting tractive effort	20,000 lb.
Maximum speed	70 m.p.h.

The extension of the Shenfield electrification to Chelmsford and to Southend has necessitated the provision of completely different rolling stock from the short-distance suburban stock described. In 1956, British Railways built thirty-two 4-car units which are very similar to the Southern semi-fast units already described (see page 158). The motor coach is, however, the third coach of the unit, a driving trailer being provided at each end. First-class accommodation is provided and there are adequate lavatory facilities. The units are not vestibule connected.

The motor coaches are equipped with pantographs for overhead collection of 1,500 v. D.C. Both bogies are motored there being four nose-suspended motors, the two on each bogie being connected permanently in series. Control is by series or parallel connection of the two pairs of motors with bridge transition.

Alternative forward positions of the reverser handle give two rates of acceleration. When wet or greasy rails are encountered, the rate is 0·85 m.p.h. per second; the normal rate on dry rail is 1·1 m.p.h. per second. Each motor has an output of 200 H.P. (one rating) on full field and 220 H.P. on weak field.

The units are fitted with electro-pneumatic brakes.

Multiple-unit stock for Liverpool–Southport line – 630 v. D.C. Another type of British multiple unit stock are the 59 motor coaches, 59 non-driving trailers, and 34 driving trailer coaches for the Liverpool–Southport line; they can be formed into three-coach units, consisting of motor, non-driving trailer and driving trailer coaches, or two-coach units consisting of a motor coach and non-driving trailer.

The new vehicles were built at Derby works, and embody the most up-to-date methods of welded steel coach construction, which enabled the tare weight of a three-coach unit to be kept down to the very low figure of 90·5 tons. Electro-pneumatic braking equipment and door gear is fitted, the latter being actuated from push-buttons in the passenger compartments, under over-riding control by the guard.

The standard schedule for the 18½-mile Liverpool–Southport section – which is, in general, level and free from severe curvature – is now forty minutes including stops at fourteen intermediate stations; the average station-to-station distance is 1·2 miles, and the overall average speed, including stops, is 27·7 m.p.h. The units are capable of a maximum speed of 70 m.p.h., although the maximum service speed is 60 m.p.h.

Four nose-suspended traction motors are provided on each motor coach; they are 4-pole series-wound machines with a one-hour rating of 235 H.P., and a continuous rating of 184 H.P., at 580 v. On each motor coach the motors are arranged in pairs for series/parallel operation. Including gear wheel, pinion and gearcase, each motor weighs only 4,484 lb. The traction motors are fitted with roller-type armature bearings, the suspension bearings being of the normal oil-lubricated sleeve type. Multiple armature windings with equalizing connections are used to ensure good commutation at the relatively high power output and speed at which these motors are required to operate.

Each motor coach incorporates two identical but independent control equipments, each equipment providing series/parallel control with automatic acceleration, also field tap control, of two traction motors. Nine series and four parallel full-field notches are provided, also intermediate and weak-field parallel notches, automatic acceleration being controlled by a current limit relay of the balanced armature type which is not affected either by vibration or train movement.

In the event of a fault occurring in any of the traction motors, the power supply to each pair can be cut off by a manually operated isolating switch on the underframe. Provision is also made for the isolation of a two-motor control equipment by means of push-buttons located in the driver's cab, which actuate control cut-out switches.

The power supply for each two-motor equipment is taken from a power bus-line running throughout the length of the train and fed directly by third-rail collector-shoes mounted on the motor coaches and driving trailers. Each two-motor equipment is connected to the power bus-line through its own isolating switch mounted on the underframe. The two-motor circuits are divided into two sections, in each of which the current passes through a traction motor and its associated resistances to a running-rail negative return; in each half an over-load relay and line-breaker are also included.

In the event of an excessive traction-motor current, the overload relay, in series with the defective motor, operates and causes the associated line-breaker to open and insert the limiting resistance in the fault circuit. The opening of this line-breaker de-energises, by interlock contacts, the second line-breaker, which interrupts the fault current which already has been reduced by the limiting resistance previously mentioned.

Each two-motor control equipment consists of seventeen electro-pneumatic contactors which arrange the traction motors in series, parallel and weak-field connections and cut out successive steps of starting resistance during acceleration.

The motor-generator set, motor-driven compressor, battery, starting resistances and equipment isolating switches are mounted on the motor coach underframe. The cab-mounted equipment is grouped into the following three assemblies – auxiliary cupboard, master controller, and control connection box.

There is a 1 kW. motor-generator set and a 36-cell alkaline battery which supply the low-voltage power for the control circuits at 52 v. D.C. The starting switch of the motor-generator set is operated by a removable key which can only be withdrawn when the switch is opened, thereby making it impossible for two sets to be operated in parallel in a train with more than one motor coach. The batteries normally work in parallel and are charged from the motor-generator set.

Main particulars are as follows:

Nominal line voltage	630 v. D.C.
Control-circuit voltage	52 v. D.C.
Over-all length of three-coach unit over buffers	209 ft 2¼ in.
Tare weight of motor coach	40·5 tons
Tare weight of three-coach unit	90·5 tons
Motor bogie wheelbase	8 ft 6 in.
Maximum passenger capacity	seated 268
	standing 352
Wheel diameter	36 in.
Traction-motor voltage	600 v.
Gear ratio	17 : 64
Maximum starting tractive effort	22,200 lb.
Maximum service speed	60 m.p.h.

Multiple-unit stock for Southern Region for 670–750 v. D.C. (third rail): This stock has already been described in Part I *Southern Railway Scheme* (page 158).

Single-phase, 50-cycle A.C. stock for 25 kV. or 6·6 kV. When the Liverpool Street outer suburban electrification scheme is completed by British Railways it will reach Bishop's Stortford on the Cambridge line and will include Hertford, Enfield and Chingford branches. These routes will use single-phase, 50-cycle A.C., at 25 kV., and the existing D.C. electrification between Liverpool Street, Shenfield, Chelmsford and Southend will be converted to A.C. (*see* page 182).

To operate these new services British Railways placed an order in 1957 with the General Electric Co. for the electrical equipment for seventy three-coach multiple-unit trains, the mechanical parts of which will be built in British Railways' own works. Each three-coach unit will consist of two driving trailers with a motor coach in between. The power equipment of the motor coach will be entirely underframe mounted and batteries, compressors, etc., will be similarly mounted beneath the second driving trailer. All coach space will therefore be available for passengers.

The motor-coach transformer and the switchgear for series or parallel connection of the primary winding to permit of operation on either 25 kV. or 6·6 kV., form a single assembly with the cam group tap-changer in the secondary circuit, for control of the motor voltage.

A.C. will be rectified by groups of single-anode rectifiers with liquid cooling by forced circulation.

The four axle-hung, self-ventilated traction motors will each be rated at 200 H.P. The motor contactors, weak-field contactors and the reverser will be electro-pneumatically operated.

Over-all protection will be by an air circuit-breaker.

Single-phase, 50-cycle A.C. stock for the Lancaster–Heysham line. The first electrification of the British Railways (London Midland Region) Lancaster–Morecambe–Heysham line was carried out in 1908 by the former Midland Railway to test the use of a single-phase, 6·6 kV., 25-cycle system for electric traction. The experimental re-equipment at 6·6 kV., 50-cycle has been carried out by the British Insulated Callender's Cables Co. and the English Electric Co. Trial running commenced in November 1952. The line was opened for public service in 1953, and the intervening period was used for trials, including tests to determine the effect on the power supply and for modifications to be made to the telecommunication and signalling system found necessary as a result of the trials.

The rolling stock comprises three three-coach units equipped to give an acceleration of about 1·3 m.p.h.p.s. and a maximum speed of 75 m.p.h., although in practice the speed on this section will not exceed 60 m.p.h.

Surplus electric stock was reconstructed at the Wolverton Works of the L.M. Region to accommodate the new traction equipment. The coach interiors have been refitted, providing seating for 146 passengers. The original Westinghouse air-brake equipment has been retained.

The new electrical equipment comprises components which were readily available and adequate for the purpose of the experiment. It is not representative of future equipment, which would be underframe-mounted to avoid loss of seating capacity.

Each motor coach includes four 215 H.P. (one-hour rating), self-ventilated, nose-suspended, standard D.C., 750 v. series motors, connected permanently two in series, a rectifier feeding each pair. The two rectifiers are supplied from one transformer.

Speed control is effected by varying the motor voltage by means of a motor-driven, cam-operated, on-load tap-changer on the two halves of the transformer secondary winding. This supplies two six-anode, sealed, air-cooled, steel tank mercury-arc rectifiers, connected so as to give full wave rectification. Some smoothing of the D.C. current is obtained by the use of a D.C. reactor consisting of two series reactors on a common magnetic core. The control gear provides four normal running positions, low, half and full voltage and weak field in both directions of running. Acceleration is automatic up to the selected notch. The three-car sets can work in multiple unit. The transformer primary is protected by a fuse, earthing contactor and lightning arrester. The rectifier anodes have individual fuses and the D.C. circuit includes the normal overload and earth fault relays. The transformer is oil-cooled with an air-blown cooler and thermostatically-controlled shutters are fitted to the equipment compartment. Electric heaters are fitted to each rectifier. All auxiliary motors are supplied at 230 v. single-phase from a tertiary winding, except the compressor, which is fed through a dry-plate rectifier from the same supply. This winding, 230–0–230 v., supplies individual transformers on each coach, giving a

24 v. lighting supply. The coach heaters are connected across the outers at 460 v. The driver's cab contains a master controller, air-brake valve, control and indication panel, and accessories. The pantograph is raised by air pressure from a reservoir or, failing this, by a foot-pump. The train is driven from the master controller in exactly the same manner as a standard D.C. equipment.

MOTOR COACH TRAINS FOR INDIA

The former Great Indian Peninsula Railway and the former Bombay, Baroda & Central Indian Railway have in use twenty-eight motor and trailer coaches, which were built by Metropolitan-Cammell Carriage & Wagon Co. Ltd; English Electric Co. Ltd and B.T.H. Ltd supplied the electrical parts. The twenty-eight units comprise seven trains, each consisting of two driving motor coaches and two trailing coaches, which form the centre part of the train.

The design incorporates a number of developments with a view to reducing the weight of the coaches by use of modern methods of construction and light-weight materials, the main body construction material being pressed or rolled steel sheet. As the coaches are to be used for suburban services, easy access is important and each coach, therefore, has six double sliding doors. They can be arranged for manual or power operation. The interior of the coaches is laid out in open saloon fashion, with transverse seats. The design comprises many unusual features, such as a heavy roof structure and extensive use of plastic materials with smooth inner and outer surfaces to facilitate cleaning. The roof consists of seven longitudinal members of high-tensile steel, together with a number of smaller members. Each coach is subdivided into sections. The outer ends of each four-coach unit are provided with side buffers and centre couplings to enable it to be worked as a complete train. The coaches themselves are close-coupled.

The coaches often have to operate in shade temperatures of 115° F., so great attention has been paid to insulation, five-ply Isoflex being used, which is held by wire mesh against the outer skin.

The bogies are of the four-wheel type and carry two motors each. The brake equipment is of the Westinghouse electro-pneumatic type with individual cylinder to each brake block.

The electrical equipment consists of four axle-hung, nose-suspended motors, which drive the 36-in. diameter wheels through single-reduction gears having a 19 : 61 ratio. The motors are 4-pole series-wound and have a one-hour rating of 175 H.P. at 700 v. The whole of the control equipment is mounted on the main frame in the high-tension compartment. There are also the usual auxiliaries, such as motor-generator sets, compressor, battery and alternator for lighting.

The main particulars of these coaches are:

Gauge	5 ft 3 in.
Length over body	58 ft 0 in.
Width over body	12 ft 0 in.
Height (to top of roof)	12 ft 6 in.
Distance between bogie centres	48 ft 0 in.
Bogie wheelbase	10 ft 0 in.
Wheel diameter	36 in.
Tare weights:	
Motor coach	53 tons
Trailer coach	31 tons
Motors per motor bogie	2
Voltage	1,400 v. D.C.
Maximum speed	70 m.p.h.
Maximum tractive effort	31,200 lb.
Seats for passengers per unit	415

MOTOR COACHES IN SWITZERLAND

Swiss Federal Railways: all-purpose motor coaches. Excellent examples of recent main-line motor coaches are the 1,600 H.P. single-phase vehicles of type CFe 4/4, Nos. 841–871.

In planning these new motor coaches, an attempt was made to produce a vehicle which would be suitable for the widest possible range of service. On many routes, particularly on secondary lines, there are services on which the seating capacity required does not normally exceed 200, and a motor coach can be used far more economically than a locomotive-hauled train. The output capacity of the motor coach must, however, be sufficient for it to work subsidiary services, such as parcels traffic, and even light freight and cattle trains. While it is especially on secondary lines that self-contained train units, made up of motor coach, driving trailer, and, when necessary, intermediate coaches, have shown their worth, such train units have also proved to be specially well adapted for fast inter-urban traffic. This has been amply demonstrated by the excellent results obtained from the shuttle-train unit consisting of an Re 4/4 locomotive, five intermediate coaches and a driving trailer, which operates between the Zurich and Lucerne terminal stations.

A further field in which the new motor coaches have been successfully used is in dealing with the steadily increasing suburban traffic of larger cities. This traffic can be operated most satisfactorily with unit trains in which the turn-round of motor coaches at terminal stations is not necessary. Several units can, of course, be coupled together as required.

The CFe 4/4 motor coaches were, therefore, designed to fulfil exacting conditions, and they were required to haul a trailing load of 246 tons at a speed of 46·5 m.p.h. up a gradient of 1·2 per cent.

In detail, they were required to be capable of:

(i) Haulage of all kinds of trains on secondary lines as single units or with attached passenger coaches and freight wagons of all existing types of construction.

(ii) Haulage of passenger trains with frequent stops on main lines and in suburban traffic.

(iii) Use as shuttle units consisting of a motor coach, a driving trailer and up to five intermediate coaches.

(iv) Use as shuttle units for suburban traffic consisting of a motor coach, intermediate coaches, a driving trailer and up to four such units coupled together in multiple-unit control.

The required output programme led to a four-axle motor coach equipped with four traction motors of a total hourly output of 1,600 H.P.

Main particulars are:

Gauge	4 ft 8½ in.
Current system Single-phase, 16 2/3 cycles, 15 kV.	
Hourly output of the motor coach at	
43·4 m.p.h.	1,600 H.P.
Tractive effort at hourly output	13,440 lb.
Maximum tractive effort	23,800 lb.
Maximum speed	62 m.p.h.
Number of seats, third class	44
Total weight	62·0 tons

The main body dimensions of the motor coach were adapted to those of the lightweight coaches of the Federal Railways, so that train units made up of motor coaches and light-weight coaches look alike. The body, which is electrically welded throughout, is carried on bogies of the same design as those of the type Re 4/4 locomotives, which are distinguished by their excellent running features and extremely low maintenance. The bogie frames are of box section.

For the drive in a number of coaches, the Brown, Boveri disc drive was chosen and the Sécheron laminated drive in the remainder. The bevel gear wheels have no springing.

The brake equipment of the motor coach comprises:

(i) the electric direct-current resistor brake;

(ii) the automatic passenger train compressed-air brake with a maximum 110 per cent braking;

(iii) the non-automatic regulator brake with a maximum 110 per cent braking;

(iv) a hand brake in each driver's cab acting on the adjacent bogie.

In the layout of the electrical equipment the aim was to cover as little useful space as possible with apparatus, whilst making the accessibility such that periodic inspec-

Oertikon Engineering Co.

PLATE VII. S.N.C.F.: freight train at Hirson hauled by electric locomotive No. CC 14003. This locomotive converts 50-cycle, single-phase, 25 kv A.C. from the power line to 3-phase A.C. for use in the traction motors.

tions can be carried out as easily and rapidly as possible. The transformer with the oil-cooling set, the converter set and the compressor were placed beneath the coach floor, while the contactors, the reversers and the remaining electric and small pneumatic apparatus, are assembled together in an apparatus box arranged near the luggage compartment. The fans for cooling the traction motors are mounted on the roof above the drivers' cabs. This arrangement provides a perfect weight balance.

The roof equipment includes a pantograph current collector above the luggage compartment, a Brown, Boveri air-blast circuit-breaker, and, distributed above the central portion, the brake resistor consisting of boxes of ribbon-resistors.

The step-transformer is of the Brown, Boveri radially laminated type and has an output of 837 kVA. It is separately cooled by a special cooling set with an Oerlikon motor-driven oil pump.

The transformer has separated H.T. and L.T. windings, of which the latter are not earthed, but only placed at earth potential through an earth indicating relay. This method of connection, already introduced in a large number of traction vehicles by the Oerlikon Engineering Co., has the advantage that an earth in any part of the L.T. network is, of course, indicated, but it produces no immediate results and allows the service to continue.

The traction motors are connected permanently in series/parallel, whereby each group has an isolating contactor and both groups together have a simple reversing switch with only two positions. The regulation of the voltage of the traction motors is made by the suitable connection of the L.T. tappings in conjunction with a triple reactor and eighteen electro-pneumatic step-contactors in eighteen running steps.

The motors are of the 8-pole, single-phase series type, and are so designed that they can be heavily overloaded. Over-refinements and forced cooling are avoided. The peripheral velocities of the motor and commutator are moderate, and the loading of the armature lies far below the usual values.

All the main current apparatus, such as collectors, isolating and step-contactors, and reversing and brake change-over switches have electro-pneumatic drive. For the control of the brake change-over, a brake change-over switch which permits the use of very simple reversing switches was developed by the Oerlikon Engineering Co.

Swiss privately-owned railways: B.L.S. high-speed twin-unit rail cars. Recent developments in high-speed motor coaches are well illustrated by three twin-coach units, type BCFe 4/8, developed in 1955 by the Swiss Industrial Co. for the Berne-Loetschberg-Simplon Rail-

way from an earlier design (1946) built for the same company.

Like all modern rail cars, these light-weight steel coaches are of integral design, entirely arc-welded and provided with a spray-asbestos insulation against variations of temperature and noise.

Both sections of the twin coach are permanently connected through a central coupling. One section contains eighteen second-class seats, a toilet, a third-class compartment, and a smoking saloon. The second section contains a luggage compartment, toilets and two third-class compartments. The smoking saloon in the first section can be converted to house a transformer to enable the motor coach to operate on a 3,000 v. D.C. overhead line. It is thus possible to use the motor coaches for rail tours into Italy where this latter electric power is used (*see page 155*). For this reason also, the dimensions of the vehicles have been made to conform to the loading gauge of the Italian State Railways. There is only one pantograph, one transformer and one Sécheron control jumping switch gear for both sections of the twin unit. The four Brown, Boveri traction motors develop an hourly output of 1,200 H.P. The entire electrical equipment is concentrated on the middle of the train, and all high-tension elements are located on the coach roof (transformer with connected switch gear, main switch). Special boxes in the luggage compartment contain the compressor group, the motor generator for excitation of dynamic brake and the magnetic cut-outs for heating. The two motor bogies are also located in the middle of the train with the advantage that only low-tension circuits have to be connected to the drivers' cabs. The coaches are provided with multiple control equipment allowing several coaches to be controlled from one driving cab. The maximum speed is 110 Km/h. (68·5 m.p.h.). A twin motor coach can haul a load of 50–60 tons up gradients of 2·7 per cent at a speed of 75 Km/h. (46·5 m.p.h.).

Special consideration has been given to the construction of the bogies, which are equipped with torsion bars. The coach body is supported on a swing bolster outside the motor bogie, and supported inside the trailing bogie by means of side bearings. The swing bolster hangs on inclined swing links in the centre of the frame to which the torsion bars are connected. Well lubricated cylindrical axle-box guides are arranged inside of the helical spring suspension. The riding qualities are excellent even directly above the motor bogies.

Main particulars are as follows:

Gauge	1,435 mm. (4 ft 8½ in.)
Length over buffers	23,700 mm. (77 ft 10 in.)
Number of seats	64

Weight of mechanical part	38 tonnes (37·4 tons)
Weight of electrical part	26 tonnes (25½ tons)
Weight in working order	64 tonnes (62·9 tons)
One-hour performance of the four traction motors	2,000 H.P.
Maximum tractive effort at the wheel rim at starting	13,000 Kg. (28,000 lb.)
Maximum speed	110 Km/h. (68·5 m.p.h.)

Swiss privately-owned railways: smaller companies' equipment. There are numerous small railways in Switzerland which have recently been re-equipped on the lines indicated. These railways usually link the smaller communities in the mountain valleys with the main lines. Most of them are electrified, and in size they range from the Rhaetian Railway which owns 276 Km. (171·4 miles) of metre-gauge track down to the Uetli-Berg Railway, near Zurich, which is but 9 Km. (5·6 miles) long and is of standard gauge and the Schoellenen Railway, 4 Km. (2·5 miles), metre gauge. Some are mountain railways, as, for example, the Jungfrau and Rigi railways.

The rolling stock used can be roughly divided into two groups:

(i) wooden coaches on steel frames which were built until recently; and

(ii) new types of vehicles, using light-weight metals, welding technique, and of pleasing air-smoothed appearance.

First, in importance, are the new motor coaches for the Swiss South-Eastern Railway, which is an important link between Eastern Switzerland and the Gotthard line. The railway is standard gauge, 49·2 Km. (30·5 miles) long, of which 18·8 Km. (11·7 miles) are in the mountains, having ruling gradients of 1 in 20. Electrification was carried out in 1939, with 15,000 v., 16 2/3 cycles, single-phase A.C., the line having been opened originally in 1877. The eight motor coaches ordered run singly, or in twin units with a trailer, the maximum speeds being 48 Km/h. (30 m.p.h.) for a motor coach with one trailer (100 tonnes (98 tons) tare weight) and 65 Km/h. (40½ m.p.h.) for the motor coach alone.

The motor coaches have four traction motors and regenerative braking.

Another type of railway is the standard gauge rack-railway, leading from Arth up the Rigi Mountain. It was electrified as early as 1907 with 1,500 v. D.C. Like all scenic railways, it has to cope with considerably increased traffic during the holiday season, and has, on occasions, handled as many as 7,000 passengers per hour. To meet these requirements, the old rolling-stock of single motor coaches had to be replaced by trains and, as

a first step, two motor coaches, capable of running with trailers, were ordered. It was required that a motor coach with two trailers and with a total weight of approximately 55 tonnes (54·1 tons) should be able to make three consecutive ascents and descents with a five minutes' turn-round at each terminus. It was calculated that this required an output of 690 H.P. The coach is 15,400 mm. (50 ft 7 in.) long, seats sixty, and weighs 26·5 tonnes (26 tons) fully laden. It rests on two four-wheel bogies, each having one fully suspended traction motor; this motor is connected by a flexible gear train to the wheels. The motors can be connected in series and parallel, and by means of electro-pneumatic contactor gear four running steps are available. When descending, there are twenty-four braking steps, the motors acting as generators. The motor coach must always be at the lower end of the train, which means that, in addition to the driver, there has to be an observer in the leading (top) coach. It is hoped to build, later, pilot coaches containing a cab in front which will house the driver, who will be able to control the motor coach by means of electro-pneumatic contactor gear, and without the help of a second man. Four braking systems are fitted: electric, hand, and two independent compressed-air brake systems. One of the air-brakes acts automatically in the following emergencies: Non-operation of dead-man's handle; no current; operation of maximum current relay; and use of alarm signal by a traveller.

Again a different kind of vehicle is required for the Forchbahn, which is an inter-urban railway of metre gauge, built as an electric railway (1,500 v. D.C.) in 1912, and having a length of 17 Km. (10·5 miles). It connects the centre of Zurich with several small communities leading to the town of Esslingen. The railway has gradients up to 1 in 40; its line in the urban area of Zurich runs together with the municipal tram lines and uses 600 v. D.C. In the country districts, however, the voltage used is 1,200 v. D.C. The ever-increasing traffic required the acquisition of additional and more modern rolling stock and, in 1945, two motor coaches were ordered. The coaches are 15,920 mm. (52 ft 3 in.) long, seat forty-four and carry, in addition, fifty-six standing passengers. The tare weight is 26 tonnes (25½ tons). There are cabs at both ends with ample space for luggage and parcels. The all-steel body is carried on two four-wheel bogies each axle of which is motored and driven through a gear train. The motors are designed for 600 v. tension. On the 600 v. section, two motors are permanently connected in parallel, and the two groups (of two motors each) can be coupled in series or parallel. On the main line section (1,200 v.) two motors are permanently in series, and can again be connected to the group in series and parallel. There are twenty-six controller positions operated by electro-pneumatic contactor gear, as well as sixteen brake positions. The cars are equipped with dead-man's handle, automatically closing doors, electric heating, electric and Westinghouse compressed air brakes, as well as rail brakes.

Among the railways which modernized their rolling stock recently were the Bernese Oberland Railways, which acquired three powerful electric motor coaches for their services radiating from Interlaken to Grindelwald and Lauterbrunnen. They are suitable for operation on rack and adhesion sections. The ruling gradients of the lines over which they are to operate are 1 in 40 on the adhesion and 1 in 8 on rack sections; the total length being about 20 Km. ($12\frac{1}{2}$ miles). The railway is built to metre gauge and uses the Riggenbach system on its rack sections (in all about 5 Km.) (3·1 miles). Maximum speeds are as follows: adhesion 70 Km/h. (43·5 m.p.h.); rack ascending 35 Km/h. (21·7 m.p.h.); rack descending 19 Km/h. (11·8 m.p.h.).

The minimum radius of curves is 100 m. (109 ft $4\frac{1}{4}$ in.), and the maximum axle load is 12 tonnes ($11\frac{3}{4}$ tons). The railway was electrified in 1914 with 1,500 v. D.C., and so far has been worked by locomotive-hauled trains, the locomotives being of uniform type with a C wheel arrangement.

The new coaches, built by S.L.M., Winterthur and Brown, Boveri Ltd, seat thirty-eight third-class and ten second-class passengers. There is also a luggage compartment and two driver's cabs. The body forms a single structure and rests on two motor bogies. The end of the main frame carries automatic centre couplers. The motors drive, through bevel and spur gears, the rack pinion as well as the adhesion axles, thus coupling permanently rack and adhesion drives. There are three separate braking systems, a Westinghouse differential pneumatic brake, a direct-acting Westinghouse pneumatic brake for the rack pinions, and a handbrake so that the wheels can be locked when the coach is stationary.

The electrical equipment comprises a number of unusual features, such as series motors with two series-connected commutators for each motor armature, necessitated by the vastly differing requirements on adhesion and rack parts, as mentioned. The four motors are normally in series/parallel but can also be used in parallel (e.g. when hauling trailers at high speeds on adhesion sections). The coaches are equipped for regenerative braking. All operations are carried out through a triple-drum master controller. The first drum has twenty notches for running with full field and two for reduced field, controlling the motors during power running and electric braking. The second drum controls reversing

and grouping of motors for rheostatic braking; and, finally, a grouping drum. This last, which controls regenerative braking, can put the motors in series for banking, and also controls battery-charging.

The ancillary equipment is the usual found in D.C. practice. During regenerative braking, a 36 v. Nife battery is charged from which all lighting and control currents are taken. The coaches are also equipped with dead-man's handles. The vehicles have been in satisfactory service for two years.

SUBURBAN TRAIN SETS FOR THE S.N.C.F.

During 1937 the S.N.C.F. put into service twenty-seven electric train sets which were the first large-scale application of the Budd process of using stainless steel in Europe. They were ordered for the Paris–Versailles– Le Mans electrification (1,500 v. D.C.). Trains are made up of the required number of two-coach articulated units (three bogies for each unit). This stock cannot be coupled to other types of coaching stock. The one-hour rating of these articulated units is 1,428 H.P., with normal speeds of up to 130 Km/h. (78 m.p.h.) and maximum speeds of 186 Km/h. ($115\frac{1}{2}$ m.p.h.). The total weight of the two-car trains is only 76 tonnes (74·8 tons), and the trains have been very satisfactory in service (Plate 50A, page 214).

The design of these trains was most carefully considered and special electrical equipment having very low weight was developed. The stainless steel used has a low carbon content; it is a nickel-chrome steel with a tensile strength of 35 Kg/mm.^{2} (22·224 tons/sq. in.). By laminating (cold forming) the steel, the tensile limit goes up to 85 Kg/mm.^{2} (53·973 tons/sq. in.). One of the most important properties of steel is that it can be welded electrically.

During and after the war the traffic problems of the Seine valley were closely studied (Paris (Quai D'Orsay) to Melun and Montereau). It was decided to use reversible train units consisting of a motor coach and trailer with an adhesive weight of 62 per cent of its tare weight, each motor bogie containing two motors. These train units can be coupled together into trains of four, six or eight coaches. The design shows an interesting compromise between express services and suburban railway requirements. The coaches have a high degree of comfort and also eight double doors at each side of a train unit, allowing for easy movement of passengers. An important problem had to be solved as the new coaches were to be used in the suburban area with its high (British type) station platforms and also in the outer districts where platforms are of the low (Continental) type. A recessed double step was the solution.

These coaches are mainly constructed of stainless steel, but the end units are made from high-resistance steel. There are eight doors in each coach, manually or air operated. Each train unit can take 265 passengers, of whom 164 are seated. The inside is fitted in coppered steel and anodized aluminium; insulation against heat and noise is made by fibreglass mattresses; lighting is by fluorescent tubes, continuously and invisibly mounted on the ceiling. The units are automatically coupled by Dellner-Scharfenberg couplings. Buffers are provided at the end of each train unit only; the two coaches of each unit are connected by a simple drawbar.

TRAIN SETS FOR THE NETHERLANDS RAILWAYS

The Netherlands Railways recently put into service forty-seven four-car train sets and thirty two-car train sets (Plate 49B, page 213).

Each four-car train set consists of three third-class coaches and one second-class coach; the coaches rest on two-axle bogies, those at each end being motored. Flexible communication gangways with double diaphragms are provided, and the train units are streamlined, with driving compartments at each end. The two-car train sets are composed of a third-class and a second–third composite car, and are identical in design and construction. Main particulars of these trains are given below:

Bogie wheelbase	3,000 mm. (9 ft 10 in.)
Length of driving coaches	24,950 mm. (81 ft 11 in.)
Length of trailing coaches	23,530 mm. (77 ft 3 in.)
Total weight of four-car train	200 tonnes (196 tons)
Maximum speed	140 Km/h. (87 m.p.h.)

The motor coaches have four motors, one on each axle, and details are as follows:

Type	series, axle-hung
Voltage	1,500 v. D.C.
Cooling system	self-ventilated
Transmission ratio	22 : 57
Power ratings (continuous):	
full field	168 H.P. at 675 v., 1,270 r.p.m.
weak field	210 H.P. at 675 v., 1,550 r.p.m.
Wheel diameter	950 mm. (3 ft 1½ in.)

The body is of all-steel, welded construction. The front nose is reinforced to provide greater protection for the crew in case of collision; doors and most fittings are in aluminium alloys. The kitchen in each four-car train set is equipped with boiling plates, refrigerator, coffee machine, and hot water boiler; tables can be placed between the seats.

Brakes used are of the Westinghouse air-brake type; this is of the non-exhaustible type and allows gradual application and release. The trains are equipped for multiple-unit control and several train-sets together with mail vans can be used. The trains are equipped with automatic couplers which include air piping and electrical connections. For uncoupling, the driver depresses a pedal whereby compressed air enters a cylinder in the coupler which uncouples the draw-gear and electric circuits, and isolates the air-brake piping. As long as the pedal remains depressed, one of the train sets can be driven away. As soon as the driver releases the pedal, the mechanical and electrical parts are ready to be recoupled, the air-brake piping remaining closed as long as the couplers are not fully connected. The trains are equipped with a thermostatically controlled air heating and ventilating system; the same system is used in summer and winter; in summer the heaters are switched off and the fresh air used for ventilation.

TRAIN SETS FOR THE SWEDISH STATE RAILWAYS

In recent years a number of electric train sets have been supplied to the Swedish State Railways and to private lines in Sweden. They include a three-car set designed for a maximum speed of 135 Km/h. (83·8 m.p.h.). One outer car contains a driver's cab and two second-class compartments, pantry and luggage-room; the second car, two third-class compartments and space for the electrical machinery; the other outer car, a driver's cab and two third-class compartments (Plate 50B, page 214). The coaches are all-welded, floor, walls and roof forming supporting elements and thus providing a firm lightweight construction. The leading cars have four traction motors, mounted flexibly in the inner bogies with one motor driving each axle. Power is transmitted from the motors to the wheels by double-acting quill drive and resilient elements (A.S.E.A. system). The bogies are all-welded and the axles are supported in Skefko roller bearings. The voltage is high-tension multi-step controlled by a servo-motor operated by a load tap changer, and the operating system is designed for multiple-unit control. Should the current fail while the train is travelling, or the driver release the dead-man's handle, a valve comes into operation which applies the brake after the train has travelled a distance of 200 m. (approximately 1/8 mile). The main transformer is installed in the underframe of the centre car and the remaining equipment is mainly placed below the cars. The control devices are installed in the apparatus room of the middle coach.

Another type recently developed is designed for electric light-railway service and built for a maximum speed of 75 Km/h. (46·6 m.p.h.). The vehicles are designed as motor coaches, trailers and driving trailers.

This stock is also of all-welded construction. In the motor coaches all axles are motored and the motors are of the nose-suspended type; the bogies are all-welded and the axles are supported in Skefko roller bearings. Voltage regulation is through special starting resistances connected in circuit from the driving controllers through contactors: multiple-unit control is used. The electrical equipment is placed under the floor of the coaches and also under the seats. There are electric-dynamic resistance brakes in combination with pneumatic brakes. The resistance brake is operated from the controller, the pneumatic brake being applied by a special brake valve located beside the driver's platform.

The cars are fitted with a central coupling of the Scharfenberg type; the interior walls and roof are finished in painted wood-fibre boards and the floor is covered with ribbed rubber matting. The walls and floor are heat-insulated with Isoflex. The entrance doors are double folding doors operated electro-pneumatically by the conductor accompanying each car. Fans mounted in the underframe circulate through the car air heated by the starting and braking resistances.

Part VI. Underground railways

Underground railways were conceived and developed in large cities and their suburbs for the purpose of relieving the surface traffic congestion.

Underground railways are of two types according to their method of construction.

(i) Those which are built by tunnelling of one means or another. The London "Tube" lines are excellent examples of this type of underground railway.

(ii) Those which are built by the "cut and cover" method in which a deep channel is dug from surface level, and then the surface replaced over it to form a tunnel. The tunnels of the Metropolitan Railway are mostly of this form of construction.

It is interesting that in London, where the first underground railway was built, the idea was mooted very soon after railways had become an accepted form of transportation. Parliamentary sanction was obtained as early as 1853 for the construction of the first underground line, although it was not until 1863 that the line was opened for traffic.

The first underground railways were, of course, steam operated, with what difficulties and discomforts can well be imagined. The advent of electric railway traction can nowhere have been more important than in the sphere of underground railways, and it was in 1890 that the first electric underground railway, the City & South London, was opened for traffic.

Many large cities now operate their own underground railways and New York, Toronto, Moscow, Berlin, Paris, Rome and Madrid may be quoted. The extent of the system and the type of construction depends largely upon the geological nature of the ground. Thus the New York tunnels which had to be bored through hard rock, in areas where "cut and cover" construction was not practicable, were an infinitely more difficult undertaking than the London tubes which are bored through clay.

LONDON'S UNDERGROUND RAILWAYS

In 1954 the London Transport Executive operated 253 route-miles of railway. On these lines ran 2,475 motor coaches, 1,553 trailers and 69 electric locomotives. In that year 671,400,000 passengers were carried.

The beginnings of this, the largest metropolitan transport undertaking in the world, were very modest. The North Metropolitan Railway connected North and West London to the City, by a $3\frac{3}{4}$-mile tunnel from Bishops Road Station, Paddington, to Farringdon Street, the line being opened for traffic in January 1863. The success of this line led to extensions to Notting Hill, Hammersmith and, later, to Harrow in the west and Aldgate in the east – the so-called Inner Circle line being completed in 1884.

The City & South London Railway was opened in 1890 as the first electric underground railway in England (*see* Part I, page 144) and the first tube railway in the world. It had its own power station and was worked on a three-wire D.C. system by locomotives and trailers; the tunnels were smaller than those of the later tubes. During 1923–4 the existing line from Clapham Common to Euston was reconstructed to conform with the rest of the Underground system, and in 1926 the system was extended to Morden, making a total length of $12\frac{1}{2}$ route-miles ($32\frac{1}{2}$ track-miles).

In 1901 the Central London Railway was opened from Shepherd's Bush to the Bank, and this line also was electrified from the beginning. It was a very severe competitor for the Metropolitan Railway, and the latter, together with the so-called Metropolitan District Railway, decided to electrify their Inner Circle line. Current used was 600 v. D.C., and the third- and fourth-rail conductor system was employed. The motor cars (four motors of 150 H.P.) weighed 11·39 tons, and had a maximum speed of 40 m.p.h.

The London Electric Railway consisted of the Bakerloo, Piccadilly and Hampstead and Highgate tube railways, covering 32 route-miles (79½ track-miles). The Bakerloo Line was opened in March 1906, from Baker Street to Waterloo, and was extended by short stages, reaching Queen's Park in 1915; a through joint-stock service is operated from Elephant and Castle to Watford Junction (L.M.R.). The Piccadilly Line was opened in 1906. The Northern (Hampstead and Highgate) Line was opened in 1907, from the Strand to Golders Green and Highgate, and as far as Hendon in 1923, and Edgware in 1924.

In 1931 the whole of London's local railway system was incorporated, together with trolley-bus, tram and bus systems, in the London Passenger Transport Board (L.P.T.B.).

It was decided to extend the underground system by continuing the Central Line westwards to Denham and eastwards to Woodford–Epping–Ongar; to electrify the Chesham lines as far as Rickmansworth, and, together with the L.N.E.R., to carry out a 1,500 v. D.C. electrification from Liverpool Street to Ilford and Shenfield. World War II interrupted the work already started and it was not completed until after the end of the war (see Part V, page 224).

In 1948, when all the railways in Britain were nationalized, the railways which formerly came under the jurisdiction of the L.P.T.B. still retained a measure of autonomy, but the managing body became known as the London Transport Executive (L.T.E.).

Electricity to supply the power for L.T.E. railways is generated as 33 1/3-cycle, three-phase A.C., at a pressure of 11,000 v. There are three power stations: Lots Road, Neasden and Greenwich, and the current they supply is fed to sub-stations situated at convenient points all over the system. In these sub-stations, the A.C. is converted to 600 v. D.C., which is supplied to the trains through positive and negative insulated conductor rails.

Except for some outer suburban trains of the Metropolitan Line which are worked by electric locomotives (see Part I, page 149), all L.T.E. passenger services are operated by motor coaches and trailers. For service on the tube lines the modern stock is that which was first put into service in 1938 and to which few modifications have since been necessary. Trains are made up of driving motor coaches, non-driving motor coaches, and trailers. Motive power is thus distributed more evenly throughout the train than is the case with only motor coaches and intermediate trailers. Each motor coach has one 168 H.P. motor on each bogie. Trains are normally made up of seven coaches comprising one three-coach and one four-coach unit coupled together. Each three- or four-coach unit has a driving motor coach each end,

with a trailer only in between them in the three-coach unit, and a trailer and a non-driving motor coach in the four-coach unit. A seven-coach train, therefore, has an aggregate of 1,680 H.P.

The latest stock has automatic centre couplers which incorporate thirty-eight electrical connections, and Metadgar control (Plate 51A, page 215).

The latest District line (surface) stock is similar to that introduced in 1939 and is known as R-Class Metadyne stock. The Metadyne system insures smooth acceleration and deceleration by controlling the rate at which current is supplied to the motors proportionally to the rate of acceleration. During deceleration the polarity of the motors is reversed and current is returned to the conductor rails at a rate proportional to the rate of deceleration. The 1939 stock is operated in three-car units (two motor coaches and a trailer between) or in two-car units (two motor coaches). The motor coaches have each bogie powered by one motor of 152 H.P. (one hour rating).

The R-Class stock has no trailers, every coach is a motor coach having one nose-suspended self-ventilated traction motor of 110 H.P. driving each bogie. This gives exceptionally good acceleration, and much of the older stock is being converted to this principle. R-Class stock is made up of 4-car and 2-car units, each with driving compartments in the end coaches (Plate 51B, page 215).

THE PARIS METRO

The first 10 Km. (6·2 miles) of underground railway in Paris was opened for traffic in 1900 and there were eighteen stations. There are now 202 Km. (125·4 route-miles) and 284 stations on the Metro. The system is still being extended.

The Paris Metro runs partly underground and partly on the surface. The underground sections are nowhere very deep but construction was carried out on the system of *galeries boisées* in preference to the cut-and-cover system.

Motor coaches with pneumatic tyres. In 1954, the Paris Metro (R.A.T.P.) decided to equip the whole of its line eleven (Place du Chatelet–Maria des Lilas) with eighteen train sets running on pneumatic wheels. Ten of these were ordered from Messrs Renault, and the remaining eight from Messrs Brissoneau et Lotz (Plate 52A, page 216).

The trains have a total length of approximately 60,000 mm. (196 ft 8 in.), and consist of four vehicles permanently coupled in the following manner: One motor coach, second class, with driver's cab; one trailer, first and second class; one motor coach, second class, and, finally, one motor coach, second class, with driver's cab. Each motor coach can carry 214 seated passengers, and it is claimed that in the rush hour the first class can take eighty-nine passengers and the second class 561. The bodies of the vehicles are completely arc-welded. On each side of the coaches are four large sliding doors which are operated by the conductor.

Each coach has two two-axle bogies and the bogies of the three motor coaches in each train are motored, being equipped with two nose-suspended motors each having a continuous output of 70 H.P. They are of welded steel construction and rest on their axles with a rubber primary suspension. The secondary suspension of each bogie consists of helical springs. Each axle has two normal running wheels with pneumatic tyres (similar to lorry tyres) and also two safety wheels which are similar to railway wheels, but lighter. The running wheels run on wooden rails (made from Azobe timber) while the safety wheels are not in contact with the rail except in case of emergency, such as the deflation of a tyre. There are also guide wheels which press laterally on to vertical guide rails. These guide rails are fixed on both sides of the track.

Power is taken from conductor rails which are fixed beneath the wooden side-guiding-rails through horizontally-mounted collector shoes. The normal acceleration is 4·25 ft per sec. per sec., and the braking rate is 4·75 ft per sec. per sec. There are twenty-three notches in the camshaft control gear, with automatic operation under either the Jeumont-Heidmann all-electric system or the C.E.M. electro-pneumatic operation. Gear ratio is 1 : 11.

It is claimed that the advantages of these pneumatic tyre trains are:

(i) They are not submitted to the shocks caused by steel wheels running on steel rails and, in consequence, they are considerably lighter.

(ii) The greater adhesion of the pneumatic tyres allows better acceleration and braking than with normal railway wheels giving characteristics which approach those of the automobile.

(iii) The safety rail can be made much lighter than the normal rail used at present.

(iv) Noise is greatly reduced.

Other particulars are as follows:

Weight of Motor coach:

With cab	22·6 tonnes	(22 tons)
Without cab	21·8 tonnes	(21¼ tons)
Weight of trailer	15·7 tonnes	(15¼ tons)
Line voltage	600 v. D.C.	

ROME UNDERGROUND RAILWAY

In 1955, Rome began to operate its first Underground railway, work on which had begun in 1938 and was interrupted as a result of the last war. Originally the line was to connect the centre of the town to the area of a world exhibition, which did not take place, and it runs from Rome Central Station to the suburb of Laurentina, a distance of 10·6 Km. (6·5 miles). There are seven intermediate stations. The line runs first as underground railway to the Porta Sao Paolo station, then in a cutting and above ground to the station of Esposizione. It then runs again in tunnel to the terminus at Laurentina. The train frequency is six per hour, and eighteen four-axle motor coaches provide the service. These have been supplied by the Italian Brown, Boveri Co. (T.I.B.B.) and have the following main particulars:

Gauge	1,435 mm.	(4 ft 8½ in.)
Total length of coach	19,100 mm.	(62 ft 9½ in.)
Distance of bogie centres	11,000 mm.	(36 ft 1 in.)
Wheel diameter	900 mm.	(2 ft 11½ in.)
Weight, empty	40 tonnes	(39¼ tons)
Number of passengers: seated		48
standing		195
Maximum speed	100 Km/h.	(62·1 m.p.h.)
Tractive effort (continuous) at 48 Km/h. (29·8 m.p.h.)	2,720 Kg.	(5,984 lb.)
Current	1,500 v. D.C.	

The electrical equipment consists of four D.C. motors with an hourly output of 117 kW., and continuous output of 88·5 kW. There is an elaborate braking system, operated by air and electrically, and in addition there are four electro-magnetic sledge brakes. The distance of 10·6 Km. (6·5 miles) is run in fifteen minutes. The motor coaches can be used in composite trains of two or four units with multiple-unit control from one cab (Plate 52B, page 216).

TORONTO SUBWAY COACHES

The city of Toronto has a population of one and a quarter millions and suffers very severely from traffic congestion, particularly at rush-hour periods. Statistics show a density of automobile ownership of no less than one car per 3·03 persons resident in the metropolitan area. This adds considerably to traffic problems and it was decided to improve public transport by building a railway below street level. The Toronto Transit Commission developed a plan to divert passengers from surface tram-cars on Yonge Street, the main traffic artery of Toronto, to a railway below or adjacent to the street. The line was to have a maximum capacity of 40,000 passengers per hour, an increase of 26,000 over that provided by the tram-car system, which would, therefore, allow the Yonge Street tram-cars to be abandoned.

Building commenced in September 1949 and was estimated to cost approximately $63,000,000. The initial double track, 4½ route-miles, has been laid under or adjacent to Yonge Street, with its present termini at Eglington in the north and Union Station in the heart of the city. About 2·7 miles of the route are in tunnel and the rest is in the open. The underground portion of the project was constructed by the cut-and-cover method. A temporary deck was placed over the excavation to allow resumption of normal street traffic, including tram-cars. Further excavation, followed by laying of the railway tracks, then proceeded underneath the decking. In the less densely built-up areas, both cut-and-cover and open-cut construction were employed; any houses directly in the way were demolished and adjacent buildings underpinned. Tunnelling was employed only where the track passes under the Canadian Pacific Railway, and where it passes north-westerly from St Clair under Yonge Street.

The twelve stations are all situated at intersections with main cross-roads. All platforms are of uniform length, 500 ft long, and are outside the tracks, with the exception of those at the terminus stations, where they are of the island type. The unusual track gauge of 4 ft 10 7/8 in. was chosen to enable the existing tram-cars to run over the railway tracks.

An order for 104 coaches was placed with the Gloucester Railway Carriage & Wagon Co. Ltd, who ordered from the British Thomson-Houston Co. Ltd electric traction and auxiliary control equipment, together with all the control and auxiliary coach wiring (Plate 52C, page 216). The vehicles are 57 ft long and 10 ft 4 in. wide, have sixty-two seats, and carry a maximum load of 240 passengers. The multiple-unit train sets are made up of one, two, three or four two-car units according to traffic requirements. Each coach of a unit has a driving cab at the outer end, which occupies only one-third of the car width. One car of each two-car unit carries the air compressor, the other carries the auxiliary motor-generator set. The superstructure, comprising the car underframe and body, is of the all-steel type. Special precautions were taken against fire, and the body panels are insulated against both noise and vibration. There are three electro-pneumatically controlled double doors on each side of a vehicle. Four cars are made of aluminium alloy, to allow investigations on comparative current consumption.

The control equipment provides for series/parallel control of the four traction motors; they are arranged in pairs permanently connected in series and mounted longitudinally in the bogies; they drive through cardan shafts and bevel gearing on the axles. The trains can be operated in multiple-unit control. All main items of traction control equipment – including two line-breakers, camshaft accelerating unit, reverser, series/parallel, transfer switch, field shunting contactors, accelerating, volt-ampere and overload relays, etc., are mounted on the main frame.

The equipment comprises a number of electro-pneumatic units including a cam-operated contactor group which controls the acceleration. Two steps of field weakening are provided. The use of a large number of accelerating points provides rapid and smooth acceleration. The two line-breakers, the reverser, and the series/parallel switch are all electro-pneumatically operated.

The vehicles contain certain electrical features which are not normal British practice:

(i) The master controller is operated by the driver's left hand and is fed from a traction control relay, the operating coil of which is energized through the door interlocks, of which there are ninety-six in series on an eight-car train. A combined door cut-out and interlock shorting switch is provided for each door. In addition, a sealed master-switch in the driving cab can be used in an emergency to cut out the traction control relay.

(ii) There is a visual starting signal.

(iii) The door circuits are designed to provide for up to four operational guards on one eight-car train. The zoning feature operates automatically and divides the train into the same number of zones as there are guards.

(iv) Both forward and rearward door open and close control push-buttons are provided.

(v) Guards can signal to each other and to the driver by buzzer, but the driver can only signal to the guard by bell.

(vi) The cars are equipped with thermostatic control of heating.

CHAPTER 4

The Reciprocating Steam Locomotive

by C. R. H. SIMPSON

This section, dealing with the construction and to a lesser extent with the design of locomotives, has been written with railway companies' locomotives in mind. Industrial and contractors' locomotives have not been included although, of course, these engines possess many features in common with those designed and built for main line use. There are, however, certain practices, of which geared drives, vertical, and water-tube boilers may be quoted, which while having had experimental application on main line engines are now restricted in use to locomotives operated by contractors or industry. Other special designs, for example the fireless boiler, have always been limited to industrial use.

Much of the practice followed by steam locomotive designers and builders, for example, bogie construction, the transmission of power by side rods, and plate frames, has to some extent been adapted to the requirements of diesel, electric, and gas-turbine locomotives.

In recent years steam locomotive design and construction has been greatly influenced by factors other than that of performance on the road. The much greater use made of welding, the difficulty in obtaining suitable labour for shed duties, and the impossibility in some areas of the world of being able to obtain good class coal at an economic price are some such widely differing factors.

Labour problems have resulted in the production of designs incorporating components intended to run from shopping to shopping with little or no shed attention. This trend, combined with designs incorporating increased accessibility of working parts, has combined to give greater availability. High availability is, be it noted, the strongest claim made by the forms of motive power which rival the steam locomotive. It must, however, be borne in mind that high availability loses much of its value if it is not accompanied by good utilization. In order to achieve such utilization there has been, in recent years, an increasing tendency to build general utility, or mixed traffic, locomotives. It is noteworthy, in this connection, that the British Railways series of standard locomotives, which may well be the last series of standard steam locomotive designs to be produced anywhere in the world, consists of mixed traffic locomotives only,

with the exception of the 2–10–0 of Class 9 intended for working heavy mineral trains.

Locomotive design and construction have taken the various forms which experience has shown to be best suited to the particular conditions prevailing on the railways concerned. It is, however, a mistake to think that particular designs, associated normally with one class of work, cannot be modified satisfactorily for another. As an example, it is customary to associate the Beyer-Garratt locomotive with a high tractive effort and a restricted axle load. As this form of motive power is extensively employed for long hauls, on single lines, where track is frequently light, it follows that speeds are not normally high. Should, however, the demand exist, this form of articulated locomotive can be produced to exert a high tractive effort at high speeds – as in fact has been done on the Algerian Railways, where speeds of over 80 m.p.h. have been attained with very smooth riding. On the other hand, some locomotive designs would never have been suitable for duties other than those for which they were originally intended. In this category fall all vertical boiler, and Shay, locomotives, to quote but two examples. The output of the former was strictly limited by size, and the latter was incapable of safe operation at high speed, although excellently suited to low-speed haulage on indifferent track.

Of recent years steam locomotive design and construction have become much more exact than formerly. It is only in comparatively modern times that limits and fits have been introduced, these having formerly been decided, often empirically, by the man on the job. Normally trouble only arose when work was put out to those unfamiliar with locomotives. Further examples of the exactness, now practised, may be found in the more scientific methods of erection, where the strained wire (Plate 60A, page 260) has been displaced, in many erecting shops, by optical lining up, and axlebox guides are ground by a machine mounted, with the frame, on a prepared bed. To some extent, closer limits have been necessitated by the introduction of ball and roller bearings. Generally there has been an increase in the amount of gauging employed in locomotive work, and in its standard of accuracy.

The metallurgist has also contributed much to locomotive development, and the use of higher-tensile steels has enabled lighter components, including boilers, to be used than would have been possible otherwise.

It was not until the 1930's that steam locomotive design became really well balanced. While progress has been made continuously, and is continuing, it is impossible to avoid the conclusion that as far as some components are concerned, for example, the piston valve, finality has been reached in design, if not in dimensions.

Part I. Construction & design – a concise encyclopædia

Adhesion. Adhesion, the link between wheel and rail, is difficult to measure and various authorities differ in the figures they quote for the co-efficient of friction or adhesion. The variation between dry rails and greasy or frosty rails is large, as may be seen from the reproduced table, due to Molesworth. It will be noted that there is relatively little difference between the co-efficients for very wet and very dry rails.

Weather conditions	Adhesion lb. per ton	Corresponding co-efficient of friction or adhesion
Rails very dry	600	0·268
Rails very wet	550	0·245
Misty weather, rails greasy	300	0·13
Frosty or snowy weather	200	0·09

The application of sand naturally affects the value of the co-efficient; for any given circumstances, generally speaking, it increases the value by 0.05.

Adhesive factor. This factor, also termed the factor of adhesion, is found by dividing the weight on the coupled wheels by the tractive effort. To obtain adequate adhesion in average conditions the factor should be not less than 4·3 for a two-cylinder locomotive and approximately 3·5 in the case of an engine with three or four cylinders. Due to the torque developed by diesel and electric locomotives being more uniform it is possible in the case of such forms of motive power to design for an adhesive factor of 3·3.

The amount of effective adhesion available depends largely on the condition of the rail; the figures quoted above for steam power are sufficient for a clean dry rail; with such a rail well sanded, a factor similar to that quoted for diesel and electric power would suffice, while with a greasy rail a factor of 6·66 would be required.

The adhesive factor does not have to be based on the starting tractive effort. In the case of locomotives intended for operating at high speed a lower factor is permissible than would be the case in engines which spend much of

their working life developing high drawbar pulls a low speeds.

Admission. This is the point at which the valve opens to allow steam to enter the cylinder. If no lead was provided admission would occur immediately the ends of the stroke were reached but in practice it is necessary to allow lead so that admission shall occur slightly before the end of the stroke and thus provide a cushioning effect, as well as advance steam admission under rapid piston and valve movements when running at high speeds (Fig. 62, page 286).

Anti-carbonizer. To prevent the formation of carbon deposit in the steam chest, on the valves, in ports, cylinders and blast pipe, an anti-carbonizer is fitted between the mechanical lubricator and the cylinders as close as possible to the steam-chest. Its function is to atomize the oil by mixture with a steam jet (*see* Atomizer).

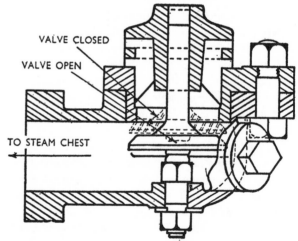

FIG. 33. Anti-vacuum valve mounted on a steam chest.

Anti-vacuum valves (Figs. 33 and 34). These valves, sometimes termed snifting valves, allow air to be drawn into the cylinders when the engine is running with the regulator closed. They are usually fitted on the superheater header as the circulation of air in this manner prevents the elements from becoming overheated. Such valves prevent the harmful results arising from ashes being drawn down the blast pipe with resultant scoring of valves, liners, pistons and cylinders. These valves are also sometimes fitted near the cylinders, and are entirely automatic, being kept seated by steam pressure when the regulator is open.

Arch. A term, sometimes used in American practice, to apply to the smokebox.

Arch-bar. The top member of the side frame in a diamond-framed truck, the member immediately below it is known as the inverted arch-bar.

Arch-bar truck. A type of truck, or bogie, used under tenders. The side frames consist of arch-bars and a tie-bar.

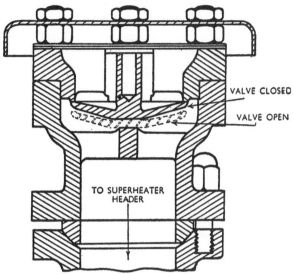

FIG. 34. Anti-vacuum valve mounted on a super-heater header.

As the spaces between the arch-bars are generally of diamond shape this form of truck is also known as a diamond truck.

Arch, brick. The purpose of the brick arch is to increase the length of the path taken by the gases on their way from the firebed to the tubes, thus providing opportunity for the admixture of supplementary air (via the firehole) which is essential for complete combustion; the rearward diversion of the gases also secures the more effective utilization of the entire heat absorbing surfaces of the firebox. The arch reduces the opportunity for the smaller particles of ignited fuel to pass directly to the tubes whilst still only incompletely burnt; its presence also assists in maintaining a high firebox temperature in general, and in imparting its reverberant heat to newly-fired coal. The arch also prevents the harmful results which would occur from air entering by the firehole passing direct to the tubes.

The arch is supported on studs in the side plates and sometimes on arch tubes. Its inclination to the tube plate is from 60° to 70°. The length varies, and a space is some-

times provided between the end of the arch and the tube-plate to prevent excessive ash accumulation on the top of the arch. While firebrick was formerly the only material used for the construction of the arch, refractory concrete is now extensively employed.

Arch tubes. Solid steel tubes, two or more in number fitted in the firebox and connected by expanding or welding, or a combination of both, to the lower portion of the firebox tube plate at their lower end and to points near the top of the firebox door plate at their upper end. In addition to increasing the heating surface and enhancing the circulation, the tubes also help to support the brick arch (Plate 58A, page 258).

Arch tube plug. As it is essential to keep the interior of arch tubes free from scale, provision is made for cleaning them out by providing these plugs in the throat plate and door plates at points opposite to the ends of the arch tubes.

Armand water treatment. Also known as the T.I.A. (*Traitement Intégral Armand*) this system of treating feed water on the locomotive has proved exceedingly successful. Briefly the process consists of dosing the feed water (which may or may not have received prior softening treatment in a stationary plant) with chemicals of the necessary kind and quantity. The apparatus consists of a simple auxiliary reservoir, devoid of valves or moving parts, which delivers the compound required to the main tank, by means of a siphon arrangement. The quantity fed is proportional to the amount of water taken on. The effect of this system of treatment is greatly to increase availability and reduce boiler maintenance (*see* Water).

Articulated locomotive. A locomotive usually having two sets of cylinders, but occasionally more, driving separate bogies, or engine units. These units may be directly connected together by an articulated joint, as in the case of the Mallet locomotive, or to a girder frame carrying the boiler, as in the Kitson-Meyer and Beyer-Garratt systems. While at one time some articulated locomotives were designed to operate as compounds, latterly almost all have been constructed for simple expansion working.

Over 100 different systems of articulation and partial articulation have been tried, some of which resembled an orthodox locomotive with provision for radial and lateral movement of coupled axles, thereby enabling a greater number of coupled axles to be incorporated than would have been possible with a rigid wheelbase. Among the better-known types of articulated locomotives are the Beyer-Garratt, Fairlie, Meyer, Shay, Mallet, Kitson-Meyer, and Golwé and their sundry modifications. While the Beyer-Garratt is capable of working trains at high speed – those supplied to the Algerian State Railways had such perfect stability that they were capable of regularly attaining speeds in the neighbourhood of 140 Km. (87 miles) an hour without difficulty on suitable track – other designs of articulated locomotives were generally suited to much lower speeds and some, for example, the Shay, to low-speed operation only. The Mallet locomotive had relatively little application outside the North American continent, Russia, and certain Continental countries. In America it assumed enormous proportions. Today the sole example being built is the Beyer-Garratt, used in many countries throughout the world, and built for all gauges, particularly the 3 ft 6 in. and narrower.

The technical reasons for employing this type of locomotive are fourfold:

(i) In many cases the maximum permissible axle-load has already been attained and the adhesive weight cannot be increased further by adding to coupled axles in a rigid wheelbase with obvious limitations. In such an instance the Beyer-Garratt system of articulation will give a high tractive effort on a restricted axle-load.

(ii) The use of an articulated locomotive of greater power than that which could be produced in non-articulated form enables train weights to be increased with a consequent increase in the carrying capacity of a railway. There are instances where single lines, operated by Beyer-Garratt locomotives, are carrying traffic which would necessitate doubling the track were rigid-frame engines employed.

(iii) The engine is extremely flexible and stable and its good riding on narrow gauges and over sharp curvatures is of great benefit to the track.

(iv) The arrangement of the boiler clear of the wheels and on a girder frame gives great freedom in design, enabling a large-diameter barrel to be incorporated with a large firebox and grate area. In addition to ensuring free steaming these features enable a low grade of fuel to be burnt, should the need arise.

While many of the components of a Beyer-Garratt are the same as those of a rigid-frame locomotive, obviously the girder frame carrying the boiler, together with the pivots connecting it to each engine unit, and the ball and expansion joints in the steam and exhaust pipes, are peculiar to it. In the designs produced in recent years the pivots are of the patented self-adjusting type, these completely eliminate the need for adjustment in service and do not even require dismantling for inspection when the locomotive is shopped for general repairs.

Additional particulars of some of the other forms of articulated locomotives may be found under the names of the types mentioned (*see* Chapter 7, part V).

Ash ejector. An arrangement to lift ashes from the bottom of the smokebox in order that they may be entrained in the blast and ejected from the smokebox. The device may be simply a perforated pipe into which steam is turned periodically to agitate any accumulation of ashes.

There is another form, which is only operated when the engine is stationary. This ejects ash accumulations from the bottom of the smokebox horizontally to the side of the line.

Ashpan. In addition to its main and obvious function the ashpan is provided to regulate, through damper doors, the amount of air entering the grate. The ashpan is sometimes attached to the foundation ring and sometimes carried by brackets on the frames; the latter method has the advantage of leaving it undisturbed when the boiler is lifted off. The shape of the ashpan varies with the width of the grate and the position of axles, in some instances it is in a number of sections. It may also be arranged in the form of a hopper which greatly facilitates emptying (Fig. 35).

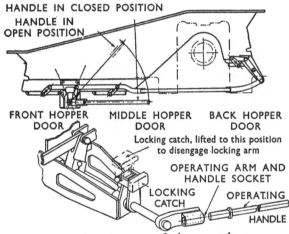

FIG. 35. Arrangement of a hopper ashpan.

Ashpan drencher. A pipe or pipes with nozzles or slots fitted inside the ashpan to enable hot ashes to be drenched when necessary, or frozen ashes thawed.

Atlantic locomotive. The name given to tender locomotives with the 4–4–2 wheel arrangement.

Atomizer (lubricating oil). A fitting situated between the mechanical or hydrostatic lubricator and the steam chests and as close to the latter as practical, the function of which is to mix the oil with steam, the resulting atomization reducing the carbonization which would otherwise occur. In some cases the steam supply to the atomizers is controlled by the cylinder drain-cock operating gear (*see* Anti-carbonizer).

Atomizer (fuel oil). A fitting in which steam and fuel oil are brought together to atomize the oil prior to combustion.

Automatic train control (A.T.C.) (*see* Chapter 7, Part III, Signalling). The term applied to various systems which provide an audible or visual indication in the cab of the position of signals. In some cases, as on those sections of British Railways where it is applied, the information provided in the cab is limited to repeating the position of the distant signal, whereas in other instances the position of all signals is indicated in the cab and the condition of the section ahead is continuously indicated; this latter system is known as Continuous Train Control (C.T.C.).

The British systems are divisible into two types:

(i) Those worked by a fixed ramp, 200 yards or so in advance of the distant signal, making contact with a shoe operating a switch on the locomotive.

(ii) Those worked by magnetic induction working a receiver on the engine.

The Great Western system falls into the first category; when the distant signal is in the caution position the ramp, situated in the four-foot way, is dead. The effect of passing over it is to lift the locomotive shoe, sound a siren in the cab and, after a short interval of time, apply the brake unless action is taken to cancel the indication by operating the handle provided. When the distant shows clear the ramp is electrified and, while passing over it raises the shoe as before, the current collected from the ramp rings a bell in the cab, in lieu of the siren, and no brake application is made (Fig. 36).

The magnetic induction system was first tried in Britain on the London, Tilbury and Southend system. In common with the Great Western system the fixed apparatus is located in the four-foot way at distant signals and gives audible warning by a horn followed by a brake application when a caution indication is concerned but additionally there is a visual indication in the form of an indicator showing black or yellow. Should the distant signal be at clear an electro-magnet is energized which cancels out the effect of the permanent magnet; as a result the audible warning is of short duration and no brake application follows.

The foregoing brief description of the L.T. & S. system is also applicable to the system selected by B.R. with the exception that in the B.R. system a clear indication is audibly indicated by a bell (Figs. 37, 38).

Axle. The shaft, running in bearings termed axle-boxes, upon which the wheels are mounted by hydraulic pressure. Axles may be solid or hollow, the latter only being used for the larger diameter driving and coupled axles, and they may be straight or cranked. In both cases the steel used is acid open hearth or electric process, the British standard specification calling for an ultimate tensile stress of 28–33 (long) tons p.s.i. in the case of solid forged crank axles, and 35–40 tons p.s.i. in the case of straight axles. Coupled and driving wheels are usually keyed on their axles in addition to the press-fitting.

Axle-box (Fig. 39). The bearing interposed between the axle and the locomotive, tender, bogie, or truck frame. It transmits the weight of the engine or tender from the spring to the axle and, in the case of driving or coupled boxes also resists forces arising from piston thrust. The side thrusts resulting from flange contact with the rail head are borne by the outside face of the box in contact with the wheel hub; collars on axles have not been perpetuated due to their causing boxes to run hot. It is necessary to provide for vertical movement of the box in the horns of the frame and it is now common practice to line the faces of the axle-boxes in contact with

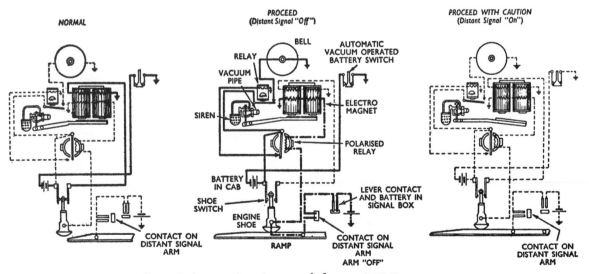

FIG. 36. Automatic train control, former G.W.R. system.

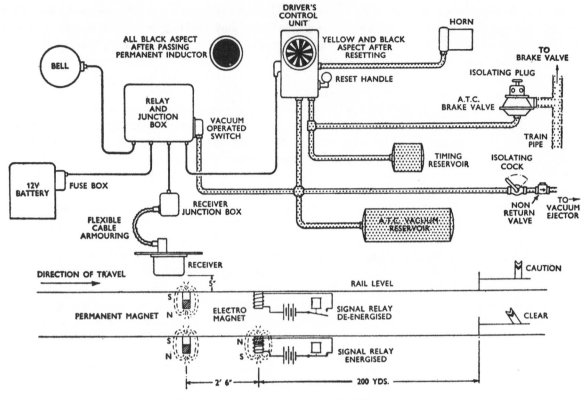

FIG. 37. Automatic train control, B.R. system.

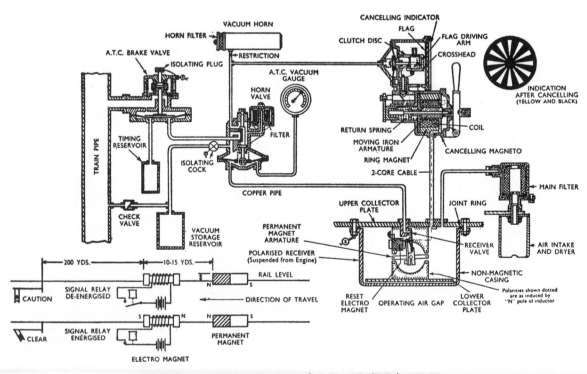

FIG. 38. Automatic train control, L.T. & S. (E.R.) system.

the horns with manganese steel to reduce the wear which otherwise occurs at this point (*see* Hornblock).

Axle-boxes may be fitted either inside or outside the wheels, the latter position has the advantage of increasing

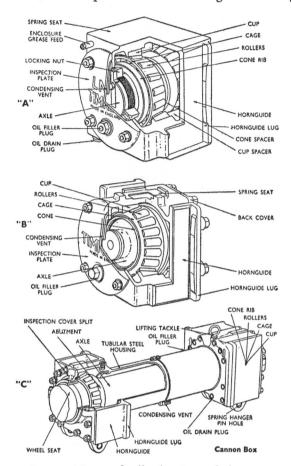

FIG. 39. Types of roller-bearing axle-boxes.

    A. Trailing truck or tender axle.
    B. Tender axle.
    C. Cast-steel box for inside journals.

the width of the spring-base but it is not now used for driving and coupled axles except in the case of some narrow-gauge locomotives. Axle-boxes may be of solid bronze or gunmetal but the general practice is to use a cast steel body with a pressed-in brass.

Ball- and roller-bearing axle-boxes have had a considerable application in recent years and in some instances have run over 2,000,000 miles. They have various advantages, among which are lower starting resistance, practical immunity from running hot, and ability to exclude dirt and retain lubricant. In some instances the boxes are individual ones, whereas in other cases the two are joined by a split tubular body with which they are integral. This split "cannon" axle-box, as it is termed,

keeps the bearings on an axle in perfect alignment and requires only one horn flange on each box.

Axle-boxes, of either plain or roller type, are normally split horizontally, the keep filling in the portion below the axle; there are cases where boxes of both types have been split vertically.

Baffle plate. This plate, also termed a deflector plate, is provided in the upper part of the firehole, its function being to deflect air down on to the fire when the door is open (*see also* Arch, Brick). Being subjected to much heat the plate requires frequent renewal.

The term is also used to refer to the diaphragm plate in the smokebox.

Balancing, *see* Masses.

Bar frames, *see* Frames.

Bed, cast-steel, *see* Frames.

Beyer-Garratt, *see* Articulated Locomotives.

Bissell truck. While this term is usually applied to two-wheeled trucks, it can relate to a four-wheeled truck the pivot of which is located outside the two-axle wheelbase (Plate 63A, page 263).

Blast nozzle. The orifice on the end of the blast pipe through which exhaust steam is discharged prior to entering the chimney. Its function is to impart velocity to the steam in order that the resulting jet may create a partial vacuum in the smokebox and entrain the gases from the tubes. Its correct siting and proportioning are matters of great importance for upon them the free steaming of the boiler will largely depend. The blast nozzle is the one feature which, in a coal-burning boiler, co-relates the boiler output with the demand, as the greater the weight of steam passed through the nozzle the more intense the blast and the consequent draught through the firebed and tubes. Nozzle size is determined by the many factors influencing the draughting of the boiler. Many forms of nozzle have been tried, among those used being ones which, viewed in plan, were oval, cruciform, star-shaped, annular, etc. The Giesl, Kylchap, and Kylala, are examples of multiple arrangements. The jumper blast nozzle, a form of variable nozzle, in common with other variable nozzles has not been perpetuated owing to the difficulty of maintaining such equipment in working order (*see* Part II, Blast Pipes, *and* Chapter 6, Draughting).

Blast pipe. The pipe or pipes connecting the exhaust pipes with the blast nozzle or nozzles attached to its upper end.

Blower. A jet or jets into which steam can be turned to induce draught up the chimney. There are two basic forms, the jet type consisting of one simple jet, rarely used, and the ring type having a number of holes and vertical jets. While ring blowers are usually mounted on the top of the blast nozzle they may also be housed in the petticoat pipe or formed integral with the chimney. If a variable blast nozzle is used the blower is generally housed in the chimney.

Blower, soot. Soot-blowers, also termed sand guns, are provided on the firebox door plate, or side plates if thermic syphons are fitted in the firebox. Their purpose is to clear deposits off the tubeplate and the tube walls and to this end they are used periodically as circumstances necessitate.

Blow-down valves (Fig. 40). These valves, also known as blow-off valves, take two forms, continuous and for washing out. The former, mounted on the back plate, functions continuously while the regulator is open or while the injectors are in operation. The object is to minimize priming by reducing the concentration of soluble salts. The latter is used for blowing down when washing out and accordingly is located on one or more of the water legs.

Bogie. Bogies carry weight while providing a flexible wheelbase. They considerably affect the guiding, and consequently the riding, qualities of a locomotive, both of these characteristics being influenced by the wheelbase. The centre pin is not necessarily on the longitudinal centre of the bogie, but is sometimes set back to increase the leverage effect of the leading axle when curving.

Bogies may be plate or bar framed, or the frame may be a steel casting. The side control springs should preferably be of the through type.

Many bogies have side control obtained by the use of various forms of pendulum links, these have the advantage that the risk of an engine becoming laterally unstable, due to a broken control spring, is obviated.

Boiler (Plates 57A, 57B, page 257). The boiler of a modern locomotive is almost always of the Stephenson form, by which is meant a firebox surrounded with water communicating with a smokebox through tubes and flues. While water-tube boilers have been tried (*see* Part II, page 310 and Chapter 9, Part II, page 464) their use today is confined to a few industrial locomotives.

Boilers may be divided into two types, those with round top fireboxes, and those with Belpaire – or modified Belpaire – boxes. Both these may again be classified as those with narrow or wide fireboxes. The former are fitted between the frames and the latter are carried over the frame, often to the maximum permissible width; it follows that the narrow box is normally deeper than the wide box.

In the case of round-top boxes the staying may be direct or they may be radially-stayed. Girder stays, or roof bars, are no longer used owing to the virtual impossibility of keeping the firebox crown free of scale in bad

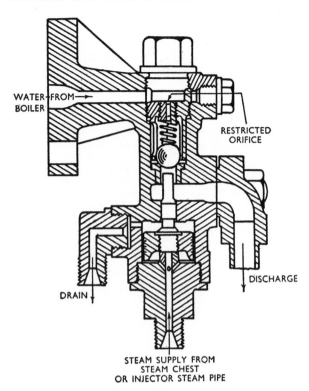

FIG. 40. Continuous blowdown valve.

water areas. Advantages of the round-top boiler are that it is relatively simple to manufacture, and that it is lighter than the Belpaire boiler both in itself and also in the amount of water it contains.

The Belpaire boiler with its flat firebox plates which are normal to each other, is a simple staying proposition and offers increased steam space and water surface, the latter being a factor reducing priming when a boiler is being forced (Plate 59A, page 259).

The width of the firebox, and hence of the grate, has a most important bearing on the boiler output and efficiency. The only way to increase steam output of a boiler is by burning more fuel and this may mean that the amount of coal burnt per square foot of grate area per hour becomes excessive; assuming that the length remains constant, the wider the grate, the greater the amount of fuel that may be fired without the rate becoming excessive – due to the greater area available. Although as much as 200 lb. or more coal may be fired per square foot of grate area per hour, anything over 100 lb. may be regarded as a high rate. There is often an enormous difference between coal fired and coal burnt. There are many factors in addition to the grate which will determine whether or not fuel is burnt efficiently; one of these is the volume of the firebox, which in turn will be influenced by grate area. Whether the boiler will steam freely is dependent upon the correct design and ratios of

Canadian Pacific Railways

PLATE VIII. Canadian Pacific Railways: Montreal–Toronto express in charge of 4–6–4, No. 2856. This fine locomotive is of the "Royal Hudson" Class, so called in 1939 in honour of the visit of H.M. King George VI and Queen Elizabeth to Canada; the Royal Train was hauled by engines of this class over Canadian Pacific metals.

the various components of the boiler. These are usually grouped together as the boiler details, the gas circuit and the draughting arrangements and it is necessary that they should all be in correct balance with each other if steam is to be produced efficiently in the desired quantities.

The boiler details will include the barrel diameter, the tube, flue and superheater element number and diameter, the length between tubeplates, the area of the grate, the heating surfaces, the water surface, and the volume of the steam space. Various ratios are also usually given in the boiler data; they include the total piston swept volume expressed as a percentage of the steam volume, the ratios of firebox volume to grate area, the evaporative heating surface, the superheating surface, etc.

The gas circuit relates to areas through ashpan dampers, air space through grate, free area through tubes and flues etc.

Draughting arrangements cover the proportions of the blast orifice, chimney, petticoat pipe, etc.

Boilers have parallel or taper barrels, the former are easier to manufacture but are heavier, do not give such good visibility and the water level is influenced more by gradients, but they carry more water for a given weight of boiler. The tapered barrel saves weight at the front end where, in any case, evaporation is less. In the case of tapered barrels the taper may be gradual throughout the barrel length or it may be embodied in the course next the firebox, the front portion of the barrel being parallel.

Barrels are invariably of steel; the firebox in British practice is usually constructed of copper but elsewhere steel is extensively employed. While all-welded boilers are seldom met with, welding is widely used on both steel and copper fireboxes. When welding is employed radiographic inspection is regarded as essential but there are varying opinions on stress-relieving.

Boiler pressures, in the Stephenson form of boiler, are not normally carried above 280 p.s.i., and are generally lower.

In many instances rocking grates are fitted. Thermic syphons and arch tubes, while met with in considerable numbers overseas, are seldom used in British practice; a copper box does not lend itself so readily to their use as a steel one.

Flexible stays have been increasingly used in the breaking zones in recent years. Plain pin stays, welded in, are also used; in some instances seal welding is employed in conjunction with screwed stays.

Additional particulars of boiler components will be found under the heading of the component concerned.

Booster. Auxiliary cylinders, driving wheels otherwise used only for carrying purposes, have been applied to many locomotives to boost the tractive effort at starting or when extra power is required. The drive is through gears which can be disengaged when desired. The additional horsepower provided is of the order of 300, the maximum cut-out speed being about 30 m.p.h. While usually applied to the trailing truck of a locomotive, on occasion the device has been fitted to tenders. It had a strictly limited application in Britain but a wide one in America. Boosters went out of favour due to high maintenance costs and relatively high steam consumption, to which the boiler capacity was not always equal (Plate 63B, page 263).

Brakes. Many types of brakes have been employed but today two systems only are in world-wide use on trains, viz., the compressed air, or Westinghouse, and the vacuum. For locomotive purposes steam brakes are frequently incorporated. They have some disadvantages but they are much easier to fit in where space is at a premium, as a considerably smaller cylinder can be used in conjunction with the higher pressure available; further, the use of steam on the engine and tender avoids encroaching on the vacuum employed on the train.

Counter-pressure brakes are sometimes used where long descents are encountered. They have the advantage that apart from saving tyre and block wear their use avoids the heating-up of tyres with consequent risk of their becoming loose on the wheel centre (see Counter-pressure, Steam, Vacuum, and Westinghouse brakes).

Bridle rod, see Reach Rod.

British thermal unit. Commonly abbreviated to B.T.U. (or B.Th.U.), this is a quantitive measurement of heat and is the amount of heat required to raise one pound of water from 62° F. to 63° F.

The calorific value of coal varies considerably; whereas that of a high quality locomotive coal will be of between 13,000 and 15,000 B.T.U. per lb., that of sub-bituminous coal will be about 11,000 B.T.U. The calorific value will be higher when the coal is dry than when in the "as received" condition. As an example a coal with an as received calorific value of 13,970 B.T.U., may have a value of 14,220 B.T.U. when dry.

The calorific value of fuel oils also varies; a good oil may have a value of over 19,000 B.T.U. per lb.

The average value of wood fuel under ordinary conditions is 5,500 B.T.U. per lb. (see Chapter 6, Part I, Coals).

Buffing gear. In addition to the buffing gear provided at the ends of a locomotive and tender, intermediate buffing gear is fitted between the engine and tender. This gear takes various forms, sometimes consisting of buffers on either side of the drawbar, and sometimes centrally located rubbing pads on the engine and tender. The

Goodall articulated drawgear consists of a drawbar having cup-shaped members at each end impinging on hemispherical blocks bolted to engine and tender respectively. This form of drawgear is able to transmit buffing forces in addition to traction forces.

Bull-ring. To reduce the wear which would result if a cast steel or forged steel piston came in contact with a cylinder wall, a bull-ring of cast iron or brass is riveted to the body of the piston head, or alternatively cast on.

Bunkers. Coal bunkers are usually constructed from steel plate, but when weight has been at a premium, aluminium has been used. Construction may be with rivets and angle, or by welding. They are preferably designed to be self-trimming to avoid the formation of dead space with the result that coal becomes stale.

By-pass valve. The function of a by-pass valve is to destroy the vacuum which can otherwise result when a locomotive with piston valves or slide valves above the cylinders is drifting with the regulator closed or almost closed. This vacuum has the most harmful effect of drawing smokebox char and gases down the blast pipe into the cylinders. By-pass arrangements may consist of a valve at each end of the cylinder which places the live and exhaust steam spaces on either side of the piston-valve head into communication, when pressure in the steam pipe falls; or it may be a single valve, placed midway on a passage connecting the cylinder ends, which when steam pressure is absent, falls from its seat and establishes connection between cylinder ends.

Cartazzi axle-box. A form of axle-box, mostly used for trailing trucks, in which the horn faces and the mating faces of the horn-blocks when viewed in plan may be plain flat ones, or arcs of curves having their centre on the longitudinal centre-line of the locomotive. Centring is controlled by inclined slippers between the top of the box and the underside of the spring, the weight acting on the inclined plane providing the centring force.

Chimney bell. The chimney bell is a downward extension of the chimney into the smokebox; its function is to equalize the draught over the tubes, as if it were absent there would be a tendency for more gases to pass through the upper tubes than the lower ones. The design of the bell is generally bound up with the proportions of the draught apparatus. An alternative to a chimney bell is a petticoat pipe, in which the downward extension is discontinuous.

Chimney dampers. Dampers on the chimney prevent cold air circulating through the tubes when the engine is cooling down. The fitting is of particular value on engines burning liquid fuel. Generally speaking, this equipment is more particularly found on European locomotives.

Circulator tube. A steel tube, in the form of an inverted T, fitted transversely across the firebox. The cross-piece is welded to the water legs and the top is welded to the crown. The object is to improve circulation and the tubes serve the added purpose of partly carrying the brick arch, and protecting the crown sheet in the event of low water. It is customary to fit washout plugs in the outer wrapper plate opposite the circulator tube ends.

Clearance, exhaust. Exhaust clearance is provided on the valves of some locomotives as it is claimed to give a freer exhaust than would result from line and line. The amount of clearance given is of the order of $\frac{1}{16}$ in. on an engine working at low speed and may be as much as $\frac{3}{16}$ in. on an express locomotive in British practice. In American practice exhaust clearance was more extensively employed and $\frac{1}{4}$ in. was a common figure on piston valve engines, an amount similar to that allowed on British compounds. Latest British practice provides no exhaust clearance.

Collector seat. A collector seat, also called mud drum, when provided, is located at the bottom of the boiler barrel. The purpose is to collect scale and sludge and allow for its easy removal.

Combining or combination lever. The member in Walschaerts valve gear, the function of which is to impart lap and lead to the valve spindle. While usually disposed vertically it can be arranged horizontally, as in the Stephenson-Molyneux system of Walschaerts gear.

Combustion chamber. An extension of the firebox into the boiler barrel, almost always now used on large boilers. It increases the firebox volume and heating surface and reduces the tube length, which is often desirable.

Compensated springing. Also known as equalizing; compensation allows the effect of track irregularity to be spread by means of levers over a group of wheels or even over all the wheels on one side of an engine. In Britain it is used more particularly for simple equalizing on bogies and is not met with on coupled and trailing truck wheels to the extent that it is encountered elsewhere.

Compounding. There have been many systems of compounding used throughout the world but today it is extinct as far as new construction is concerned. While magnificent work was performed by many compounds, notably those in France, and by the Mallet and Smith-Johnson-Deeley engines, latterly the streamlining of ports and passages, and alteration of valve events, had so improved the simple engine with long-lap valves that the additional expense of constructing and maintaining compounds was not generally considered to be worth while. There was the added disadvantage that in the large modern locomotive the piston associated with the low-pressure cylinder – sometimes as much as forty-eight inches in diameter – became exceedingly heavy.

Compression. The pressure in the cylinder resulting from the closing of the valve before the piston has reached the end of its stroke. Apart from the thermal advantages of this arrangement it provides a cushioning effect for the reciprocating parts (Fig. 62, page 286).

Conjugate valve gear. A valve gear in which the motion for a third cylinder, or third and fourth cylinders, is derived through a series of levers from the two sets of gear operating the remaining valves. The most widely applied conjugate gear is that of Sir Nigel Gresley but this was preceded by that evolved by H. Holcroft, and earlier by the Prussian State Railway (*see* Valve Gears).

Connecting rod. The rod transmitting thrust from the cross-head to the crankpin. These rods are sometimes of special steels and may be fitted with roller bearings at one or both ends or plain or floating bearings. It is interesting to note that the greatest forces to which the rod is subjected are those arising from inertia (Plate 62B, page 262).

Counter-pressure brake. Where inclines are severe and lengthy, counter-pressure braking has been widely used to relieve brake blocks and tyres from the wear and tear to which they are otherwise subjected.

In operation the engine is brought into mid-gear and later into reverse, the cylinders then acting as compressors and therefore absorbing power. When air is used this has the disadvantage of generating heat on compression. This form of counter-pressure braking is known as the Riggenbach type and may still be found on Continental locomotives, including small contractors' engines.

The Le Châtelier system, at one time the most extensively employed counter-pressure brake, had provision for admitting hot water from the boiler into the cylinders. When the water was subjected to pressure it became vaporized.

The braking effect was determined in both Riggenbach and Le Châtelier systems by the position of the reverse. In the case of the Riggenbach system it was customary to discharge the air to atmosphere through a silencer. Both systems were open to mishandling and as a partial protection against this, ample pressure-relief valves were fitted on the cylinders (*see* Chapter 9, Part V, page 475).

Couplers. Automatic couplers take several forms but all have to comply with the basic requirements of automatically coupling on impact, and the ability to be uncoupled without the need for staff passing between the vehicles. Such couplers also handle buffing loads.

Couplers are represented in a simplified form by the Norwegian hook, where the coupling is effected by vertical movement of the coupler hook. In the knuckle type of coupler, which in various guises is the most extensively used form of coupler, the necessary movement controlling coupling and uncoupling is a horizontal one.

The automatic coupling of vehicles is not limited to drawgear alone, but may in some cases include the connection of brake pipes, electrical connections, etc.

On railways where screw coupling drawgear and buffers are in use on some vehicles, and central couplers on others, it is necessary to provide a coupler which is hinged on the drawbar so that it may be brought into use when required. In such instances it is also necessary for the side buffers to be retractable as all buffing loads are transmitted through the central couplers.

Coupling rods. Also termed side rods, serve to spread the tractive effort over the coupled wheelbase. Roller bearings, plain bearings, or floating bearings are all used. In an engine with inside cylinders only, the crank pins upon which the rods are mounted are usually opposite the corresponding inside cranks but they sometimes occupy the position corresponding to the cranks. This latter position although reducing axle-box wear by lessening the internal reversal of forces, increases the gross weight needed for balancing purposes.

Covers, cylinder. Are almost invariably made of cast iron and are ribbed or dished. In the case of the hind cover, if the slide bars are supported on it, integral brackets are cast-in. As a safeguard against the result of an accumulation of water, or a crosshead gudgeon pin, or big-end bearing failure, it is customary to arrange for the cover or cover studs to break comparatively easily to save more expensive damage resulting. This is effected by weakening the cover by the provision of a groove at the point where the flange runs into the spigot, or alternatively waisting the studs. The hind cover is sometimes cast integral with the cylinder.

Crank axle. While numerous engines still survive which have inside cylinders only, for many years it has been customary to design two-cylinder locomotives with the cylinders outside to obviate the use of a crank axle which, where width between frames is restricted, is always difficult to design with adequate strength. In the case of multi-cylinder locomotives it is, of course, necessary to provide a crank axle for the one or two inside cylinders. While forgings were at one time universally employed, later it became usual, particularly in large engines, to build up the crank axle.

Crank pins. These are pressed into the wheel centres. To retain the coupling or connecting rod on the pin, a split brass is sometimes employed in conjunction with a shoulder integral with the pin; alternatively a washer may be employed which is held on the pin end by a bolt or stud; another form of fixing met with is a collar locked by a pin passed diametrically through the crank pin, the collar being generally screwed. Where return cranks are used they suffice to retain rods on the pins concerned.

Crank spacing. In a two-cylinder locomotive the cranks will be spaced at 90°, the engine being termed right- or left-handed according to which crank is leading the other. In a three-cylinder locomotive the cranks will all be spaced at angles of 120° to each other, subject to slight alteration to allow for differences in cylinder inclination. In a four-cylinder locomotive the cranks may be spaced at 90° to each other or an intermediate spacing may be used. In the former case four power impulses per revolution will result, the outside back and inside front, and the outside front and inside back sequences coinciding. If an intermediate spacing is employed, with settings spaced 90°, 45°, 90° and 135° respectively, eight regularly-spaced impulses will result.

While it was at one time claimed that three- and four-cylinder locomotives, with six or eight impulses, and hence exhausts per revolution, produced a more even draw upon the fire it is now considered that this point has more theoretical than practical significance.

Crosshead. The function of the crosshead is to guide the piston rod, thus relieving it of the vertical forces arising from the angularity of the connecting rod, and to house the gudgeon pin connecting the piston rod to the connecting rod. It may also carry the arm connected to the union link on an engine fitted with certain forms of valve gear, and it is also utilized to provide the drive to vacuum pumps on engines fitted with crosshead-driven pumps.

The crosshead may take many forms, largely depending upon the arrangement of the slide bars. Thus the crosshead may

(i) surround a single bar;

(ii) slide midway between an upper and a lower bar – known as an alligator crosshead;

(iii) slide below two bars – the Laird crosshead;

(iv) be T-shaped, sliding below three bars – one wide bar above the top, and one on each side below the ledges. All the slide bars are above the centre line of the crosshead. This is the Dean type.

(v) have blocks on either side, sliding between two bars on each side.

(vi) be of the multiple ledge form – also a Dean type. The number of faces may be as many as nine.

There are other forms but they are not commonly met.

While the piston rod is normally fitted into the cross-head on a taper and secured by a cotter it may also be fixed by machining a series of annular recesses in the cross-head, which is split on the vertical longitudinal centre, and providing shoulders on the piston rod which mate with these recesses when the two halves of the crosshead are bolted together. In relatively few instances the piston rod is drawn in by a nut on the end of the rod.

The shoes – or slippers – may be white-metal lined cast iron or bronze with babbitt inserts. The gudgeon-pin bush is usually of bronze but roller bearings are sometimes used; the body is a steel casting (see Slide Bars).

Cross-stays, or cross-ties as they are known in America, run transversely from one side of the frame to the other to impart rigidity. While they are usually steel castings they may also be pressings, or built up from angle and plate, or may be fabricated.

Crown bar. A girder running along the crown sheet, and secured thereto by bolts, being slung by stays from the outer wrapper. This method of supporting the firebox crown was discontinued due to difficulties arising from impeded circulation and obstruction to cleaning.

Crown stays. The stays supporting the firebox crown and connecting it with the outer wrapper. The method of fixing is similar to that of other stays (q.v.). Crown stays are also known as roof stays.

Cut-off. The position in the piston stroke at which the admission of steam to the cylinder is cut off by the closing of the valve. The point of cut-off may be of the order of 80 per cent at starting, and in the case of a modern locomotive with poppet valves may be as little as 5 per cent when running at speed (Fig. 62, page 286).

Cylinders. Heat and pressure energy present in steam are converted into mechanical work in the cylinders. The minimum number used other than on unorthodox locomotives is two and the maximum four. Cylinders are cast in close-grained high-grade iron, or in steel; if steel is employed cast iron liners are fitted. Sizes vary considerably and in the larger cylinder sizes securing by turned bolts does not necessarily ensure permanent freedom from movement on the frame. This trouble has led to the employment of shear strips, as on the British Railways locomotive *Duke of Gloucester*, or to the manufacture of the cylinder integral with the frame, as in the cast-steel bed (Plates 60A, 60B and 104A, pages 260 and 380).

Dampers. Dampers are usually fitted on the ashpan or pans to regulate the amount of air drawn through the fire. They are not universally employed and their omission can result in control of the fire being difficult (see Ashpan).

Deflector plate, also known as baffle plate, q.v.

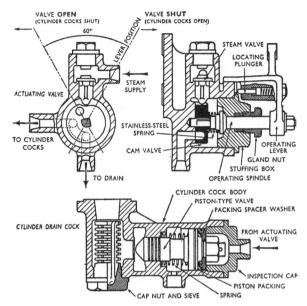

FIG. 41. Steam actuating valve and cylinder drain cock for clearing condensate.

Diagram factor. The ratio of the actual mean effective pressure (q.v.) to the theoretical M.E.P. The factor will vary with piston speed and may lie between about 0·8 and 0·6. With Walschaerts valve gear the indicator diagrams obtained from full fear to 40 per cent cut-off at lower speeds are very nearly perfect. At high speeds and early cut-offs this gear, even with long-lap valves, has deficiencies. These are:

(i) excessive throttling during admission;
(ii) too early a release;
(iii) too high a compression.

As a result of these conditions the diagram factor at 20 per cent cut-off and five revolutions per second may be only 41·1 per cent, and the steam rate 16·60 lb./I.H.P.Hr. A rotary poppet valve gear at 10 per cent cut-off and the same speed gives a diagram factor of 69 per cent and the same horsepower is developed at a steam rate of 15·93 lb./I.H.P.Hr. (*see* Chapter 6, Part I, page 392).

Diaphragm plate, or baffle plate, is the plate in the smokebox which – when fitted – is placed between the tube plate and the chimney to deflect sparks downwards and to equalize draught over the tubes.

Die block. The block in any form of link motion which is mounted in the link and to which the valve spindle is connected through the medium of an extension or intermediate link.

Dome. An extension, usually but not always cylindrical, on top of the boiler shell. Its purpose is to collect dry steam and, when space permits, it usually houses the regulator. A second dome is sometimes provided in connection with top-feed arrangements.

Dragbox. The cross brace at the rear end of the locomotive and the front end of the tender to which the ends of the intermediate drawbar are attached. Formerly cast, it is now customary to fabricate it by welding or build it up from plate and angles.

Drifting valve. A valve provided in the regulator, or elsewhere, to admit a small amount of steam to the steam chest when the engine is drifting. The position the regulator handle must occupy to bring this valve into operation is usually marked on the regulator sector plate or quadrant.

Drop grate or dumping section. A section, or sections, of the grate so hinged that they may be rotated to a near-vertical position. Control is normally by lever from the footplate or worm and wheel from ground level. The equipment is of use for removing large masses of clinker when fire cleaning and is also of value should it be necessary to drop the fire in an emergency.

Drop plug. A form of fusible plug, used in American practice, in which the fusible alloy is replaced by a steel centre held in the body by an annular film of fusible metal. Should this plug drop, the whole of the way through the bronze body is immediately opened to steam.

Dry pipe, *see* Pipe, Main Steam.

Eccentric. This device for translating rotary movement into reciprocating motion is almost always used to impart movement to valve gears deriving their drive inside the frames, unless the drive is derived from the connecting rod, as in Joy's gear, or is a cross-driven one, e.g., Deeley's or Young's. While eccentrics are also to be found, particularly on older Continental locomotives, providing motion for gears placed outside the frames, their use in this position has been largely superseded by return cranks.

Ejector, ash, *see* Ash Ejector.

Ejector, brake. On steam locomotives fitted with, or equipped for working, the vacuum brake, the vacuum is invariably created by an ejector although it may be maintained by an air pump driven from a crosshead. Most ejectors contain large and small cones, the former or large ejector as it is known, is for creating the vacuum initially or for releasing the brake after an application; the small ejector is used continuously, unless a pump is fitted, to maintain vacuum against the small amount of leakage always present (*see* Vacuum Brake).

Element, *see* Superheater.

Equalizer. The lever used for compensated springing, q.v.

Equalized brake rigging. A layout of rigging in which the brake-block pressure per wheel is maintained equal and independent of brake-block or tyre wear.

Events, valve. The points of admission, cut-off,

release, and compression during one complete revolution of the wheels. These events will all vary according to the point of cut-off, always expressed as a percentage of the stroke, and be interdependent, unless the gear is a poppet one. Poppet valve gear provides that the valves always open practically fully, and release occurs late at whatever cut-off is selected.

Expansion line. The line running from the point of cut-off to release on an indicator diagram.

Fairlie. Some Fairlie articulated locomotives had a single swivelling engine unit underneath the boiler, in conjunction with a trailing carrying bogie. Other Fairlies had two power bogies under a double boiler, and carrying wheels were dispensed with. While there was no necessity that it should be so, the wheel arrangement was usually 0–4–0+0–4–0 or 0–6–0+0–6–0.

The design gave latitude in the design of the firebox. At the end of last century two separate boilers were fitted, which considerably simplified the arrangement of piping, etc.

In the modified Fairlie, which has a single boiler, the coal and water supplies are transferred from the power units on to the superstructure, on which the boiler is mounted. This arrangement increases the overhang on curves and also has the disadvantage, compared with a Beyer-Garratt locomotive, that the considerable weight of the coal and water supplies has to be borne by the pivots.

Feed, boiler. Boiler feed is most usually by means of injectors but pumps are also extensively used, especially when feed-water heaters are employed. Delivery in modern practice is usually through clacks – check-valves – placed on the top of the barrel, sometimes on to trays housed in the steam space to collect impurities. Of recent years much attention has been given to the treatment of feed water which treatment may be effected in a lineside plant, or on the locomotive. Increasing numbers of locomotives are being fitted with the T.I.A. system of feed water treatment (*see* Armand Water Treatment).

Attention paid to the correct treatment of boiler feed water has considerably increased engine availability in addition to reducing the cost of boiler maintenance.

Feed water heaters have been applied to a small extent in Britain and to a large extent elsewhere. In areas of bad (hard) water, serious trouble may be experienced with feed heaters which become scaled up and it may even be necessary for them to be cleaned out on every day the engine is in steam.

Where feed is by injector the water delivered is always warm or hot.

Ferrules. To protect ends of boiler tubes ferrules are sometimes driven into the tube end and they may be beaded over and seal welded to the tube plate. Copper ferrules may also be fitted over the end of the tube, where it passes through the tube plate, to assist in obtaining a water-tight joint. This practice, however, is not universally accepted as it results in electrolytic action.

Firebox (Plates 58A, 58B, page 258). The firebox of the Stephenson boiler remains surrounded by water on the top and all sides, despite the many efforts that have been made to dispense with the water-legs. Where British practice is followed the inner box is nearly always copper but elsewhere steel is usually utilized. The joints between the plates are frequently welded and, in the case of steel boxes, welding may be used for the stays which may otherwise be screwed into the plates. Flexible stays are frequently employed in the breaking zones. The Belpaire box, or modified forms of it, is extensively used, due largely to the simplicity of the roof and cross-staying which is direct; another advantage is the increased surface are on the water line. In the case of large round-topped boxes it is customary to provide welded seatings for the cross stays.

Fireboxes intended for oil-burning have a flash wall, constructed of firebrick at the end opposite the burner and it is usual to line the lower part of the walls and cover the bottom with firebricks to protect the firebox and ashpan plates.

On large modern locomotives it is general practice to extend the firebox at its front into a combustion chamber. In addition to the brick arch, which is almost universally used with coal fuel, the firebox may incorporate arch tubes, circular tubes, or thermic syphons.

In the case of small locomotives, more especially those for industrial use, circular fireboxes have been employed; this permits of extremely easy renewal, but it is quite unsuitable where the steaming rate is high.

Where water-tube boilers have been used the firebox walls have consisted of banks of vertical tubes.

The width of fireboxes is governed by the class of fuel to be burnt in addition to the size of the boiler.

Firehole. In addition to its obvious function the firehole is, in many boilers, the only means of admitting secondary air. The firehole may be formed:

(i) by interposing a ring between the holes in the inner and outer firebox plates.

(ii) the plates may be flanged to butt each other and be welded.

(iii) the plates may lap each other and be riveted.

(iv) the plates may be dished and riveted.

A protection plate is usually fitted in the lower half of the firehole to protect it against the abrasion of the shovel and fire-irons, and a deflector plate is normally fitted in the top half of coal-burning locomotives. In oil-burning locomotives the firehole may be semi-permanently closed and an inspection window provided in the cover. On locomotives with large fireboxes and wide grates two, and even three fireholes may be provided.

Firebox ring. The term used in American practice to describe the foundation ring (q.v.).

Firedoor. Firedoors may be hinged vertically in one piece (or in two or more sections) or horizontally. They may otherwise be divided into two parts which slide apart. The "Ajax" firedoor is operated by steam, or compressed air; the two half-doors are pivoted at the top and are connected by a toothed quadrant which keeps the doors in phase as power is applied to one half-door only. Control is by a pedal, rendering operation of the doors between shovelfuls a simple matter. Other power-operated doors in use are the Franklin, which has butterfly doors arranged to open horizontally, and the Shoemaker radial firedoor in which the joint between the two doors is also arranged horizontally. Almost all doors have some provision for the entry of secondary air through holes or shutters provided in the back of the door. (Plate 64A page 264).

Flange lubricators. To cut down the heavy wear which results from flange friction it is sometimes the practice, where curvature is severe, to provide for a film of lubricant to be introduced between the flange and the side of the rail-head. This may be effected by lubricators installed on the track, or fixed on the locomotive. In the past, water has been used for flange lubrication, but today, oil or grease are customarily employed. In some applications lubrication is continuous when the locomotive is in motion, whereas, in others it is pendulum controlled and only comes into action when the engine is nosing or negotiating a curve.

It is most usual to feed the lubricant on to the flanges of coupled wheels but there are exceptions to this practice. Flange lubricators have been fitted on the trailing trucks of tenders, in which position they also lubricate flanges of vehicles in the train.

The saving which flange lubrication makes in tyre life can be considerable and an increase of up to 300 per cent in mileage between tyre turning is claimed for one form of flange lubricator.

Flues. While at one time it was customary to refer to all firetubes as tubes or flues it is usual, since the advent of superheating, to use the term flues specifically for the large-diameter tubes which house the superheater elements.

Foaming. The formation of foam on top of the boiler water is due to grease or dirt on the surface which prevents the bubbles of steam from bursting. It is not to be confused with priming (q.v.), although the two are fre-quently regarded as one and the same trouble. In the event of a boiler foaming it should be remembered that the water level as shown in the gauges will be a false one, the actual level being lower than that recorded.

Foundation ring. The foundation ring, also known as a mud ring, unites the lower edges of the inner and outer firebox sheets. Forgings were at one time used; later, steel castings found favour, while in recent years the ring has been formed from a U-section pressing welded to the plates.

Frames. Frames take three basic forms, plate, bar, and bed. The plate frame (Plates 59B and 60A, pages 259, and 260) is more particularly suited to good tracks and is associated with British and Continental practice, although for many years some Continental locomotives have had bar frames. Plates are of the order of one to two inches in thickness, bar frames are usually from four inches to seven inches thick. Bar frames may be flame cut or machined from rolled-steel slabs or they may be steel castings.

Cast steel beds may be single castings consisting of the complete frame, all necessary stretchers, cylinders, smokebox saddle, and even air reservoirs, etc. The cast steel bed became necessary owing to inability to keep large cylinders tight on the frame. The production of cast-steel beds is a limited one but locomotives for export embodying these have been constructed in Britain employing beds imported from America (Plate 60B, page 260).

One advantage of the bar frame over the plate frame is that the centre line of the frame coincides with the centre line of the axle-box. This, however, can be arranged on plate frames as was originally done on the Bulleid "Pacifics" on the Southern Railway and later on certain of the standard locomotives produced by British Railways. The location of the springs above the axle-boxes of bar-framed locomotives is also a simple matter and the necessary brackets and hangers are arranged symmetrically about the frame. The considerable width of the frame enables the hornstays to be adequately proportioned.

A disadvantage of the bar frame is the reduced width between frames; this is particularly apparent when a between-frames firebox has to be fitted, but is overcome by the provision of a trailing truck. The bar frame is normally heavier than its plate counterpart.

Fusible plugs. Also known as lead plugs, give warning of dangerously low water when the lead alloy core melts and allows steam to escape into the firebox. In order that they may remain reliable they must be replaced frequently and should be examined at washouts. A dropped plug is normally accompanied by a buckled crown sheet.

Gibson ring. This form of tyre fastening consists of the provision on the outer edge of the tyre of an integral lip and on the inner edge a groove which houses a specially formed ring. The ring is placed in the groove and protrudes above it locating the tyre laterally relative to the wheel centre. The ring is secured in position by the lip of the tyre being closed down upon it.

Golwé locomotive. An articulated locomotive having two motor bogies, connected by a girder frame carrying the boiler, cab, and the water tank and fuel. No fuel or water are carried at the front end, all supplies being behind the cab. In order to reduce overall length the pivots are located at the outer end of each unit.

In this locomotive the firebox design is comparatively unrestricted but the boiler layout is hampered by the presence of the front engine unit below the barrel. While the wheel arrangement of both units is sometimes the same, e.g., 2–6–0+0–6–2, on other occasions it differs, e.g., 2–6–0+0–6–0.

Goodfellow tips. V-shaped projections fitted in the blast orifice to sharpen the blast. This enables the diameter of the blast nozzle to be increased, while still retaining the same pull on the fire, combined with a reduction in back pressure.

Grate. The grate may be horizontal, inclined, or partly horizontal and partly inclined. Today it is usual to use inclined grates as, apart from the necessity to adopt this form to clear a coupled axle, it assists the maintenance of an even depth of fire. The grate may take many forms but construction is always of cast iron. Bars are normally employed which may be parallel when viewed in plan, or made up of a series of small castings mounted on carrier bars. Their axes may be parallel to the longitudinal centre line of the locomotive, as in the Waugh grate, or at right angles to it as in the Hulson grate. Today, rocking grates are almost universally used on modern coal-burning locomotives. The rocking may be done manually or by steam cylinders.

Round-hole or slotted-table grates are not now used to the extent that they were, having been largely replaced by firebars.

Grease separator. A grease separator is fitted in the pipe conveying exhaust steam to an exhaust injector to prevent the carriage of cylinder lubricant into the boiler. It is designed to impart a rotary motion to steam entering the separator, any particles of oil or water being separated from the steam by centrifugal action and drained through an automatic drip valve.

Hornblock. Almost always a single steel casting bolted or riveted to the frame, in which the axle-box is free to rise and fall. Wedge adjustment is sometimes provided, provision occasionally being made for this adjust-ment to be automatically effected. Common modern practice is to fit manganese steel liners which cuts down the maintenance required; bronze liners are also used. Cross stays are sometimes provided from the bottom of one horn to that on the opposite side, to provide additional resistance to side thrusts. Single castings have been employed combining the hornblocks on opposite sides united as by a brace; such practice has been confined mainly to narrow-gauge locomotives (see Axle-box).

Hornblocks may also be fabricated from plate and located with their centre line on that of the frame. This practice, which is used both in the Bulleid 4–6–2 locomotives of the former Southern Railway and on the standard 4–6–2 locomotives of British Railways, relieves the frames of the eccentric loading which occurs when the hornguides are offset. Hornblocks are required only on plate frames.

Hornguides. Hornguides are, in effect, hornblocks minus the top portion. They are widely used and may be fitted with manganese liners in the same way as hornblocks. They lack the reinforcement which is imparted to the frame by the top of the hornblock and for this reason are not regarded favourably by some designers. There are instances on trailing coupled axles of engines having between-frames fireboxes where hornguides are necessarily used instead of hornblocks.

Horsepower, see Pressure, Mean Effective; and Thermal Efficiency.

Injectors (Figs. 42–48). Injectors, in British practice, are almost universally used for boiler feeding. Elsewhere they are widely, but not so extensively employed. They take two forms, those operated by live steam and those functioning on exhaust steam; the two types are basically similar but due to the much lower pressure in the exhaust injector the cones are necessarily of different design. It was at one time customary to arrange injectors vertically on the boiler backhead, when they were combined with the check valve. Later they were arranged horizontally below the footplate. The vertically disposed injector, also below footplate level, which was previously used in America, has now been adopted in Britain as a result of war-time experience with imported locomotives. It is now usual to arrange both injectors on the fireman's side.

Due to the combination of hot feed water and steam at boiler pressure, the delivery – always hot in the case of any injector – is, in the case of a hot water injector, above boiling point at atmospheric pressure. For this reason the overflow passage on hot-water injectors must be kept closed.

The sizes of injectors are quoted in the size of the

Beyer, Peacock and Co. Ltd

PLATE 57A. Boiler for Beyer-Garratt locomotive – firebox end, showing firehole, boiler fittings and flexible stay heads in the braking zones.

Beyer, Peacock and Co. Ltd

PLATE 57B. Boiler for Beyer-Garratt locomotive – smokebox end, showing superheater header and elements, front tubeplate and small firetubes.

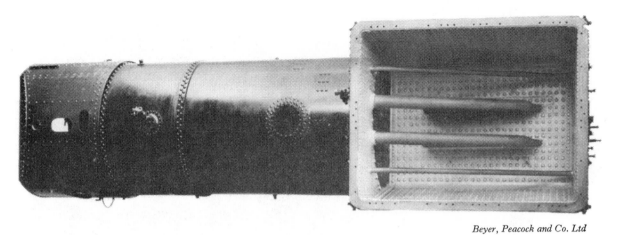

Beyer, Peacock and Co. Ltd

PLATE 58A. Firebox viewed from underside, showing thermic syphons and arch tubes.

Alco Products, Inc.

PLATE 58B. Firebox showing security circulators.

P. Ransome-Wallis

PLATE 59A. Belpaire boiler, showing staying between inner
and outer wrapper plates, and back tubeplate.

British Railways

PLATE 59B. Plate frames and buffer beam with smokebox saddle. Shows levers for Gresley-Walschaerts conjugated valve gear.

British Railways

PLATE 60A. Plate frames ready for lining up cylinders.

General Steel Castings Corporation

PLATE 60B. Cast steel engine beds for Mallet locomotive.

Beyer, Peacock and Co. Ltd

PLATE 61A. Walschaerts valve gear for outside cylinders.

P. Ransome-Wallis

PLATE 61B. A modern application of Stephenson link-motion for outside cylinders: L.M.S. class 5.

P. Ransome-Wallis

PLATE 62A. Reidinger rotary cam poppet valve gear applied to British Railways 2–6–0 locomotive.

Delaware and Hudson Railroad

PLATE 62B. Roller-bearing big-end applied to an American locomotive.

Beyer, Peacock and Co. Ltd

PLATE 63A. Two-wheel trailing truck for metre gauge locomotive. The wheel centres are wider than normal and protrude outside the tyres in order that conversion to 3 ft 6 in. gauge may be made at a future date. Timken roller bearing axle-boxes are fitted.

Lima-Hamilton Corporation

PLATE 63B. Trailing four-wheel truck with 2-cylinder booster engine.

PLATE 64A. British Railways. Footplate of "Coronation" Class 4-cylinder 4–6–2 locomotive.

British Railways

PLATE 64B. Cab view of South African Railways Class GMAM, Beyer-Garratt locomotive, showing engine and conveyor for mechanical stoker.

Beyer, Peacock and Co. Ltd

delivery cone measured in millimetres, those in common use vary from 5 to 13.

Inside admission. A piston-valve is said to have inside admission when live steam is admitted to the steamchest between the piston-valve heads. The advantage of this arrangement is that the valve spindle is submitted only to

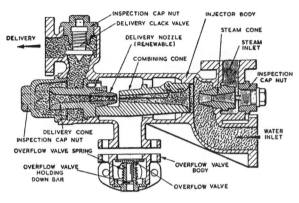

FIG. 42. Injector having combining cone with hinged flap.

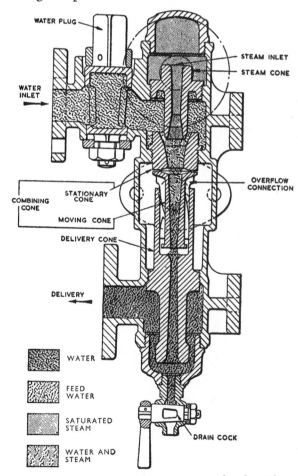

FIG. 43. Injector combining cone, fitted with moveable portion.

the low pressure of the exhaust steam where it passes through the steamchest cover (*see* Outside Admission).

Lap, exhaust. When the valve is centrally situated, relative to the ports, if the exhaust edges of the valve overlap the edges of the exhaust ports, the amount of overlap is termed the exhaust lap. The lap is "negative" if both ports are open slightly. Should neither positive nor negative lap be provided the exhaust is termed "line and line".

Lap, steam. Lap is essential to expansive working and is obtained by increasing the length of the valve face in order that the admission edges of the ports may be overlapped at each end when the valve is in mid-position.

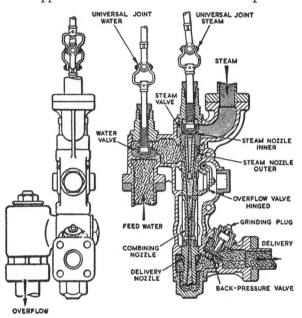

FIG. 44. Live steam injector Monitor type.

It is customary to provide about one inch lap on an engine with slide valves, whereas one with piston valves would have about 50 per cent more lap. The provision of long laps obviously necessitates using a valve gear giving long travel.

Lead. The amount which the valve is open to steam when the piston is at the end of its stroke. It enables the cylinder to take steam more readily than would be the case if the valve did not begin to open until after the stroke commenced, and it has the added advantage that it provides a cushioning effect to the piston, piston rod, crosshead, and to a lesser extent the connecting rod, at the end of the stroke. Within limits, benefit would be derived from admission occurring earlier at high speeds; this effect is to some extent obtained with Stephenson's valve gear, where lead increases as the cut-off becomes earlier when the eccentric rods are open. If the rods are crossed the lead decreases as the gear is notched-up. The

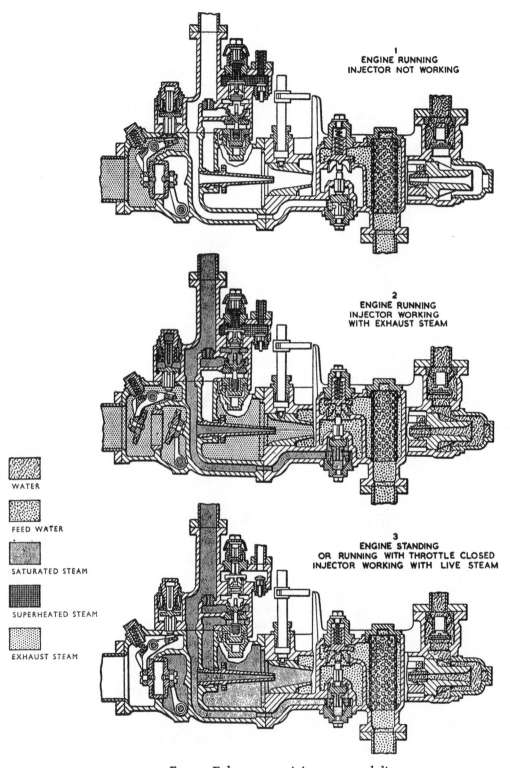

FIG. 45. Exhaust steam injector control diagrams.

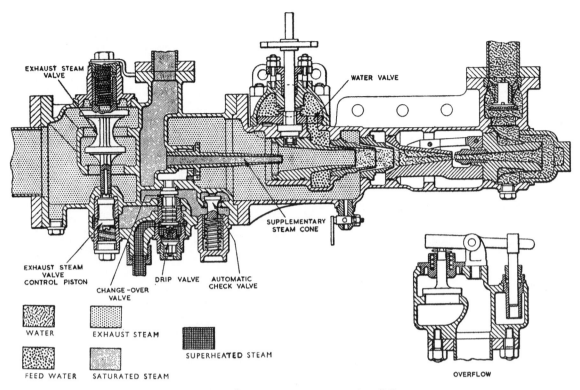

FIG. 46. Exhaust steam injector, Class "J".

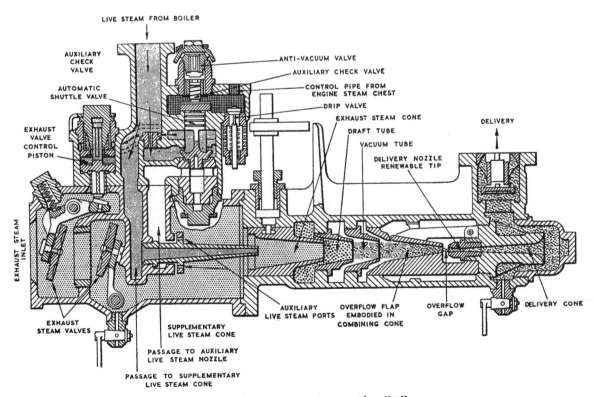

FIG. 47. Exhaust steam injector, Class "H".

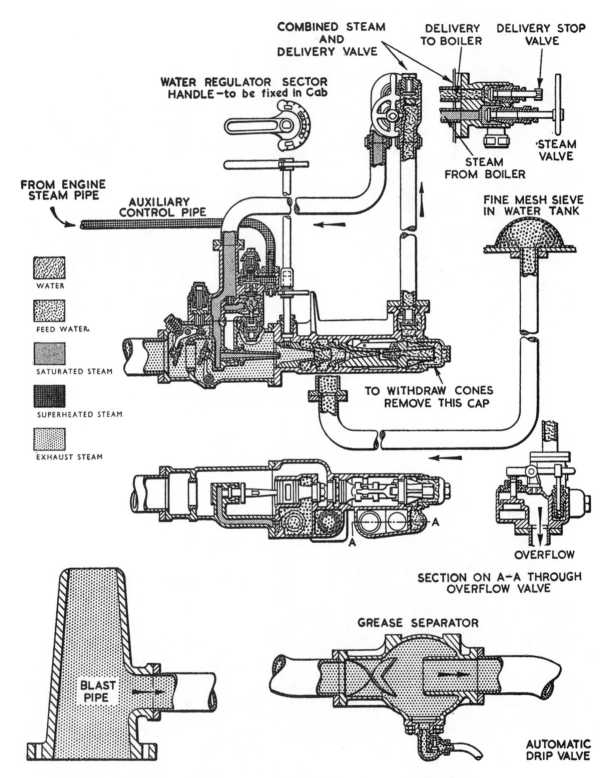

FIG. 48. Exhaust injector arrangement and control, Class "H".

amount of lead now provided is of the order of $\frac{1}{16}$ in., increasing to $\frac{1}{4}$ in.

Lens joint. A form of joint, sometimes used for steam pipes, consisting of a ring with spherical faces introduced between seatings having ground mating faces held together with flanges. The arrangement is one capable of accommodating a certain amount of mal-alignment.

Liners for cylinders and valves are frequently used in cast-iron cylinders, and always employed in cast-steel and fabricated cylinders. In the case of piston-valve liners, the ports are provided in them and the liner is machined outside to tolerances necessary to ensure a press-in fit. Alternatively the liner may be frozen in. It is necessary accurately to locate the liner axially to ensure the position of the ports being correct. To assist in valve setting some form of inspection opening is often provided.

Low-water alarm. These alarms are met with in American practice, examples being the Barco, Ohio and Nathan. The first-mentioned device is actuated by the height of the water over the crown-sheet, the second employs a fusible plug, and the last-named relies upon temperature change. In each instance a whistle is sounded to draw the attention of the engine-men or firelighters to the position.

Lubrication (Figs. 49–52). Basically lubrication may be effected in two ways, by oil or by grease. Various grades of oil are employed and both hard and soft grease are used. Graphite – an excellent lubricant – was at one

period employed for cylinder lubrication but due to the relative difficulty of feeding, it is not now used. Tallow, at one time extensively used, has fallen into disfavour because it becomes decomposed when heated, and the fatty acids so produced cause corrosion.

For the three forms of motion met with on a locomotive, i.e., rotating, rocking, and sliding, oil may be fed in various ways. For items where only intermittent lubrication is required it may be applied by hand when the engine is prepared. Where lubrication must be continuous, it may be fed by worsted pad or syphon, as for example to axle-boxes; restrictor plugs may be used, as on revolving parts, where the plug limits the amount of oil able to pass to the bearing. Cane plugs are sometimes employed; by restricting the amount of air entering the oil chamber they also limit the amount of oil that can flow out. Fountain type lubricators, where the amount of oil fed is also restricted by the amount of air allowed to enter, have a limited application for axle-boxes.

Lubrication to axle-boxes and many motion details, crossheads, link trunnions, etc., is often effected mechanically. Practice in this connection varies considerably and an item such as a die block which on the majority of locomotives is lubricated by the driver with an oilcan, is, in other instances, lubricated mechanically. In the case of piston rods and valve spindles lubrication may be by mops, from oil boxes, or by mechanical lubricator.

Cylinder lubrication on superheated locomotives is effected mechanically, or hydrostatically by sight feed

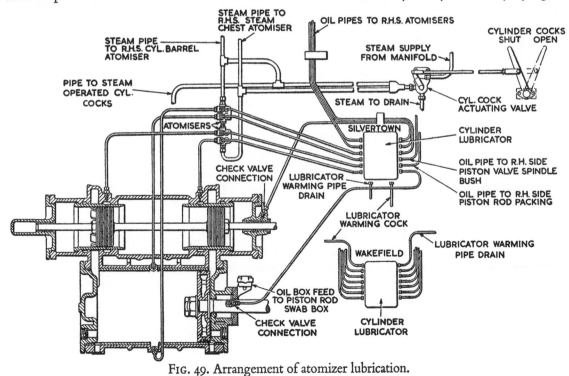

FIG. 49. Arrangement of atomizer lubrication.

lubrication (q.v.); locomotives using saturated steam also use hydrostatic lubricators, where steam condenses and the resultant water displaces an equivalent amount of oil. Well-known examples of mechanical lubricators are

The lubrication of big ends and side-rod bushes may also be effected, as in recently constructed Chinese locomotives, through drilled axles and crankpins, as in automobile crankshafts.

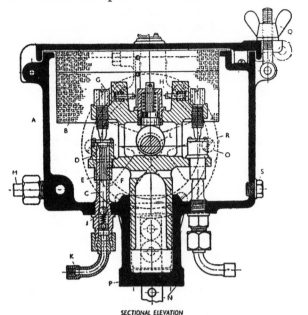

SECTIONAL ELEVATION

A. OIL RESERVOIR
B. WIRE GAUZE STRAINER
C. PUMP BARREL
D. PUMP PLUNGER
E. SLEEVE VALVE
F. OIL PORTS
G. OIL REGULATING PLUG
H. OIL REGULATING LOCKING PEG
J. NON-RETURN VALVE
K. OIL OUTLETS
L. DRIVING ECCENTRIC SHAFT
M. OIL WARMING PIPE
N. DRIVING ARM
O. RATCHET DRIVE AND GEAR CASE
P. FIXING LUGS
Q. FLY BOLT TO SECURE LID
R. FLUSHING WHEEL
S. DRAIN PLUG

NOTE.—No. 7 (a) pattern operates in a similar manner to the above, but the oil outlets K lead out from the sides of the oil reservoir, level with the centre of the driving eccentric shaft L.

FIG. 50. Wakefield mechanical lubricator.

Friedmann, Silvertown, and Wakefield; hydrostatic lubricators often met with are the Detroit and Eureka.

With both mechanical and hydrostatic lubricators it is customary to employ an atomizer.

While in British practice the use of grease lubrication is limited to ball and roller bearings, motion parts, brake and water pick-up gear, reversing screws, etc., in overseas practice it is used most extensively; parts so lubricated including axle-boxes, hub faces of wheels, coupling and connecting rods, gudgeon-pins, etc. Grease is applied by means of a gun capable of exerting, if necessary, pressure of the order of 2 to $2\frac{1}{2}$ tons per sq. in. While grease is applied intermittently in the case of the components mentioned, such items as valve spindle crossheads and hornblocks may be arranged to receive continuous lubrication from an automatic lubricator.

A lubrication feature which may be encountered in diesel locomotive practice but not in steam locomotives, is mechanical lubrication of the crank pins through the medium of return cranks.

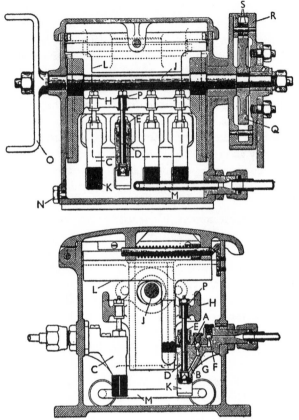

A. CAVITY
B. PASSAGE
C. SUPPLY PUMPS
D. PUMP PLUNGER
E. PACKING
F. BALL VALVE
G. BALL VALVE
H. DRIVING FRAME
J. SHAFT
K. SMALL SIEVE
L. FINE-MESH SIEVE
M. WARMING PIPE
N. DRAIN PLUG
O. DRIVING SHAFT HANDLE
P. THIMBLES
Q. DRIVING WHEEL
R. FIXED WHEEL PLATE
S. PAWL

FIG. 51. Silvertown mechanical lubricator.

The method of lubrication affects the design of bearings.

Main rod. The American term for the connecting rod.

Mallet locomotive. Mallet's name is associated both with compounding and with a form of articulated locomotive. Strictly, only compound engines built on this system of articulation should be referred to as Mallets. Four-cylinder single expansion articulated locomotives have, however, come to be referred to as "Mallet Simples" (*see* Chapter 5, North America – Articulated).

The Mallet articulated locomotive had two motor units, the rear unit being arranged integrally as in the case of a normal rigid-frame locomotive, while the front unit is free to adapt itself to both horizontal and vertical

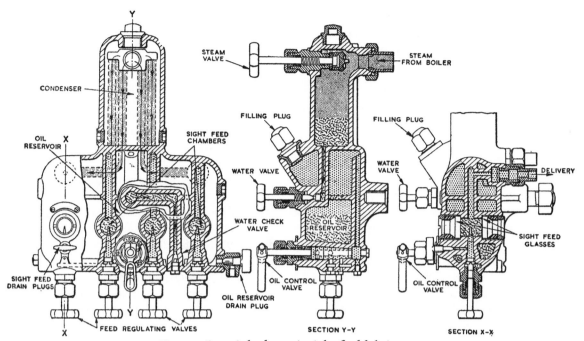

FIG. 52. Detroit hydrostatic sight feed lubricator.

curvature through the medium of a hinged joint between it and the hind unit. The boiler is supported on the front unit by a saddle with provision for translation. Centring devices are usually, but not always, provided.

While compounding has been tried on other forms of articulated locomotives, the Mallet is the only system of articulation where the use of compounding has been extensive.

The use of compounding involves heavy reciprocating parts and the engines in that form are unsuited to high-speed operation. Later, when single-expansion Mallets were built, running was considerably accelerated and there were rare instances where the design permitted speeds of up to 60 m.p.h. When compounding went out of fashion the demand for Mallet locomotives fell off. In compound Mallets, the high-pressure cylinders were located on the fixed unit which meant that flexible steam connections were subjected to low pressure only.

Normally separate tenders are used in conjunction with Mallet locomotives but some have been constructed as tank engines.

Mallet locomotives with tenders differ from most other forms of articulated locomotives in that they are suited for operation in one direction only, i.e., with chimney leading.

When curving, the boiler of a Mallet locomotive assumes a position tangential to the curve, and in certain cases the front of the locomotive can laterally overhang considerably. Unless a single boiler bearing is used on the front unit, corresponding to the centre of gravity of

the ordinary rigid engine, the centre of gravity, when the engine is on a curve, takes up a position nearer the outer rail.

The Mallet locomotive is still met with in many countries throughout the world but it had its greatest application in the United States of America.

Manganese-steel liners. Liners fitted to hornblocks and horns to reduce the wear to which the faces are normally subjected by pounding and movement of the axle-boxes. These liners may also be fitted to the corresponding faces of the axle-boxes. The liners may be of manganese-steel throughout or may be formed of a manganese-steel plate attached by riveting and welding to a mild steel backing plate (see Axle-box and Hornblock).

Manifold, see Turret.

Masses. The masses on a steam locomotive take two forms, rotating and reciprocating. To ensure steady running of the locomotive, and safe progress along the track, it is essential that these masses be balanced as far as practicable. The out-of-balance rotating masses are the crank pins, eccentrics, coupling rods, and part of the weight of the connecting rod. The balancing of the rotating masses presents no special problem and is carried out by the provision of balance masses in the wheels. These masses may be cast in, added by providing lead in cored-out holes, or may be made up from segmental plates riveted on and filled with lead as may be necessary. It is possible for the lead to become loose, and subsequently disappear when it has been pulverized, unless it is suitably alloyed with antimony.

The balancing of reciprocating masses is a much more difficult matter. These consist of the pistons, piston rods, crossheads, and part of the weight of the connecting rod.

Practice varies largely and is much influenced by the ratio of the weight of the reciprocating parts to the total engine weight: it is also influenced by piston speed and hence by wheel diameter. It was the practice on American locomotives to leave reciprocating masses unbalanced to the extent of one four-hundredth of the locomotive weight (not including the tender) on each side of the engine. The provision of balance masses on the crank axle – if fitted – or in the wheel centres, will go far to reduce disturbances in the horizontal plane but it is not practicable completely to balance reciprocating parts. To do so would cut out unbalanced horizontal forces, and the engine motion associated therewith, but it would introduce exceptional variations in axle loading in each revolution and result in bad hammer-blow.

In practice a compromise is usually arrived at to avoid bad surging and excessive hammer-blow. The proportion of reciprocating mass which is balanced varies from 0–66 per cent.

To minimize hammer-blow it is customary to spread reciprocating balance over the coupled wheels but not necessarily in equal proportions.

Four-cylinder locomotives having the inside and outside adjacent cranks set at 180° and the right- and left-hand cranks at 90°, practically balance themselves. Wheels revolving at far higher speed than that at which they are balanced may give rise to serious damage to track and bridges due to excessive hammer-blow; the same harm may result from operating an engine, even at low speed, if side rods are removed as the balancing will thereby be affected. In bad cases of hammer-blow the wheel will be lifted clear of the track at each revolution. (see Chapter 7, Part I, Route Availability, page 414).

Metals. The metals used in the manufacture of locomotives form the subject of specifications which are constantly under review. In British practice the British Standards Institution gave early attention to the subject of materials for British Rolling Stock, the principal specifications concerned being contained in B.S. Report 24. This report, which is in several sections, covers materials for most of the components of a locomotive, and in addition to giving particulars of the process and chemical composition of the material, includes details of the physical properties and the bend or other tests concerned.

Meyer locomotive. The Meyer form articulation has several varieties. Originally the two engine units, mounted below a single boiler, had the cylinders at the inner end of each unit. As these units carried the buffing and drawgear, and were joined to each other by draw-gear, no buffing or traction stresses were transferred to the boiler or its supports. Later modifications (see Meyer Kitson), included the cylinders being positioned at the rear of each bogie. In some instances compounding has been employed, the term Mallet-Meyer being used to describe such engines.

Meyer Kitson locomotive. An articulated locomotive based on the Meyer locomotive, with modifications, introduced by Kitson & Co. The boiler and fuel and water supplies are carried on a girder frame mounted on two power bogies. A separate tender is sometimes used. The cylinders may be arranged at the outer ends of each unit, or they may be at the leading or trailing end in each case. As the boiler barrel is located over one engine unit the designer has freedom in the layout of the firebox.

Monel metal. A nickel–copper alloy containing approximately two-thirds nickel and one-third copper. Its physical properties, plus the fact that it resists attack by many organic and inorganic acids and alkaline salts over a wide range of operating conditions, have led to its widespread adoption as a staybolt material. It is also used for firebox plates.

Motion. A term synonymous with valve gear.

Motion bracket. A bracket, made up of plate and angle, fabricated, pressed, or a steel casting, to which the bearings carrying the link of a fixed link motion are attached. The same bracket is frequently employed to carry the rear ends of slide bars.

Motion girder. A girder running parallel to the main frame and supported upon a motion plate at the front end and by a bracket at the rear, on which the bearings carrying the link trunnions are mounted.

Motion plate. The cross-stay of an inside-cylindered locomotive which carries the slide bars and, in some instances, valve-rod guides. At one time built up from angle and plate, annealed steel castings have been employed for many years. In the case of engines fitted with Joy's gear it is customary to attach the anchor links also to the motion plate. In an outside-cylindered engine the bracket carrying the rear end of the slide bars is sometimes termed the motion plate.

Mud doors. Oval-shaped doors, making a joint with the inside face of the hole cut in the boiler, and held against the face by tightening up a screwed stem passing through a cross-member. Mud doors are provided for washing-out purposes and give a larger opening than that through a mud plug.

Mud drum. A receptacle of shallow depth placed in the bottom of the boiler barrel. Provided with a bolted-on cover, its purpose is to enable the scale and sediment which collects in it to be readily removed. Mud drums are also called collector seats.

Mud ring. A term used, more especially in American practice, to denote the foundation ring.

Multiple-valve regulator, *see* Regulator.

Needle roller bearing. These bearings are sometimes provided for all pin connections of valve gears except that of the return crank and the eccentric rod, where a ball or roller bearing would be employed.

Nosing. A transverse oscillation of a locomotive about a vertical axis. The result of this is that the path pursued along the track is a sinuous one. It arises as a reaction from the alternating piston thrusts in combination with slackness in the axle bearings and is to some extent affected by the balancing of the reciprocating parts.

Oiling rings. Rings fitted over a bush and into counterbores to deal with side thrusts, as on little-end bearings. Such rings have blind holes drilled in the thrust faces to carry lubricant.

Oil firing. When oil firing is employed steam is used to atomize the oil, which issues in this form through a burner situated at the end or in the centre of the firebox, modified as necessary (*see* Firebox).

In order that the oil may flow freely by gravity to the burner, it may be necessary to warm it by means of a heating coil in the tank. Steam for lighting up purposes is obtained from a connection to an outside steam supply.

With oil firing it is necessary for the fireman to anticipate the demand for steam as the automatic relation of supply to demand provided by the blast pipe on a coal-fired locomotive will not apply with oil-burning.

It is usual to fit shorter superheater elements on locomotives burning oil than on engines burning coal, to protect the element ends from the intense flame.

Oil tanks. These tanks may be cylindrical or rectangular and are mounted on top of the water tank in the case of a tender or in the rear bunker, in tank engines. Filters are provided inside the filler and the filler cover must be a good fit and designed to be tightened down. Heating coils are provided so that the oil may be warmed, when necessary, to allow it to flow readily; they have the further advantage that slight warming has the effect of separating any water present in the oil. In order to obtain this effect the temperature must not be raised to such an extent that vaporization takes place. Additional heating chambers are usually provided in the pipe-line from the tank to the burner.

Oscillation. Movement of a locomotive in any plane.

Outside admission. A valve having outside admission is one where the outer edges of the valve and the outer edges of the ports control the admission of live steam to the cylinder. In such cases the outer edges are called the "steam edges", and as a corollary the inside edges of the ports and valve are known as the "exhaust edges". Whereas all slide valves give outside admission, piston valves with inside admission are in the majority (*see* Inside Admission).

Packing. Various forms of packing are in use, largely dependent upon the purpose for which they are employed. Packings formerly used for piston rods, such as hemp soaked in tallow, have been rendered obsolete by increasing pressures and consequently increasing temperatures. Sundry asbestos packings have been used and while this material is excellent, particularly in graphited form, for such items as the spindles of screw-down valves and certain types of water-gauge cocks, for piston rods cast-iron, white metal, or lead–bronze packings are now generally employed.

Metallic packings may be classified as:

(i) those where the seating effect is derived by pressure applied on the end of the packing through the medium of studs and nuts.

(ii) those similar in principle, but deriving the necessary end-wise pressure from a coil spring;

(iii) those in which pressure is obtained from a garter spring. Lubrication of rods where they pass through glands may be effected mechanically, by mop, or from an oil box.

Pads, boiler. Steel pads are welded or riveted to the firebox wrapper or backhead in order that the boiler fittings may be mounted upon them. When rivets are employed they must be countersunk as the pad must be faced to receive the flange of the fitting concerned. The largest pads, e.g., those provided for the safety valves, are of a relatively complicated nature. In addition to the ways through them, they must be suitably machined for the reception of the studs holding the fitting on, for the rivet holes securing the pad to the plate, and also to clear any stay heads.

Pads, lubricating. Pads are sometimes employed where it is necessary to wipe oil on to a bearing. Worsted pads are used for some axle-boxes, the pad being located in the keep and wiping oil, supplied to it by tail feeders from a reservoir in the keep, on to the underside of the journal. Pads made of felt are also sometimes fitted in suitably machined grooves provided in coupling and connecting rod bushes where they wipe oil delivered from the oil box, via a plug trimming or needle, over the crankpin.

Pick-up gear. The water scoop may be hand or power operated and is connected to the pick-up tube running vertically to the delivery box which is suitably shaped to deflect the water into the tank. In recent years it has been customary to fit a deflector in front of the scoop; this rises and falls with the scoop and banks the water up, thus reducing loss by splashing. Water scoops, when fitted on tank engines, are arranged for operation when working in either direction.

Pipe, breeches. The pipe used to bring together the exhaust from the cylinders and connected to the blast pipe at its upper end.

Pipe, main steam. The main steam pipe, or dry pipe, as it is sometimes termed, is usually constructed from copper in British practice, but steel is generally used elsewhere. When copper is employed external strengthening bands may be brazed on. The joint between the pipe and the front tube plate may be made by expanding the pipe and fitting a steel ferrule. When a steel pipe is used, the joint may be made by welding or brazing on a flanged sleeve. It is customary to provide wrought iron or steel supports for the dry pipe.

The dry pipe may take steam from the boiler, via an elbow terminating in the dome, or through a series of slots in the top of the pipe where the dome is either shallow or non-existent.

The dry pipe is sometimes arranged externally.

Piston (Fig. 53). Pistons must be designed as lightly as is consistent with strength, in order to minimize the disturbances arising from partly balanced masses. The piston head is nearly always mounted on the rod but is sometimes forged integrally with it or welded. When separate, the head is fitted on the rod either on a taper drawn up by a nut, or it may be screwed on by a parallel or slightly tapered thread. Cast iron and cast steel are materials also used for heads, the former material being most generally employed. Cast steel, or forged, heads incorporate a cast-iron bull-ring to reduce the wear which would result from the use of steel alone, this bull-ring may be cast on or riveted on. Brass has also been used for bull-rings.

Pistons may be dished, in which case the contour of the cylinder covers will correspond with that of the piston head, partially to equalize steam volume where no tail-rod is used and also to keep down clearance volume. Pistons may also be of box form in which case they will be cored out. A compromise, where the piston is dished and

a steel plate is inserted to complete the box form, is sometimes used. A box piston lends itself to housing a spring-loaded bronze pad, as used on British Railways standard locomotives and formerly on the London, Midland & Scottish Railway. This pad is situated at the bottom of the piston and is easily, and relatively inexpensively, renewed when worn. Wear liable to occur on the bottom of a piston may also be reduced by increasing the width of the piston head to present a larger area in contact with the wall on the underside of the piston. Pistons usually carry two or three rings, those now used being about $\frac{5}{16}$ in. wide, which is narrower than those formerly advocated.

Piston rods. Piston rods in British practice are usually separate from the piston heads and the crosshead but they may be forged integrally with either, or attached to the head by welding. The connection to the crosshead is usually by means of a taper, the rod being drawn in by a nut or more frequently a cotter. Occasionally the crosshead is split along its vertical centre line and machined, with a number of annular recesses engaging with corresponding shoulders machined on the rod.

While at one period considerable use was made of piston rods with an extended tail these have now passed out of fashion in Britain as the additional complication involved did not prove worth while.

Piston valves, *see* Valves.

Plug, fusible, *see* Fusible Plug.

Pony truck, *see* Truck.

Poppet valves, *see* Valves.

Port. The cavity in the valve face and the steam passages leading to the cylinders, in the case of a locomotive with slide valves, or in the liner in the case of an engine with piston valves, through which steam is admitted to and exhausted from the cylinder as the port is opened or closed by the valve. The port is now always rectangular in the case of a slide valve engine though circular ports have occasionally been used. In the case of a piston valve engine the ports may be rectangular, triangular, rhomboidal, etc. The shape of the port makes an appreciable difference to the port opening for a given valve movement. The greatest opening is obtainable with the rectangular port, but ring wear is better distributed with ports of other forms.

Port bar. That part of the cylinder of an engine fitted with slide valves which divides the exhaust port from the admission port.

Port bridge. The pieces of metal situated between the ports in a piston valve liner. Apart from other functions these bridges prevent the rings on the valve head from springing into the port.

Pressure, back. The pressure on the exhaust side of

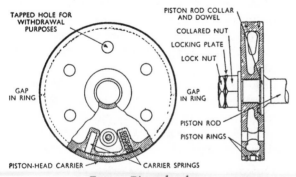

TAPPED HOLE FOR WITHDRAWAL PURPOSES

PISTON ROD COLLAR AND DOWEL

COLLARED NUT

LOCKING PLATE

LOCK NUT

GAP IN RING

GAP IN RING

PISTON ROD

PISTON RINGS

PISTON-HEAD CARRIER — CARRIER SPRINGS

FIG. 53. Piston head.

the piston; this may be considerable and is ascertainable from an indicator diagram. In the case of an engine fitted with poppet valve gear it is at a minimum, as the exhaust valves are always fully open and their timing is independent of other valve events.

Pressure, mean effective. The mean effective pressure acting on the piston is sometimes taken as 85 per cent of the boiler pressure and sometimes as 75 per cent, the higher percentage applying generally to engines using superheated steam. The M.E.P., as it is usually known, is made use of in formulae for calculating horsepower developed, and tractive effort. Horsepower is calculated from the well-known formula:

$$H.P. = \frac{apln}{33,000}$$

where

a = mean effective pressure in p.s.i.

p = length of stroke, measured in feet.

l = area of the piston in square inches.

n = number of piston strokes per minute, i.e., two per cylinder, multiplied by the number of revolutions of the driving wheels per minute.

As this formula takes no account of steaming capacity, its primary value is as a basis of comparison.

Mean effective pressure is largely dependent, in practice, upon the layout of the steam pipes and passages, the degree of regulator opening, the valve events, the heat loss occurring in the cylinders, the rate of cut-off, and the piston speed. There is not as much difference between the mean effective pressure on engines working with superheated or saturated steam as might be supposed. Lawford Fry showed that at a cut-off of 20 per cent, the M.E.P. was higher with a saturated steam engine than with a superheated engine; at 40 per cent there was practically no distinction between superheated and saturated engines, but at 60 per cent the M.E.P. of the superheated engine was considerably higher.

To secure the closest possible agreement between theoretical M.E.P. and actual M.E.P. the boiler must be able to supply the cylinders adequately at all times, the steam circuit must be properly proportioned, the valve events must be satisfactory, and both valves and gear designed to give efficient distribution (*see also* Diagram Factor *and* Chapter 6, Part I, page 392).

Primary air. Air admitted to the firebox, for combustion purposes by way of the firebars and the firebed (*see* Chapter 6, Combustion).

Priming. The carrying over of water with steam. It may be due to too high a water level, sudden opening of the regulator, resulting in a local reduction of pressure on the surface of the boiler water, bad boiler design, or working the boiler above its capacity. Priming is always

bad and may be very bad for an engine. At the least it washes oil from cylinders and valve liners, while at worst it may break pistons, cylinder covers or cylinders, and bend piston or connecting rods. A superheated engine when it "gets the water" will take longer to clear itself than a saturated one; there is the added trouble when priming occurs on a superheated engine that scale is deposited in the elements.

Pull, drawbar. The drawbar pull which a locomotive is required to develop, and the speed at which this must be developed, will largely determine the design. When calculating such power it is customary to neglect the power required to propel the locomotive and tender, and produce a result which is the force available at the wheel rim, i.e., the tractive effort or tractive force (q.v.).

The drawbar pull results obtained in practice are dependent partly on the ability of the boiler to supply steam, the front-end design generally, the rate of cut-off, the degree of opening of the regulator, etc., some of these factors being particularly important at speed. Whether the power developed can be fully employed at low speeds will be determined by the adhesion.

Pulverized fuel. The burning of pulverized fuel in locomotive boilers has had little application outside Germany. While coal is the fuel normally associated with such working, peat has also been employed. The subject is treated in Part II (page 315) and in Chapter 9, Part I, page 463.

Pump, air. Air compressors are provided whenever compressed-air brakes are used.

On steam locomotives the pump is steam driven; on diesel or electric locomotives it is belt or motor driven.

Air supplied by a compressor on the locomotive may be employed for the operation of dumping-wagons, and also used on the locomotive to operate sanding gear, water scoop, firehole doors, grease guns, etc., and even to operate the whistle-valve.

The size and capacity of the air pump depends upon the duties it is required to perform. It may be quite small – a single-stage compressor cylinder and one steam cylinder – or large with three-stage compressors operated by two cross-compounded steam cylinders. While normally one air pump suffices to supply requirements there are instances where two are employed. Steam-driven air pumps are invariably arranged with the cylinders vertical.

Pump, feed. An important difference between feeding a boiler by a pump or by an injector is that the former normally feeds cold water; for this reason it is frequently the practice to utilize a feed-water heater in conjunction with the pump.

Reciprocating pumps may be operated by a steam

cylinder, in which case they are disposed either vertically or horizontally, or they may be plunger pumps driven from some part of the locomotive, e.g., from a crosshead or an eccentric. A feature of all modern steam-driven reciprocating pumps is the considerable working range; this enables the pump to be set to operate continuously.

While reciprocating pumps are better for handling hot feed water, and their use is almost universal when feed water heaters are employed, they have the disadvantage that the valves are subjected to considerable "hammering", necessitating fairly frequent attention.

Centrifugal pumps are employed in American practice; compared with reciprocating pumps they have the advantage that as the water flow is continuous, less attention to valves is required.

Pump, vacuum. With a view to saving steam consumption when a train is running, vacuum pumps, operated from a crosshead, are sometimes fitted. In such cases use of the ejector is only required initially to create vacuum, lift the brakes, and help maintain vacuum at low speeds. Diesel or electric locomotives equipped with vacuum braking equipment, have belt- or motor-driven exhausters (see Vacuum Brake).

Pusher, coal. A device, operated by a steam cylinder, to push coal forward from the back of the tender. The piston rod is connected to a T-shaped crosshead, or main pushing member, which has a supplementary pushing member connected by links in front of it. The device is brought into operation when coal requires bringing forward to the shovel plate, in the case of a manually-fired locomotive, or when the conveyor screw is not receiving sufficient coal in the case of a mechanically-fired engine.

Apart from the obvious advantages, a coal pusher reduces corrosion of tank plates, as coal is not left stagnant, and it is particularly valuable when coal is liable to become frozen. It is important that care be exercised not to use the pusher when sufficient coal is already at the front of the tender as this may cause damage to the coal gates.

The pusher should be operated at the end of a journey and before taking coal, to bring forward that already in the tender and prevent its becoming stale.

Whether the installation of a coal pusher is warranted or not naturally depends on many factors, including the size of the tender. The first application in Britain was on tenders of Pacific locomotives of the London, Midland & Scottish Railway having a coal capacity of ten tons.

Quarters. The four quarters are when the crankpin is on the front dead centre, the back dead centre, and the top or bottom centres, each of the latter being spaced at 90° to the back and front centres. The crank positions

mentioned are bisected at 45° by positions known as angles. Viewed from the right-hand side and reading in a clockwise direction from the top, the positions, each 45° apart, are: top quarter, top front angle, front quarter, bottom front angle, bottom quarter, bottom back angle, back quarter, top back angle.

Radial axles. Radial axles are those in which the boxes are housed and move in radial guides. They were first used successfully by Webb and employed on a number of British railways. Later the Cartazzi box was extensively used. Whereas in the Webb form the two axleboxes on one axle were connected, in the Cartazzi arrangement the boxes were independently mounted. Both Webb and Cartazzi layouts permitted the axle to swing sideways in an arc as with a conventional form of pony truck, but dispensed with a radial arm, the controlling influence of which was replaced by the curved guides. Other forms of radial truck have been employed, as, for example, on the Western Australian Government Railways, where the radial boxes are carried in a common frame.

Radial bogies. In a radial bogie the straight slot in the bogie stretcher is replaced by a radially-machined one, having its centre of curvature corresponding to that of a pony truck.

Radial truck (Plate 63A, page 263). A truck, having two or four wheels, attached to a radius arm, or bar, and free to turn about the point of attachment of the arm on the centre line of the frame. A well-known example of a radial truck is the Bissell. Whereas radial trucks may be used at either end of a locomotive, by common usage the terms Bissell and pony truck have come to be used for two-wheeled leading trucks only. The arc on which a radial axle or pony truck swings sometimes has had its centre at the middle of the rigid wheelbase; this reduced "hunting".

Radiation. To reduce loss of heat by radiation, lagging is necessary on the boiler, cylinders, and outside steam pipes. Still air is an efficient insulator and the provision of an air space alone between the boiler and the cleading gives excellent results if the air contained can be kept still. As this is a difficult matter, especially so on a locomotive, it is usual to lag the boiler with asbestos, spun-glass, or slag-wool, etc., the lagging being covered by cleading secured by bands, which are sometimes of stainless steel. When asbestos is employed it may be in the form of tailored blocks, mattresses, or foil, or it may be spread on. When glass wool is used it is more than ever necessary that the weight of the cleading, normally carried by crinoline bands, should not fall upon the lagging which depends for its efficiency upon minute air pockets. This necessity has, in the past, produced D-shaped cleading, supported on the bottom legs.

Radius rod. The rod, in Walschaerts valve gear, which is connected to the die block at its hind end and to the combination lever at its front end. In American practice this rod is also called a radius bar.

Reach rod. The rod connecting the reversing gear, or lever, in the cab to the reverse shaft arm. In British practice this detail is sometimes called the Bridle Rod.

Reciprocating masses, *see* Masses.

Regulator (Figs. 54–56). The regulator, controlling the supply of steam to the cylinders, may be located in the dome on the end of the drypipe, when space permits, or in the smokebox. In the latter case it may be on the saturated or superheated side of the header. The valve, if sliding, may be a simple flat one, disposed vertically or

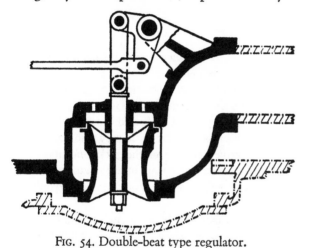

FIG. 54. Double-beat type regulator.

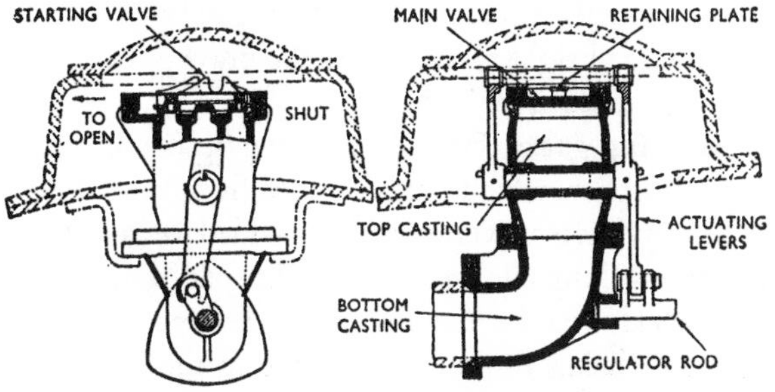

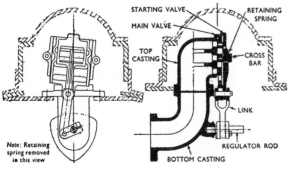

FIG. 55. Regulator valve, horizontal-dome type.

FIG. 56. Regulator valve, vertical-dome type.

horizontally, or may be of the grid type. If rotating, the valve may be of the butterfly type. When lifting valves are employed it is usual to employ either a number of small mushroom valves, as in the MeLeSco multiple-valve regulator, or a double-beat form, as the Lockyer regulator. The Joco regulator has three valves arranged vertically and concentrically.

With the ordinary sliding regulator it is customary to provide a pilot – or first valve – besides the main valve. In some forms of compound locomotive the position of the regulator determines whether the engine is working simple or compound.

In American practice the regulator is termed the throttle.

Reidinger valve gear. A form of poppet valve gear (*see* Valve Gear) (Plate 62A, page 262).

Release. The point at which the valve opens to exhaust. It is dependent upon other valve events in the case of slide and piston valve engines, but not in the case of locomotives with rotary poppet valve gear, this being one of the features of that gear. Release is arranged to occur, on a slide or piston valve engine, before the dead centre is reached, in order to minimize back pressure. On an engine with poppet gear, delayed release may be employed, thereby obtaining more work from the steam, as the exhaust valve opens very rapidly (Fig. 62, page 286).

Release valve. A spring-loaded valve, fitted to each cylinder cover, to release any water which may enter and become trapped in the cylinder. These valves supplement the ordinary drain cocks and are provided to protect the cylinder covers, etc., against damage resulting from the presence of water in the cylinder.

Reservoir, brake, *see under* Vacuum Brake *and* Westinghouse Brake.

Return crank. An arm secured to the outer end of the crankpin to impart movement to the valve gear, lubricator, speedometer drive, etc. In rare instances return cranks are employed to convey lubricant to crankpins from a mechanical lubricator.

Reversing gear (Fig. 57). Reversing is normally effected in the cab by a wheel and screw, arranged vertically or horizontally, a lever moving in a sector plate, or by means of power. In the case of wheel and screw, gearing may be interposed between the wheel and screw so that the wheel – or more particularly the drum attached thereto on which the positions of cut-off are marked – is readily visible to the driver. Means of locking are usually arranged on the wheel or lever, as the case may be, but can also be located on the reversing shaft by means of a vacuum-operated clasp brake working on a drum secured to the shaft. A quick-nut, attached to a lever, has been used in conjunction with a wheel and

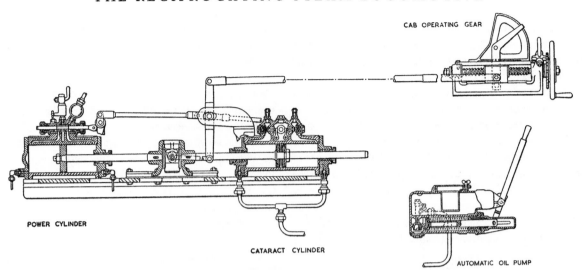

FIG. 57. Hadfield power reverser.

screw. In such a case, a screw of elongated barrel-shape may be employed.

Power reverses have employed steam and air pressure. While the advantages of power reverse are considerable, especially on a large locomotive, it has not had a wider application due to the amount of difficulty experienced owing to "creeping". This difficulty has been overcome in the Hadfield power reverser, where selected cut-off may be attained accurately and without effort. The provision, in the gear mentioned, of specially arranged valves in the cataract cylinder, and the method of automatically recuperating that cylinder, have entirely eliminated creep – for so many years the greatest enemy of power reversers.

Riding. The riding of a locomotive is a product of the condition of the track, the condition of the locomotive, and the design of the locomotive. A most comprehensive survey of the disturbing forces having their origin in the locomotive is to be found in the *Pacific Locomotive Committee Report* (*see* For Further Reading, page 508). The following definitions of movements which may arise are taken from this source.

HUNTING. Nosing and rolling rarely occur separately, but are generally found acting together in varying proportions. The resulting oscillation is described as hunting.

LURCH. One semi-amplitude of movement in the action of hunting.

OSCILLATION. An inclusive term used to describe movements of a locomotive in any plane.

PITCHING, also referred to as galloping, consists of the fore and back ends of the locomotive rising and falling about a transverse horizontal centre line.

ROLLING. Transverse oscillation of the locomotive on its springs, about a longitudinal centre line.

SHUTTLING. Oscillation in a fore and aft direction parallel to the track.

TRIMMING. A settling down of the locomotive, either at the front or rear end.

Some of these forces are common to diesel and electric locomotives although their causes may be different. As an example, rolling could be initiated in a steam locomotive by slidebar pressure whereas it may also arise, in any form of motive power, from the track condition.

Shuttling, having its origin in uneven torque and lack of balance in reciprocating masses, is associated only with steam locomotives.

Nosing is particularly apparent on some types of electric stock with axle-hung motors, while trimming may be associated with any type of motive power where the horizontal centre line of the driving axles does not fall at the same height above rail as the draw gear.

Ring, piston, *see* Piston.

Rocking grate (Fig. 58). Rocking grates consist of hinged firebars which can be rocked from the cab when necessary. Two movements are provided for, a restricted one to break up clinker when on the road, and a greater one which enables the fire to be dropped when the engine is being disposed of. While in the majority of cases the rocking is effected manually, some American locomotives are fitted with steam cylinders which supply the movement.

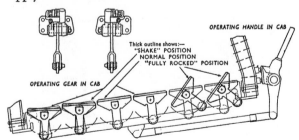

FIG. 58. Arrangement of rocking grate.

When dropping fires it is essential that the ashpan doors are opened prior to dropping as otherwise the ashpan will be damaged by the heat.

Rocking lever. A lever transferring movement from one plane to another. Rocking levers have frequently been employed to transfer motion from an inside valve spindle to an outside one, in which case the lever may be a simple one, or it may take the form of two pendulum arms on a single shaft, when the whole is termed a rocker. The use of vertical rocking levers to transmit power from cylinders to wheels, through the agency of a connecting rod attached to the lower end of the lever, is rare. It has been employed on tramway engines; the largest locomotive to which this form of drive is known to have been fitted is the Russian 2–10–6 OR class.

Rods, piston, *see* Piston Rods.

Roller bearings. Roller bearings have, of recent years, had a wide application and have proved the answer to many hot box problems. Some of the advantages of roller bearings also apply to ball bearings, but as the latter naturally bear over a restricted area they are unable to carry the loads which the increased bearing area of the roller makes possible. Nevertheless, for applications where the loading is not so high, e.g., for return cranks or to deal with thrust loads in axle-boxes using parallel rollers, ball bearings are exceedingly satisfactory.

Hoffman axle-boxes, which employ parallel rollers to carry the load and ball bearings for endways location, have been fitted to many carrying axles of locomotives and to tenders with outside boxes. The same principle is employed in the box, manufactured by this company, for mounting inside on driving and carrying axles.

S.K.F. boxes, utilizing barrel-shaped rollers capable of carrying end and radial loads, are used for axle-boxes in all situations. Some S.K.F. bearings, for locomotive coupled axles, are housed in axle-boxes which are split vertically.

Timken roller bearings, embodying tapered roller bearings, are able to carry both heavy radial load and considerable end thrust. In addition to having supplied many individual axle-boxes, Timken have placed in service in America, Britain and overseas, over 17,500 cannon boxes. This design of axle-box comprises a rigid tubular housing of cast steel in which are mounted two single Timken bearings. This form of construction provides great transverse stiffness as well as maintaining complete bearing alignment under conditions of track irregularity (*see* Axle-box.)

Individual roller-bearing axle-boxes have run over 2,000,000 miles.

Roller bearings have also had a considerable application on connecting rods, coupling rods, etc.

Needle roller and ball bearings have been used for motion (Plate 62B, page 262).

Ross "Pop" safety valve, *see* Safety Valve.

Rotary cam gear, *see* Valve Gear.

Rubber springs. Rubber may be used as a spring at various points. It is met with in draw and buffing gear. It is also used as an auxiliary spring between spring hangers and brackets, and between hangers and equalizing levers.

Safety links. These are fitted between engine and tender as a precaution against the effects of drawbar failure.

Safety valves (Fig. 59). The number of safety valves which are fitted is dependent upon the size of the boiler.

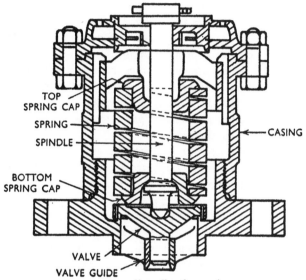

FIG. 59. Ross "pop" safety valve.

Most modern locomotives are fitted with a valve which uses a helical spring in compression. The design of the valve is such that when it lifts the area subjected to pressure is considerably increased and the valve rapidly lifts to allow ample release. An example of this type of valve is the Ross, which occupies comparatively little space, compared with the older Ramsbottom valve and the difference of the former between lifting and shutting down pressure is about 1 to 2 p.s.i., against the 5 p.s.i. (or more), of the Ramsbottom valve.

Ramsbottom valves are now seldom met with. They had the merit that compared with valves formerly in use they were difficult to tamper with, but boiler explosions sometimes occurred due to incorrect assembly and incorrect machining tolerances.

Latterly, direct-loaded valves have always been fitted, examples met with including, Ashton, Coale, Crosby, Consolidated and Knorr.

The location of safety valves has been governed

largely by consideration of space. For this reason, their position in earlier days, on the top of the dome, has been abandoned. This position was also a bad one in that as the regulator is housed in the dome there was always the likelihood of water entering the dry pipe when the safety valves lifted. In many modern locomotives there is even insufficient space above the firebox and this has necessitated locating the safety-valves on the barrel.

Sand guns. Sand guns are mounted on the boiler back-plate and when operated discharge sand, entrained in a steam jet, over the tube plate and into the tube ends. The gun is normally operated for a period of about thirty seconds at intervals of an hour, but the time and frequency will be dependent upon the quality of the fuel. Sand consumption is about two pounds at each operation and clean tubes are indicated by the gases from the chimney clearing. For obvious reasons the gun should only be operated when the engine is steaming hard and never, in any circumstances, when the regulator is closed, as sand would enter the blast-pipe and get drawn into steam-chests and cylinders. To obviate this possibility, the steam supply to the sand gun is sometimes obtained from the steam chest.

The shortened life of fireboxes and tubes, which results when oil fuel is employed, is contributed to by the necessity for using a sand gun more frequently with this form of fuel.

Sand guns are useful for cleaning superheater flues and elements which cannot be swept out.

Sanding gear. Sand is used between the tyre and the rail to increase adhesion when necessary. Such sand is stored on the locomotive in sandboxes which are arranged either on the frames or on the top of the boiler, where the radiated heat assists in keeping the sand dry. So far as is practicable, sand boxes should be located in positions from which any sand spilled while filling will not fall on to working parts of the locomotive. They are not, in modern practice, incorporated with splashers or built into smoke boxes, though it is always of importance that the sand should be kept perfectly dry. An exception to this is if the system of wet sanding (q.v.) is employed.

To deliver the sand where it is needed, between tyre and rail, four forms of sanding gear are in common use.

(i) *Gravity sanding:* The sand runs down suitably located sand pipes from the sand boxes to the rails just in front of the driving wheels. The flow of sand is controlled by valves which may be operated manually, or by vacuum or compressed air.

(ii) *Steam-sanding gear:* An ejector is fitted near the lower end of each sand pipe. When steam passes through the ejector, a current of air is induced in the sand pipe, and this draws sand from the sand box and forces it between the wheel and the rail. In order to prevent sand running out by gravity when the gear is not in operation, a trap is provided. There is an air inlet in the top of this trap so that when sand is being withdrawn by the action of the ejector, it will flow freely from the trap.

(iii) *Compressed-air sanding gear:* This operates on the same principle as steam-sanding gear. A more simple sand trap and ejector is adequate, when air is used, and the ejector need not be placed near to the exit end of the sand pipe, thus obviating a considerable amount of piping.

(iv) *Wet-sanding gear:* The sand is flushed from the sand box by way of a trap, to the sand pipe. This system has the advantage that damp sand does not cake in the box or pipe as sometimes happens with other methods of sanding; it is nevertheless desirable that water does not enter the apparatus when it is turned off and accordingly an automatic drip valve is incorporated.

The number of wheels for which sand is provided will vary, but in the average six-coupled engine, gravity sanding will be provided to one axle and power-assisted sanding on another. It is also usual to provide at least one axle of such an engine with equipment for operation when reversed.

It is no longer customary to provide sanding gear on tenders for backward running.

As sand left on the rail head after the driving wheels have passed over it is injurious, in that it increases train resistance and does not provide good contact for track-circuiting purposes, rail washing equipment is sometimes fitted. This consists of a valve, situated on the side of the firebox in the water space, which is normally closed. When the sanding gear is operated the valve opens and delivers hot water, through a suitably located pipe, on to the rail head behind the trailing coupled axle.

Saturated steam. Steam in the "as generated" condition is termed saturated as in this state it is wet, containing minute particles of water. Saturated steam is now used only on some old locomotives and on shunting locomotives. For many years it has been customary on all new construction, with the exception of shunting locomotives, to use superheated steam. The temperature and volume of saturated steam are dependent upon its pressure (*see* Steam).

Secondary air. Air required for combustion purposes which is supplied above the fire, i.e., through the firehole door or through orifices provided in the water legs. Normally it is not possible in a coal-fired engine to supply from below all the air required for the efficient combustion of the volatile gases, which are generated in the firebox. To attempt to do so, save in the case of the very thinnest fire, would necessitate uneconomical spacing of the firebars.

The quantity of secondary air necessary to ensure complete combustion varies with different coals, those high in fixed carbon and low in volatile matter, as, for example, good Welsh steam coal, will require less secondary air but more primary. If the position is reversed and the volatile content is high, as with good steam coal from some Yorkshire seams, more secondary air is necessary.

The quantity of secondary air needed is also determined by the amount of coal fired, and the frequency of firing. The best results are obtained by firing little and often. Insufficient air results in incomplete combustion, of which smoke is always a visible sign, and excessive secondary air can also cause considerable loss, as it involves uselessly heating up a large column of air. In the last instance there will be no visible indication and to save loss in this way it is best to so adjust the supply of secondary air that a light grey haze is just visible at the chimney.

As a thinner fire can be carried with a stoker-fired engine than by one which is fired manually, it follows that on a locomotive with a mechanical stoker the secondary air requirements will be much less than on a hand-fired one (*see also* Baffle Plate, Firehole, Fire Door *and* Chapter 6, Combustion, page 390).

Self-cleaning smokebox, *see* Smokebox.

Shay locomotive, *see* Chapter 9, Part I, page 462.

Side rods, *see* Coupling Rods.

Sight feed lubricators (Fig. 51). Hydrostatic or displacement lubricators are those cylinder lubricators in which the drops of oil ascend in a glass tube filled with water, making it a simple matter to check the rate of feed, which will vary according to the type of work upon which the engine is engaged. In British practice a rate of two to three drops per minute per feed is representative.

Modern practice embodies the whole in one self-contained fitting in which the frequency of the ascending oil-drops can be observed through what are termed "bulls-eyes" in the body of the casting.

Simple expansion. Those locomotives where the cylinders always receive steam from the boiler and exhaust direct to atmosphere are termed simple expansion. This is in contra-distinction to compounds.

Slide bars. The function of slide bars is two-fold, viz., to guide the piston rod, through the medium of the cross-head, and to resist the thrusts in the vertical plane, arising from the angular positions of the connecting rod. With steam "on", the thrust is always upwards when travelling in fore gear and always downwards in back gear. The forces may be of considerable magnitude, depending on the piston thrust, the length of the connecting rod, and

the stroke; they are the causes of rolling, or shouldering as it is sometimes termed.

Slide bars are sometimes attached to a bracket, carried on the hind cylinder cover, at their front end, and to a motion plate or slide bar bracket at their rear end. It is also common practice to support them instead, on a substantial bracket having its transverse centre line on that of the bar. This method of attachment also offers the advantage that if desired the piston rod can be lengthened and the slide bars placed farther back than would otherwise be possible. If this is done the piston rod may be arranged to pass through a bushing situated at the front end of the slide bars.

Slide bars may vary from a simple rectangular bar, shrouded by the crosshead, to a single multi-ledge bar; they may be of rectangular, I or T cross-section. The latter sections are particularly employed when the bar overhangs from the fixing bracket. Multiple bars may also be used.

It is customary to attach slide bars rigidly at their back end, but provide for some longitudinal movement at the front end, when they are attached to the cylinder cover, to allow for the expansion which occurs when the cylinders are heated up (*see* Crosshead).

Slide valves, *see* Valves.

Smokebox (Fig. 60). The smokebox is a forward extension from the boiler barrel, through which the products of combustion pass before being ejected from the chimney, or smokestack.

Smokeboxes may be divided into those which are extended and those which are non-extended. An extended smokebox is a cylindrical box whose length exceeds that of the saddle. The cylindrical shape is now always used, but there are still many old engines with D-shaped boxes. The box is built up of plate and angle, or the front may be a flanged plate, pressed ring, or of cast steel in American practice.

It is customary for the front tube plate to be of the same outline as the smokebox front; this is the chief reason for using a cylindrical box, as it enables a drum-head tubeplate to be employed.

Smokeboxes are of two types, those which are self-cleaning and those which are not. In both instances the box houses the blast-pipe, blower, superheater header (if fitted), chimney bell or petticoat, steam pipe or pipes, and, in the case of a self-cleaning box, the diaphragm and table plates, and wire screens. The products of combustion in a self-cleaning smokebox have to pass under the table plate and then pass through the screens before being ejected to the atmosphere (Fig. 60). Normally the larger cinders, or char, in a box which is not self-cleaning, fall to the bottom of the box where they remain

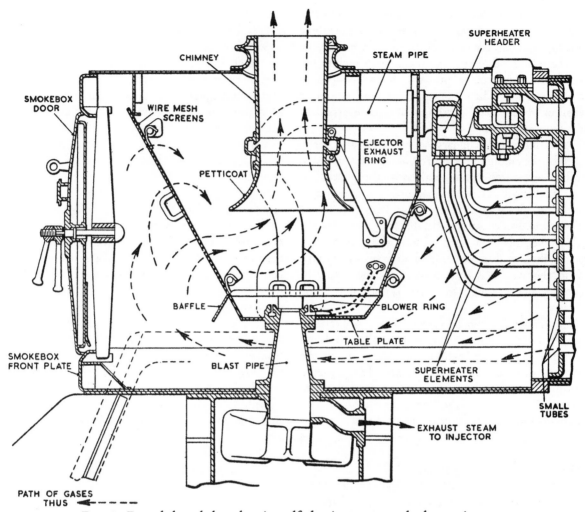

FIG. 60. Extended smokebox showing self-cleaning screens and other equipment.

until removed. In a self-cleaning box most of the larger cinders are broken up by contact with the wire screens and reduced in size sufficiently to allow them to be ejected.

Smokebox door. The door on the front of the smokebox. It is secured by a dart and crossbar, arranged horizontally or vertically, by dogs, or by a combination of both. It is provided to allow the smokebox to be cleaned out when necessary, and to allow access to the smokebox fittings, boiler tubes, flues, and superheater elements for maintenance purposes. In American practice it is customary to provide a relatively small door and allow for the front of the smokebox to be readily removable, when easy access is required inside the box. In British practice the smokebox door ring is riveted in and is a fixture, while the door provided is a large one.

The successful steaming of the boiler depends upon the creation of a vacuum in the smokebox. It is necessary, therefore, for the smokebox to be airtight and it is

important that the door should be a good fit. To ensure this, it is customary to provide an asbestos ring (smokebox front door washer) for it to bed against.

A smokebox door liner is generally fitted inside the door, but held by distance pieces away from it, to protect it from overheating. This plate may also be set farther away from the door at the bottom so that char does not fall out when the door is opened.

Smokebox netting. The spark arrester netting used in a self-cleaning smokebox. This has a mesh varying from $\frac{1}{4}$ in. to $\frac{3}{8}$ in. square. This netting may also be used in the form of a vertical cylindrical sieve above the blast pipe; in this case it serves as a simple spark arrester.

Smokebox saddle. The seat upon which a cylindrical smokebox is mounted. The saddle may be a casting or may be fabricated. When a casting is employed it is sometimes integral with the inside cylinder, or it may be cast in one with all the cylinders. It is usual in American practice to cast a cylinder with half the smokebox saddle which is divided vertically, the two halves being bolted together through flanged joints.

The saddle acts as a strong frame stay and in addition to carrying the weight of the smokebox, also carries part of the boiler weight (Plate 59B, page 259).

Smoke stack. The American term for chimney.

Soot blower. A device, mounted on the back plate, and passing through the rear water leg, capable of providing a steam jet which can be moved through an angle, to blow soot off the tube plate, combustion chamber, tubes, flues, superheater elements, etc. The device is operated by a wheel which rocks the jet by the amount necessary to sweep the area involved.

In the case of locomotives fitted with thermic syphons, soot blowers are fitted on each side of the firebox, a housing tube passing through each firebox leg.

Spark arrester. Spark arresters may take the form of baffles, wire netting, or perforated plate, the purpose being either to break up large cinders into smaller ones, or alternatively, to prevent their being ejected from the chimney. Size of sparks is a most important factor in the prevention of line-side fires as it is only the larger cinders which remain incandescent for the period occupied in being thrown into the air and returning to earth. When deflectors are used they may be arranged to subject the cinders to centrifugal action in order to break them up and extinguish them.

It is not always easy to keep spark arresters clean, especially if the fuel produces tarry deposits, but it is essential to do so if the draught is not to be hindered.

Spectacle plate. Same as Motion Plate (q.v.).

Spindle, valve. The valve spindle, or valve stem as it is termed in American practice, is usually connected to the valve spindle crosshead at its end remote from the valve. There have, however, been instances where the valve spindle has been of sufficiently large diameter to constitute a trunk guide.

Springs. The suspension of a steam locomotive is effected by springs which may be plate (or laminated), helical, or volute. A combination of one or more of these forms may be employed. Springs may be arranged above or below the axlebox, and are termed overhung or underhung respectively. In the case of carrying wheels, springs may be used in pairs, one on either side of the horn, the weight being transferred from the axlebox through a spring beam; in such a case they are of helical form. It is desirable that the coupled wheel springs should not all have the same periodicity. While helical and volute springs are normally employed in pairs, there are instances where nests are employed.

Rubber pads, or auxiliary rubber springs as they are termed, are often interposed between the spring hanger and the spring bracket; they do much to lessen vibration. Helical springs are also sometimes introduced between the brackets and the hangers of laminated springs.

If the springing is compensated, horizontally arranged helical springs are sometimes employed in the compensating gear.

While torsion bar springing is not used on steam locomotives, it has been employed on railcars, and in electric motor bogies (q.v.).

Stand, steam, see Turret.

Stays, boiler (Plates 57A, 59A, pages 257 and 259). The locomotive boiler, by reason of its flat surfaces, requires considerable staying, the pitch and size of the stays being dependent upon the working pressure, the material, and the thickness of the plates. Stays take several forms:

(i) *Direct stays* are of bar and may be screwed in and riveted over or provided with nuts, alternatively they may be drilled down and drifted. In some cases they are of plain bar welded in to plain holes. Direct staying is employed for the firebox sides and ends, and generally the crown.

(ii) *Girder stays* were formerly used for firebox crowns, the girder being bolted or stayed to the crown plate and slung by means of sling stays from the outer wrapper. This arrangement provided for easy vertical breathing of the crown, but rendered satisfactory washing out difficult. In earlier days boiler explosions occurred due to crown sheets collapsing, owing to the girders terminating short of the top edges of the vertical firebox plates.

(iii) *Longitudinal stays* run the whole length of the boiler, being screwed and nutted at the ends, and are sometimes used for top corners of Belpaire fireboxes.

(iv) *Transverse stays* are of similar type and, in the case of round-topped boxes, require suitably shaped pad pieces under the nuts.

It is customary for both longitudinal and transverse stays – being steam space stays – to be of steel.

(v) *Water-space stays* are either *rigid*, sometimes called direct, or *flexible*, sometimes called articulated. Rigid stays if screwed are reduced in diameter between the threaded portions in the plates; one of the reasons for this is to give some degree of flexibility, the term rigid being merely a relative one as considerable bending stresses arise from the difference in temperature between the inner and outer firebox plates – particularly when lighting up. The shouldered-down portion may be either parallel or waisted. Water-space stays may also be *hollow* or *hollow-ended*. In the first instance a coaxial hole is drilled through the stay from end to end. Hollow-ended stays have holes drilled down for a short distance from each end. While the primary function of the hollow stay is to give indication of cracks developing, it has the added advantages that it possesses greater flexibility and the provision of holes is an aid to fixing, either by drifting or by setting explosively. For the last-mentioned method it is claimed that less skill and time are required to produce satisfactory results. Steel is employed for water space stays in steel fireboxes and also in some copper boxes, while the use of copper is restricted to some boxes of the same material. Monel metal is a well-known nickel-copper alloy widely employed for water-space stays in both copper and steel fireboxes. This is rendered possible by the absence of galvanic corrosion. In the case of copper boxes the Monel metal is sufficiently close to copper in the galvanic series to render it troublefree, whereas in the case of a steel box the area of steel is so much greater than that of the Monel metal employed that galvanic corrosion does not arise.

Stays may be seal-welded or the outer plate may be caulked around the stay; nuts are usually provided at the inside end and in the case of screwed stays to protect the stay end from the action and heat of the fire.

In the breaking zones of the firebox, i.e., those areas in which the greatest movement takes place and consequently the largest number of stays is likely to break, it is customary to fit flexible stays to accommodate the upwards and outwards movement of the inner box. These stays, which may also be solid or drilled for the greater part of their length, have a ball formed at one end which seats in a spherical socket either screwed into the outer plate or welded to it. A cover is provided which may take the form of a dome, welded on to the plate, or a plug screwed into the end of the housing. Variation in the method of attachment is met with, but the basic principle remains the same.

(vi) *Gusset stays*, manufactured from plate, are sometimes used to assist in tieing the backhead to the outside wrapper plate, or the front tube plate to the first course. In the former location they are also termed backhead braces in American practice.

(vii) *Palm stays* are used to give support to the firebox tube plate at a point between the lowest smoke-tubes and the uppermost water-space stays. Palm stays are steel forgings and are fixed to the bottom of the boiler shell by riveting, the connection to the firebox tube plate is made by means of direct-screwed stays or bolts.

(viii) Although not classed as stays, *the smoke-tubes and flues* have a considerable staying effect upon both the front tube plate and the firebox tube plate.

Steam. Steam is the medium through which the heat energy in the coal is converted into mechanical energy. It has the property of being exceedingly elastic which enables cut-off to take place early at high speeds and work to be obtained from it as it expands in the cylinder. At atmospheric pressure steam occupies some 1,700 times the volume of the water from which it was produced, or, expressed another way, one cubic inch of water will produce one cubic foot of steam. The volume is governed by the pressure; steam temperature is constant for any given pressure. The foregoing remarks apply to saturated steam.

Superheated steam is steam to which further heat has been added after its primary evaporation in the boiler, i.e., in the case of superheated steam the temperature is not dependent upon pressure alone. Three advantages are derived from superheating, the most important of which is economy. Compared with a saturated engine, a superheated one will produce economies on water of the order of 35 to 40 per cent, and on coal of 25 to 35 per cent. The power of the locomotive is increased as the volume of steam available is greater, and the steam is capable, compared with an equal volume of saturated steam, of producing more work due to its containing a greater amount of heat, no water, and not being subject to the initial condensation of saturated steam. These attributes have sometimes led designers of superheated locomotives to lower steam pressures and to enlarge the cylinders; such a procedure enables the same amount of work to be performed at less cost, and boiler maintenance is reduced, but it lowers the full effectiveness of the locomotive which would otherwise accrue.

The properties of steam, i.e., pressure, temperature, total heat B.T.U. per lb., volume, and density, are to be found in steam tables. In the case of tables relating to superheated steam the percentage increase in volume of superheated steam over saturated steam is given for various degrees of superheat, in addition to the particulars already enumerated (*see* Chapter 6, Cylinder Performance, page 392).

Steam brake. Steam brakes are frequently fitted on locomotives and tenders, the operating cylinders being single acting and taking steam at boiler pressure. Return of the piston is accomplished by a pull-off spring attached to the rigging. Due to the initial condensation which takes place, particularly when the equipment is cold, it is important that application should be made earlier than is customary with other types of brake.

On railways where it is usual to work vacuum-braked trains with locomotives fitted with steam brakes, it is usual to provide a vacuum-operated, graduable steam-brake valve which has the effect of controlling the application of the steam brake on the locomotive as the vacuum brake is applied on the train; similarly steam is exhausted from the steam brake cylinder as vacuum is restored in the train pipe. Even if the locomotive is fitted with steam brakes it will be necessary to provide it with an ejector to operate vacuum-braked trains.

Steam brakes have a feature in common with air brakes, i.e., a relatively small cylinder (usually eight inches to ten inches in diameter in the case of the steam brake) can be used. They have the advantage over vacuum brakes in that, if necessary, brake cylinders may be placed in close proximity to the firebox or ashpan. Deterioration of packing would rapidly occur in such conditions in the case of vacuum cylinders (Fig. 61).

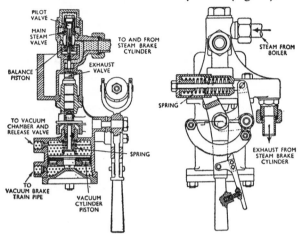

FIG. 61. Vacuum-operated graduable steam brake valve.

Steam Chest. The steam chest houses the valve and may or may not be integral with the cylinder; in the case of slide valve cylinders the chest is almost always cast with the cylinder, a separate cover being provided on the top, bottom, or side, according to whether the chest is above, below, or at the side of the cylinder. In the case of piston valve cylinders the chest is always integral save in the few instances where conversion from slide to piston valves has taken place. Where piston valves are concerned the steam chest is cylindrical, provided with a liner or liners, and has a cover at each end.

Steam collector, see Dome.

Steam drier. A device fitted at the steam space end of the dry – or internal main steam – pipe, to remove moisture from the steam. The principle upon which the tangential steam drier works is that of steam flowing at high speed being suddenly subjected to change of direction. The particles of moisture, being relatively heavy, continue in a comparatively straight direction and are precipitated and ultimately returned to the boiler.

Steam space. That portion of the boiler volume above the level of the surface of the water.

Stoker, mechanical. While mechanical stoking is frequently associated only with large grates its use is by no means confined to such applications, as substantial benefits are to be derived from the use of stokers, even if they are not rendered essential on the ground of large

steam output. The limit for hand firing is about 4,000 to 5,000 lb. per hour, and this amount can be sustained for a short time only. It is wrong to conclude that the main factor influencing the decision whether or not to fit a mechanical stoker is the grate area. The decision must be based on the sustained drawbar horse-power, which is the factor determining the firing rate. For this reason it may well be that hand-fired locomotives which have been able to perform satisfactorily the work they have been called upon to do must be fitted with stokers if the load is increased and/or the timing cut down. Such a situation is envisaged in Great Britain under the modernization plan, more especially in the case of speeded-up freight trains.

The Standard mechanical stoker has been selected for illustration. It may easily be fitted to existing locomotives designed originally for hand firing. The screw-to-jet feature is incorporated. Coal is carried from bunker, or tender, by the screw located in the bunker conveyor unit, which carries the fuel to the telescopic intermediate unit, located between the bunker, or tender, and the cab. From the intermediate unit it passes up the elevator pipe, which is rigidly attached to the backplate, and then is fed to the distribution table from which it is blown by steam jets to the desired parts of the firebed. Rotation of the conveyor screw is effected by a two-cylinder, double-acting reversible steam engine (Plate 64B, page 264).

A stoker-fired locomotive is never short of steam providing that its operation is properly controlled; it is erroneous to refer to a mechanical stoker as being automatic, for it is fully under the fireman's control. In addition to providing ample steam a mechanical stoker has other important advantages:

(i) the ability to use efficiently run-of-mine coal and coal of poor quality;

(ii) the greater capacity and utilization of the locomotive which are no longer limited by the fireman's physical ability;

(iii) the higher operating efficiency resulting from a thin fire and even firebox temperature which combine to reduce back pressure and firebox maintenance.

(iv) the fireman is relieved of manual work and is able to devote more time to maintaining a lookout.

Stretchers, see Frame.

Stroke. The distance traversed by the piston in the cylinder. Reference to the tractive effort formula will show that the force developed by the locomotive is directly proportional to the stroke. *Piston speed*, which may be plotted against mean effective pressure or indicated horsepower, is equal to:

r.p.m. × twice the length of the stroke in feet per minute.

It is unusual in orthodox locomotives for the piston to average more than 1,000 ft per minute, although it may rise to almost double that figure at very high speed (Fig. 62).

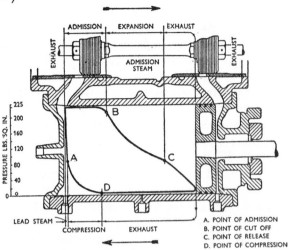

FIG. 62. The distribution of steam on one side of the piston for a double stroke.

Stuffing box. A form of steam-tight seal in which hemp rings soaked in tallow, or asbestos packings, were housed in a recess around a piston rod or valve spindle and tightened as necessary by screwing down nuts on the studs securing the gland to the cylinder cover or valve chest. Higher pressures rendered this method of maintaining steam-tightness obsolete, its place being taken by metallic packing.

Superheater. The purpose of the superheater is to add further heat to the steam after it has left the boiler steam space. The temperature of superheated steam may be 200° to 400° F. above the temperature of saturated steam, the final steam temperature may thus be as high as 850° F. Steam at this temperature is at the red heat of iron and so adequate and regular lubrication of valves and pistons assumes great importance.

Many forms of superheater have been tried (see Part II, page 316) but the firetube superheater is the only one which need now be considered.

The superheater consists of a cast header, and elements. The header, which is situated on the front tube plate, at the end of the main steam pipe, is divided into compartments some of which contain saturated, and others superheated, steam. Steam enters the saturated steam compartments from the boiler and passes into the elements. The steam returns from the elements to the compartments collecting superheated steam and passes thence to the cylinders. Headers sometimes incorporate multiple-valve regulators. The elements may be expanded into the header but more usually are connected

thereto by hemispherical seatings, machined in the header, into which the ball ends of the elements are clamped (Plate 57B, page 257).

Elements are constructed of solid-drawn weldless steel tubing, usually $1\frac{1}{2}$ in. outside diameter $\times$ 9 S.W.G., and are housed in flues of about $5\frac{1}{4}$ in. outside diameter. Whereas at one time elements consisted of a single loop, modern practice is to use four lengths of tubing connected by three forged return bends, providing for four passes of steam. The forged return bends are sometimes fitted with sheet metal protectors to prevent cinder cutting and it is customary to use rather shorter elements on oil-fired locomotives to keep the ends away from the intense heat. The formation of deposits on the bends will also result in overheating and consequently shortened life (see Steam and Chapter 6, Part I, pages 392–394).

Superheater dampers. Dampers have been fitted in the smoke-box to reduce the flow of hot gases through the flue tubes when the regulator is closed, in order to protect the elements from overheating. The dampers can be operated automatically, being held open by a steam cylinder taking steam from the main steam pipe, and hence closing whenever the regulator is shut; or they may be manually controlled. The fitting of superheater dampers has now been rendered unnecessary (i) by placing the regulator in the smokebox, between the superheated side of the header and the cylinders. This allows the elements always to be kept full of steam; (ii) by the provision of anti-vacuum valves, situated on the saturated side of the header, which enable air to circulate through the elements when the engine is running with the regulator closed.

Swing link bogie. A bogie depending for its side control on swing links. The weight on the bogie is carried through links, free to swing as pendulums, which always seek to return to their normal position. Each link is attached, by one or two pins at its upper end, to a cross member, and by one pin at its lower end, to the bogie. This arrangement can result in a twist being imparted to the main frame whenever the bogie is deflected and to overcome this two sets of links, arranged one above the other, with an intermediate member, have been employed, their angularities cancelling out. Swing links are also used on trucks.

Syphon, Nicholson thermic. (Plate 58A, page 258). The Nicholson thermic syphon is a flattened funnel, the upper edge of which is flanged and attached to the crown sheet, while the neck is connected to the lower portion of the tubeplate. The flange at the top is sufficiently wide to accommodate a row of crown stays on either side and the flat sides of the syphon are also fitted with stays.

While the principal function of syphons is to increase firebox heating surface they have the added attractions that they increase boiler efficiency by some 10 per cent; they may be used to support the brick arch, and the upward flow of water through the syphon greatly assists circulation and helps to protect the crown sheet in the event of low water. Cases are on record where the crown sheet has remained covered even when the water level had fallen several inches below it.

The number of syphons, and consequently the increase in the heating surface, will depend upon the size of the firebox; from one to three syphons may be used and if a combustion chamber is provided a second bank of syphons may be fitted therein. When Duplex thermic syphons are used the syphons have each two necks, one of these is usually attached to the tube plate and the other to a water leg. Should a Duplex syphon be used between two other syphons the necks are connected to the floor of the combustion chamber (Fig. 63).

Tail rod. A tail rod is an extension of the piston rod which is carried through the front cover. Today they are seldom used in Britain. The object is to support the weight of the piston on the rod with a view to reducing wear of both piston and cylinder. The tail rod is either carried in a bush, periodically partly rotated to accommodate wear, or alternatively a small crosshead, known as a tail rod guide, is provided.

Tandem main rod drive. The big end of the connecting rod in this drive is forked and a steel bush inserted which is the full width of the forked end. The coupling rod from the driving axle to the adjacent trailing axle is carried at its leading end on the bush. A concentric floating bush, inside the steel bush, transmits the drive to the main crank pin and carries the trailing end of the coupling rod to the axle ahead of the driving axle. This arrangement has the advantage that the main crank pin is relieved of a considerable proportion of the load it normally carries; the amount of relief will depend upon the number of coupled wheels, the position of the driving axle, etc., but in the case of a twelve-coupled locomotive can be as much as 50 per cent.

Tanks. Water tanks may be in the form of side, saddle, pannier, well or tender tanks; they may also be located in or under the rear bunker of tank locomotives. Tanks used for oil fuel are situated on top of the tender water tank or, in the case of a tank engine, in the space normally occupied by the rear bunker.

Construction may be by riveting or by welding. Normally steel plate is employed but there have been exceptional cases in which, due to restrictions of weight, aluminium has been used. It has the further advantage of being freer from corrosion than mild steel.

Equipment for feed water treatment may be provided in tender or side tanks, in cases where feed water treatment takes place on the locomotive (*see* Tenders).

Tenders. Tenders have much in common with locomotives where general construction is concerned, but this is somewhat lighter. For example, where plate frames are incorporated they have a thickness of about $\frac{7}{8}$ in. Roller bearing axle-boxes are widely used, these being in some instances of similar size and type to those used for locomotive carrying axles.

Steel castings are sometimes used for bogie frames and the tender frame. The tender frame may be an open bottom one, on which the tank is mounted, or it may be a water bottom one, the casting constituting both the frame and the tank bottom.

Tanks may be rectangular, cylindrical, or U-shaped; in the last-mentioned instance the tank may form the frame, having suitable mountings for the bogies attached to it, as in the German war locomotives.

Tender bogies do not require provision for translation and merely rotate about their pivots.

It is now common practice to provide equipment for treating feed water in the tender, a well-known example being the Armand system, T.I.A. (q.v.).

The coal space is usually on top of the tank and slopes downwards to the firing step, the latter being slightly

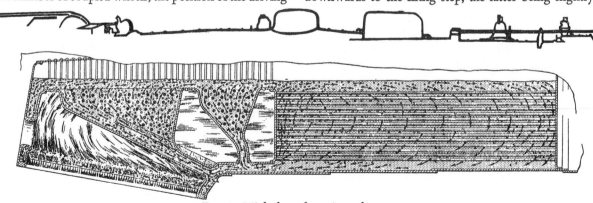

FIG. 63. Nicholson thermic syphons.

higher than the firehole. While the slope results in what is known as a "self-trimming" tender, the term is only a relative one, and sometimes a coal pusher (q.v.) is fitted. The type of tender in which the coal space is U-shaped, the legs of the U being part of the water tank, with no water below the coal space, is now seldom met with.

Many tenders are provided with a well tank between the frames; apart from increasing the water capacity this improves the riding when the tender is partly empty.

Equipment for picking up water (see Pick-Up Gear) is often fitted.

Gauges are usually attached to show the water level. Many modern locomotives are fitted with shut-down valves which enable the tender strainers to be removed for cleaning, after the valve is closed, and without the necessity of draining the tank.

The brake equipment of a tender is generally similar to that of a locomotive, but it is customary to incorporate a hand-operated brake in addition.

Test cocks. These cocks, also referred to as try cocks, are sometimes fitted in order that the water level may be ascertained approximately in the event of the water gauges failing. The lower cock is fitted a few inches higher than the crown sheet. The difference between water or steam issuing from the cocks is most easily distinguished by ear.

Thermal efficiency. This is sometimes called the overall efficiency and is the heat equivalent of the work done relative to the heat supplied. As the efficiency is expressed as a percentage it equals:

$$\frac{\text{B.T.U. per hour equivalent to 1 H.P.}}{\text{Actual B.T.U. of coal per d.b.h.p.hr.}} \times 100$$

The mechanical equivalent of heat is 778 ft lb. per B.T.U. and 1 H.P. equals 2,545 B.T.U. per hour.

The overall efficiency is determined by the boiler efficiency, the cylinder efficiency, and the machine efficiency.

Boiler efficiency will vary with the efficiency of the grate, superheater, and boiler arrangements generally, and will be dependent upon the rate of combustion. It is represented by the total heat transferred divided by the thermal value (B.T.U.) of the coal fired, both expressed on an hourly basis. The boiler efficiency of a modern locomotive may be of the order of 80 per cent. It will be understood that in the event of the locomotive being fitted with a feed water heater the heat added by this per hour must be added when calculating the efficiency.

Cylinder efficiency. The steam consumption per I.H.P. hr is a measurement of cylinder efficiency. It is usually found by ascertaining the indicated horse power and the amount of steam supplied to the cylinders (arrived at by measuring the amount of water fed to the boiler), deduc-

ting the steam requirements of auxiliaries, and dividing I.H.P. by the steam consumption. Cylinder efficiency is seldom above 12 per cent and more usually about 8 per cent. On occasion it has risen to about 15·75 per cent, as on the British Railways Class 8 locomotive *Duke of Gloucester*, fitted with poppet valve gear, but this is exceptional. The cylinder efficiency of a locomotive is low compared with that obtained in other practice. This is primarily due to the limited initial pressure and the absence of a condenser. The Stephenson form of boiler does not readily lend itself to pressing beyond 300 p.s.i., and attempts to use steam at increased pressure, derived from other types of boilers, have resulted in the thermal advantages being more than offset by increased boiler complications. Condensers have been used but not normally with a view to raising thermal efficiency; the subject is referred to in Part II, page 314.

Machine efficiency. This efficiency, also termed the mechanical efficiency, expressed as a percentage, is obtained by dividing the drawbar horse-power by the indicated horse-power and multiplying the result by 100. The difference between the drawbar horse-power and the indicated horse-power is the amount of power absorbed in internal friction by the engine, and tender where fitted, plus the work done in overcoming air resistance.

The three efficiencies referred to, combine to produce a thermal, or overall, efficiency of from about 3 to 12 per cent. The former figure would apply to a locomotive with high standby losses, while the latter figure relates to a locomotive in continuous operation under optimum conditions. It is a mistake to judge the steam locomotive as a thermo-dynamical machine, particularly so in the days when fuel was considerably cheaper than it is now. While thermally the performance may be poor it is an excellent traffic machine which should be one of the determining factors when reviewing motive power (see Pressure, Mean Effective).

Thermic syphon, see Syphon.

Throat plate. The plate at the front of the outer firebox uniting it to the boiler barrel.

Throttle. The American term for the regulator.

Top feed. In modern practice it is customary to introduce feed water into the boiler through check valves fitted on the top of the barrel. It is arguable that from the point of view of circulation a better place for it to enter would be at the bottom of the boiler barrel, but this would be an impossible position for the maintenance of check valves. When top feed is employed, trays can be introduced into the steam space below the delivery pipes, and scale deposited upon these may be periodically cleaned out. Delivery through the steam space also

Stan Kistler, Pasadena, California

PLATE IX. Southern Pacific Railroad: 4–8–4, No. 4455, Class GS–4. These locomotives were among the finest of their type ever built, and were used to haul the Southern Pacific Daylight Expresses, often said to have been the most beautiful trains in the world.

facilitates the dispersal of any excess oxygen sometimes present in the feed.

Torque, *see* Turning effort.

Tractive effort. Tractive effort, tractive power, and tractive force are to a large extent used synonymously, but strictly speaking tractive effort and tractive force only apply to the effort which can be exerted in moving the train from a state of rest. Power, which is the rate at which work is performed, involves a time factor and for this reason a locomotive hauling a train of any given weight, and on a given gradient, will produce a tractive power varying with speed.

Formulae exist for arriving at the tractive effort of various forms of compound locomotives and also locomotives fitted with boosters (*see* Pressure, Mean Effective, *and* Pull, Drawbar, *also* Chapter 6, Cylinder Performance and Traction Relations, pages 392 and 408).

Travel, valve. Long-travel valves have been employed since early in this century and their use has been universal for all new construction for many years. The long travel is necessitated by the use of long laps. Piston valves are much better suited to long-travel than are slide valves, due to their reduced friction. Long-travel valves usually have about 7 in. movement but up to about 9 in. is sometimes met with. Above $8\frac{1}{2}$ in. difficulty is encountered due to excessive angling of the link; this trouble is pronounced with Walschaerts gear and difficulties also arise with Baker gear.

Valve travel is naturally dependent upon the rate of cut-off. For example an engine with $7\frac{3}{4}$in. travel at 80 per cent cut-off has $5\frac{1}{32}$ in. at 50 per cent, $4\frac{1}{16}$ in. at 20 per cent, and $3\frac{7}{8}$ in. at 7 per cent.

The maximum travel of both slide and piston valves is normally twice the steam lap plus twice the port opening. There are instances where the valve, even at full travel (i.e., in full gear), never fully uncovers the steam port; in such cases the provision of the uncovered width of the port is to give extra area for the steam when exhaust takes place.

Minimum travel is not less than twice the lap plus twice the lead. The latter, in the case of an engine with a gear giving variable lead, will be measured in mid-gear (*see* Chapter 7, Valve Design, page 420).

Trick ports. Trick ports, used in both slide and piston valves, have the effect of increasing the port area open to steam; they consequently permit of a shorter valve travel being employed for any given opening.

Trofinoff piston valve. An inside-admission piston valve in which the two heads are free to slide on the spindle. When the regulator is open the heads are held by steam pressure against collars on the valve spindle and the valve functions normally. When the regulator is

closed the two heads slide towards each other and automatic by-passing takes place. Carbonized oil on the valve spindle seriously interferes with the operation of this form of piston valve.

Truck. A truck is provided to spread the load, and the weight it carries has no adhesive value unless the truck is fitted with a booster which device has, with very few exceptions, been applied only to trailing and tender trucks. Locomotive trucks normally have two or four wheels but six-wheeled trucks have sometimes been used, as in the Pennsylvania Railroad S-1 Class, where the 6-4-4-6 wheel arrangement embodied both leading and trailing trucks of this type. For tenders four, six, and eight-wheeled trucks are used in America.

Trucks may have cast-steel frames or the frame may be constructed of plate and angle. In recent years the use of roller bearing axle-boxes on trucks has become widespread. In the case of tenders, trucks are merely pivoted, but in the case of locomotives, side movement is also provided for. Side movement may be in the form of:

  (i) radial movement;
  (ii) swing link;
  (iii) sliding, with centring controlled:
    (*a*) by coil;
    (*b*) laminated springs;
    (*c*) by weight acting through inclined planes, as in the Cartazzi axle;
    (*d*) through swing links.

In instances where the axle-boxes are provided with movement in the horizontal plane, relative to the frame, it is not necessary to allow also for movement of the truck framing relative to the locomotive main frame.

A pony truck is a single-axle one, having provision for swivelling, and used at the front end.

In American practice it is customary to refer to a leading truck as an engine truck; a trailing truck is so called or alternatively designated a rear truck or trailer (*see* Radial Truck) (Plate 63A, page 263).

Trunk guide. In place of the usual valve spindle guide the valve spindle may be increased in diameter to form a trunk guide.

Trunk type piston-valve. A piston valve consisting of two heads joined by a cylindrical trunk, through which steam can pass in certain valve and steam-chest arrangements.

Tubes, *see* Boiler.

Turning effort. The turning effort of a steam locomotive of the Stephenson form, i.e., one where the drive is direct from the piston to the crank pin without the intervention of gearing, varies through wide limits in each revolution. In addition to the variation arising from the position of the crankpin, and hence its effective length as a lever, torque varies according to the force exerted on the piston and therefore is dependent upon the rate of cut-off and the mean effective pressure. A torque diagram, or crank effort diagram as it is variously termed, is usually prepared from indicator cards, or in their absence, from a Zeuner diagram.

The effort is zero when the crankpins are on the back and front centres and at its maximum near the top and bottom centres. To obtain the mean torque it is necessary to superimpose the diagrams got from the number of cylinders with which the locomotive is provided. The humps on the diagram obtained for the mean torque will considerably exceed – probably by about 50 per cent – the valleys on the diagram, and this clearly demonstrates a potential source of slipping in steam locomotives.

Turret. The turret, also termed the steam stand, manifold, or steam fountain, provides steam for a variety of purposes; among fittings usually obtaining steam from this source are injectors, ejector, steam brake, carriage warming apparatus, sand gun, and the sight-feed lubricator. Important advantages deriving from the use of a turret are that the steam supply to the auxiliaries it serves necessitates cutting but one hole in the boiler, with the provision of only one pad. As it is usual to provide a stop valve isolating the steam supply to the turret it is also possible to give attention to any of the fittings concerned when the boiler is in steam.

Turrets may be located inside or outside the cab. In the former position they have the advantages that radiation loss is less and, if necessary, some attention may be given to the fittings when the engine is in motion. The external position has the attractions that the cab remains cooler whilst accessibility is much greater when the engine is stopped.

Tyres. Tyres are now almost always shrunk on to wheel centres, but may be fastened by bolts, rivets, or by retaining rings. In one design the wheel centre is machined with a lip on either side of the tyre seat which allows a wider area of contact on the wheel centre than is possible when retaining rings are used.

Until recently an enormous number of tyre profiles, with most variation occurring in the shape and dimensions of the flange, were in use. In recent years much has been done to standardize these. The steels used form the subject of Standard Specifications and include high-tensile alloy-steel to take advantage of the wearing qualities.

The fitting of bolts radially through the wheel centre and tyre has been discontinued. Theoretically in the event of a tyre breaking the pieces were retained on the centre by the bolts. In practice, cracking generally arose through the bolt holes. With tyres secured by shrinkage alone no trouble normally arises from the tyre slipping on the rim. This method of attachment may, however, prove unsatisfactory on long descents, where heavy and continuous brake applications are required, resulting in the tyres becoming overheated.

Flangeless tyres, or tyres with thinned flanges, are used when the wheelbase, considered in relation to the curvature of the track, calls for some easing when negotiating curves. When the flange is omitted and the tyre has the tread extending to its full width and beyond, the tyre is said to be "blind".

The tread is usually coned, being slightly less in diameter at its outer edge. This coning approximates to the angle of inclination on the railhead and keeps the wheels central with the track and the flanges out of contact with the railhead on tangent track. To an approximate extent it also provides a differential effect when negotiating curves.

To restore worn treads, and particularly flanges, to their original contour, re-turning is necessary from time to time. There has been a limited application of devices to restore tyre profiles without dropping the axle. In the case of steam locomotives this has been done by toolholders, containing stationary tools affixed to the track, or to brake hangers, the wheels being rotated as necessary. In the case of other stock, rotating cutters have been employed.

There is a limit to the safe radial thickness of the tyre; this is often indicated by a groove machined on the outside face to show the minimum size to which re-turning may be carried.

Flange forces are difficult to measure, are not uniform, and vary with the wheel arrangement, the weight of the locomotive, the control arrangement for side play, the speed, and with other factors. In the case of a rigid-frame locomotive with coupled wheels leading, the forces at the leading wheels will be of a very high order as there is no "build up", as is the case with bogies and leading trucks. As soon as the initial clearances have been taken up, the whole mass of the locomotive will bear against the rail through the leading flange. Flange forces of the order of twelve tons have been recorded and they may well be much higher in certain circumstances.

Vacuum brake (Figs. 64, 65, 66). The vacuum brake depends for its power upon atmospheric pressure, which measured at sea level is 14·7 p.s.i. In practice there are limits to the amount of vacuum which can be created and maintained; 14·7 p.s.i. is equivalent to thirty inches of mercury, but in practice, a partial vacuum of between twenty to twenty-four inches is used. The power available for braking is, therefore, about 10 to 12 p.s.i.; the higher figure is exceptional, and it is usual to calculate on the basis of the lower. It will be noted that the pressure in

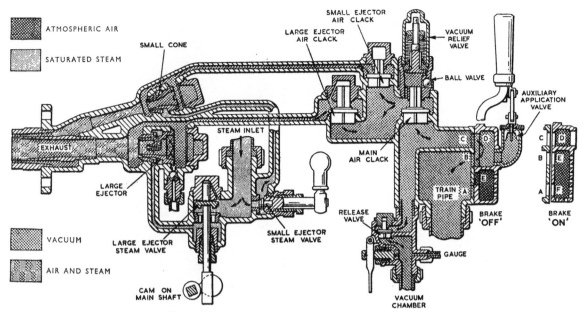

FIG. 64. "Dreadnought" vacuum ejector and driver's brake valve.

p.s.i. available in the cylinder may be taken as half the vacuum in inches of mercury. The force developed in the cylinder is greatly increased by the levers through which it is connected to the brake rigging.

Vacuum is created on a steam locomotive by an ejector and on diesel and electric locomotives by an exhauster, which may be belt driven from the diesel engine, or may be electrically operated. In the case of steam locomotives vacuum may be maintained, when running, by a pump driven off the crosshead.

Ejectors are commonly of the combination type, incorporating a large ejector for creating initial vacuum and for rapidly releasing the brake, a small ejector to maintain vacuum and which is continuously in use when running, and the driver's application valve.

The driver's control handle has three positions: "off", "running" and "on".

(i) In the "off" position the large ejector is in operation together with the small ejector, to create rapidly vacuum in the train pipe and in the connected reservoirs, or chambers.

(ii) In the "running" position the small ejector remains in operation but the large ejector is cut off.

(iii) In the "on" position air enters the brake pipe, the degree of application depending upon the amount of vacuum destroyed.

With some layouts, placing the handle in the "on" position also places the small ejector in communication with the vacuum chambers. This has the combined advantage of maintaining brake power and reducing the size of the chambers.

Separate ejectors are frequently used, the ejector being mounted outside the cab or even on the smokebox side. In such an arrangement the "running" position of the driver's valve is replaced by two stop valves, controlling steam to the large and small ejectors respectively, and the control valve merely has "on" and "off" positions, although, of course, application may be made progressively.

Vacuum brake cylinders are now always mounted on trunnions, although this has not always been the practice. When vacuum is created the top side of the piston is exhausted through the ball valve provided, while the underside is exhausted direct. Applying the brake, by admitting air to the train pipe, results in admitting air to the underside of the piston, but the presence of the ball valve retains vacuum above the piston, which is forced upwards by the atmospheric pressure.

The brake is released by restoring a state of equilibrium in the cylinder on both sides of the piston. This may be done by re-creating vacuum, or, if the engine is dead, by admitting air on both sides.

A rolling ring or band, interposed between the piston and the cylinder wall, allows freedom of movement while at the same time keeping air tightness. A packing ring prevents air leakage up the piston rod.

The equipment fitted to the train differs from that on the locomotive in that the type of cylinder employed is often of the combined type, in which the vacuum chamber is incorporated with the cylinder. It is also the practice to fit braked vehicles with direct admission valves. These admit air to the brake cylinders throughout the train in exact relation to the reduction made in the

293

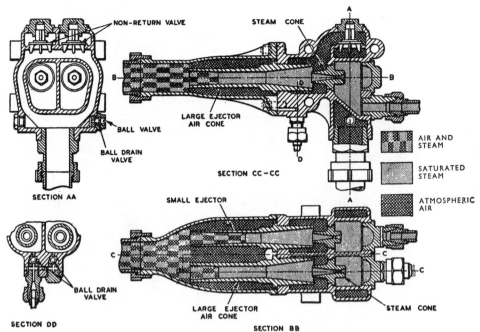

FIG. 65. S.S.J. Ejector for vacuum brakes.

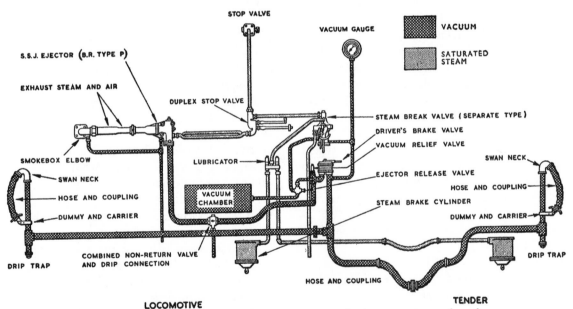

FIG. 66. General arrangement of vacuum automatic brake on engine and tender.

train pipe vacuum. The admission of air locally reduces the time which would otherwise elapse if air were only admitted by the driver or guard and had to travel along the train.

A special driver's brake valve is sometimes incorporated which gives independent control of the brakes on the locomotive; this enables the brake on the latter to be held on while permitting the train brakes to be released (*see* Ejector, *and* Pump, Vacuum).

Valve events (Figs. 62, 67). The points of admission, cut off, release, and compression constitute the valve events. Their correct occurrence is of great importance, for upon them the cylinder performance will largely depend. They are determined by the design of the valve gear in conjunction with the rate of cut-off. Except in the case of rotary gears for operating poppet valves, and gears of the Berthe, Bonneford and Marshall types, the valve events are always interdependent.

Valves for steam distribution (Figs. 68, 73). The admission of steam to, and its exhaustion from, cylinders may be controlled by slide, piston, or poppet valves. Rotary valves, or those having a partially rotating movement, have had such small application that they need not be considered here.

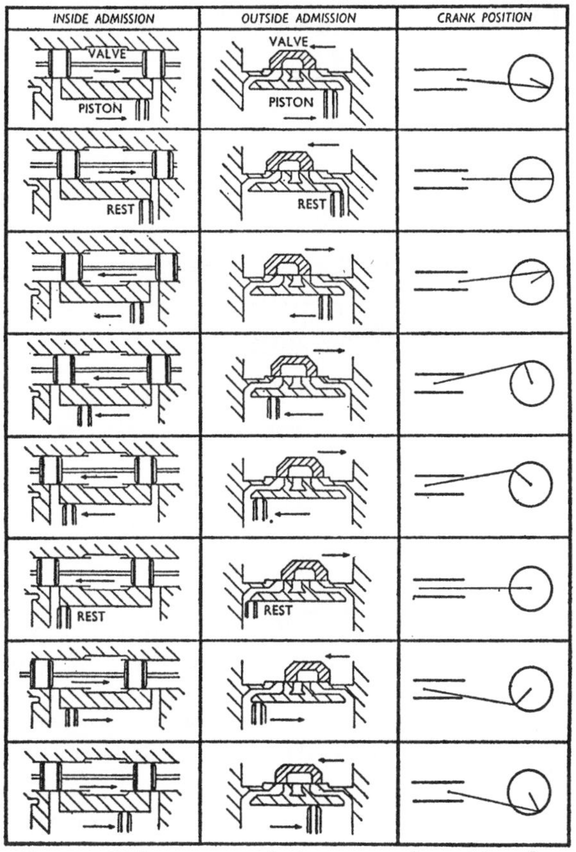

FIG. 67. Valve events relative to crank position for one revolution.

Slide valves. In the simple "D" box form the slide valve is the simplest type of valve to manufacture. Generally speaking the valve distributes steam to the two inlet ports and one exhaust port; exceptionally, two valves have been provided on a common spindle, each valve controlling steam flow through ports (one inlet and one exhaust) close to the cylinder end. Such an arrangement has the merit of keeping the steam passages short, which cannot be done with one centrally situated slide valve. Slide valves necessitate the use of packing on the valve spindle, which is always subjected to high pressure where it passes from the steam chest; they have the added disadvantage, in the simple form, that they have high friction, whenever the regulator is open, and furthermore, they are difficult to lubricate adequately when temperatures are high. For this reason they are not suitable for use with superheated steam.

In an attempt to overcome the high friction of the normal slide valve, balanced slide valves were introduced; in these the back of the valve is relieved of steam pressure by means of a plate above the valve. Spring-loaded packing strips are inserted in the valve to prevent the entry of steam between the plate and the valve back.

The slide valve, in both its balanced and unbalanced forms, is able to move away from the port face sufficiently to release accumulation of water from the cylinder. In its simple form, when placed at the side of, or below the cylinder, it also has the merit of leaving the seat sufficiently to permit of a by-pass action taking place, when the engine is running with the regulator closed.

In conjunction with simple D slide valves, trick ports have sometimes been provided. The provision of trick ports has the effect of increasing the port opening, but they may be rendered ineffective, or partly so, by carbonized oil obstructing the passages.

Piston Valves. A piston valve consists of two pistons, or heads as they are more usually termed, on a common spindle. Piston valves are usually of inside admission type, the live steam entering the valve chamber between the two heads; thus the pressure acting on the two inner faces of the valve heads places the valve in balance. The valve gear is, therefore, only called upon to impart to the valve spindle the force required to overcome friction. This is only about one-sixth of the work needed to actuate an unbalanced slide valve on an engine of similar size.

The amount of steam which a piston valve can pass at any given pressure will depend upon its diameter, travel, shape and area of ports, etc.

Various types of piston valve are in use (*see* Trunk valve, Trofinoff valve, *and* Willoteaux valve). Trick ports are sometimes employed, with the same aim and possible difficulty as that experienced with these ports on slide valves. In recent years the rings used on piston valves have been fewer in number and narrower than was formerly the case. Originally plug valves without rings were used, but the steam leakage which could occur was considerable. The steam edge of a piston valve may be on the head itself, or the head may be so machined that it presents the edge of the ring as the steam edge. In the latter instance the ring viewed in cross-section may be rectangular or L-shaped; where the last-mentioned rings are employed two are provided, arranged back to back, with a wide spacer ring between them.

Poppet valves. Considerable advantages result from the use of poppet valves, and they have been widely used. The valves may be arranged vertically or horizontally. As double-beat valves are always used, they are in near-equilibrium and require comparatively little power to

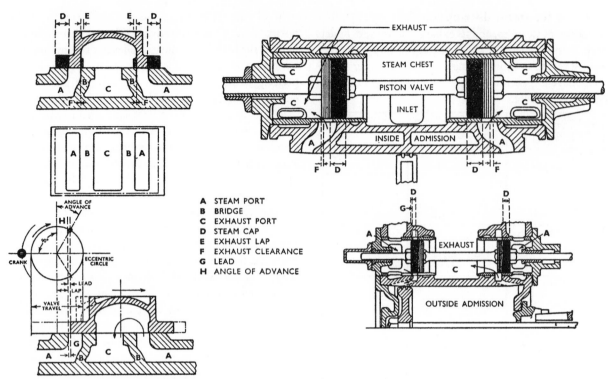

A STEAM PORT
B BRIDGE
C EXHAUST PORT
D STEAM CAP
E EXHAUST LAP
F EXHAUST CLEARANCE
G LEAD
H ANGLE OF ADVANCE

FIG. 68. Sections of slide valve (left) and of piston valve (right).

operate, enabling the components of the valve gear to be lightly constructed. The valves, for a given movement, provide a much greater opening than other types of valve. As the valve lifts from the seating, the lubrication required by the valve is limited to the spindle. The valves remain steam-tight over long periods and require little attention, having no rings. With any other type of valve, exhaust opening is always related to remaining valve events; with poppet valves it is possible for the exhaust events to be dissociated from other valve events and for the exhaust always to open fully. The steam valves may be held on their seatings by steam pressure, in which case when the regulator is closed, the valves will move to the fully-open position and remain there, providing excellent by-passing. Normally two inlet and two exhaust valves are provided for each cylinder.

Some of the advantages of poppet valve arrangements are derived from the particular type of gear used to operate the valves (*see* Valve Gears *and* Chapter 7, Valve Design, page 314).

Valve gears. Many types of valve gear have been evolved to impart the desired movement to slide valves, piston valves and poppet valves.

For some years past, so far as engines with slide and piston valves are concerned, new construction has been limited to Stephenson gear, Walschaerts gear, and variations of the latter. Gears such as Allan's and Joy's are still

to be found in use on locomotives built some time ago. It is not sufficient that a gear should provide good distribution; additionally it must be easily fitted in, be easy to maintain, and must not throw strains on other parts of the engine. The last-mentioned factor was an important consideration where Joy's gear was concerned, the drive being derived from a point on the connecting rod which was subjected to considerable loading. Valve gear providing reciprocating motion has been applied to the use of oscillating cam poppet valves; modern applications of poppet valves are of the rotary cam type, the motion for which is derived from a specially designed gear.

Valve gears may be classified under three main headings, link motions, radial gears, and poppet valve gears. The best known link motions are the Stephenson, Allan, and Gooch gears but only the first-mentioned of these has had any large-scale use in the last fifty years. Of radial gears easily the most extensively employed is Walschaerts and many other gears which are generally looked upon as complete entities are variations of this gear; in fact only two radial basic gears which had any appreciable application did not fall into the category of Walschaerts variants. These were the Strong gear and the Joy gear.

STEPHENSON VALVE GEAR (Fig. 69, and Plate 61B, page 261) is operated by two eccentrics or return cranks for each cylinder; it is met with in a number of variations. The links may be of the launch type, the rods should be "open" but are sometimes "crossed", and the drive from the die block to the valve spindle may be direct or through the medium of a rocking shaft. A feature of the gear, when the rods are open, is the increase in lead which takes place as the gear is notched up, due to the control of the back gear and fore gear eccentrics varying according to the position of the die in the link. This increase has value at high speeds. Should the rods be crossed the lead in similar circumstances will decrease.

The layout and construction of the gear is clearly shown in Fig. 69. It will be noted that reversing and variation in rates of cut-off are obtained by moving the link relative to the die block.

Disadvantages of Stephenson's gear are the considerable weight and the excessive axle rigidity resulting when four eccentrics are applied on an axle.

In the case of an engine with outside admission and direct drive, the eccentrics are keyed on the axle at an angle of 90°, plus the angle of advance, ahead of the crank. This setting applies to each direction of travel. When inside admission is employed, assuming direct drive, the angle of the eccentrics, relative to the crank, will be 90°, less the angle of advance, behind the crank. When, on an inside admission engine, the drive is indirect, due to the provision of a rocking shaft, the eccentrics will be set as for outside admission.

Some of the other link motions which have found application in locomotive practice are:

Alfree. A Stephenson link motion with an additional movement derived from the crosshead. A rocking shaft is employed which is connected to the valve rod by an eccentric pin. A pinion, mounted on the pin end, mates with a segmented rack on the rocking shaft, the rack being actuated by a rod from the crosshead.

Allan. The straight link of this motion is connected to back and fore gear eccentrics, as in Stephenson gear. Whereas in the latter gear the die block remains stationary in the vertical plane, in Allan's motion, when reversing or notching up, both block and link move in opposite directions. Steam distribution is good and the increase in lead which takes place on notching up is less than that with Stephenson gear with open rods. An objection to this gear is the additional joints required.

Anderson. A form of Stephenson gear in which the movement normally derived from eccentrics is replaced by that from a return crank fitted with double pins.

Gooch. In this gear the link is arranged with its concave side towards the valve-chest and is stationary in the vertical plane, the die block at one end of the radius-rod or valve-rod being moved as necessary. Two eccentrics impart motion to the link which, to minimize slip, is usually suspended with its centre line rather below the centre line of the motion, with the back gear eccentric rod lengthened and the angular advance of the eccentric concerned reduced. Due to the angularity of the eccentric and valve rods, the gear suffers from indirect thrusts.

Marshall. J.T.Marshall's gear was a double eccentric one with the link arranged as in Gooch's motion, the valve rod being movable vertically. One eccentric, at

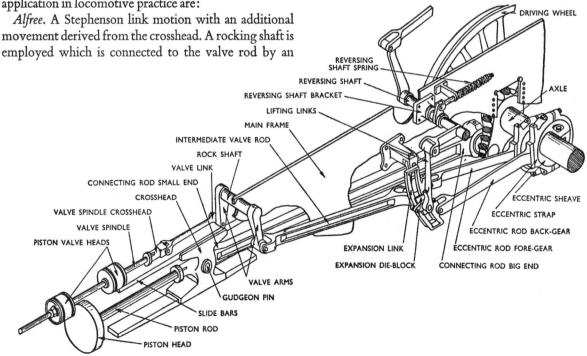

FIG. 69. Stephenson valve gear with outside admission piston valves, direct motion.

180° to the crankpin, was connected to the centre of the link and provided lap and lead. The second eccentric, which followed the crank by about 90°, operated a bell-crank the function of which was to oscillate the link. The distribution was, in some respects, similar to Walschaerts gear.

WALSCHAERTS GEAR (Plate 61A, page 261) is relatively light and is so arranged that each component has but one duty to perform, e.g., lap and lead are derived from the combining lever, the eccentrics or return cranks imparting the steam port opening only, This gear has had a very wide application since the beginning of this century and has appeared in various guises.

Fig. 70 shows the gear arranged for inside admission: when applied to a locomotive with outside admission the radius rod is connected to the combining lever at a point below the connection of the valve spindle.

Reversing and variation in the rate of cut-off are obtained by moving the die block relative to the link which swings on trunnions mounted in fixed bearings. Due to the return crank being 90° out of phase with the crosshead, the link will be approximately in the centre of the arc through which it swings when the crankpin is on the back and front centres. As the combining lever, which imparts lead, is entirely dependent for this function on the position of the crosshead, it follows that the lead will always be constant, irrespective of the rate of cut-off.

When inside admission is employed the return crank, or eccentric, follows 90° behind the crank; in the case of outside admission the settings will be 90° ahead of the crank.

Some of the modifications of the Walschaerts radial valve gear which have found application in locomotive practice are:

Baguley. A gear with some features of Walschaerts gear but deriving movement from the main crank pin. A system of levers superimposes lap, lead, and valve travel, lap and lead being obtained from horizontal displacement of the link trunnion which is carried in an eccentric bush.

Baker. Introduced in America, this gear is very similar to Walschaerts except that the link is replaced by a crank-shaped rocker. As no sliding surfaces are used, restoration to new condition is possible by fitting new pins and bushes. It has the disadvantage of requiring more space laterally than Walschaerts gear.

Beames. A Walschaerts form of gear, applicable to locomotives with inside cylinders, in which the combining lever derives the movement, normally obtained from the crosshead, from a forward extension of a coupling rod. A rocking lever transfers the movement from the top of the combining lever to the valve rod. Obtaining movement for the combining lever from the coupling rod had previously been done by Golsdorf.

Berthe. A form of Walschaerts gear having two separate links for admission and exhaust piston valves respectively. This makes provision for independent notching-up.

Bulleid. A Walschaerts gear in which the motion to rock the link is obtained from a chain-driven counter-shaft. A connection from this shaft also actuates the combining lever (Fig. 71).

Deeley. A form of Walschaerts gear in which the link movement is obtained from the crosshead on the other side of the engine. The combining lever is of the usual Walschaerts arrangement. A feature of such cross-driven valve gears is that the engine will be completely disabled if it is necessary to disconnect one side.

Fidler. A Walschaerts type of gear. A horizontal arm, forming part of the link, is moved in the vertical plane by a return crank set at 180° to the main crankpin. A double lever is mounted on the valve spindle, in place of the usual combining lever, and the radius rod is connected to the top of this, while the bottom connection is made to a rod pinned to an intermediate point on the driving rod from the return crank.

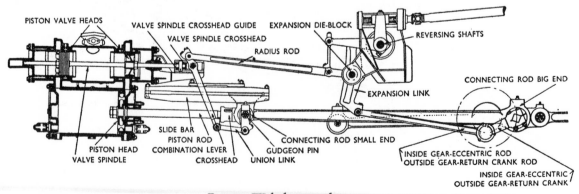

FIG. 70. Walschaerts valve gear.

Jones. A Walschaerts gear having an additional link fitted between the combining lever and the valve rod. The function of this link, which is connected to the reversing lever, is to give variable lead.

Kingan-Ripken. A form of Walschaerts gear in which the crosshead connection to the combining lever is dispensed with, and a connection made instead to the connecting rod, at a point near the small end. The claim made for this form of gear is that it allows release, compression, and lead to take place later for any given cut-off than would otherwise be the case.

Kitson. A Walschaerts gear in which the link derives its movement from the coupling rod.

Pilliod. A modification of Walschaerts gear in which all sliding components were dispensed with, pin joints only being employed. The necessary movements, as in Walschaerts gear, were derived from a return crank and the crosshead.

Stephenson-Molyneux. A Walschaerts gear in which the difficulty of arranging the combining lever for the inside set of motion of a three-cylinder locomotive was overcome by arranging the lever either horizontally or at an angle.

Young. A cross-connected Walschaerts gear. The link is driven from the crosshead and the combining lever receives its movement from a connection to the top of the link on the opposite side of the engine.

RADIAL GEARS, OTHER THAN WALSCHAERTS, have been very widely used in locomotive practice. Originating with Hackworth, and still being in use, they have covered the life of the steam locomotive.

For the last decades, however, their use in new construction has been very limited. While various radial gears were patented, the two best known are Joy's and the Southern gear.

Joy. While the basis of this gear was Hackworth's, important differences were the use of a curved slide by Joy, and the provision of correcting motion. In the days when inside cylinders were the rule, Joy's gear had the great attraction that, as the movement was derived from a point on the connecting rod, it was possible to dispense with the eccentrics and the space so saved permitted journal and crank-web dimensions to be increased.

The disuse of the gear was due to connecting rods fracturing through the pin-hole from which the drive to the gear was taken. A characteristic of gears driven from such a point is that the valve setting is influenced by the fluctuating height of the framing relative to the rail, and hence distribution was to some extent influenced by the camber of the springs and the amount of water in the boiler. This susceptibility resulted in some locomotives equipped with Joy's gear having especially stiff springs to cut down the vertical movement of the engine.

Southern. The Southern gear was the last form of Hackworth gear to be used and was virtually the Strong gear reintroduced on the Southern Railway of America. A fixed horizontal link was incorporated in connection with a vertical radius bar. Movement to the valve spindle was imparted by a bell crank, being initially obtained from a return crank in the Southern form and an

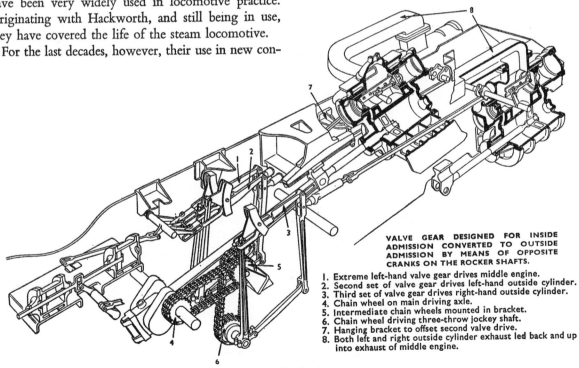

VALVE GEAR DESIGNED FOR INSIDE ADMISSION CONVERTED TO OUTSIDE ADMISSION BY MEANS OF OPPOSITE CRANKS ON THE ROCKER SHAFTS.

1. Extreme left-hand valve gear drives middle engine.
2. Second set of valve gear drives left-hand outside cylinder.
3. Third set of valve gear drives right-hand outside cylinder.
4. Chain wheel on main driving axle.
5. Intermediate chain wheels mounted in bracket.
6. Chain wheel driving three-throw jockey shaft.
7. Hanging bracket to offset second valve drive.
8. Both left and right outside cylinder exhaust led back and up into exhaust of middle engine.

FIG. 71. Bulleid valve gear.

eccentric in the Strong version. The distribution which this gear gave was good and the lead was constant, as is always the case with gears having a fixed link. The failing, which caused it to disappear, was lateral weakness.

CONJUGATED VALVE GEARS are based on the principle, enunciated by H.Holcroft, designer of the first such gear to be fitted to a locomotive in Britain, that if the harmonics of two valve gears of different phases are combined, a third harmonic is produced of another phase. The savings in weight resulting from the use of such a gear are considerable.

Gresley (Fig. 72 and Plate 59B, page 259). Applicable only to three-cylinder locomotives this gear obtains its movement from extensions of the two outside valve spindles, one of which is connected to a 2:1 lever joined at its other end to a floating lever, the ends of which are pinned to the valve spindles of the other outside and the inside cylinder. Locomotives fitted with this gear have performed magnificent work, including attaining a speed of 126 miles per hour. Gresley's gear is susceptible to poor maintenance, any play being both cumulative and subject to multiplication. A further peculiarity is that spring in the levers, added to any slackness, results in overtravel of the inside valve, particularly at high speed. In this connection it may be noted that when applied to engines in North America the maximum speed recommended was 35 m.p.h. Although usually attributed to Gresley, valve gear of similar layout was fitted to express locomotives of the Prussian State Railway at least eight years before it was applied in Britain.

Holcroft. This gear may be applied to three-cylinder locomotives, including those with unequal crank spacing, or may be adapted, by an additional arm on the floating lever, to four-cylinder engines with the inside and outside cranks spaced at 135°. In addition to effecting a considerable saving of weight this gear, compared with other conjugate gears having a 2:1 ratio, has the advantage that the levers may be kept short and the pivots carried on the frames. These features eliminate the cross-stay otherwise necessary to carry the pivot and considerably reduce the likelihood of whip occurring in the cross lever.

POPPET VALVE GEARS basically take two forms, those in which the cams oscillate and those in which the cams rotate. The cams are sometimes in contact with the valve spindles and on other occasions the movement is imparted to the valves through levers. The early oscillating cam poppet valve gears were operated by conventional Stephenson or Walschaerts gear and the cylinder performance could not be expected to be much higher that that normally attained with such gears, with the valve events remaining inter-related. Furthermore, the weight of the gear and the power which it absorbed remained constant, the only saving resulted from lighter valves. The introduction of rotary cam gears enabled independent timings to be given to inlet and exhaust valves and a big reduction to be made in the weight of gear and

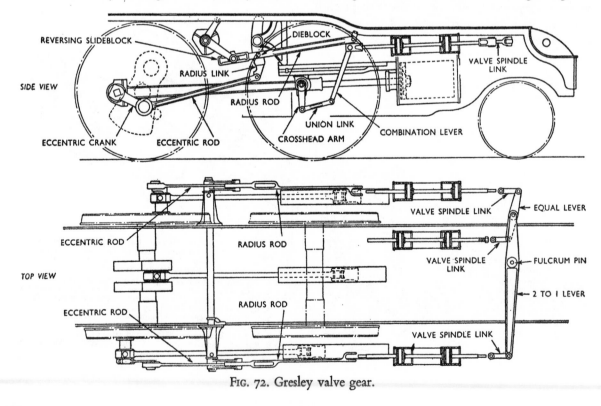

FIG. 72. Gresley valve gear.

the power required to drive it; this in the case of a rotating cam gear may be about 4 H.P. against 25 H.P. for piston valves and Walschaerts gear. Oscillating cam gears are still extensively used in France, Spain and Italy.

The Lentz is a well-known example of an oscillating cam gear while rotary cam gears include British Caprotti, Dabeg, Franklin, "R.C." and "R.R." gears.

British Caprotti (Figs. 73 and 74). One form of this gear obtains its rotary drive from worm gearing enclosed in an axle-mounted box between the frames. This drive is taken to a small gearbox midway between the frames, whence it is transmitted through bevel gearing to cam boxes located above each cylinder. In the latest form a return crank gearbox is used. One steam cam governs the opening, and another controls closing, of the steam valve located at each end of the cylinder.

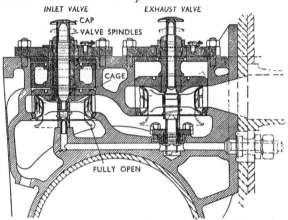

FIG. 73. British Caprotti valve gear, section through inlet and exhaust valves.

The exhaust cam always gives full opening, but the periods when cut-off and release take place vary according to the position of the reversing handwheel which has sufficient locking teeth for the gear to be regarded as infinitely variable. Rates of cut-off and reversing are determined by the position of the scrolls, which are moved endways relative to the helically-splined shaft upon which they are mounted. Axial movement thus results in angular displacement of the cams. This gear makes it possible to obtain constant periods of release and compression for any cut-off. Both this and the later form of gear are variable within limits.

Dabeg. This form of gear, as developed on the French Railways, was available in both oscillating and rotating forms. In the former a Walschaerts gear was used, while in the latter a shaft on which were varying cam contours was moved relative to the cam followers.

Franklin. In the Franklin Type A system of steam distribution there are two inlet and two exhaust valves at each end of the cylinder, all valves being horizontally arranged. The valve gearbox, which provides independent control of the valve events through multiple links, receives its oscillating movement from the two crossheads. These provide two distinct motions per cylinder, one for the exhaust and the other for the inlet valves.

The Franklin Type B system of steam distribution employs rotating cams having a continuous contour. The valves and the method of moving them, are the same as for the Type A, having oscillating cams, but the valves in the Type B gear are located in one horizontal plane. Drive is derived from a return-crank gearbox on each side; rotation of the camshafts, housed in separate cam boxes, is synchronized with the axle.

The Woodard gear is similar to the Franklin but the drive is crossed, that for the right-hand cam box being derived from the left-hand side and vice versa.

Lentz. The Lentz gear is a type of oscillating gear which has been fitted in conjunction with orthodox link motions, as well as with bevel drive from an axle.

In one form, a concentric valve arrangement is employed, the valves and their operating cams being located in housings which can be inserted in place of piston valves after the valves, liners, and steam chest covers have been removed. In this layout the exhaust valves, located at the outer ends, have hollow spindles in order that the steam valve spindles may pass through.

A modified form, the *O.C.* valve gear, had the four valves arranged in one plane, the horizontal arrangement being retained. In this and various other layouts of this gear it is not possible to interchange the valve with piston valves and special cylinder castings are necessary. In the O.C. gear the location of the valve spindles on the centre line of the camshaft was discontinued and intermediate levers have been introduced. These levers, which have their fulcra at the lower ends and engage the valve spindles at their upper ends, carry rollers which ride on the cams, at their centres.

R.C. This gear also has two inlet and two exhaust valves per cylinder, these being arranged horizontally. The drive may be taken between the frames from a coupled axle or from a return crank gearbox. The camshaft, which rotates at engine speed, is moved axially to engage the desired admission cams. The exhaust cams are arranged as stepped cams. The arrangement of the intermediate levers is generally similar to that of the O.C. gear. Pre-determined valve events only are available and the gear is not infinitely variable.

In a typical application to a three-cylinder locomotive a single camshaft was arranged transversely, being carried in a housing integral with the cylinders and driven at one end, through bevel gearing, from a return crank gearbox on one side only.

ELEVATION AND PLAN

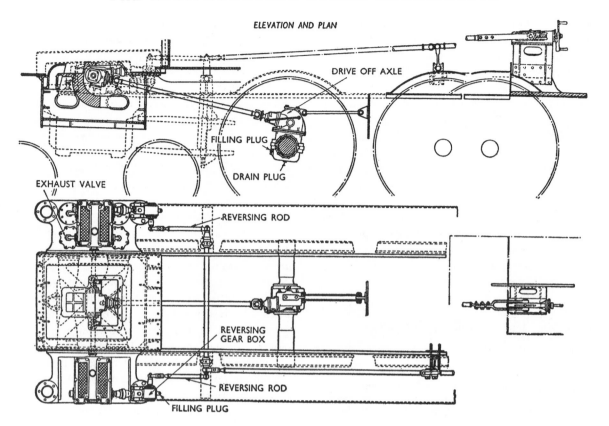

OUTSIDE AND END VIEW

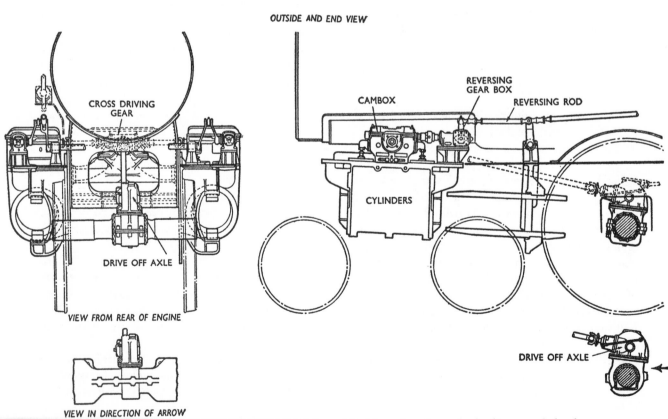

Fig. 74. Arrangement of Caprotti valve gear, with drive taken from the leading coupled axle.

Reidinger valve gear (Plate 62A, page 262). Also known as the "R.R." valve gear, this gear has the unique feature of being infinitely variable, although the independent events which are required are naturally obtained by effecting the necessary adjustment in the shed, such alteration being outside the scope of the driver.

Figs. 75, 76 and 77 show its application to a three-cylinder locomotive; the camshaft is located above the cylinders and runs across the locomotive. The plan view (Fig. 76) shows how the camshaft receives its drive from a worm gearbox on the return crank, through the medium of gears on one end of the shaft. The shaft may be housed in cavities integral with the cylinders or in separate boxes bolted to the cylinder, and is capable of

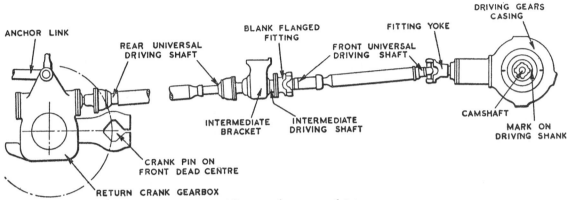

FIG. 75. Reidinger valve gear – driving gear.

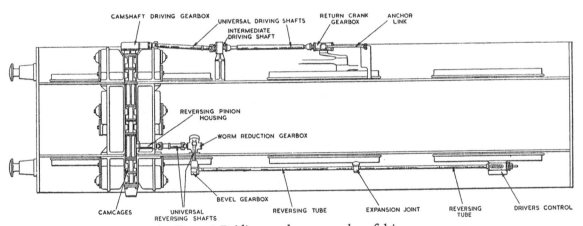

FIG. 76. Reidinger valve gear – plan of drive.

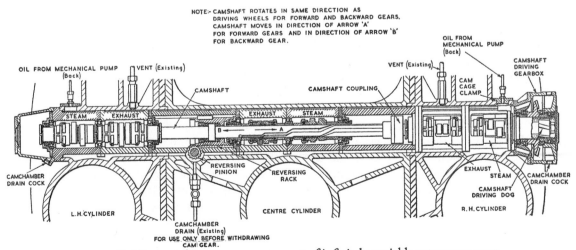

FIG. 77. Reidinger valve gear – arrangement of infinitely variable rotary cam gear.

303

being withdrawn as a unit. Each cylinder has two inlet and two exhaust valves, operated by their own cams. In the case of the admission valves the cams provide a differential action which is transmitted to the valve spindles through intermediate levers, fitted with swing beams and follower rollers.

The differential action provided by the steam cams allows the cut-offs from full-gear to mid-gear in both fore and back directions to be infinitely variable. It also allows the steam valve to open fully at all ranges of cut-off, from full-gear to about 12 per cent, whilst at lower cut-off the valve opening is fully adequate, being 90 per cent of the maximum valve area at 10 per cent cut-off.

Differential action is also employed for the exhaust cam, or cams if more than one is used; while the intermediate levers are fitted with one roller per cam, the swing beams are dispensed with. This is due to the fact that the exhaust events do not alter to the extent of the admission events, but varying action is also employed for the exhaust cams. The exhaust cams give full valve openings at all cut-offs. A special feature of the gear incorporated on a British Railways locomotive is a separate cam-adjusting device for resetting the angularity of the cams relative to the main crank pins of the locomotive. This provision makes possible individual adjustment of the valve timings for pre-admission, release, and compression for any rate of cut-off. Over 1,000 different sets of valve events may be obtained without any replacement of parts being involved, but the adjustment must be made in the shed. Adjustments on the road are limited to those resulting from the use of the reversing control in conjunction with the particular settings for which the gear has been set. The possibility of co-relating infinitely variable cut-off and the points of pre-admission, release, and compression, enables tests to be readily carried out with a view to obtaining the optimum setting from the aspects of thermal and mechanical efficiencies. After the optimum valve events have been obtained the separate vernier adjustment may be dispensed with.

Reversing and variation of cut-off are obtained by transverse movement of the camshaft which, it will be noted, has specially formed keyways in it which house the driving keys of the cam-operating driving members. When it is desired to alter the valve events the driving keys are brought into engagement with a different position of the camshaft keyways. When the reversing control is operated, the combined action resulting from moving the camshaft through the cam bores and driving members adjusts the angular position of the cams relative to the driving axle. This, combined with the variation in angularity resulting to the helically-splined driving

dog from lateral movement of the camshaft, provides the desired setting of the cams for either fore or back gear.

Cossart valve gear. This gear has cam-operated piston valves, separate admission and exhaust valves being provided. These are arranged vertically and it was claimed to have the advantage that, compared with poppet valves, the piston valves were travelling at high speed when they opened and closed the ports. Three cams per cylinder are used; these are an admission cam controlling the opening of the steam valve, an expansion cam governing the closing of the steam valve, and an exhaust cam opening and closing the exhaust valve. The camshaft derives its motion from a return crank; valve events are determined by varying the angularity of the cams.

Volume, clearance. The clearance is the volumetric space between the valve and the piston at the end of its stroke. As this volume, which must be kept at a minimum, includes the volume of the ports it is necessary that these should be kept as short as possible. Clearance volume varies considerably between engines, between cylinders located differently on a locomotive, and sometimes between front and back ends of a cylinder. It is also influenced by the diameter of the cylinder, and the type and location of the valves. Expressed as a percentage of the piston-swept volume, it may range from 4 per cent to 12 per cent, an average for piston valve cylinders being between 7 and 10 per cent. Clearance volume is always greater in compounds where, in the high-pressure cylinder it may be of the order of 25 per cent.

Wash-out plugs. Also termed mud plugs, these are provided in order that, when washing out a boiler, jets of water may be brought to bear on the inside surfaces as necessary. They also enable rods to be inserted to loosen scale, and finally an inspection may be made, through them, of the interior of the boiler.

Wash-out plugs have a fine, slightly-tapered thread, of about $1\frac{5}{8}$ in. diameter. A square head permits the use of a box spanner. Wash-out plugs are always fitted to give access to the firebox crown and the water legs. Other locations will be decided by the size and type of boiler and the quality of the feed-water.

Flanges or pads are generally fitted to afford additional length of thread for the wash-out plugs.

Water. The presence of an ample supply of water suitable for locomotive purposes will largely determine whether steam locomotives continue in use or are replaced by some other form of motive power.

Some water is suitable in its raw state for locomotive use whereas other water must be treated prior to use. A water suitable for a boiler pressed to, say, 175 p.s.i., is not necessarily suited to one carrying a higher pressure.

Of recent years increasing attention has been paid to the treatment of feed-water, with remarkable results. It is probable that had some of these developments been made ten years earlier, alternative forms of motive power would not have made the progress that they have. The boiler has always been the Achilles heel of the locomotive, with the necessity of frequent wash-outs. Wash-outs not only involve the engine being out of traffic – although the time can be greatly cut down if a hot-water system is used – but also consume a lot of water.

Bad waters take two forms, those forming hard scale and those which, while forming a non-incrusting scale, produce corrosive effects or foaming. A further but more complex trouble, occurring in certain conditions of boiling water, is caustic embrittlement. Certain salts, in solution, will give rise to electrolytic action, a serious cause of wastage of plates and tubes.

The Armand system of feed-water treatment (q.v.) is extensively used in France, where it is known as T.I.A., and also in Britain. It is reported that with this water treatment, deterioration due to scaling and corrosion has been much reduced. For certain locomotives the distance covered between general and boiler overhauls exceeds 200,000 Km. (125,000 miles), and in many cases runs of over 1 million Km. have been made between shoppings.

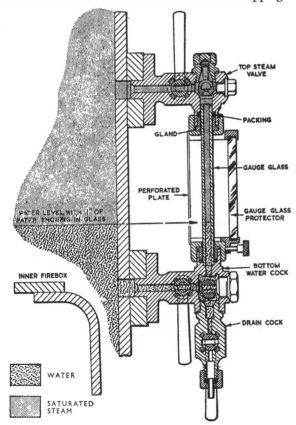

INNER FIREBOX

TOP STEAM VALVE

PACKING

GLAND

GAUGE GLASS

PERFORATED PLATE

WATER LEVEL WITH 1" OF WATER SHOWING IN GLASS

GAUGE GLASS PROTECTOR

BOTTOM WATER COCK

DRAIN COCK

WATER

SATURATED STEAM

FIG. 78. Water gauge.

As an example of the benefits to be derived from the increased availability, records made at Nice shed showed that in 1955 with T.I.A. in operation forty-one locomotives performed almost as much work as that for which 106 engines were required prior to the introduction of T.I.A. As a natural corollary to maintaining boilers in a cleaner condition, the fuel consumption fell, in the case of the engines referred to, by 12·6 Kg. per 1,000 tonne/kilometres.

Water gauges. One or more gauges are fitted on the backhead to show the level of water in the boiler. The gauge takes the form of a glass tube, or a rectangular box having a plate-glass front, the top of which is connected to the steam space while the bottom fitting is located at such a height, that when water is in sight in it, the crown of the firebox is covered by a pre-determined amount of water. A protector is fitted to the gauge glass and, as an added safeguard, restrictor or ball valves are sometimes incorporated in the top and bottom fittings to cut off the escape of steam and water should the glass break. The handles operating the top and bottom cocks may or may not be coupled (Fig. 78).

In American practice the gauge frame takes a different form, the top being connected to the steam space through a vertical pipe, entering the wrapper plate higher up. Test- or try-cocks may or may not be incorporated in the gauge fittings.

Water space frame. An American term for the Foundation Ring (q.v.).

Westinghouse brake (Fig. 79). This compressed-air brake is supplied with air by a compressor on the locomotive. In the case of a steam locomotive the compressor will be a steam-driven one, having either a single- or a three-stage compound air compressor. In the case of diesel locomotives it may be either belt-driven off the engine or, as in the case of an electric locomotive, motor-driven (see Pump, Air).

The compressed air is stored in the main reservoir on the locomotive and fed, via the driver's brake valve, into the main brake pipe (also termed the train pipe) and, through the triple valves, into the auxiliary reservoirs located on each vehicle. The size of these auxiliary reservoirs is determined by the weight of the vehicle concerned.

The air pressure is fairly high, usually 90 p.s.i. in the main reservoir and 70 p.s.i. in the brake pipe, which permits of the parts being relatively small. Generally, pipes and couplings have an internal diameter of one inch, and the brake cylinders, which may be of either vertical or horizontal types, vary from six inches to fourteen inches in diameter. A duplex pressure gauge is provided on the locomotive; the red hand indicates

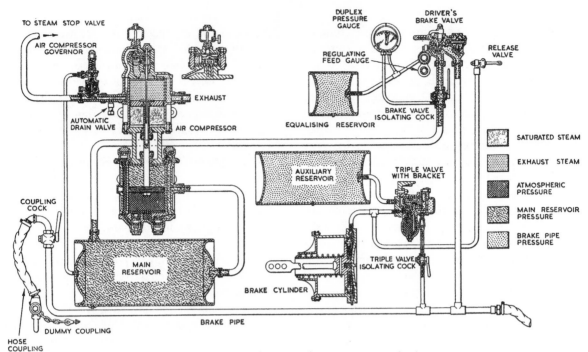

FIG. 79. Diagram of Westinghouse automatic brake.

main reservoir pressure and train pipe pressure is shown by the black hand.

Reduction in the pressure in the brake pipe has the effect of moving the triple valves, which allows air under pressure in the auxiliary reservoirs to enter the brake cylinders, which are normally free from pressure, with the result that the blocks are brought to bear upon the wheels with a force depending upon the leverage provided in the brake gear and the degree of the application. A return spring, housed within the cylinder, returns the piston to the "off" position when normal pressure is restored in the train pipe and the application pressure from the auxiliary reservoir is released.

The driver's brake valves are of different types. An example is the widely-used No. 4 valve which has five principal positions:

(i) the position for charging the train and releasing the brakes;

(ii) the position whilst running;

(iii) the lap position;

(iv) the position for moderate brake applications, and

(v) the emergency application position.

The No. 6 valve has four principal positions. Numbers 1 to 3 inclusive are as for the No. 4 valve, but intermediate positions between the third and fourth result in brake applications increasing in severity until the fourth, or emergency application position is reached.

Important components of the Westinghouse brake are the slide valve feed valve, and triple valve.

The slide valve feed valve is attached to, or works in conjunction with, the driver's valve; its function is to ensure the brake pipe pressure being automatically and correctly maintained, irrespective of the main reservoir pressure.

The triple valve, of which various types exist, some electrically operated, is provided on engines, tenders, and all vehicles fitted with the air brake; it is of the utmost importance to the working of the brake and is automatically operated by air pressure. Its three functions are:

(i) to charge the auxiliary reservoirs (fitted on each vehicle);

(ii) to apply the brakes, and

(iii) to release the brakes.

The presence of this valve makes the brake much quicker acting than would be the case if air had to travel to each vehicle from the driver's valve, it also renders the brake automatic, i.e., it is not rendered inoperative in the event of a mishap to the train pipe, which would apply the brake.

Wheel arrangements. The wheel arrangement of a locomotive is determined by:

(i) the weight of the engine;

(ii) the weight of the rail in use;

(iii) the number of axles which can be coupled;

(iv) the curvature over which the locomotive will be required to work, and

(v) the length of existing turntables.

The maximum permissible axle weight will be

Canadian National Railways

PLATE x. Canadian National Railways: 4–8–4 locomotive, No. 6401, Class U–4a. This streamlined version of a famous class is a fine example of the ultimate development of the express steam locomotive in Canada.

governed by the Civil Engineer's requirements and will be dictated by rail weights, sleeper spacing, and strength of bridges. The last-mentioned consideration, in addition to limiting axle loading, will also affect the total engine weight and the weight per foot-run.

Within the above-mentioned confines, the weight of the complete locomotive will be determined by the tractive effort it is expected to produce and maintain, for this will lay down the amount of adhesive weight required. The two simple methods of obtaining adhesion by increasing axle loading, or increasing the numbers of coupled axles, have their limitations, and when these have been attained in a rigid-frame steam locomotive there are only two alternatives:

(i) the use of double-heading, or

(ii) the employment of an articulated locomotive.

An articulated locomotive is the more economical proposition as one engine crew is saved and there is only one boiler to maintain.

Guiding- and carrying-wheel requirements will be determined by the class of work the locomotive is expected to perform and the amount of weight which cannot be accommodated on the coupled wheels. It is sometimes necessary, if the rigid wheelbase would be excessive, to allow for lateral translation of coupled axles. Various methods exist for providing this side movement, well-known systems being Zara and Kraus-Helmholtz, where a carrying axle and a driving axle constitute a bogie. Bergiot levers are another example of arranging for side movement of coupled axles; there is also the simple Dewhurst-Cartazzi axlebox, effectively used for the outer axles of long coupled wheelbases.

In the Henschel arrangement the outer coupled axles have side play but to overcome the difficulties associated with providing this while retaining coupling rod drive, the latter type of drive is replaced by gearing on the longitudinal centre line of the locomotive.

In the Golsdorf axle the lateral play of the outer axle necessitates lengthened crank pins to allow for side play of the rods, while in the Klien-Lindner system, a hollow axle connects the wheels which are free to turn about a pin connecting the hollow axle to an inner axle which always remains parallel with the axles to which it is coupled. A similar arrangement is attributable to Heywood.

Hollow axle arrangements have, in the main, been applied only to locomotives working over light tracks.

The engine units of a Beyer-Garratt locomotive are, in effect, bogies and have proved to be particularly easy on track, even when lightly laid.

A locomotive is said to have "tapered" axle loading when the highest axle loading is on the centre, or two centre axles, and becomes progressively less towards the outer axles.

This is now considered, by some railways, to be an essential requirement to obtain the best results having regard to rail stress and sleeper loading.

Wheel centres. Driving-wheel centres are now almost always high-quality steel castings, but some cast-iron centres are in use. The diameter of the wheel centre will be determined by the type of work the engine is expected to perform. At one time an engine was regarded as having a maximum speed equal in miles per hour to the tyre diameter measured in inches. In recent years speeds have increased and this ruling no longer applies. The size of wheel is to a large extent determined by the gauge, and cases are met with where it has been necessary to dish the boiler barrel on narrow-gauge locomotives to accommodate wheels of the desired diameter.

Spoked steel wheel centres may have elliptical or channel-section spokes, the latter being met with in the Scoa-P wheel centre. In a cast-iron wheel centre of simple pattern the spokes are usually of H-section. Spokes should always be increased in section where they merge into the boss, and a large radius is desirable where they join the rim, which in modern practice is frequently of triangular cross-section. In the case of the Scoa-P centre, the rim is of channel section. Wheel centres often have the spokes arranged in umbrella-dished form, this enables the length of the axlebox bearing area to be somewhat increased.

Disc wheels have had extensive application, well-known examples are the Scullin, Boxpok, and B.F.B., in addition to the normal plain dished single-plate disc type.

Willoteaux valve. A form of piston valve having double admission and double exhaust ports. Viewed from the aspect of steam consumption the results given by this type of valve have been satisfactory, but it is not always easy to incorporate the valve in the design, due to its size. As it gives better port openings for a given size and travel it has been applied in France to low-pressure cylinders on compounds.

Part II. Steam locomotive experiments

The experiments which have been made to improve the steam locomotive are legion. They commenced with the advent of the locomotive and they still continue, although in a very diminished degree; in fact so far as some components are concerned, experiments may be said to have ceased. An example of this is to be found in the piston valve; an immense number of experiments have been conducted to improve the piston valve but a stage has now been reached where it appears that nothing further can be done with this form of valve to enhance its performance.

Every component of the locomotive has formed the subject of experiment, and this has been rendered much easier of evaluation in recent years by the greatly improved testing facilities available. An enormous amount of experimentation and testing has been devoted to valve gears and the possibilities of compounding. Despite all the time, effort, and money spent on the work, the valve gears still in greatest use, Stephenson and Walschaerts, have been used for very many years, and compounding is not embodied in the modern locomotive.

The many attempts to improve performance, however, have not been wasted. Many of the ideas which have been tried, whilst in themselves abortive, have led to other developments, or alternatively, have focused research on particular aspects of locomotive design which has ultimately led to improvement.

A general criticism can with justification, however, be levelled against the attempts to improve upon the steam locomotive, for inventors have had the thermal aspect out of focus. Far too much attention has been devoted to saving infinitesimal quantities of fuel, when the first requisite of the locomotive is that it shall be a good traffic machine. In this capacity a high degree of reliability is essential. Many of the ideas produced, and even many of the ideas tried, had the weakness that they constituted possible causes of failure. This is a particularly important point on any railway, but when some overseas railways were concerned, where the nearest locomotive depot may have been several hundred miles away, it took on an added significance.

This short survey is devoted to attempts to improve upon component parts of steam locomotives; essays to improve upon the Stephenson conception as a whole are dealt with in Chapter 9.

Also included here are those developments which became established in some cases to a very considerable extent, but which were unable ultimately to retain their place in current practice.

Experiments reached their zenith in the years between World Wars I and II when much was expected, and realized, from steam power. No new developments were introduced in this period to account for the high performances obtained, but rather was the success due to the correct proportion or balance between the components, coupled with excellent maintenance. It is interesting to note that the more highly developed the steam locomotive became, the greater were the further attempts made to improve upon it. This is contrary to experience in other fields of engineering. The steam locomotive also is peculiar in that in its basic form it has not only survived but vastly progressed. When assessing the value of motive power, it is essential to include such aspects as first cost, ease of maintenance, reliability, and intrinsic characteristics. So far as steam power is concerned it has not proved possible, for general service, to improve upon the Stephenson form of locomotive, by which is understood one with a multi-tubular firetube boiler, reciprocating pistons, and direct connection from cylinder to wheels without the intervention of gearing. This in spite of the numerous attempts made in many parts of the world to introduce locomotives differing in part, or in toto, from the Stephenson conception.

BLAST PIPES
(*see* Chapter 6, Draughting, page 394)

Blast pipes of almost every form have been tried, varying from square to star-shaped, in conjunction with many arrangements of petticoat pipes and chimney chokes evolved from many formulae. Today arrangements seem to have settled down to the ordinary circular blast nozzle, of fixed diameter, double chimneys above two circular nozzles, and the Giesl blast-pipe where a chimney of elliptical section is used in conjunction with a number of jets arranged fanwise on the longitudinal centre line. Variable nozzles, of which the "jumper cap" type was very widely used, have disappeared from locomotive practice, due to the difficulty of maintaining them in working order. The blast arrangements play a most vital part in the draughting of the boiler and there have been many instances where very slight adjustments to the blast pipe relative to the other factors involved, have had a phenomenal effect upon the rate of steam production.

Among the multiple-jet blast pipes, which have had a wide application, may be mentioned Kylala, Kylchap, and Lemaitre.

BOILERS

Attempts to improve on the Stephenson form of boiler may be divided into three categories:

(i) those in which the complete boiler was replaced by a boiler of another form;

(ii) those in which the so-called improvement was limited to replacing the orthodox firebox with a firebox of a different type; and

(iii) those where the designer was content merely to add something to the Stephenson form.

Complete departures from the orthodox boiler took the form of water-tube boilers, and while these lent themselves to the use of high pressures, such pressures were not invariably used. When high pressures of above 300 p.s.i. were utilized, water-tube boilers, which dispensed with flat stayed surfaces, were almost always employed. There were, however, a few instances where the Stephenson boiler was pressed as high as 325 p.s.i. On the other hand, some water-tube boilers were designed for pressures as low as 220 p.s.i., as, for example, the Brotan boiler produced for the Swiss Railways in 1908. Compared with fire-tube boilers those of water-tube form had serious shortcomings, which were much greater than the troubles which they set out to overcome. On the credit side the water-tube boiler had no stay troubles, and in some forms these boilers were very rapid steam raisers. On the debit side, the cost of manufacture was high, it was an exceedingly difficult boiler to keep clean internally and externally, it lacked strength as a beam, and in many forms it depended for its satisfactory working upon a host of accessories, all of which were sources of potential failure. Almost without exception the complete water-tube boiler (i.e., where both firebox and flues were replaced by water tubes) was used with locomotives of special types.

Water-tube fireboxes gave the designer a freer hand as far as higher pressures were concerned and did away with what – in American practice at least – had become an enormous number of stays to maintain. In at least one instance, the Woolnough boiler, the water-tube firebox not only replaced the firebox of the Stephenson boiler but the barrel as well. The Brotan water-tube firebox, used in conjunction with either drums or a single shell, was used extensively in some European countries but was little known outside Europe.

Many attempts have been made to dispense with the water legs of the Stephenson boiler. The advantages being lower first cost, reduced maintenance and the possibility of increasing pressure. Alterations to almost anything in the engineering world are accompanied by items on the debit side as well as on the credit side. Suppression of the legs, and their substitution by firebrick, removes as much as half of the heat-absorbing potential of the boiler and always results in excessively hot footplate conditions. In the case of the boiler in the "Leader" class locomotive (see Chapter 9, Part I, page 463), the water legs were in effect replaced by thermic syphons. Such an alteration has the advantage that considerable heat absorption still takes place in the firebox and the water circulation is actively maintained. While the circulation in water legs does not necessarily always follow the same course, being largely dependent upon the condition of the fire, the water legs are of primary importance to water circulation.

Circular fireboxes have been fitted, of both cylindrical and corrugated form, with and without a wet back.

These forms of box have certain advantages, among which are ease of replacement and low first cost, particularly when the back is dry, but their use on main-line locomotives proved unsatisfactory. The last large-scale application was by the German railways during World War II and the distortion troubles previously experienced were again a source of trouble. While cylindrical fireboxes normally consisted of one drum, there was no reason why the number should be so restricted, especially when oil burning was in use, but connection to the tube plate, unless the drums ran into a common chamber in front of that plate, presented a problem. In the case of the Player boiler the firebox consisted of three tubes.

The replacement of the ordinary inner wrapper of the Stephenson boiler by one of corrugated form was tried in the Jacobs-Shupert firebox. Whereas in the cylindrical corrugated boxes the corrugations were rolled in and the wrapper sheet was in one piece, in the Jacobs-Shupert firebox the wrapper was built up from a number of channel-section pressings, riveted together. Theoretically a corrugated firebox is able easily to accommodate lengthwise expansion but in the Jacobs-Shupert application it proved excessively rigid and gave much trouble due to leakage at the joints.

While some locomotive engineers were generally content with the Stephenson form of boiler they sought to modify it by the introduction of auxiliaries such as extra air inlets, water tubes, and even the replacement of the tube nest by a single flue having crossed water tubes.

Extra air inlets were sometimes fitted through the water legs and, on occasion, provided through the throat plate. As applied to large boxes, where a large volume of secondary air was required, they could give rise to excessive noise. A further trouble arose from distortion, as there was the likelihood of considerable differences arising in the temperature of the plate in the neighbourhood of the opening.

Water tubes, especially when a flue with crossed water tubes was employed, were difficult and very costly to maintain. Much more satisfactory are arch tubes and thermic syphons, which were later employed, and which are easy to maintain and economical in service.

Serve tubes, having a number of internal gills, have been used in locomotive practice, particularly in French-built locomotives. The advantages, however, are apparent rather than real, for they are excessively rigid, difficult to expand, and not easy to clean. Fluted tubes and tubes having a shallow helix from end to end, have also been used. In each instance proper cleaning presents great difficulty, but in the case of the "ESS" tube, a modification in the design produces a swirling motion of the gas passing through the tube which is sufficient to keep it clean. These tubes have the further advantage of breaking up cinders but at the expense of increased resistance to gas flow which in turn leads to intensification of the blast.

While there has been a considerable application of welding to boilers, more particularly to fireboxes, the all-welded boiler is a comparative rarity, partly due to the difficulty in stress-relieving where this is considered essential; in countries where stress-relieving is not insisted upon, there still remains the need to subject welds to radiographic examination. An objection to an all-welded boiler is that the firebox is not easily removed and for this reason some boilers which are otherwise welded have the firebox joined to the barrel by riveting.

The very large volume of air required for combustion by a large boiler, well over two million cubic feet per hour, plus the lack of space on a modern locomotive, have combined practically to rule out the pre-heating of air for this purpose. Attempts have generally been confined to special types of locomotives, e.g., the L.N.E.R. high-pressure locomotive No. 10000 or the Ljungström turbine (see Chapter 9). When attempts to pre-heat air were made with orthodox boilers it was done with boilers of relatively small diameter where the loading gauge allowed fittings to be applied to the smokebox externally.

BOOSTERS

Like so many of the attempts which have been made to improve the locomotive the booster in theory was an excellent idea, for it enabled extra tractive effort to be available at starting or when working hard. Boosters had two cylinders of the order of 10 in. × 12 in., and could produce a very useful addition to the locomotive's power. This, assuming that the boiler was able to generate the extra steam required, could be as much as 300 H.P. at speeds of from 10 to 15 m.p.h., while the increase in tractive effort, at starting, could be about 15,000 lb.

Normally the location of boosters was on the trailing truck but on occasion they were fitted on tender trucks, in which case the truck wheels were coupled and the term auxiliary locomotive was applied. All later boosters were cut in or out through the medium of gears, which proved their undoing due to the resulting high maintenance. It is of interest to note that in experimental applications of boosters, at the turn of the century, the booster – or auxiliary engine as it was then termed – was in the form of an axle which was retracted when not required.

COMPOUNDING

The schemes which have been produced for compound working cover every possible cylinder arrangement, and some which proved to be quite impossible; an example of the latter was Johnstone's annular system where an annular low-pressure piston, with rods at diametrically opposite points, surrounded the high-pressure piston. All rods for both high- and low-pressure cylinders were connected to a common crosshead.

Compounding commenced in earnest about 1880 and persisted for about fifty years, for much of the time it only survived because the Locomotive Superintendent or Chief Mechanical Engineer had an enthusiasm out of touch with reality, for unless the cost of coal was high and that of labour low, it was impossible to make out a good case for it. This applied particularly when engines were allocated to individual crews and the mileage covered was much less than is now customary. With coal at the prices then reigning, even assuming that the savings claimed had been attained, it is extremely doubtful whether the additional first cost, maintenance, and running expenses would have been met. Sponsors of the various systems had no difficulty in producing statistics showing that their particular system resulted in a lower fuel consumption than the equivalent simple engine – what they omitted to disclose was the fact that sometimes the fuel used by compounds was of higher calorific value than that fed to simples, nor did they reveal the initial and upkeep costs. Finally the advent of superheating did much to discourage interest in compounding.

When coal was costly and maintenance cheap it was, in the state of the art which then prevailed, a simple matter to state a case for the compound. The results derived in stationary or marine practice could never be obtained by locomotives. For one reason, it was often impossible to fit large-diameter low-pressure cylinders inside the frames or within the limits of the loading gauge. There was also the necessity to have the exhaust at a sufficiently high back pressure to provide the necessary blast. In spite of these limitations and other difficulties, good results were achieved. Particular success attended compounding in France, where the results obtained, usually on very indifferent fuel, were of a high order. Good results were also forthcoming from Mallet compounds, where speeds were low and space existed for really large low-pressure cylinders. The Mallet compound did not lend itself to high-speed working any more than did any other form of locomotive where separate wheel groups are driven by the high- and low-pressure cylinders. Slipping of the high-pressure unit must lead to choking the receiver. While large low-pressure cylinders have some advantages they possess the drawbacks of large port dimensions and heavy reciprocating parts, particularly when the cylinder diameter went as high as forty-eight inches, as it did on the Virginian Railway 2–10–10–2 Mallet freight locomotives. These engines had a tractive effort of no less than 147,200 lb. The gain in thermal efficiency on Mallets which resulted from compounding was something between 2 and 3 per cent. While the thermal efficiency of compounds in their most highly developed form was, for a steam locomotive, high (it was 12·8 per cent measured on an I.H.P. basis in the case of a Chapelon 4–8–0 compound, and some 2 per cent higher than a comparable simple), the compound ultimately disappeared from the scene as the additional complication was not considered worth while. It must also be remembered that the introduction of long-lap valves did much to improve the efficiency of simple locomotives, which were developed to such an extent between the two World Wars that their coal consumption fell by more than 30 per cent.

As has already been stated, compounds took many forms not only in the layout of the cylinders but also in the arrangements made for the supply of steam to the low-pressure cylinder. In some cases provision was made for independent adjustment of the cut-off to the high- and low-pressure cylinders, whereas in others variation of cut-off to the high-pressure cylinders only was possible, the low-pressure events being constant.

For starting purposes steam could be turned into the low-pressure cylinder either at boiler pressure, or, dependent upon the design, at reduced pressure. In some instances whether the engine worked simple or compound was determined by the position of the regulator, which covered or uncovered ports allowing steam to pass to the low-pressure cylinder; in other cases an automatic starting valve was fitted. While this allowed boiler steam to pass to the low-pressure cylinder at starting it closed automatically when the pressure of the exhaust steam from the high-pressure cylinder had risen sufficiently for it to perform work in the low-pressure cylinder or cylinders. Such valves effected a rapid changeover to compound working and came into operation after only a few revolutions of the wheels. Accordingly it was not possible for an engine so fitted to be worked as a simple, which could be done with locomotives having the change-over arrangement incorporated in the regulator. There were exceptions to these layouts where the automatic change-over could be cut in or out, whereas, in some cases, the engine was able to work compound only. Provision for turning high-pressure steam into a low-pressure cylinder did not necessarily mean that drivers would avail themselves of the facility. Where low-pressure cylinders of thirty-inch diameter were involved there was the consideration of keeping passengers seated!

The ratio of the volume of the high-pressure cylinders to that of the low-pressure was usually 1:2. There were cases where it ranged from 1:1·7 to 1:3; in the case of the lower ratio it was due to the engine having no change-over arrangement, and the consequent necessity of its being able to exert sufficient power in the high-pressure cylinders alone to start the train. Generally the proportions were determined by the need of equalizing, or approximately equalizing, the amount of work performed by high- and low-pressure cylinders. This was not so much of importance on three- or four-cylinder engines as it was on two-cylinder compounds. It was general for high- and low-pressure cylinders to have a common stroke but there were a few exceptions to this.

Two-cylinder systems. Two-cylinder systems had some vogue as they enabled compounding to be indulged in while retaining almost all of the simplicity of the simple locomotive. The two cylinders, which always drove the same axle, were arranged either between or outside the frames. Large numbers of such locomotives were constructed, some being of considerable size. It is of interest to note that some two-cylinder compounds, which had only one reverser and hence no means of notching-up high- and low-pressure cut-offs independently, had link-hangers of different lengths; the effect of this was that the cut-offs in high- and low-pressure cylinders varied, in one application in the ratio of 4:5. The best-known two-cylinder non-automatic systems were those of Worsdell and von Borries, and Mallet. The two first-mentioned designers were also responsible for a two-cylinder automatic system, in which work they were associated with Lapage. Vauclain and Rogers also produced systems of compounding in the same category.

In the Lindner system the intercepting valve was dispensed with, but boiler steam could be supplied direct to the low-pressure cylinder. This system was termed a semi-automatic one.

Three-cylinder systems. Three-cylinder systems have had a considerable use. In some instances, e.g. Smith, there were one high- and two low-pressure cylinders, in the case of Webb's system there were two high- and one low-pressure cylinders. The Smith system readily lent itself to the use of three cylinders of equal size, whereas the Webb system involved the use of a very large cylinder between the frames and bad surging was sometimes experienced. With three-cylinder systems the drive was sometimes on a single axle and sometimes divided between two axles. In addition to Smith's system other three-cylinder compounds were attributable to Webb, Rickie, Sauvage, and the Swiss Locomotive Works.

Four-cylinder systems. Much use has also been made of four-cylinder compounds, with which Continental railways are particularly associated. A four-cylinder locomotive having four cranks set at 90° to a large extent balances itself. Accordingly, four-cylinder compounds, with such a crank phasing, are often referred to as balanced compounds, irrespective of whether the drive is on one axle or divided. In all four-cylinder balanced systems, either both low-pressure cylinders are between the frames or both are situated outside. Well-known systems of this type are Webb, Golsdorf, von Borries, and Maffei.

Other four-cylinder systems are those in which two cylinders are connected to a common crosshead and only two cranks are employed; these systems have the attraction of simplicity. They are of two forms which have either tandem or superimposed cylinders. Tandem systems were particularly associated with France, Hungary and Russia, and had the high- and low-pressure pistons, and their respective valves, mounted on common rods; in some cases the cylinders were mounted directly end to end, whereas in other instances space was provided between them. An exception to the common piston rod arrangement was made in the case of the Du Bousquet (Woolf) tandem system, where a simple valve of special design fed both high- and low-pressure cylinders and the low-pressure piston had two rods, outside the high-pressure cylinder. Mallet's tandem system was applied in Russia.

In the Vauclain compound, associated more particularly with American practice, the high- and low-pressure cylinders were superimposed, one valve and one crosshead sufficing for the pair of cylinders. Locomotives designed for passenger service normally had the high-pressure cylinder on top, the cylinder positions being reversed in the case of freight engines.

Triple-expansion. Triple-expansion on locomotives is exceedingly rare. The last occasion when this was tried was on the Delaware and Hudson Railroad in 1933. Four cylinders were incorporated, arranged in pairs at opposite ends of the engine, high-pressure and intermediate pressure being at the rear end and the two low-pressure at the front. All cylinders drove on to one axle.

CONDENSING
(*see* Chapter 9. Condensing Turbine locomotives, and Condensing Tenders, pages 465 and 472)

Condensing has been applied to many locomotives but only very rarely with a view to bettering the performance. The most usual reasons are to improve visibility in tunnels – which has led to its use on sections of the London underground railways – or to conserve water where it is either difficult to obtain, or alternatively where the water available is of very bad quality.

Condensing to aid tunnel conditions is effected by simply turning the exhaust into the water supply, but if this is done for prolonged periods it may result in the feed-water being heated to such an extent that it is difficult to handle it, even with pumps. This has sometimes led to pre-arranged changing of the water in the tank at regular intervals, which practice has been known to cause much dislocation of busy suburban traffic.

To conserve water, condensing has been applied in the Argentine and is currently the practice in certain districts in South Africa, where Henschel condensing tenders are fitted to some of the Class 25 locomotives. It will be apparent that if the exhaust is condensed it is not available in the blast pipe. For this reason a blower, operated by air or steam, must be used to obtain the necessary smoke-box vacuum. On the particular locomotives referred to, the exhaust steam is passed through a turbine actuating a blower fan in the smokebox. From this turbine the steam is exhausted, through an oil separator, to a turbine located in the tender. The purpose of this second turbine is to draw air through the condensing elements into which the steam passes before finally returning as condensate to the tender. The saving in water which results, compared with a similar locomotive with orthodox tender, is said to be 90 per cent (Plate 123C, page 481).

Condensing to improve performance has taken two forms:

(i) it may be used to create vacuum on the exhaust side of the piston, as is done in marine and stationary practice; or

(ii) it may be utilized to heat feed-water, thereby slightly improving the thermal efficiency.

(i) The use of condensing in the first form is another good example of the futility of pursuing the thermodynamic aspect of the steam locomotive to an illogical degree. Theoretically, advantages to be derived are large, and in the case of a boiler working at 250 p.s.i., an increase in efficiency of over 70 per cent is available by providing a condenser.

In practice the following disadvantages are manifest:

(*a*) very little space is available unless a special tender is provided;

(*b*) the equipment is necessarily bulky and heavy;

(*c*) it is not cheap in first cost or maintenance;

(*d*) it is unlikely to be able to produce constant results; and

(*e*) alternative draughting must be provided.

In short, the only simple factor involved is that of otherwise absorbing any saving effected on the fuel bill!

(ii) Where feed-water heating is concerned:

(*a*) the heat may be imparted to the water by directly introducing steam into it. This necessitates adequate separation of lubricant from the condensate.

(*b*) the heat may be imparted to the water by the use of a heater in which the steam flows over the outside of a pipe containing the water to be heated.

The two systems are referred to as open and closed

respectively. In neither case do they constitute condensing in the accepted sense of the word, as only a proportion of the exhaust is utilized and the main body of exhaust steam is discharged up the chimney in the usual manner.

Full condensing, provided that the separation of lubricant is efficiently carried out, offers the great advantage associated with a closed circuit, the reduction in maintenance. To obtain this advantage and also save fuel, heat conservation equipment was provided on a locomotive of the Southern Railway of England (No. A 816). All the exhaust was led to a cooler where the heat content of the steam was slightly lowered. Much of the latent heat was retained and the steam, in what was described as a supersaturated condition, was returned to the boiler. The experiments were discontinued because of fan trouble, an illustration of the manner in which the additional complication associated with most departures from the Stephenson form easily proved their undoing.

CYLINDERS

Cylinders during the history of the locomotive have moved through an angle of 90°, and additionally have been transposed from one end of the locomotive to the other. Valves in relation to their cylinders have also moved completely round, having begun on top and having now returned to that position.

While various attempts to improve cylinder performance have been made by alterations in the cylinder itself, e.g., by the use of jacketting, or by the uniflow system, ultimately such improvements as were attained were due to superheating and better valve events and passage arrangement in conjunction with higher pressures.

The layout of cylinders has at times presented difficulties on account of their size. This has been overcome in some rigid-frame locomotives by arranging for cylinders to drive different wheel groups. Occasionally opposed pistons have been employed and there are a few instances, of which the Russian "OR" class 2–10–6 is an example, where the drive has been through the medium of rocking levers. Such an arrangement has increased weight and complication but, on the credit side, reciprocating balance is simplified.

Multiple cylinders were tried to a small extent but never became popular. They were usually incorporated in engines having individual axle drive. A good example was the 1–DO–1 locomotive constructed by Henschels in 1941. A V-twin engine was fitted to each driving axle, the engines being designed to work at speeds up to 800 r.p.m., giving a rail speed of about 110 m.p.h. (*see* Chapter 9, Part I, page 461).

The additional complication of multiple-cylinder loco-motives, with individually-driven axles, plus the added possibilities of failure, proved sufficiently strong deterrents to prevent them ever seriously competing with the orthodox locomotive.

To simplify front-end layout various locomotives had cylinders with crossed-ports, enabling one valve to feed two cylinders. It cannot be emphasized too strongly that engineering is a matter of compromise and any change almost always brings some disadvantage in its wake. In this case large clearance volume almost inevitably arose. The magnitude of this shortcoming ultimately depended upon the clearance and the work upon which the locomotive was engaged.

FUELS

(*see* Chapter 6, Coals, page 389 *and* Chapter 7, Fuel, effect of, page 411)

Coal and oil. Coal and oil have predominated as locomotive fuels for many years and oil-burning in steam locomotives is now making progress at the expense of coal. Whereas a number of railways have converted to oil firing it is relatively seldom that conversion from oil to coal occurs. The decision to burn one or the other fuel is usually determined by price, a fact which led to the large-scale ignominy in Britain in 1946, when, after deciding to convert a very large number of locomotives to oil burning, oil was not available at an economic price. This was probably the most extensive and expensive fiasco ever made in connection with locomotive operation.

Colloidal fuels. Colloidal fuels, manufactured from varying mixtures of coal dust, creosote, pitch, fuel oil, smokebox char and coke breeze in various combinations were tried in Britain by J.G.Robinson and were said, at the time, to have attained considerable success; but the amalgamation of the Great Central Railway into the London & North Eastern Railway ended the experiments. In connection with these experiments Robinson designed a firebox where the water-legs were extended downwards and inwards to join in the form of a U. This form of construction, which dispensed with the foundation ring, had some similarity with the firebox introduced experimentally by F.W.Webb on the London & North Western Railway, which is said to have had exceedingly bad circulation.

Pulverized coal. Pulverized coal has had a fairly large application, particularly in Germany where the A.E.G. Locomotive Works and Henschel & Son developed and produced the necessary equipment. These experiments were largely influenced by the desire to use brown coal for locomotive purposes, which fuel did not readily lend itself to any other method of firing. The theoretical advantages of being able to utilize a cheaper grade of fuel, and one which did not store without disintegrating, were more than offset by the difficulties which arose:

(i) there was much unburnt fuel ejected from the chimney;

(ii) the fuel was apt to freeze or cake above the outlet from the tender;

(iii) slag formed on the tubeplate;

(iv) there was a source of danger in the form of explosion if the apparatus was not handled properly;

(v) working was exceedingly dirty and led to public complaint.

Many ideas which have proved successful in stationary steam practice have not been enthusiastically received in the locomotive world, one of the factors contributing to unsatisfactory results being lack of room. Combustion, particularly so when the fuel is pulverized, requires space, and with that available on a locomotive, the rate of heat absorption necessary would have had to have been many times higher than in a stationary boiler. It is only right to add that at the time most of the pulverized fuel experiments were conducted, fireboxes and combustion chambers were not as large as they subsequently became, but the difficulty still remained even if in less degree.

Wood and peat. Wood has all but disappeared from use for locomotive purposes although for many years it was of necessity widely used where no alternative was available. The same remarks apply to peat, which while capable of producing great heat – there are cases on record where the smokebox became bright red on peat-burning locomotives–has the disadvantages also common to wood, i.e., considerable bulk, excessive ash formation, and the ability to absorb much water. Peat, like coal, could be adapted to pulverized firing, but only when it was dry (*see* Chapter 9, page 463).

STREAMLINING

In the decade before World War II the fashion of locomotive streamlining or air-smoothing became worldwide. Aerodynamically, it benefited locomotives little, its chief value being a publicity one. It was only at higher speeds that any appreciable saving in the horsepower required to propel the locomotive became apparent and many of the locomotives to which streamlining was applied spent little of their time running at speeds (over 50 m.p.h.) at which it showed advantageously, and some never reached it at all. The occasions on which streamlining saved 200 H.P. or more must have been few; against this spasmodic saving was the additional first cost, increased weight, and higher maintenance charges. The weight was considerable, sometimes of the order of eight tons or more, and the transport of such a load during the lifetime of a locomotive is costly. Higher maintenance charges arose in two ways, the difficulty of obtaining access to parts cased in by fairing, and the need of some of these for more frequent attention. There was a tendency on some streamlined locomotives for parts to run warm which had previously kept cool when there was an ample flow of air around them.

The resistance resulting from head-on air pressure is slight compared with that resulting from a wind blowing at such an angle that it impinges on the end of every vehicle in a train, or is broadside-on, when flange friction is greatly increased.

Streamlining was abandoned for new construction after about 1942 and subsequently it has been removed from many locomotives to which it was originally fitted.

SUPERHEATERS

When the large benefits to be derived from superheating became recognized much experimental work had still to be carried out before the fire-tube superheater established its supremacy. Many of the earliest superheater designs were of doubtful effectiveness, and maintenance of some of them was bound to be difficult and expensive. Alternative positions, in which superheaters were located, were the smokebox, the boiler barrel, and the firebox.

Smokebox superheaters. The smokebox superheater was the favourite form of early superheater, and Baldwin, Drummond, Phoenix, Schmidt, and Vauclain were among the types which were housed in the smokebox. It is, perhaps, flattering to Drummond's device to include it in a list of superheaters for in the designer's own estimation it was a "steam dryer", the heat imparted over and above that of saturated steam being but some 20° F.

Boiler barrel superheaters. Superheaters housed in the boiler barrel took three forms. In Aspinall's superheater, the front tubeplate was set well back in the barrel, the smokebox being, in effect, extended backwards.

The second form was used on large articulated locomotives and consisted of a superheater in the form of a chamber, with flues passing through it, equal in diameter to the boiler barrel. This chamber, as applied on the Santa Fé, was divided vertically, one part serving as the superheater for steam passing from the dome to the high-pressure cylinders, and the second portion acting as a reheater for steam exhausted from the high-pressure cylinders before entering the low-pressure cylinders. This type of superheater is known as the Buck-Jacobs or Santa Fé.

Reheaters, besides being thus incorporated on compounds with superheaters, were also arranged separately. The experiment of superheating receiver steam only was carried out by the Eastern Railway of France.

The Clench superheater is usually regarded as a smoke-box superheater but, in fact, it consisted of a chamber in the front end of the barrel, the flues being extended to pass through the second tubeplate. Superheating took place as the steam came in contact with the sections of the flues between the two tubeplates.

The third form, represented by the Pielock superheater, was fitted to some German and Swiss locomotives. It was housed in the boiler barrel and comprised a closed drum through which the flues passed, and in which the steam took up superheat from the flues. Compared with the Clench superheater it had the advantage that it was located much nearer to the firebox, and was stated to provide a high degree of superheat. However efficient such a device might be in imparting heat, it was doomed to failure by reason of its inaccessibility. It had one unique feature in that it never gave trouble from leakage as the pressure both inside and outside was the same.

Cusack-Morton superheater. This superheater, applied on the Midland Great Western Railway of Ireland, was located on top of the boiler, deriving its heat from firebox gases. This, like many other devices introduced in the days of relatively small locomotives, even had it proved successful, would have disappeared from the scene, at all events in its original form, as space could not be found for it on most locomotives now running.

Fire-tube superheaters. Fire-tube superheaters have taken three basic forms:

(i) those in which the header is at the top, as in The Superheater Company's very successful designs;

(ii) those in which two side-headers were used, as in the Horwich superheater and some American applications of the Schmidt superheater;

(iii) those where a comb-header is used, as in the Vaughan-Horsey superheater, where saturated and superheated headers are separated.

The best-known fire-tube superheaters are the Schmidt, Robinson, Swindon, Vaughan-Horsey, and those produced by The Superheater Company.

While the advantages of superheating steam have been recognized from very early days, designers preferred to obtain increases in output by raising pressure (which also raised steam temperature) rather than to impart superheat. Superheating introduced great problems of lubrication, particularly with the slide valves then in use. The combination of piston valves and positive lubrication arrangements enabled superheated steam to be used to full advantage. There have been a few cases where superheated locomotives have had slide valves but these were usually cases of converted engines where the original valves had not been replaced.

VALVES

The slide valve has served steam engines of all types well, but it had the serious objection, in its simple form, of excessive friction, which obviously increased as the steam pressures became higher, and as cylinder – and hence valve sizes – increased. It was natural, therefore, that designers should turn their attention to balancing, and various slide valves were introduced where the pressure on the back was reduced, if not removed.

Balancing took the form of preventing steam pressure having access to the back of the valve, which was achieved by providing spring-loaded strips bearing against a plate, parallel with the valve face. The provision of these additions, plus clearance to allow the valve to lift should water be present in the cylinder, inevitably added to the height required in the valve chest, which, except in the case of very small inside cylinders, ruled out balanced valves for application between or at the sides of the cylinders. Balanced valves took various forms, of which the Richardson was, perhaps, the best-known example. Some balanced valves were trick-ported, while the Wilson balanced valve also had additional exhaust ports.

The American Fay-Richardson valve was a balanced valve having a hollow body, which allowed the exhaust steam to pass directly through the back of the valve to the blast pipe.

While normally a single slide valve controlled the inlet and exhaust arrangements to both ends of a cylinder there were instances where two valves were mounted on a single stem, each controlling the steam flow through ports situated close to the end of the cylinder. Such a layout enabled short and direct steam passages to be used and greatly reduced the clearance volume.

Apart from the normal development of piston valves, which ranged from the use of a plug-type valve without rings, moving in a steam-jacketted liner, to the wide-ringed valve, and ultimately to the valve with narrow rings, relatively little has been done to alter the valve form. Many layouts of ports have been tried but no investigations have been carried out of the value of ports located in the semi-circle remote from the cylinder.

Trick ports were applied to piston valves. Other special forms, such as Trofinoff and Willoteaux, are described in Part I of this Chapter.

For use on balanced compounds the Baldwin Locomotive Works produced a piston valve having three heads on one spindle, which consequently required but one set of valve gear. This valve controlled the steam distribution to both the high- and low-pressure cylinders; it had a fairly extensive application on railways in the Western States of America.

A feature usually associated only with poppet valves is

the provision of separate inlet and exhaust valves. The Nadal layout of piston valves, however, provides one valve to control live steam and another for exhaust. This arrangement was fitted on the French State Railways. Among other gears having separate valves for admission and exhaust were the Berthe and Bonneford gears, also applied in France, and J.T.Marshall's gear tried by the Southern Railway of England.

Valves which rotated, or partly rotated, have had a very limited application to locomotives as they were developed for use only on locomotives of unconventional form (see Chapter 9, page 463).

Corliss valves have been tried on a small scale both in America and France but operation was by established locomotive valve gears, modified Stephenson and Gooch respectively.

In the early years of this century many ideas, including methods of steam distribution, after progressing to some extent, were dropped, as superheating and compounding, separately or in combination, gave greater promise of efficiency and economy.

Compared with the slide valve, the ordinary piston valve had the disadvantage that water trapped in the cylinder or clearance space was unable to escape. To overcome this objection both Smith and Robinson produced valves with provision for trapped water to be freed. In the case of the Smith valve, collapsible segments were employed, whereas in the Robinson valve, pressure release rings were incorporated. Ball relief valves were also used for this purpose of relieving any excessive pressure.

The average steam locomotive has a life which varies from fifteen to twenty years in America to forty to fifty years or more in Britain and many other countries. Many railways, therefore, endeavour to keep their motive power up to date by incorporating, as far as is economically practical, ideas more recent than those in vogue when the engine was built. To meet such a demand, valves have been produced which can be fitted to cylinders originally designed for an earlier form of distribution.

The Universal valve chest was produced to allow piston valves to be fitted to locomotives originally having slide valves. The valve chamber was bolted on to the cylinder in place of the original steam chest and various styles were available to suit various positions of the steam pipe. Valves of various diameters were also provided, with allowance for vertical adjustment of the valve

spindle. As the valves, which were replaced, were of the outside admission type, the replacements were always similar.

Later, when the piston valve in its turn was challenged by the poppet valve, the Dominion Engineering Company supplied housings which could be bolted on to the ends of piston valve chambers to allow poppet valves to be fitted.

VALVE GEARS

A tremendous number of valve gears have been evolved and well over one hundred different gears have been fitted. Those which have been used at all widely have been described on page 296 (Valve Gears). An experimental valve gear worthy of mention was the Meier-Mattern which was an hydraulically-operated gear, working poppet valves. While this was not the first use of hydraulically-operated gears, D.Joy having had such a gear fitted to an engine on the London, Brighton & South Coast Railway, it was probably the last one to be applied. Some thirty years have elapsed since the writer was on the footplate of the Dutch 4–4–0 locomotive fitted with the Meier-Mattern gear, but the impression still remains of everything coming up solid – a feature of many hydraulic mechanisms. To stand a reasonable chance of success all modifications, improvements and new ideas must be understood by shed personnel. The more complicated a mechanism is, the less are its chances of succeeding generally on locomotives. Not only is it usually more prone to failure but having failed it is not so readily rectified. Thus an ingenious conjugated valve gear for three-cylinder locomotives due to Pickersgill failed because of its complication and its fourteen pin joints. It failed to fulfil the essential attribute of steam locomotive practice – to be a good traffic proposition. The same may be said of Marshall's valve gear which had separate valves for inlet and exhaust. It was too heavy and too complicated.

In the last thirty years, especially, availability has become a factor of great importance, and the higher the first cost of a locomotive becomes, the more importance it assumes. Today, the availability of a modern steam locomotive may be very high if maintenance is good, and at least one British Railways locomotive class has an availability well over 90 per cent. Such availability has been obtained by keeping things relatively simple and accessible, and emphasizes the reasons why so many experimental ideas have never become established practice.

CHAPTER 5

Illustrated Survey of Modern Steam Locomotives

by H. M. LE FLEMING

In the following pages one hundred examples of modern steam locomotives are described and illustrated. Some of the examples described have already been scrapped, but they are included because they illustrate the summit of steam power achieved in that particular country or region. This is, of course, particularly true of locomotives of the United States where the steam locomotive reached proportions never equalled in any other part of the world. Such motive power will never be seen again; its nearest rivals in size and power are in the U.S.S.R., but unfortunately little opportunity exists to see these machines and exact descriptions are difficult to obtain.

In the hundred examples chosen, the caption number, the picture number and the number in the tables coincide, so that reference to all available details may readily be made.

Notes on the tabulated dimensions. With the different systems of measurement and formulae in use it is necessary to adopt a common standard for comparative purposes. For example, there are three tons: the long (2,240 lb.), the metric (2,204 lb.), and the short (2,000 lb.). Weights are therefore given to the nearest pound, and where tons are quoted in the text, long tons of 2,240 lb. are meant. Lengths are given in feet and/or inches and decimals of an inch. Areas are given in square feet and decimals, and capacities in imperial gallons.

Dimensions marked * have been scaled from drawings and diagrams or computed from data.

Wheelbase, coupled. Where Krauss-Helmholz or similar trucks are fitted, the end coupled axle or axles have considerable side-play. Such examples are denoted by (K). Total wheelbase refers to engines with tender.

Tractive force. For non-compound engines this is based on 85 per cent boiler pressure, but for engines with limited cut-off a lower figure is often quoted officially, since the larger figure would show too low a factor of adhesion. In such cases the official figures are shown in

the tables and the 85 per cent in footnotes for comparative purposes. With compounds the official figures have been used for the basis and are marked (V).

Heating surface. Unless otherwise stated, the figures are for the water side, but in some countries, notably Germany, the fire side, denoted by (f), is used. These are some 7 to 9 per cent less than the water side figures. Fireboxes containing arch tubes (whose heating surface is included) are denoted by (a), circulators by (c) and siphons by (s). (p) indicates a preheater closely integrated with the boiler, the additional heating surface being shown in a footnote.

Maximum axle load. Different types of tender are often attached to the same class of engine and in some cases have heavier axle loading than the engine. Those of the engines are shown in the tables, and the tenders' (where heavier) in footnotes.

Maximum speed. This is not the maximum of which the engine is capable. On many railways the maximum is officially laid down, either for the class of engine or for the route and type of traffic, and the design is based on such requirements. Where there is no such ruling the figure given in parenthesis indicates a figure likely to be reached in normal working. Where two figures are quoted, the lower is for freight working, the higher for passenger.

Minimum track radius. On main lines sharper curves are subject to speed restrictions, but to reach sheds or sidings the engine must often traverse sharp curves at slow speed and the minimum radius is quoted where known. This depends on the gauge widening which is customary, and is therefore not truly comparative, but it illustrates the comparatively narrow limits found throughout the world. As an example, the large American 2–10–4 (No. 8) will pass over a 430-ft curve with the gauge widened by 0·75 in., but a 1,000-ft radius is the limit on track laid strictly to gauge; (m) indicates the radius which can be traversed at moderate speed where the lower figure is not available.

Part I. Standard gauge: North American

1. **Chesapeake & Ohio Railway** (Plate 65A, page 325).
Class L-2. 4–6–4. Built Baldwin, 1948.

Earlier engines of this fine class had Baker valve gear, but the engine shown is fitted with Franklin poppet valves. Designed to haul fifteen-car trains at speeds up to 95 m.p.h., their high-speed boosters could remain in operation up to 33 m.p.h. They were amongst the last and the heaviest 4-6-4's to be built in America. The C. & O. was the second railway in the U.S.A. to adopt the "Pacific" type, in 1902. The L-2's were the successors to a long line of handsome and beautifully kept "Pacifics", which for many years, resplendent in polished metal work, headed the named trains over the more level stretches.

2. Canadian National Railways (Plate 65B, page 325).
Class U-1f. 4-8-2. Built Montreal, 1944.

These neat engines were designed for high-speed service between Montreal and Chicago and Montreal and Vancouver, where loads were less than those handled by the larger 4-8-4's.

3. Atchison, Topeka & Santa Fe Railway (Plate 65C, page 325).
Class 3765. 4-8-4. Built Baldwin, 1937-44.

The Santa Fe pioneered long-distance runs without change of engine, crews being changed *en route*, and its great oil-burning 4-8-4's, amongst the largest built, were used on continuous runs of some 1,780 miles. The main line between La Junta, Colorado, and San Bernadino, California, rising to over 7,500 ft, includes the steepest (1 in 28½) and the longest sustained grades (126 miles at 1 in 70 or steeper) on the main lines of North America. Banker assistance is provided for the steeper sections, but long spells of high power output are necessary to maintain 50-60 m.p.h. with 1,350-ton trains. The I.H.P. was 6,300 with 60 per cent limited cut-off. Notable features were the light-alloy steel rods of very deep section, aluminium crosshead slippers and 15-in. piston valves. The telescopic chimney could be raised when on open stretches of line. It was operated by compressed air from the cab and was a characteristic of larger Santa Fe steam locomotives.

4. New York Central System (Plate 65D, page 325).
Class S-1b. 4-8-4. Built Alco, 1945-46.

These very neat and compact engines were used for heavy passenger work, hauling twenty-car passenger trains at speeds up to 85 m.p.h. The enormous tenders with a 41-ton coal capacity (equal to the total weight of many European tenders) enabled refuelling stops to be reduced to the minimum, and special rapid-loading facilities were installed at depots to deal with them. On the other hand, the water capacity was less than usual, since water pick-up apparatus suitable for scooping at up to 80 m.p.h. was fitted. The success of these time-saving measures is reflected in the high engine mileages; No. 6024, for example, reached 228,849 miles in eleven months. Extensive comparative tests were made with these locomotives against diesel power.

5. Norfolk & Western Railway (Plate 66A, page 326).
Class J. 4-8-4. Built N. & W. Railway, 1941.

Mountain grades of the main line include one of 30 miles at an average of 1 in 120, with the last five miles averaging 1 in 71. Class J were introduced to operate 900-ton passenger trains over this route. The maximum of 5,100 H.P. is reached at 40 m.p.h., but they are also capable of 100 m.p.h., which was exceeded with a 1,350-ton test train on a 30-mile stretch of straight track. In service they average 15,000 miles a month and 238,000 between heavy repairs, and no doubt these figures could be exceeded with longer runs.

6. Great Northern Railway (Plate 66B, page 326).
Class O-8. 2-8-2. Built G.N.R., 1944-46.

The "Mikado", for so many years the freight contemporary of the "Pacific", was by no means always superseded by larger classes. These fine modern examples, built at the company's shops, had Belpaire type fireboxes, for many years characteristic of the G.N.R. and of the Pennsylvania Railroad. Amongst the heaviest 2-8-2's ever built, they were rated to haul 2,000 tons on a grade of 1 in 100 at 10 m.p.h., and were permitted to run at speeds up to 55 m.p.h.

7. New York Central System (Plate 66C, page 326).
Class A-2a. 2-8-4. Built Alco, 1948.

These handsome engines were noteworthy as the last steam locomotives to be built by the American Locomotive Co. The first of the 2-8-4's was the famous "A-1", built by the Lima Locomotive Works in 1925 and regarded as the forerunner of modern American steam power. The engine illustrated was built to operate 12,000-ton freight loads at up to 35 m.p.h., on the Pittsburgh & Lake Erie lines of the N.Y.C. System. These engines could, however, operate fast passenger trains at up to 70 m.p.h. when called upon to do so.

8. Bessemer & Lake Erie Railroad (Plate 67A, page 327).
Class H1g. 2-10-4. Built Baldwin, 1944.

The first of the B. & L.E. 2-10-4's came out in 1929 and the last was built in 1944. Used entirely for very heavy freight work, trains of up to one hundred loaded iron-ore cars weighing over 10,000 tons were worked by two of them between Lake Erie and the Pittsburgh area. In the opposite direction equally heavy coal trains were operated, one engine being coupled at each end in both cases. Some of them have been sold to the D.M. & I.R. for their heavy ore traffic.

9. Pennsylvania Railroad (Plate 67B, page 327).
Classes J-1, J-1a. 2-10-4. Built P.R.R., 1942-44.

War Production Board regulations prevented the

development of redundant locomotive designs and the P.R.R. chose the Chesapeake & Ohio Railway's T-1 class as a basis for their 2–10–4's. Altogether 127 of the J-1 and J-1a classes were built, the difference being only in the material used for certain parts. The engines differ from the C. & O. only in minor details, such as chimney and cab, but the tenders are P.R.R. pattern. Capable of hauling loads of 10,000 tons and over, they proved a most successful and trouble-free heavy freight design.

10. **Canadian Pacific Railway** (Plate 67C, page 327).

Class T-1c. 2–10–4. Built Montreal, 1944.

This type was introduced in 1929 for service between Revelstoke and Calgary in the Canadian Rockies and the illustration shows one of the last batch. The climb includes some eight miles at 1 in 45, over which the T-1's were rated to take 1,000 tons. They are oil-burners, which eased operation over this arduous route.

Since the dieselization of most of the trains over the Rockies, these engines have been used on freight train duties on other sections of the C.P.R.

11. **Chesapeake & Ohio Railway** (Plate 67D, page 327).

Class C-16. 0–8–0. Built Lima, 1930–48.

The C-16 class are typical of the American eight-wheel switcher, a universal and indispensable type to be found on practically all railroads. Superheaters are general, but the valves are very generously proportioned for working at long cut-offs. On these engines Baker valve gear operates 14-in. diameter piston valves with a travel of $8\frac{1}{2}$ in., similar to those of much larger main line engines. The design remained practically unchanged from 1930 to 1948.

12. **Canadian Pacific Railway** (Plate 68, page 328).

Class F-1a and F-2a. 4–4–4. Built Montreal, 1936.

Before the War there was an unexpected revival of the four-coupled locomotive for light high-speed trains in North America. After a lapse of twenty years or more, some "Atlantics" were built for the Chicago, Milwaukee, St Paul & Pacific Railroad. The Canadian Pacific developed two classes of 4–4–4 designed to haul light trains of 200–250 tons at speeds up to 90 m.p.h. The class illustrated (F-1a) have coupled wheels 75 in. driving on the rear axle. Class F-2a have slightly larger wheels and drive on the leading axle.

13. **Duluth, Missabe & Iron Range Railway** (Plate 69A, page 329).

Class S7. 0–10–2. Built Baldwin, 1936.

Originally built for the Union R.R. transfer service in the Pittsburgh area handling iron-ore traffic, and hence

the name "Union" given to this type. In 1949 they were purchased by the D.M. & I.R., whose principal traffic is also iron ore. Capable of taking very heavy loads on comparatively short runs, they have proved to be most useful engines. At one time they were used on eighty-five-car trains for the seven-mile run from Proctor to the Duluth ore docks. They represent the final development of the American non-articulated switcher.

14. **Union Pacific Railroad** (Plate 69B, page 329).

Class 9000. 4–12–2 (3-cylinder). Built Alco, 1926–30.

Eighty-eight engines of this class were built and form the most impressive examples of American 3-cylinder engines built during the 1923 to 1930 period, when such engines were in vogue in the U.S.A. In later designs a preference was shown for three sets of valve gear in place of the Gresley arrangement, and seven of the Union Pacific engines were so altered. The aim in adopting three cylinders was to obtain greater power without resorting to Mallet compounds, but in the meantime, simple articulated locomotives became increasingly general and avoided the large and inaccessible inside big-ends. The 9000 class operate freight loads of 3,800 long tons at average speeds of 35 m.p.h., with a ruling grade of 1 in 122.

15. **Pennsylvania Railroad** (Plate 69C, page 329).

Class T-1. 4–4–4–4. Built P.R.R., 1945–46.

Tests showed that these very distinctive engines could reach a speed of 100 m.p.h., with loads of 1,000 tons or more, and produce 6,100 d.b.h.p. By dividing the driving mechanism into two parts, revolving and reciprocating masses were much lighter than with the conventional 4–8–4 type, but the wheelbase was longer. The chisel form of streamlining was impressive and much more pleasing than most other forms. The whole of the engine framing with the four cylinders was a single steel casting. Typical of the P.R.R. were the enormous tenders running on two eight-wheel bogies; their high coal capacity obviated refuelling stops, and water pick-up was fitted. The engines were equipped with Franklin oscillating-cam poppet valve gear.

16. **Pennsylvania Railroad** (Plate 69D, page 329).

Class Q-2. 4–4–6–4. Built P.R.R., 1944–45.

As on the last class the driving mechanism was divided to reduce forces on the rails, but the Q-2's were freight engines. They were larger and more powerful than contemporary 2–10–4's and were possibly the most powerful non-articulated engines ever built.

Part II. Standard gauge: North American articulated

17. **Chesapeake & Ohio Railway** (Plate 70A, page 330).

Class H-6. 2–6–6–2(c). Built Baldwin, 1949.

The American 2–6–6–2 compound first appeared in 1906 on the Great Northern Railway, and large numbers

PART I. STANDARD GAUGE: NORTH AMERICAN

EXAMPLE NO.	units	1 (Plate 65A, page 325)	2 (Plate 65B, page 325)	3 (Plate 65C, page 325)	4 (Plate 65D, page 325)	5 (Plate 66A, page 326)	6 (Plate 66B, page 326)	7 (Plate 66C, page 326)	8 (Plate 67A, page 327)	9 (Plate 67B, page 327)	10 (Plate 67C, page 327)	11 (Plate 67D, page 327)	12 (Plate 68, page 328)	13 (Plate 69A, page 329)	14 (Plate 69B, page 329)	15 (Plate 69C, page 329)	16 (Plate 69D, page 329)
RAILWAY		C & O	CNR	AT & SF	NYC	N & W	GN	NYC	B & LE	PRR	CPR	C & O	CPR	DM & IR	UPRR	PRR	PRR
Class		L-2	U-1f	3765	S-1b	J	O-8	A-2a	H-1g	J-1a	T-1c	C-16	F-1a	S-7	9000	T-1	Q-2
Type		4-6-4	4-8-2	4-8-4	4-8-4	4-8-4	2-8-2	2-8-4	2-10-4	2-10-4	2-10-4	0-8-0	4-4-4	0-10-2	4-12-2	4-4-4-4	4-4-6-4
Cylinders (no.) dia. × stroke	in.	(2) 25×30	(2) 24×30	(2) 28×32	(2) 25½×32	(2) 27×32	(2) 28×32	(2) 26×32	(2) 31×32	(2) 29×34	(2) 25×32	(2) 25×28	(2) 16·5×28	(2) 28×32	(1) 27×31 (2) 27×32	(2) 18¾×26 (4)	(2) 19·75×29 (2) 23·75×29
Coupled wheel dia.	in.	78	73	80	79	70	69	63	64	70	63	52	75	61	67	80	69
Wheelbase, coupled	ft in.	14' 0"	19' 0"	21' 3"	20' 6"	18' 9"	18' 3"	16' 9"	22' 4"	24' 4"	22' 0"	15' 0"	7' 2"	22' 0"	30' 8"	25' 4"	26' 4·5"
Wheelbase, total	ft in.	91' 8·25"	86' 10·75"	108' 2"	97' 2·5"	95' 4·75"	83' 6·63"	95' 5·3"	95' 2·5"	104' 0·5"	87' 5·13"	54' 0·5"	64' 9·5"	67' 2·75"	91' 6·5"	107' 0"	107' 7·5"
Tractive force (85%)	lb.	52,100	52,315	66,000[1]	61,570	80,000[2]	77,330	67,130	96,700[3]	93,750	76,900	57,200	26,000	90,900	96,650	58,300	100,800
(booster)	lb.	12,600	—	—	—	—	—	—	13,235	15,000	12,500	—	—	17,150	—	—	15,000
Boiler pressure	p.s.i.	255	260	300	275	300	250	230	250	270	285	200	300	260	220	300	300
Grate area	sq. ft	90·2	70·2	108·0	101·0	107·7	98·5	90·3	106·5	121·7	93·5	46·9	45	85·2	108·25	92·0	121·71
Heating surface	sq. ft	4,178(c)	3,584(a)(s)	5,312(s)	4,819(a)	5,271(a)	4,726(c)	4,276(a)	5,904(a)(s)	6,568	4,590(a)	2,600(s)	3191	4,777(s)	5,832(a)	5,209(c)	6,725
Superheating surface	sq. ft	1,785	1,570	2,366	2,073	2,177	2,110	1,881	2,487	2,930	2,055	637	900	1,360	2,560	1,537	2,930
Weight full, engine	lb.	443,000	355,700	511,000	471,000	494,000	425,540	426,000	523,600	574,730	449,000	244,000	230,130	422,000	515,000	510,870	619,100
tender	lb.	393,000	281,840	458,055	420,000	378,600	327,000	352,800	385,120	411,500	283,500	158,400	149,480	240,000	310,500	442,500	422,000
Coal capacity	lb.	60,000	36,000	—	92,000	70,000	—	44,000	52,000	59,800	—	24,000	24,000	28,000	88,000	85,200	75,000
Oil capacity	imp. gal.	—	—	5,830	—	—	4,850	—	—	—	4,100	—	—	—	—	—	—
Water capacity	imp. gal.	17,500	5,000	20,400	15,000[1]	16,650	14,360	16,650	19,150	17,500	12,000	6,860	7,700*	10,000	15,000	16,000[4]	16,000[4]
Maxm axle load	lb.	72,500	60,000*	73,500	70,000*	72,000	81,250	70,000*	75,984	77,270	62,500*	63,200	55,625	70,000*	59,000	69,620	79,780
Overall length	ft in.	104' 8"	93' 0".	122' 0"*	115' 5·6"	109' 2·25"	95' 9·9"	108' 3·6"	108' 2·25"	117' 8"	97' 10·63"	70' 7·9"	76' 11"	86' 6·5"	103' 1·63"	122' 9·75"	124' 7·13"
Overall height	ft in.	16' 9·5"*	15' 1·8"	16' 0"	15' 2·75"	16' 0"	16' 1·75"	15' 2·75"	16' 2·25"	16' 5·5"	15' 7"	15' 0"	14' 11·5"	15' 11"	16' 1·5"	15' 10"	16' 5·5"*
Overall width	ft in.	10' 9"	10' 9"	11' 0"	10' 7·5"	11' 2"	10' 10"	10' 2"*	11' 2·75"	11' 3"	10' 8"	10' 4"	10' 8"	10' 9"	11' 2"	10' 0"	10' 0"
Normal maxm speed	m.p.h.	95	(90)	—	85	78	55	35/70	—	(45)	65	—	90	35	67	80	—
Minm track radius at slow speed	ft	288	360	573	310	477	303	310	430	400	320	280*	288	360	360	385	400

Tractive efforts with limited cut-offs quoted at lower figures: [1] At 70% cut-off – with 85% B.P. would be 79,970. [2] At about 80% cut-off – with 85% B.P. would be 85,000. [3] At 68·75% cut-off – with 85% B.P. would be 106,350. [4] Water pick-up fitted.

were built in the next twenty years; in World War I it was adopted as the U.S.R.A. light standard Mallet. Soon superseded by the 2–8–8–2 for the heaviest work, it retained a large sphere of usefulness on secondary work. The C. & O. H-6 class were used on the various colliery branches where the shorter wheelbase was advantageous. Most of them were built in 1920–23 and when more were ordered in 1949 the design was sufficiently satisfactory to remain practically unchanged.

18. Norfolk & Western Railway (Plate 70B, page 330).
Class Y-6b. 2–8–8–2. (c). Built N. & W. R., 1948–52.

At one period the universal heavy freight locomotive of America, its subsequent development was only pursued on the N. & W.R. Since the Y-2a's of 1918, eight successive enlargements have culminated in the present series, and they all bear a strong family likeness. Boiler pressure has always been high, which accounts largely for their great success. In the later engines, low-pressure piston valves of no less than 18-in. diameter are fitted. To clear these on curves the smokebox is short, and to avoid the superheater header the exhaust and chimney axis is inclined 10° forward. The chimney top is of course horizontal, this unusual feature being only perceptible from the side. With loads of up to 13,000 tons on the level, and at about 25 m.p.h., the Y-6b's develop 5,500 H.P., with about 60 per cent cut-off in the high-pressure cylinders and 55 per cent in the low pressure. Up to 4 m.p.h., a simpling valve admits boiler steam to the L.P. cylinders at reduced pressure, at the same time by-passing the H.P. exhaust direct to the chimney. During World War II a number of the Y series were loaned to other systems, where they did excellent service.

19. Great Northern Railway (Plate 71A, page 331).
Class R-2. 2–8–8–2. Built G.N.R., 1929–30.

The 2–8–8–2 was the first simple articulated type to be generally adopted and these Great Northern engines with their distinctive grey-green boiler and cylinder jackets are fine examples. They were rated to take 4,000 tons up 1 in 100 grades at 10 m.p.h. in the Rocky Mountains on the Continental Divide, and could tackle enormous loads on the level. The G.N.R. was a pioneer in the use of simple articulated locomotives and inaugurated in 1924 a large conversion programme of compound Mallets to simples.

20. Norfolk & Western Railway (Plate 71B, page 331).
Class A. 2–6–6–4. Built N. & W.R., 1936–50.

The possibilities of high speed with simple articulated engines were demonstrated on the Baltimore & Ohio Railroad in 1930 with two 2–6–6–2's. Other railroads followed with 2–6–6–4's, amongst which the Norfolk & Western A class are unsurpassed, both in performance and appearance. The maximum d.b.h.p. of 6,300 is reached at 45 m.p.h., but they have hauled 6,700 long

tons at 64 m.p.h., and can exceed 70 m.p.h. on passenger work. As in nearly all modern American locomotives, they have cast-steel engine beds, the two units combining 66 major and 634 minor parts, the section at the side sills being 7¼ in. The boiler contains 4,925 stays, normally holds 8,100 gallons of water, and expands over 1 in. from cold to working-pressure temperature. Evaporation reaches 116,000 lb. per hour. Roller bearings are fitted on all axles and needle bearings for motion parts. Baker valve gear operates 12-in. diameter valves with a maximum travel of 8½ in. There are 238 lubrication points; special sheds are provided at depots for rapid servicing.

21. Chesapeake & Ohio Railway (Plate 72A, page 332).
Class H-8. 2–6–6–6. Built Lima, 1941–49.

The C. & O.R. traverses the largest bituminous coal-field in the world and has transported upwards of fifty million tons of coal annually. Much of this is over difficult territory and on single track, which in this Allegheny area reaches maximum track occupation of any railway. Consequently, it is not surprising to find an abundance of massive motive power: 160-car trains 1⅓ miles in length and up to 11,500 tons are operated at speeds up to 45 m.p.h. Later classes showed an increase in coupled wheel diameter to increase average speeds, culminating in the 528-ton H-8 class. Sixty of these great engines were built for the C. & O., and eight similar engines for the neighbouring Virginian Railway. They are the only locomotives with six-wheel trailing trucks, apart from some experimental types and tank engines.

22. Duluth, Missabe & Iron Range Railway (Plate 72B, page 332).
Class M-4. 2–8–8–4. Built Baldwin, 1941–43.

Iron ore produces the most concentrated loads on railroads. Ore-cars, though short, often exceed 100 tons loaded weight. On the D.M. & I.R.R. this traffic originates in the Vermilion and Missabe ranges, where the largest open-cast mines in the world are situated, and some forty million tons a year are transported to Lake Superior. On the Missabe division the heaviest trains consist of 185 ore cars, 75 to 80 per cent of 70-ton capacity, and the balance 50-ton cars. Speed is restricted to 30 m.p.h. with these trains, which total between 12,500 and 15,500 tons and are the heaviest regular train loads operated by steam in the world.

23. Southern Pacific Co. (Plate 73A, page 333).
Classes AC-11, AC-12. 4–8–8–2 (cab in front). Built Baldwin, 1943–44.

The ruling grade over the 140-mile Sierra Nevada section on the Southern Pacific lines is 1 in 41. Heavy working through numerous tunnels and snowsheds on this route often completely obscured the view from cabs in the normal position. This defect was first overcome in

1910 by arranging the engine to run cab-first, and such engines became a universally recognized characteristic of the Southern Pacific. As they were oil-burners, the fuel was pumped from the tender, but, when working up steep grades, 5 lb. air pressure in the tank is necessary to force the oil to the front end. Prior to 1928, Mallet compounds were used, but since that year 195 simples have been built, one of the later batches being illustrated. Unassisted, these engines handled the heaviest sixteen-car transcontinental trains over this route, and with three engines, 4,000-ton trains of 100 refrigerator cars with perishables on fast schedules were operated.

A unique feature of these engines is the "smoke splitter", a hinged frame over the chimney, operated by compressed air from the cab. This divides the exhaust and prevents the pressure at the centre shattering the roofs of the many snowsheds in the region.

24. Union Pacific Railroad (Plate 73B, page 333).

Class 4000. 4-8-8-4. Built Alco, 1941-44.

The steam locomotive reached its peak of development in the famous "Big Boys" of the Union Pacific, weighing 540 long tons and designed for speeds up to 68 m.p.h. For freight work these twenty-five great engines are rated at 5,360 tons, on lines with a ruling grade of 1 in 122. Only in the U.S.A. have locomotives of this size been built, a total weight of 500 tons being exceeded for the first time by the Northern Pacific 2-8-8-4's of 1928-30.

25. Union Pacific Railroad (Plate 74, page 334).

Class 3900. 4-6-6-4. Built Alco, 1942-44.

The "Challenger" type was introduced on the Union Pacific Railroad in 1936 and, designed for an operating speed of nearly 70 m.p.h., created a wonderful impression of speed and power. The type was soon adopted on other railroads and established the suitability of simple articulated engines for fast running. One hundred and five of these great engines were built for the U.P.R.R., in two classes, one of the later engines being illustrated. For freight work they are rated at 3,700 long tons with a ruling grade of 1 in 122. Engines of the same design were built for the Denver & Rio Grande Western Railroad.

PART II. STANDARD GAUGE: NORTH AMERICAN ARTICULATED

EXAMPLE NO. RAILWAY		17 (Plate 70A, page 330) C & O	18 (Plate 70B, page 330) N & W	19 (Plate 71A, page 331) GN	20 (Plate 71B, page 331) N & W	21 (Plate 72A, page 332) C & O	22 (Plate 72B, page 332) DM & IR	23 (Plate 73A, page 333) SP	24 (Plate 73B, page 333) UP	25 (Plate 74, page 334) UP
Class		H-6	Y6b	R-2	A	H-8	M-4	AC-12	4000	3900
Type		2-6-6-2	2-8-8-2	2-8-8-2	2-6-6-4	2-6-6-6	2-8-8-4	4-8-8-2	4-8-8-4	4-6-6-4
Cylinders (no.) dia. × stroke	in.	(2) 22×32 (2) 35×32	(2) 25×32 (2) 39×32	(4) 26×32	(4) 24×30	(4) 22·5×33	(4) 26×32	(4) 24×32	(4) 23·75×32	(4) 21×32
Coupled wheel dia.	in.	56	58	63	70	67	63	63	68	69
Wheelbase, coupled	ft in.	10' 0"	15' 9"	16' 6"	12' 4"	11' 10"	17' 3"	16' 11"	18' 3"	12' 2"
Wheelbase, total	ft in.	88' 6·75"	103' 8·25"	107' 0·4"	108' 3·25"	112' 11"	113' 4·4"	107' 2·5"	117' 7"	106' 8"
Tractive force (85%)	lb.	(s) 98,700 (c) 78,250	(s) 152,206 (c) 126,388	146,710	125,900	110,200	140,000	124,300	135,375	97,350
Boiler pressure	p.s.i.	210	300	240	300	260	240	250	300	280
Grate area	sq. ft	72·2	106·2	126	122	135	125	139	150	132·2
Heating surface	sq. ft	4,830 (a)	5,115 (c)	7,834 (c)	6,639 (a)	6,795 (s)	6,758 (s)	6,505	5,755 (c)	4,817 (c)
Superheating surface	sq. ft	991	1,478	3,515	2,703	2,922	2,770	2,300	2,043	2,085
Weight full, engine	lb.	434,900	611,520	686,400	573,000	751,830	699,700	657,900	772,250	633,500
tender	lb.	210,000	378,600	372,780	378,600	431,710	438,335	393,300	436,500	436,500
Coal capacity	lb.	32,000	60,000	—	60,000	50,000	52,000	—	56,500	56,500
Oil capacity	gal.	—	—	4,850	—	—	—	5,080	—	—
Water capacity	gal.	10,000	18,320	18,320	18,320	20,800	20,800	18,240	20,800	20,800
Max. axle load	lb.	65,000*	68,600	76,000	72,000	85,480	71,000*	69,100	68,500*	68,000*
Over-all length	ft in.	96' 9·75"*	114' 10·5"	119' 11·25"	121' 9·25"	125' 7·9"	127' 8"	111' 9"	132' 9·9"	121' 10·9"
Over-all height	ft in.	15' 0·13"	16' 0"	16' 0·13"	16' 0"	16' 7"	16' 3"	16' 5·5"	16' 3·5"	16' 3·5"
Over-all width	ft in.	10' 6"*	11' 0"	11' 0"	11' 2"	11' 1"	11' 3"	10' 6·9"	10' 10"	11' 2"
Normal max. speed	m.p.h.	45	45	40	78	60	35	60	68	69
Min. track radius	ft	—	260'	360'	260'	320'	—	320'	288'	288'

Dimensions marked* have been scaled from drawings and diagrams or computed from data.

The Baldwin Locomotive Works

PLATE 65A. No. 1: Chesapeake and Ohio Railway (*see* page 319).

Montreal Locomotive Works Ltd

PLATE 65B. No. 2: Canadian National Railways (*see* page 320).

The Baldwin Locomotive Works

PLATE 65C. No. 3: Atchison, Topeka and Santa Fé Railway (*see* page 320).

New York Central System

PLATE 65D. No. 4: New York Central System (*see* page 320).

Norfolk and Western Railway

PLATE 66A. No. 5: Norfolk and Western Railway (*see* page 320).

Great Northern Railway

PLATE 66B. No. 6: Great Northern Railway (*see* page 320).

New York Central System

PLATE 66C. No. 7: New York Central System (*see* page 320).

Bessemer and Lake Erie Railroad Co.

PLATE 67A. No. 8: Bessemer and Lake Erie Railroad (*see* page 320).

Pennsylvania Railroad

PLATE 67B. No. 9: Pennsylvania Railroad (*see* page 320).

Canadian Pacific Railway Co.

PLATE 67C. No. 10: Canadian Pacific Railway (*see* page 321).

The Baldwin Locomotive Works

PLATE 67D. No. 11: Chesapeake and Ohio Railway (*see* page 321).

Canadian Pacific Railway Co.

PLATE 68. No. 12: Canadian Pacific Railway (*see* page 321).

The Baldwin Locomotive Works

PLATE 69A. No. 13: Duluth, Missabe and Iron Range Railway (*see* page 321).

Union Pacific Railroad

PLATE 69B. No. 14: Union Pacific Railroad (*see* page 321).

Pennsylvania Railroad

PLATE 69C. No. 15: Pennsylvania Railroad (*see* page 321).

Pennsylvania Railroad

PLATE 69D. No. 16: Pennsylvania Railroad (*see* page 321).

The Baldwin Locomotive Works

PLATE 70A. No. 17: Chesapeake and Ohio Railway (*see* page 321).

Norfolk and Western Railway

PLATE 70B. No. 18: Norfolk and Western Railway (*see* page 323).

Great Northern Railway

PLATE 71A. No. 19: Great Northern Railway (*see* page 323).

Norfolk and Western Railway

PLATE 71B. No. 20: Norfolk and Western Railway (*see* page 323).

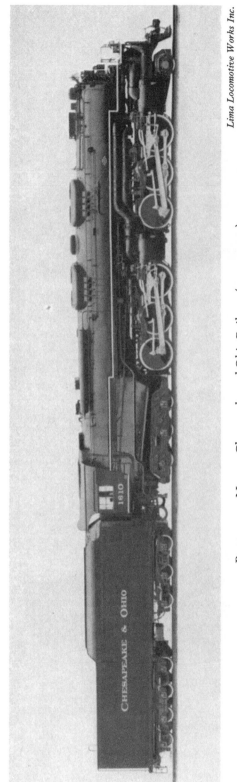

Lima Locomotive Works Inc.

PLATE 72A. No. 21: Chesapeake and Ohio Railway (*see* page 323).

The Baldwin Locomotive Works

PLATE 72B. No. 22: Duluth, Missabe and Iron Range Railway (*see* page 323).

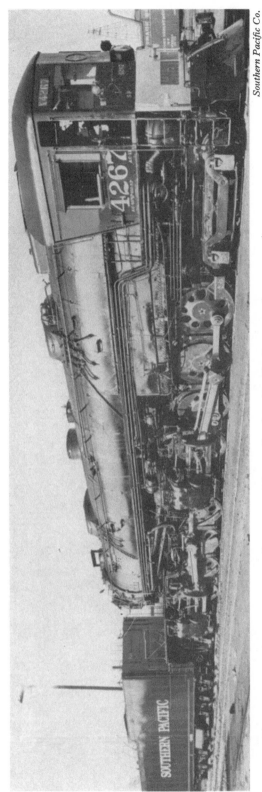

Southern Pacific Co.

PLATE 73A. No. 23: Southern Pacific Company (*see* page 323).

Union Pacific Railroad

PLATE 73B. No. 24: Union Pacific Railroad (*see* page 324).

Union Pacific Railroad

PLATE 74. No. 25: Union Pacific Railroad (*see* page 324).

British Railways

PLATE 75A. No. 26: British Railways (*see* page 341).

British Railways

PLATE 75B. No. 27: British Railways (*see* page 341).

British Railways

PLATE 75C. No. 28: British Railways (*see* page 341).

British Railways

PLATE 76A. No. 29: British Railways (Western Region) (*see* page 341).

British Railways

PLATE 76B. No. 30: British Railways (Western Region) (*see* page 341).

British Railways

PLATE 77A. No. 31: British Railways (Southern Region) (*see* page 341).

P. Ransome-Wallis

PLATE 77B. No. 32: British Railways (Southern Region) (*see* page 343).

British Railways

PLATE 77C. No. 33: British Railways (Southern Region) (*see* page 343).

British Railways

PLATE 78A. No. 34: British Railways (Eastern Region) (*see* page 343).

British Railways

PLATE 78B. No. 35: British Railways (London Midland Region) (*see* page 343).

New South Wales Government Railways

PLATE 79A. No. 36: New South Wales Government Railways (*see* page 343).

New South Wales Government Railways

PLATE 79B. No. 37: New South Wales Government Railways (*see* page 343).

S.N.C.F.

Plate 80a. No. 38: French National Railways (S.N.C.F.) (*see* page 343).

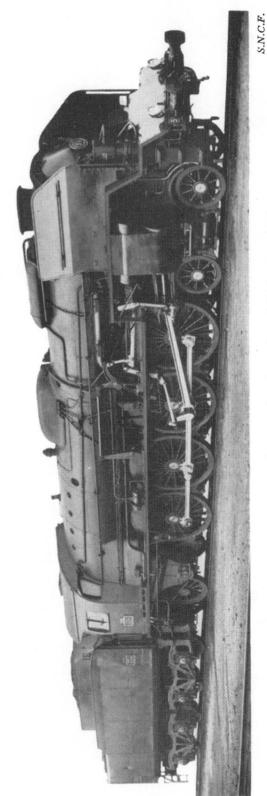

S.N.C.F.

Plate 80b. No. 39: French National Railways (S.N.C.F.) (*see* page 344).

Part III. Standard gauge: British and Australian

26. British Railways (Plate 75A, page 335).

Class 5. 4–6–0. Built Derby, 1956.

In spite of the advent of "Pacifics" the 4–6–0 is still universal in Great Britain, where it is used far more extensively than elsewhere. Although conditions are generally cramped and distances short by other standards, track curvature is easy and the mixed-traffic 4–6–0 is a most useful engine. That illustrated is one of thirty fitted with the latest form of the British-Caprotti poppet valve gear. In this, separate camshafts are actuated by external worm drives on each side of the engine and the valves are closed by steam pressure. Many improvements over earlier designs have been incorporated, including a greater variation in lead and compression.

27. British Railways (Plate 75B, page 335).

Class 9. 2–10–0. Built Crewe, 1955.

Great Britain was the last of the principal European countries to introduce the ten-coupled freight engine. The engine illustrated is one of ten equipped with boilers incorporating Crosti pre-heaters. When running, the smokebox gases are diverted through the pre-heater under the barrel and pass out through a vent on the right-hand side. In so doing, the waste heat is utilized in raising the feed-water temperature to just below that of the boiler. Feed-water pre-heating is applied in many forms on different railways, but usually not so closely integrated with the boiler. The Crosti design is a modification of the Franco-Crosti pre-heater used extensively on the Italian State Railways. In this there are two pre-heaters, one on either side of the boiler. Limitations of the loading-gauge preclude the use of this design in Britain.

28. British Railways (Plate 75C, page 335).

Class 4. 2–6–4T. Built Brighton, 1958.

The multiplicity and variety of suburban services around London and the big towns still require suitable steam power, in spite of growing electrification. For the last thirty years the 2–6–2T and 2–6–4T types have replaced earlier 0–6–2T, 0–6–4T, 4–6–2T and 4–6–4T, and of course four-coupled tank engines. Combining the virtues of the earlier types, they avoid the flange wear of the first two types and the bulk of the last two. The 3-cylinder arrangement has been discontinued on account of high construction costs and the example illustrated represents the final form of steam locomotive for this class of work in Britain.

29. British Railways (Western Region) (Plate 76A, page 336).

Class 60xx ("King"). 4–6–0 (4-cylinder). Built Swindon, 1927–30.

In the years 1902 to 1907 extensive trials were carried out with 2- and 4-cylinder simple and compound "Atlantics" and 2- and 4-cylinder 4–6–0's under Mr. G.J.Churchward. As a result, the 4-cylinder simple 4–6–0 with divided drive was adopted, and with successive enlargements remained the standard form on the Great Western Railway. The 30 "Kings" were built in 1927–30 and are remarkable for a T.E. of 40,300 lb. The original medium-degree superheater has been replaced by a much larger apparatus and an improved smokebox arrangement with double chimney has been fitted. Eight engines are to be rebuilt with new frames and roller-bearing axles. Capable of hauling 800-ton passenger trains, and of speeds of 100 m.p.h., they illustrate the extent to which the 4–6–0 has been developed in Britain.

30. British Railways (Western Region) (Plate 76B, page 336).

Class 94xx. 0–6–0T. Built Swindon and contractors, 1947–56.

The former Great Western Railway covered a multitude of duties with inside cylinder 0–6–0 tank engines. It was by far the greatest user of this most useful type. For nearly a century (1860–1956) just under 2,400 were built to G.W.R. design, and for sixty years there have been over 1,000 in service, in addition to those taken over from absorbed lines or with outside cylinders. The present example is one of a class of 210 engines and bears a strong family likeness to the prototype of 1860. A similar class, the 57xx, with domed boilers, totalled some 863 locomotives. Superheaters have been fitted from time to time but not generally adopted. Many remarkable feats have been accomplished and speeds of over 60 m.p.h. attained.

31. British Railways (Southern Region) (Plate 77A, page 337).

"West Country" class. 4–6–2 (3-cylinder). Built Brighton, 1945–47.

Seventy of these engines followed twenty heavier "Merchant Navy" class and were intended for mixed traffic. Designed under Mr. O.V.S.Bulleid, they are an interesting departure from normal practice. The three sets of valve gear are enclosed in an oil bath and are chain-driven. Welding is used extensively on the boilers and frames. The combination of many new features in one design, at a period when maintenance difficulties have been severe, led to the decision to rebuild them on normal lines. Nevertheless, remarkable performances have been achieved in their unorthodox state.

Part III. Standard Gauge: British and Australian

EXAMPLE NO.		26 (Plate 75A, page 335)	27 (Plate 75B, page 335)	28 (Plate 75C, page 335)	29 (Plate 76A, page 336)	30 (Plate 76B, page 336)	31 & 32 (Plate 77A & 77B, page 337)	33 (Plate 77C, page 337)	34 (Plate 78A, page 338)	35 (Plate 78B, page 338)	36 (Plate 79A, page 339)	37 (Plate 79B, page 339)
RAILWAY		BR	BR	BR	BR(W)	BR(W)	BR(S)	BR(S)	BR(E)	BR(LM)	NSWGR	NSWGR
Class		5	9	4	60xx	94xx	WC	V	A-1	8	AD-60	D-58
Type		4-6-0	2-10-0	2-6-4T	4-6-0	0-6-0T	4-6-2	4-4-0	4-6-2	2-8-0	4-8-4+4-8-4	4-8-2
Cylinders (no.) dia. × stroke	in.	(2) 19×28	(2) 20×28	(2) 18×28	(4) 16·25×28	(2) 17·5×24	(3) 16·4×24	(3) 16·5×26	(3) 19×26	(2) 18·5×28	(4) 19·25×26	(3) 21·5×28
Coupled wheel dia.	in.	74	60	68	78	55·5	74	79	80	56·5	55	60
Wheelbase, coupled	ft in.	15' 6"	21' 8"	15' 4"	16' 3"	15' 6"	14' 9"	10' 0"	14' 6"	17' 3"	14' 9"	15' 9"
Wheelbase, total	ft in.	52' 1"	55' 11"	36' 10"	57' 5·5"	15' 6"	57' 6"	48' 7·25"	62' 5·25"	52' 7·75"	97' 8"	87' 6·13"
Tractive force (85%)	lb.	26,120	39,667	25,515	40,300	22,515	27,720	25,130	37,397	32,438	59,560	55,000
Boiler pressure	p.s.i.	225	250	225	250	200	250	220	250	225	200	200
Grate area	sq. ft	28·7	40·2	26·7	34·3	17·4	38·25	28·3	50	28·65	63·4	65
Heating surface	sq. ft	1,650	1,432 (p)	1,366	2,013	1,347	2,122 (s)	1,766	2,461	1,649	3,041 (a)	3,390 (a)
Superheating surface	sq. ft	358	411	240	473	—	545	283	680	241	748	773
Weight full, engine	lb.	170,240	200,608	194,096	199,360	123,984	192,640†	150,304	235,648	161,504	570,696	310,772
tender	lb.	119,280	114,800	—	104,608	—	95,424	94,976	136,416	120,176	—	199,136
Coal capacity	lb.	20,160	15,680	7,940	13,440	7,940	11,200	11,200	20,160	20,160	31,360	31,360
Water capacity	gal.	4,725	4,725	2,000	4,000	1,300	4,500	4,000	5,000	4,000	9,200	9,000
Max. axle load	lb.	44,128	35,728 (1)	40,208	50,400	43,120	42,000†	47,040	50,064	35,840	35,840	51,324
Over-all length	ft in.	62' 7"	66' 2"	44' 9·5"	68' 2"	33' 4"	67' 4·75"	58' 9·75"	72' 11·7"	63' 0·75"	108' 7"	87' 6·13"
Over-all height	ft in.	13' 0"	13' 1"	13' 0"	13' 5·25"	12' 5·5"	12' 11"	13' 0"	13' 1"	12' 10"	13' 11·5"	14' 3"*
Over-all width	ft in.	8' 10·13"	9' 0·5"	8' 9·25"	8' 11·25"	8' 7"	8' 7·67"	8' 5·43"	9' 0"	8' 7·7"	9' 8"*	9' 7"*
Normal max. speed	m.p.h.	(80)	(50)	(75)	(95)	(40)	80	(85)	(95)	45/60	40	40
Min. track radius	ft	297	330	297	462	297	352	297	330	297	396	528 (m)

(p) Preheater heating surface 1,078 sq. ft.　(1) Tender axle load greater at 38,304 lb. maximum.

†These weights refer to the unrebuilt engines. Dimensions marked* have been scaled from drawings and diagrams or computed from data.

32. British Railways (Southern Region) "West Country" class as rebuilt (Plate 77B, page 337).

The weight of the engine has been increased in re-building by 4 tons 1 cwt and the maximum axle load of the rebuilt engine is 44,240 lb.

33. British Railways (Southern Region) (Plate 77C, page 337)

Class V ("Schools"). 4-4-0 (3-cylinder). Built Eastleigh, 1930–35.

The 3-cylinder "Schools" class must rank amongst the finest 4-4-0's built. In traffic they have equalled and often surpassed the performance of larger 4-6-0's. At one time they worked to schedule on a non-stop run of 108 miles in 116 minutes with 350-ton trains, the 4,000 gallons of water carried in the tender being adequate. About half the forty engines were fitted with Lemaitre multiple-jet exhausts and large chimneys, but, as with other initially excellent designs, the alteration has not shown sufficient gain to justify further adoption.

34. British Railways (Eastern Region) (Plate 78A, page 338).

Class A-1. 4-6-2 (3-cylinder). Built Doncaster, 1949.

The famous Scotch expresses of the East Coast route had for many years been worked by "Atlantics" when, in 1922, the late Sir Nigel Gresley introduced the first of a long line of 3-cylinder "Pacifics". All three cylinders drive the centre axle and two sets of Walschaerts gear drive all three valves, the middle valve being driven through conjugated valve gear mounted in front of the valves. In 1935, *Mallard*, one of an enlarged class, attained the record speed of 126 m.p.h. After Gresley's death in 1941, divided drive with three sets of valve gear was adopted, and the engine illustrated shows the last design of the former London & North Eastern Railway.

35. British Railways (London Midland Region) (Plate 78B, page 338).

Class 8F. 2-8-0. Built Crewe, 1939.

It is possible that more 2–8–0's have been built than any other type, and it is also remarkable that it has varied within comparatively small limits of size and weight in different parts of the world. In both World Wars it was adopted as a standard operational type in Britain and America. These 2–8–0's of Class 8 were the basis for British Standard Types in 1939–45 and have, therefore, seen service in various parts of the world. Capable of 1,000 d.b.h.p., they normally work freight or mineral traffic at speeds up to 45 m.p.h., but may work up to 60 m.p.h. on passenger service if required.

36. New South Wales Government Railways (Plate 79A, page 339).

Class AD-60. 4-8-4+4-8-4 (Beyer-Garratt). Built Beyer Peacock, 1952.

The N.S.W.G.R. is the largest standard gauge system outside North America and Europe. Beyer-Garratt locomotives have only been used to a limited extent in Britain, where conditions generally do not warrant their use. It is therefore interesting to see what has been accomplished elsewhere on this gauge. Class AD-60 are employed on the very heavy traffic on the main lines radiating from Sydney, where an axle load of 22·5 tons is permissible. However, when the electrification of these lines is completed, they will be transferred to other parts of the system and have therefore been built to a 16-ton limit. They are the first Beyer-Garratts to have cast-steel engine beds. The firebox is very wide and the short barrel of 7 ft 3 in. diameter gives ideal proportions, for which the type is so well suited. Mechanical stokers are provided and either arch tubes or thermic siphons are fitted in the firebox.

37. New South Wales Government Railways (Plate 79B, page 339).

Class D-58. 4-8-2 (3-cylinder). Built Everleigh, 1950.

The heavy freight locomotive illustrated is based on an earlier Class D-57, of which twenty-five were built in 1929–30 with Gresley conjugated valve gear. On the D-58's a most interesting rack and pinion arrangement has been adopted to prevent "whip" or lost motion. The right valve spindle terminates in a rack which oscillates a transverse shaft by means of a toothed quadrant. At the other end of the shaft a small pinion is geared to a second pinion which reverses the motion and engages with a sliding rack. To this is attached a crosshead bearing the pivot of the small lever. The large 2 : 1 lever is thus replaced by a light oscillating shaft.

Part IV. Standard gauge: French

38. French National Railways (S.N.C.F.) (Plate 80A, page 340).

Class 242.A. 4–8–4 (3-cylinder compound). Reconstructed, 1946.

No. 242.A.1 is the largest French express passenger locomotive constructed to the ideas of M. Chapelon and, unfortunately, it is probably also the last. It is a 3-cylinder compound and steam at 206 lb. pressure is admitted to the L.P. cylinders when starting, giving a starting tractive effort of 47,000 lb. Designed for 650-ton expresses, it can maintain 59 m.p.h. on 1 in 125 banks, and 40 m.p.h. with 600 tons up grades of 1 in 71. The maximum of 4,200 H.P. was obtained with 611 tons at 56 m.p.h. on a grade of 1 in 90. The normal speed limit in service is 75 m.p.h., but 81 m.p.h. is permissible on certain stretches and this engine reached 100 m.p.h. on trials. The corresponding 2–10–4 type freight engine was not built owing to progress with electrification. Other advanced designs were also abandoned and a most interesting phase of development was unfortunately cut short.

39. French National Railways (S.N.C.F.) (Plate 80B, page 340).

Class 241.P. 4–8–2 (4-cylinder compound). Built Schneider, 1947–49.

This engine was developed from No. 241.C.1 of the former P.L.M.Rly., which introduced the "Mountain" type in 1925. A unique characteristic of P.L.M. design lay in placing the inside cylinders to the rear of the outside, the outside L.P. cylinders driving the second, and the inside H.P. cylinders driving the third coupled axle in this case. The practice of mounting inside cylinders over the coupled wheels was adopted by M. Chapelon in the production of an experimental 6-cylinder 2–12–0, No. 160. A.1.

40. French National Railways (S.N.C.F.) (Plate 81A, page 349).

Class 232.U. 4–6–4 (4-cylinder compound). Built Corpet, Louvet & Cie, 1949.

The first 4–6–4 tender engines were put into service on the Northern Railway of France in 1911, and since 1890 this line has been noted for the fine performance of its de Glehn compounds. No. 232.U.1 is the final development of this long pedigree and differs from previous types in having two sets of valve gear. The valves are exceptionally large, but of light construction. The H.P. valves are of 11·89 in. diameter with a maximum travel of 9·17 in.; the L.P. of 16·61 in. diameter and a maximum travel of 11·89 in., believed to be the longest in existence. With bar frames and roller-bearing axle-boxes, the engine is noted for its beautiful riding qualities. It is capable of developing 4,000 H.P., and before the electrification of the Paris–Lille main line, regularly worked 570-ton express trains between the two cities.

41. French National Railways (S.N.C.F.) (Plate 81B, page 349).

Class 231.E. 4–6–2 (4-cylinder compound). Built Homecourt, 1937.

No factor could more clearly show the genius of M. Chapelon than this achievement in doubling the power output of existing locomotives by improvements rather than by construction of new designs. The Paris–Orleans "Pacifics" of 1907 – the first of the type to run in Europe – had a maximum of 1,800 H.P., later increased to 2,100 H.P. when superheated. His chief improvements may be summarized: (1) new steel firebox with Nicholson siphon; (2) greatly enlarged superheater, later with Houlet type elements; (3) all steam passages greatly enlarged and streamlined; (4) piston and slide valves replaced by oscillating cam-operated double beat valves; (5) double blast pipe of improved design with enlarged smokebox; (6) integral type A.C.F.I. feed-water heater. This transformation raised the maximum H.P. attainable into the region of 3,500. By converting further engines to the 4–8–0 type with increased pressure (290 lb.) and larger cylinders, 3,500 H.P. was obtained continuously, with a maximum of 4,000 H.P. This represents a power–weight ratio of 58·53 lb. per I.H.P., the best figure obtained from a steam locomotive of normal design. Since 1935 these engines have been operated on test at speeds up to 108 m.p.h.

42. French National Railways (S.N.C.F.) (Plate 82A, page 350).

Class 141.P. 2–8–2 (4-cylinder compound). Built Schneider, 1942–47.

These engines were the first new standard design to be built in numbers after nationalization. The "Mikado" type was extensively adopted for mixed traffic in France and was in service on three of the principal railways prior to 1914. The 141.P's have the traditional 4-cylinder compound arrangement and were designed to operate either 550-ton passenger or 1,200-ton freight workings. They develop around 3,900 H.P., and the design is based on class 141.E of the P.L.M.

43. French National Railways (S.N.C.F.) (Plate 82B, page 350).

Class 150.P. 2–10–0 (4-cylinder compound). Built La Chapelle, 1947.

These very remarkable mixed traffic engines were introduced on the Northern Railway of France in 1933. Phenomenally silent and smooth-running at high speeds, even up to 70 m.p.h., they are rated to haul a 554-ton passenger train up 1 in 200 grades at 55–60 m.p.h., developing 2,200 d.b.h.p. Alternatively, they can operate 2,214-ton freight loads at 31 m.p.h. Such versatility, involving engine speeds of nearly 400 r.p.m., is attained by using exceptionally large valves and steam passages with very light moving parts and with an adhesive weight of over 90 tons. The receiver pressure is limited to 116 p.s.i. Frames are of plate 1·4 in. thick and cast steel is extensively used in construction. The Northern was one of the few railways to retain the long, narrow grate (in this case 11 ft 6 in. × 3 ft 3·4 in.) with advanced modern designs. These later French standard engines have greater heating surface and slightly larger tenders.

44. French National Railways (S.N.C.F.) (Plate 82C, page 350).

Class 151.TQ. 2–10–2T. Built Corpet, Louvet & Cie, 1940.

The 2–10–2 tank type was introduced on the Eastern Railway of France in 1913 for heavy iron-ore traffic. The standard class illustrated was based on the more compact design used on the Belt (Ceinture) Railway of Paris, where short hauls with heavy loads form the principal traffic. They are employed on heavy suburban passenger and freight work.

PART IV. STANDARD GAUGE: FRENCH

EXAMPLE NO.		38 (Plate 80A, page 340)	39 (Plate 80B, page 340)	40 (Plate 81A, page 349)	41 (Plate 81B, page 349)	42 (Plate 82A, page 350)	43 (Plate 82B, page 350)	44 (Plate 82C, page 350)
RAILWAY		*SNCF*	*SNCF*	*SNCF*	*SNCF*	*SNCF*	*SNCF*	*SNCF*
Class		242.A.	241.P.	232.U.	231.E.	141.P.	150.P.	151.TQ.
Type		4–8–4	4–8–2	4–6–4	4–6–2	2–8–2	2–10–0	2–10–2T
Cylinders (no.) dia.×stroke	in.	(1) 23·6× 28·35	(2) 17·56× 25·6	(2) 17·56× 27·56	(2) 16·5× 25·6	(2) 16·14× 27·56	(2) 19·3× 25·2	(2) 24·8× 26
Ditto. (L.P.)	in.	(2) 26·8× 29·9	(2) 26·5× 27·56	(2) 26·77× 27·56	(2) 25·6× 25·6	(2) 25·2× 27·56	(2) 26·77× 27·56	—
Coupled wheel dia.	in.	76·77	79·5	78·74	76·77	64·96	61	53·15
Wheelbase, coupled	ft in.	20′ 2·12″	20′ 8″	14′ 4·8″	13′ 5·4″	18′ 4·47″	22′ 10·8″	19′ 8·2″
Wheelbase, total	ft in.	75′ 7″*	75′ 5·9″	74′ 1·56″	66′ 10·55″	66′ 4·85″	62′ 9·5″	36′ 8·95″
Tractive force (85%)	lb.	47,077 (v)	45,084 (v)	46,958 (v)	28,440 (v)	48,510 (v)	59,650 (v)	50,706
Boiler pressure	p.s.i.	290	290	290	246·5	290	261	203
Grate area	sq. ft	53·8	54·4	55·7	46·6	45·8	38·0	38·75
Heating surface	sq. ft	2,720 (s)	2,351 (a)	2,098	2,123	2,174	2,071	1,798
Superheating surface	sq. ft	1,285	1,166	941	861	936	884	958
Weight full, engine	lb.	326,283	289,687	284,396	234,871	246,136	230,714	257,280
Ditto tender	lb.	167,728	182,102	185,092	177,472	186,952	187,833	—
Coal capacity	lb.	25,133	19,842	25,353	20,282	26,455	19,842	10,582
Water capacity	gal.	7,200	7,630	7,710	8,160	7,208	8,056	3,000
Max. axle load	lb.	46,297	44,974[1]	50,706	41,888[2]	41,888[3]	41,006[4]	38,581
Over-all length	ft in.	87′ 3·8″*	90′ 8·9″	87′ 4·6″	77′ 7″	77′ 10·25″	73′ 8·25″	50′ 3·5″
Over-all height	ft in.	14′ 0·5″	13′ 11·5″	14′ 0·5″	14′ 0·5″	14′ 0·5″	14′ 0″	13′ 11·8″
Over-all width	ft in.	10′ 2″	9′ 8·9″	9′ 9·8″	9′ 10″	9′ 10″	10′ 6·3″	10′ 4·5″
Normal max. speed	m.p.h.	87	75	87	87	65	65	43
Min. track radius at slow speed	ft	403	460	198	441	345	345	330

(v) Figures from the S.N.C.F. diagrams for compound working.
Some modern French tenders have heavier axle loading than the locomotives: [1] 45,535 lb., [2] 44,643 lb., [3] 47,620 lb., [4] 47,840 lb.
Dimensions marked * have been scaled from drawings and diagrams or computed from data.

Part V. Standard gauge: Austrian, Czechoslovak and Scandinavian

45. Austrian Federal Railways (Ö.B.B.) (Plate 83A, page 351).

Class 12 (old 214). 2–8–4. Built Floridsdorf, 1936.

As in earlier years the Austrians had adopted the 2–6–4 in place of the "Pacific", the tradition was continued in the preference for the 2–8–4 to the "Mountain", since the Krauss-Helmholz truck was widely used. A 2-cylinder engine was built in 1928 and a 3-cylinder the following year. After trials the 2-cylinder type was adopted and was claimed to be the most economical passenger engine in Europe. Six more were built in 1931 and six in 1936 (one of which is illustrated), and the design was adopted in Rumania, where seventy-nine were built. The great skill and elegance of Austrian design is shown in producing such a large and powerful machine with such a low weight. The connecting rods are the longest used in modern practice (just under 14 ft centres) and the expansion link is exceptionally large. Lentz oscillating cam valves are fitted, as they are to all modern Austrian designs. The trailing bogie is pivoted over its front axle. The plate frames are only only 1¼ in. thick, but are braced over the coupled wheelbase by a continuous steel casting. Designed to operate 550-ton trains at 37 m.p.h., on a grade of 1 in 91, 3,000 H.P. has been reached. The service maximum is 75 m.p.h., but 96 m.p.h. has been attained. After World War II some were fitted for oil burning, with very economical results.

46. Austrian Federal Railways (Ö.B.B.) (Plate 83B, page 351).

Class 33 (old 113). 4–8–0. Built State Works, 1923–28.

For heavy, semi-fast passenger trains the 4–8–0 has proved a very popular type in Austria, Hungary and Spain, and eighty engines similar to the Austrian class illustrated were built for Poland. They are the principal passenger class on the southern main lines that are still steam-operated. Although the oldest machines described here, practically no modifications have been found necessary or desirable, with the exception of the Giesl ejector, which is now replacing the normal exhaust arrangement. This greatly improved draught arrangement first used five and later seven nozzles in line. For light working a damper is fitted, which cuts off the lower tubes and concentrates the gas flow over the superheater elements.

47. Austrian Federal Railways (Ö.B.B.) (Plate 84A, page 352).

Class 78 (old 729). 4–6–4T. Built Florisdorf, 1931–38.

The final type evolved from a long line of 2–6–2T's and 4–6–2T's. The bogie pivot centres are placed inward of the centre line by 4¾ in. on the front, and by 2 in. on the rear truck. This may account for their remarkably steady riding qualities, often noticeably absent on 4–6–4 tank engines.

48. Austrian Federal Railways (Ö.B.B.) (Plate 84B, page 352).

Class 297. 2–12–2T. rack and adhesion. Built Florisdorf, 1941.

The twelve-mile section between Eisenerz and Vordernberg, on the line reaching the Styrian iron ore area, has maximum gradients of 1 in 38 with adhesion, and 1 in 14 with Abt rack system. Since 1890, combined rack and adhesion 0–6–2 tank engines have worked the line, with three 0–12–0 tank engines added in 1912. The two large engines here described are the largest of their type and can haul 400 tons up the 1 in 14 grade at 9 m.p.h. Hard-wearing surfaces are reinforced with manganese steel. In addition to vacuum and compressed-air brakes, the Riggenbach counter pressure system (*see* Chapter 9, page 000) and a special mechanical emergency brake are fitted.

49. Austrian Federal Railways (Ö.B.B.) (Plate 84C, page 352).

Class 3071 (old Dt-1). 2–4–2T. Built Florisdorf, 1935–40.

These remarkable little engines were designed for long-distance, light-weight passenger trains and have a wider range than railcars. Oil-fired, and operated by one man, in a cab with an all-round view, the guard's compartment is also incorporated in the engine. The cylinders have Lentz oscillating cam valves, and part of the tank is used for feed-water heating. With a three-coach, 110-ton train they can run at 60 m.p.h. or more, and on a 1 in 100 grade at 40 m.p.h. Long runs, such as Vienna-Klagenfurt (208 miles) include the climb over the Semmering (2,940 ft, with grades of 1 in 40), but an average speed of 40 m.p.h. could be maintained. Speeds of 74 m.p.h. have been recorded, up to 84·5 m.p.h. reported, and trains of 160 tons worked at times. By reducing the oil and water capacity the axle load can be reduced to twelve tons for working light branch lines. Similar one-man locomotives with coal fed from a hopper were in use in South Germany many years ago. In recent years the 3071 class have been relegated to branch line work with a crew of two, and altered for coal burning.

50. Czechoslovak State Railways (C.S.D.) (Plate 85A, page 353).

Class 498.1. 4–8–2 (3-cylinder). Built Skoda, 1950.

The very modern design of these engines includes bar frames, roller bearings on all axles, mechanical stoker, arch tubes, thermic siphon, and Kylchap blast pipe. At the same time, the typical elegance and economy of Czechoslovak design is apparent. The three cylinders are operated by three sets of valve gear, the drive for the inside set being derived externally on the right side from a crank on the coupled axle immediately behind the driving axle. This well-spaced-out arrangement is standard practice. With alternative spring gear the axle load can be reduced from 18·6 to 16 tons. The engines are beautifully finished in blue with an aluminium band; wheels and framing are red.

51. Czechoslovak State Railways (C.S.D.) (Plate 85B, page 353).

Class 477.0. 4–8–4T (3-cylinder). Built Skoda, 1956.

Very fine examples of the 4–8–4 tank are to be found in Czechoslovakia where the type was introduced in 1934 for running equally well in both directions. There are several classes with both two and three cylinders and some having tanks at the back end only. The class illustrated is the latest of the side tanks and the three sets of valve gear are arranged as in the preceding paragraph. They can be used on all types of work, from expresses on steeply graded sections to short-distance heavy mineral and

PART V. STANDARD GAUGE: AUSTRIAN, CZECHOSLOVAK AND SCANDINAVIAN

EXAMPLE NO.		45 (Plate 83A, page 351)	46 (Plate 83B, page 351)	47 (Plate 84A, page 352)	48 (Plate 84B, page 352)	49 (Plate 84C, page 352)	50 (Plate 85A, page 353)	51 (Plate 85B, page 353)	52 (Plate 85C, page 353)	53 (Plate 86A, page 354)	54 (Plate 86B, page 354)	55 (Plate 86C, page 354)	56 (Plate 86D, page 354)
RAILWAY		ÖBB	ÖBB	ÖBB	ÖBB	ÖBB	CSD	CSD	NSB	SJ	DSB	DSB	VG
Class		12	33	78	297	3071	498.1	477.0	49c	E10	E	H	—
Type		2-8-4	4-8-0	4-6-4T	2-12-2T	2-4-2T	4-8-2	4-8-4T	2-8-4	4-8-0	4-6-2	2-8-0	2-6-0T
Cylinders (no.) dia. × stroke	in.	(2) 25·6 × 28·35	(2) 22 × 28·35	(2) 19·7 × 28·35	(2) 24 × 20·47	(2) 11·4 × 22·4	(3) 19·7 × 26·8	(3) 17·7 × 26·8	(2) 17·3 × 25·6	(3) 17·7 × 24	(2) 16·5 × 26	(3) 18·5 × 26·4	(2) 14·17 × 21·26
Ditto (L.P.)	in.	—	—	—	(2) 15·75 × 19·7 (r)	—	—	—	(2) 25·6 × 27·56	—	(2) 24·8 × 26	—	—
Coupled wheel dia.	in.	76·4	68·5	62	40·5	57	72	64	60·2	55·12	74·6	55·27	43·3
Wheelbase, coupled	ft in.	20' 4·5" (k)	18' 2·5"	10' 9·7"	22' 9·6"	10' 6"	19' 0"	19' 0·25"	16' 7·2"	15' 2·9"	12' 11·5"	20' 8" (k)	9' 6·17"
Wheelbase, total	ft in.	61' 7"	56' 4·14"	38' 11·7"	37' 6·8"	25' 0·5"	71' 4"	45' 11"	61' 1·3"	47' 6·9"	59' 8·5"	54' 5·5"	18' 0·5"
Tractive force (85%)	lb.	44,030	34,066	28,040	51,904 (r)	9,860	41,670	44,750	36,145 (c)	45,336	25,073 (c)	35,580	10,072
Boiler pressure	p.s.i.	213	213	185	227·6	227·6	227·6	227·6	242	199	185	170	170
Grate area	sq. ft	50·8	47·1	38·2	41·9	8·9	52·2	46·7	53·8	30·0	38·7	28·0	12·9
Heating surface	sq. ft	3,000	2,345	1,983	2,268	457	2,459 (a) (s)	2,150 (f) (s)	2,765 (f)	1,481	1,987	1,732	550
Superheating surface	sq. ft	979	641	441	780	177	794	813	1,098	576	684	592	245
Weight full, engine	lb.	255,923	187,746	240,920	277,421	92,595	250,230	292,743	218,477	163,583	194,888	179,235	83,775
Weight full, tender	lb.	132,277	96,651	—	—	—	178,130	—	119,270	92,394	121,695	125,663	—
Coal capacity	lb.	17,300	15,432	8,820	7,716	2,425	33,070	15,432	18,520	15,432	14,330	17,637	1,984
Water capacity	gal.	6,435	5,950	3,700	2,025	1,035	7,714	3,306	6,116	3,637	5,510	5,950	1,102
Max. axle load	lb.	37,597	32,760	35,705	36,150*	28,660	41,664	38,530*	34,172	28,219	39,683	37,478	22,046
Over-all length	ft in.	72' 0"	67' 8·2"	49' 2"	48' 6·6"	36' 10"	83' 10·6"	56' 9·5"	73' 1"	59' 4"	69' 9·2"	63' 1"	32' 5·75"
Over-all height	ft in.	15' 3"	15' 3"	14' 10·3"	15' 2·7"	13' 5·4"	14' 8·77"	14' 7"	14' 0·3"	14' 2·5"	14' 1·3"	14' 1·3"	12' 8·4"
Over-all width	ft in.	10' 4"	10' 0·7"	10' 0·5"	10' 0"	9' 6"	10' 0"*	10' 0"	10' 6·77"	10' 1·7"	10' 4"	10' 4"	9' 8·5"
Normal max. speed	m.p.h.	75	53	65	18·5 (r)	62·5	75	62·5	62	52	68	50	28
Min. track radius at slow speed	ft	492	394	492	590	492	492	492	394	394	590 (m)	623 (m)	230

(r) ÖBB class 297 has two rack pinions driven at a ratio of 2 : 1, the tractive force being 69,705 lb. and the maximum speed allowed when in operation is 15·5 m.p.h.

suburban trains. In view of the high output required, they are equipped with mechanical stokers, rare in tank engines.

52. Norwegian State Railways (N.S.B.) (Plate 85C, page 353).

Class 49c. 2–8–4 (4-cylinder compound). Built Krupp, 1940.

The Norwegian main line between Dombaas and Trondheim surmounts an altitude of 3,415 ft, over the Dovre Mountain, with many grades of 1 in 46–56 and curves of 650–980 ft radius, the axle load being limited to 15½ tons. Great skill has been shown in the design of these 4-cylinder compound 2–8–4's, built to operate 300-ton trains under these conditions and to maintain 37 m.p.h. on the 1 in 56 grades. Weight reduction was a major problem, the frames being only of 1 in. plate but strongly braced and the platform is of aluminium. Earlier engines of the same general design were built in Norway in 1935–36, and known as the Dovregubben (Dovre Giant) class.

53 Swedish State Railways (S.J.) (Plate 86A, page 354).

Class E 10. 4–8–0 (3-cylinder). Built Nydqvist & Holm 1947.

Swedish locomotives are notably distinctive, combining simplicity with the most advanced practice and elegant design. The finish is also excellent, with black paintwork, planished steel clothing and polished motion. Although electrification has been proceeding steadily for many years, there are still some hundreds of steam locomotives on the State railways. The 3-cylinder 4–8–0 mixed traffic engine illustrated has bar frames and three sets of Walschaerts valve gear. The tender with semi-circular tanks has long been a Swedish characteristic.

54. Danish State Railways (D.S.B.) (Plate 86B, page 354).

Class E. 4–6–2 (4-cylinder compound). Built Frichs, 1947.

This design originated in Sweden, where the engines rendered excellent service. With the extension of electrification they became redundant and were transferred to Denmark in 1937. Subsequently additional engines were built in Denmark. All four cylinders drive the centre coupled axle at the same inclination, the low pressure being outside. Only two valves are fitted and an ingenious by-pass arrangement enables the engine to be worked as a 4-cylinder compound or a 2-cylinder simple on either the high-pressure or low-pressure cylinders alone. Balancing is exceptionally good. The plate frames are 1·18 in. thick, the rear part being of cast steel. Class E are capable of working fast trains of 650 tons on grades up to 1 in 143, speed being limited to 68 m.p.h.

55. Danish State Railways (D.S.B.) (Plate 86C, page 354).

Class H. 2–8–0 (3-cylinder). Built Frichs, 1941.

The 3-cylinder freight engines of Class H have three separate sets of valve gear, the inside motion being derived from an additional return crank on the left side, as shown in the illustration. By placing these cranks on the third axle a well-set-out gear can be fitted and cramped spacing avoided. Fine examples of the medium European freight engine, they are remarkably free from slipping, and can take loads of 900 tons on grades of 1 in 100.

56. Varde-Grinsted Railway (Denmark) (Plate 86D, page 354).

2–6–0T. No. 5. Built Frichs, 1926.

The number of the once very numerous small private railways of Europe has been steadily diminishing over the last twenty-five years as a result of road competition. Others now use railcars, but there are still some steam locomotives in service. Small tank engines of great variety and often interesting design have been employed. The example illustrated is for a standard gauge private railway in Denmark. Normally limited to 28 m.p.h. it is capable of up to 40 m.p.h., and is rated to take 250 tons on the steepest grade of 1 in 70.

Part VI. Standard gauge: German and South-East European

57. German Federal Railway (D.B.) (Plate 87A, page 355).

Class 10. 4–6–2 (3-cylinder). Built Krupp, 1956.

Two engines have been built with a view to greatly reduced maintenance and repair costs. The frames are of I-section, prefabricated and welded to cross-members and to the cast-steel cylinder block. The whole assembly resembles a cast-steel bed, but is considerably lighter. Cast steel is used throughout for fittings, which can be welded to the frames or the boiler, also of all-welded construction. Roller bearings are used wherever possible and the crank axle is formed in two pieces only, which can be dismantled and reassembled by oil pressure. Surfaces requiring attention or renewal have been reduced to the absolute minimum. Streamlining has only been applied where it is of proven value. The cab is roomy and very conveniently arranged, all lubrication points being centred in it, and even footwarming is provided. The tender is of all-welded construction with arched bottom and sliding roof to the coal bunker. The first engine has supplementary oil firing to relieve the fireman of up to 30 per cent of his work, whilst the second is fully oil-fired.

S.N.C.F.

PLATE 81A. No. 40: French National Railways (S.N.C.F.) (*see* page 344).

S.N.C.F.

PLATE 81B. No. 41: French National Railways (S.N.C.F.) (*see* page 344).

S.N.C.F.

PLATE 82A. No. 42: French National Railways (S.N.C.F.) (*see page* 344).

P. Ransome-Wallis

PLATE 82C. No. 44: French National Railways (S.N.C.F.) (*see page* 345).

P. Ransome-Wallis

PLATE 82B. No. 43: French National Railways (S.N.C.F.) (*see page* 344).

P. Ransome-Wallis

PLATE 83A. No. 45: Austrian Federal Railways (Ö.B.B.) (*see* page 345).

P. Ransome-Wallis

PLATE 83B. No. 46: Austrian Federal Railways (Ö.B.B.) (*see* page 346).

Austrian Federal Railways

PLATE 84C. No. 49: Austrian Federal Railways (Ö.B.B.) (*see* page 346).

Austrian Federal Railways

PLATE 84A. No. 47: Austrian Federal Railways (Ö.B.B.) (*see* page 346).

Austrian Federal Railways

PLATE 84B. No. 48: Austrian Federal Railways (Ö.B.B.) (*see* page 346).

Skoda Locomotive Works

PLATE 85A. No. 50: Czechoslovak State Railways (C.S.D.) (*see* page 346).

Skoda Locomotive Works

PLATE 85B. No. 51: Czechoslovak State Railways (C.S.D.) (*see* page 346).

P. Ransome-Wallis

PLATE 85C. No. 52: Norwegian State Railways (N.S.B.) (*see* page 348).

P. Ransome-Wallis

PLATE 86B. No. 54: Danish State Railways (D.S.B.) (*see* page 348).

A. S. Frichs

PLATE 86D. No. 56: Varde-Grinsted Railway (Denmark) (*see* page 348).

Swedish State Railways

PLATE 86A. No. 53: Swedish State Railways (S.J.) (*see* page 348).

P. Ransome-Wallis

PLATE 86C. No. 55: Danish State Railways (D.S.B.) (*see* page 348).

Fried Krupp

PLATE 87B. No. 58: German Federal Railway (D.B.) (*see* page 357).

Henschel & Sohn

PLATE 87D. No. 60: German Federal Railway (D.B.) (*see* page 357).

Fried Krupp

PLATE 87A. No. 57: German Federal Railway (D.B.) (*see* page 348).

Fried Krupp

PLATE 87C. No. 59: German Federal Railway (D.B.) (*see* page 357).

355

Henschel & Sohn

PLATE 88A. No. 61 : German Federal Railway (D.B.) (*see* page 357).

Krauss-Maffei

PLATE 88B. No. 62 : German Federal Railway (D.B.) (*see* page 357).

Fried Krupp

PLATE. 88C No. 63 : East Hanover Railway (O.H.E.) (*see* page 357).

58. German Federal Railway (D.B.) (Plate 87B, page 355).

Class 23. 2–6–2. Built Krupp, 1953.

Of the 3,850 4–6–0's of the Prussian State Class P 8, built 1906–24, some 2,975 were taken into the German State Railway in 1924. These once ubiquitous and useful engines, with 69-in. coupled wheels, have, like old soldiers, been gradually fading away. For their replacement the Class 23 2–6–2's were designed, with welded I-section frames and the latest refinements of German practice. The all-welded tender with arched tank bottom and tall bunker will be noted. So far this is the most numerous of the post-war replacement classes.

59. German Federal Railway (D.B.) (Plate 87C, page 355).

Class 44. 2–10–0 (3-cylinder). Built Krupp, 1938.

The standard German heavy freight locomotive which was developed from experience with a great many engines of this type. A 3-cylinder 2–10–0 design was decided on in 1915 and in 1917 a change was made to bar frames with a much higher pitched boiler. In 1927, both 2- and 3-cylinder engines were built, but the former were discontinued in spite of cheaper first cost. Between 1913 and 1924 over 1,800 3-cylinder engines were built in Germany, with various forms of conjugated valve gear for the inside valve. Thereafter practice progressed in the opposite direction, with three sets of valve gear, first with a third return crank outside, next with an inside eccentric and, finally, as on these engines, with an inside crank. Class 44 have 4-in. bar frames, Krauss-Helmholz leading trucks and, at slow speed, can traverse curves of 460 ft radius. They haul over 2,000 tons on level track and can take 550 tons up 1 in 40 at slow speed.

60. German Federal Railway (D.B.) (Plate 87D, page 355).

Class 50. 2–10–0. Built Henschel, 1938–44.

The heavy engines of Class 44 were restricted to certain routes, and a much lighter design with only 15½-ton axle load was brought out to replace the old 0–10–0's of Class G-10, which were suitable for practically all lines. The 50's also subsequently formed the basis of the German war-time austerity classes: 50-UK, 52 and the larger 42. Altogether, some 10,650 engines of this basic design were built (including 900 of Class 42) the most numerous steam locomotive group in the world. During the war it was estimated that the capacity under German control could, if undisturbed, produce 12,500 such engines annually. Further engines with welded frames are now built in East Germany. The previous largest groups were the Class G-8[1] Prussian State 0–8–0's, which totalled some 5,260 engines, and the U.S.S.R. Group E 0–10–0's.

61. German Federal Railway (D.B.) (Plate 88A page 356).

Class 66. 2–6–4T. Built Henschel, 1956.

The Class 66 tank engines are the most modern in Germany and are intended to replace earlier 2–6–2 and 4–6–4 tanks, as well as covering the lighter duties of the 4–6–0's of Class P8. They therefore have a very wide range of service. Their construction is contemporary with the Class 10 4–6–2's and designed to reduce maintenance to the minimum. The first two axles are combined in an improved form of Krauss-Helmholz truck. An entirely new feature is introduced in linking the leading truck and trailing bogie by a system of levers, to equalize flange wear. The setting of this differs for foward and backward running, the change-over being automatically effected by compressed air when reversing.

62. German Federal Railway (D.B.) (Plate 88B, page 356).

Class 65. 2–8–4T. Built Krauss-Maffei, 1951.

The 2–8–2 tank has been adopted extensively on the Continent for working suburban passenger services and Class 65 has been developed to replace former German engines of this type. The adoption of a trailing bogie disposes of the large overhang at the back and heavy loading on a single axle. Welded construction is used in both frames and boiler.

63. East Hannover Railway (Ö.H.E.) (Plate 88C, page 356).

Class 84. 2–10–2T. Built Krupp, 1952.

The 2–10–2 tank type was introduced in 1913 for the heavy ore traffic on the Eastern Railway of France. In 1920 the same type on the Halberstadt-Blankenburg Railway successfully worked on 1 in 16 grades and enabled the rack system to be dispensed with. Subsequently both 2- and 3-cylinder classes were built for the Prussian State and German Federal railways for service on heavily graded sections. For industrial systems a number of very fine designs were brought out, some of them approaching 140 tons weight in working order, thereby ranking amongst the heaviest tank engines built. The modern example described is of medium weight, having bar frames of 3·15-in. thickness. It is capable of hauling a 500-ton train at 50 m.p.h. on the level and maintaining 25 m.p.h. on a grade of 1 in 100.

64. Hungarian State Railways (M.A.V.) (Plate 89A, page 365).

Class 303. 4–6–4. Built Mavag, 1950.

The locomotive illustrated is the 6,000th built at the Mavag works. German rather than the Austrian influence of former times will be noticed in this post-war design. The totally enclosed cab is very neat and a Standard

PART VI. STANDARD GAUGE: GERMAN AND SOUTH-EAST EUROPEAN

EXAMPLE NO.		57 (Plate 87A, page 355)	58 (Plate 87B, page 355)	59 (Plate 87C, page 355)	60 (Plate 87D, page 355)	61 (Plate 88A, page 356)	62 (Plate 88B, page 356)	63 (Plate 88C, page 356)	64 (Plate 89A, page 365)	65 (Plate 89B, page 365)	66 (Plate 90, page 366)	67 (Plate 91A, page 367)	68 (Plate 91B, page 367)	69 (Plate 91C, page 367)
RAILWAY		DB	DB	DB	DB	DB	DB	OHE	MAV	BDZ	BDZ	TCDD	TCDD	CEH
Class		10	23	44	50	66	65	84	303	11	46	46.051	57.01	Mα
Type		4-6-2	2-6-2	2-10-0	2-10-0	2-6-4T	2-8-4T	2-10-2T	4-6-4	4-10-0	2-12-4T	2-8-2	2-10-2T	2-10-2
Cylinders (no.) dia. × stroke	in.	(3) 18·9 × 28·35	(2) 21·6 × 26	(3) 21·6 × 26	(2) 23·6 × 26	(2) 18·5 × 26	(2) 22·4 × 26	(2) 24·4 × 26	(2) 21·6 × 27·5	(3) 20·5 × 27·5	(3) 21·6 × 25·6	(2) 25·6 × 26	(3) 22·4 × 26	(2) 26 × 29·5
Coupled wheel dia.	in.	78·7	69	55	55	63	59	53	78·7	57	52·75	·68·9	55·12	63
Wheelbase, coupled	ft in.	15' 1"	13' 1·5" (k)	22' 3·7" (k)	21' 7·8" (k)	12' 1·6" (k)	17' 2·7" (k)	21' 0" (k)	14' 9"	17' 4·6" (k)	25' 9·36" (k)	18' 8·4" (k)	21' 3·7" (k)	22' 8·4" (k)
Wheelbase total	ft in.	72' 9¼"	57' 9·9"	62' 11·5"	61' 11·7"	36' 3"	39' 3·45"	37' 8·45"	71' 0·75"	75' 9·6"	48' 2·9"	63' 10·1"	41' 0"	71' 3"
Tractive force (85%)	lb.	37,037	33,870	63,834	50,770	27,254	37,400	49,400	34,400	58,700	65,980	44,968	68,847	63,350
Boiler pressure	p.s.i.	256	227·6	227·6	227·6	227·6	199	199	256	227·6	227·6	227·6	227·6	256
Grate area	sq. ft	42·6	33·46	48·9	41·96	21·0	28·6	36·6	59·18	52·7	52·4	43·1	43	60·3
Heating surface	sq. ft	2,221 (f)	1,681 (f)	2,550 (f)	2,057	940 (f)	1,508 (f)	1,845	2,797	2,411 (f)	2,407 (f)	2,401	2,233	3,359
Superheating surface	sq. ft	1,137	794	1,076	682	481	677	624	839	860	860	1,138	846	1,346
Weight full, engine	lb.	263,452	184,306	242,508	196,432	206,683	250,224	226,635	250,886	241,626	343,800	230,048	299,828	295,680
tender	lb.	189,600	136,637	161,378	134,040				165,347	154,334		139,888		141,120
Coal capacity	lb.	19,850²	17,640	22,050	17,637	11,020	9,921	8,820	28,660	28,660	22,050	17,584	11,023	26,455
Water capacity	gal.	8,820	6,830	7,500	5,730	3,086	3,086	2,425	5,510	6,170	3,967	6,380	3,196	5,500
Max. axle load	lb.	48,500	41,890²	42,550	34,061	34,723	37,480	33,800	38,140	37,479	39,700*	40,300	44,092*	44,092
Over-all length	ft in.	86' 11·4"	69' 11·56"	74' 2·5"	75' 3·14"	48' 4·7"	50' 9·25"	49' 8·45"	83' 6·16"	75' 9·6"	59' 6·76"	75' 0"	54' 3·57"	82' 0·25"
Over-all height	ft in.	14' 11"	14' 11"	14' 11"	14' 9"	14' 11"	14' 11"	13' 11·65"	14' 11·5"	15' 0·3"*	14' 3"	14' 0·5"	14' 0·5"	14' 9·55"
Over-all width	ft in.	10' 0"	10' 0"	10'· 2"	10' 2"	10' 2"*	10' 2"	10' 2"	10' 0·3"	9' 11"	10' 2"*	10' 4"	10' 0·4"	—
Normal max. speed	m.p.h.	87	68	50	50	62	53	53	75	47	40	62	43·5	56
Min. track radius at slow speed	ft	460	460	460	460	460	460	328	590	460	560	492	460	985 (m)

¹ Also 1,000 gallons oil fuel. The second engine has 2,645 gallons of oil only – no coal.

² Can also be adapted for lighter track with a maximum of 37,480 lb.

mechanical stoker is fitted. All carrying axles are equipped with Isothermos axle-boxes. The engine is designed to operate 500-ton passenger trains at speeds up to 75 m.p.h.

65. Bulgarian State Railways (B.D.Z.) (Plate 89B, page 365).

Class 11. 4–10–0 (3-cylinder). Built Henschel, 1941.

The first 4–10–0 was the famous *El Gubernator* of the Central Pacific Railroad, built in 1884. Very few engines of this wheel arrangement have materialized, although they have been proposed and the type was given the name "Mastodon". These engines are well suited to Balkan conditions, with long hauls and frequent grades, often in very severe weather.

66. Bulgarian State Railways (B.D.Z.) (Plate 90, page 366).

Class 46. 2–12–4T. (3-cylinder). Built Berlin, 1943.

Twelve-coupled tank engines are rare and only on two railways have they been extensively used in normal traffic. Between 1912 and 1921 about forty were built for the 3 ft 6 in. gauge Indonesian State Railways, in Java and Southern Sumatra. In Bulgaria 0–12–0 tanks were introduced in 1919, and the 2–12–4T type with two cylinders in 1931, for the heavy but short-distance coal traffic from the coalfields to Sofia. The 3-cylinder engines here described first appeared in 1943, from German builders, and are the heaviest non-articulated tank engines so far built. The wheel tyres of two of the coupled axles are flangeless.

67. Turkish State Railways (T.C.D.D.) (Plate 91A, page 367).

Class 46.051. 2–8–2. Built Henschel, 1927.

Fine examples of the European 2–8–2 intended for passenger work, these engines operate trains of 400–500 tons between Haydarpasa and Ankara.

68. Turkish State Railways (T.C.D.D.) (Plate 91B, page 367).

Class 57.01. 2–10–2T. Built Henschel, 1951.

Four engines of this class are used for banking over the ten miles between Bilecik and Karakoy on the main line between Haydaparsa and Eskisehir. The ruling gradient is 1 in 33·5 up from Bilecik. The 2–10–2 tank is well suited for this work and is widely used thereon, particularly in Germany, where a number of classes, both 2- and 3-cylinder, have been built.

69. Hellenic State Railways (C.E.H.) (Plate 91C, page 367).

Class M α 2–10–2. Built Breda & Ansaldo, 1954.

Considerably larger than any previous locomotives used in Greece, they are intended for passenger and freight work between Salonika and Athens. They have plate frames, and about half of the class of twenty engines are equipped for burning oil fuel.

Part VII. Broad gauge: 5ft 6 in. – 5ft 0 in.

70. Spanish National Railways (R.E.N.F.E.) (Plate 92A, page 368).

Class 242–2001. 4–8–4. Built Maquinista, 1955.

Since World War I, large locomotives have been constructed in Spain and for many years most of the new work for the National Railways has been carried out there. As a result, modern Spanish engines have a distinctive style, being massive without appearing heavy, with clean lines and a fine finish. The class shown has two cylinders and oscillating cam valves operated by Walschaerts gear. The spokes of the coupled wheels are of "U" section and both sides of the wheels are braked. A four-wheeled pony truck is mounted at the rear and provided with Isothermos axle-boxes, as are the tender bogies. A double chimney and blast-pipe is fitted and the engines are oil-burners. They are used on heavy passenger trains where the long grades of the Spanish railways call for high sustained power output.

71. Spanish National Railways (R.E.N.F.E.) (Plate 92B, page 368).

Class 151–3101. 2–10–2 (3-cylinder). Built Maquinista, 1942.

Oscillating cam-operated valves are fitted to these heavy Spanish freight engines, but there are three cylinders each with a separate set of Walschaert motion. The first two axles are united in a Krauss truck which reduces the rigid wheelbase to 16 ft 4 in. At 12 m.p.h. they are rated to take 2,300 tons on a grade of 1 in 200, down to 600 tons on 1 in 40. The maximum speed permitted is 56 m.p.h.

72. Argentina (F.N. General Roca) (Plate 92C, page 368).

Class 15B. 4–8–0. Built Vulcan Foundry, 1949.

The broad gauge of the Argentine allows room between the 1-in. plate frames for a firebox of considerable size and in more recent years preference has been shown for the 4–8–0 as against the 2–8–2 type. The 15B class, with a maximum axle load of only 15·75 tons, are lighter than earlier engines of Class 15A. The tender oil tank can be removed and 7 tons of coal carried if a change-over is desired. The wooden cowcatcher and

PART VII. BROAD GAUGE: 5ft 6in.—5ft 0in.

		70 (Plate 92A, page 368)	71 (Plate 92B, page 368)	72 (Plate 92C, page 368)	73 (Plate 93A, page 369)	74 (Plate 93B, page 369)	75 (Plate 93C, page 369)	76 (Plate 93D, page 369)	77 (Plate 94A, page 370)	78 (Plate 94B, page 370)	79 (Plate 94C, page 370)
EXAMPLE NO.											
RAILWAY		RENFE	RENFE	FN	IGR	IGR	VGR	SAR	USSR	USSR	USSR
Class		242-2001	151-3101	15B	WG	WU	R	520	6998	P.36	LV
Type		4-8-4	2-10-2	4-8-0	2-8-2	2-4-2T	4-6-4	4-8-4	4-6-4	4-8-4	2-10-2
Gauge	ft in.	5' 6"	5' 6"	5' 6"	5' 6"	5' 6"	5' 3"	5' 3"	5' 0"	5' 0"	5' 0"
Cylinders (no.) dia. × stroke	in.	(2) 25·2× 28	(3) 22·4× 29·5	(2) 19·5× 28	(2) 21·9× 28	(2) 13× 26	(2) 21·5× 28	(2) 20·5× 28	(2) 26·4× 30·3	(2) 22·64× 31·5	(2) 25·6× 31·5
Coupled wheel dia.	in.	74·8	61·4	68	61·5	61·5	73	66	86·6	72·8	59
Wheelbase, coupled	ft in.	21' 9·9"	22' 5·3" (k)	18' 6"	17' 1"	9' 9"	12' 10"	17' 9"	16' 0·9"	19' 2·3"	21' 3·9"
Wheelbase, total	ft in.	75' 3"	72' 9"	60' 7·75"	68' 6"	27' 0"	67' 0"	77' 3·63"	86' 3·4"	86' 6·5"	71' 0·7"
Tractive force (85%)	lb.	46,865	72,050	30,000	38,890	12,750	32,080	32,600	44,150	46,923	59,185
Boiler pressure	p.s.i.	227·6	227·6	225	210	210	210	215	213	213	199
Grate area	sq. ft	57	57	32·6	45	18·5	42	45	75·75	72·63	69·2
Heating surface	sq. ft	3,160	3,160	1,752	2,237	742	2,243	2,454 (s)	2,668 (a)	2,617	2,549 (a)
Superheating surface	sq. ft	1,125	1,516	348	683	182	462	651	1,280	1,536	1,469
Weight full, engine	lb.	313,500	308,647	181,688	228,144	141,695	240,016	249,312	304,238	297,624	267,861
Weight full, tender	lb.	149,914	138,890	148,092	160,832	—	178,528	200,200	279,987*	209,436	180,780*[2]
Coal capacity	lb.	—	17,637	15,430	40,320	6,720	13,440	21,840	48,500	51,810	39,683
Oil capacity	gal.	3,135	—	2,887	—	—	—	—	—	—	—
Water capacity	gal.	6,170	6,170	6,000	5,000	1,515	9,000	9,100	10,800	10,050	6,170
Max. axle load	lb.	41,888	46,299	34,888	41,440	36,960	43,680	35,392	47,400*	39,460	43,341*
Over-all length	ft in.	88' 0·7"	84' 7·75"	69' 7·25"	78' 4"	36' 8"	77' 3·25"	87' 4·13"	96' 4·5"	97' 9·6"	81' 1·8"*[2]
Over-all height	ft in.	14' 6·65"	14' 6·65"	14' 5·3"	13' 10·5"	13' 6"	13' 11·97"	13' 6"*	16' 4·46"	16' 4·46"	16' 8·8"*[2]
Over-all width	ft in.	10' 9·9"	10' 4·4"	10' 1"*	10' 6"	10' 0"	9' 8·5"	10' 5·75"	—	—	10' 7"*
Normal max. speed	m.p.h.	93	56	—	—	—	70	70	(Note 1)	78	50
Min. track radius at slow speed	ft	558	558	328	573	573	—	462	436*	436*	410*

(1) Designed for 112 m.p.h. but normal maximum speed not known.
(2) The tender illustrated is larger than that specified which would increase tender weight and overall length.
Dimensions marked * have been scaled from drawings and diagrams or computed from data.

hinged buffers, features of Argentine locomotives, will be noted.

73. Indian Government Railways (Plate 93A, page 369.)

Class WG. 2–8–2. Built North British Loco., 1951.

The 2–8–2 superseded the 2–8–0 as the Indian broad gauge standard heavy freight engine many years ago. The recent example illustrated has a larger firebox and superheater than previous engines and bar frames 4½ in. thick, in place of plate. Provision is made for fitting mechanical stokers if desired. These locomotives haul trains of some 1,850 tons on the level and 1,300 tons on a grade of 1 in 150. Many parts are interchangeable with the WP class "Pacifics" (Colour Plate XIV, page 444).

74. Indian Government Railways (Plate 93B, page 369).

Class WU. 2–4–2T. Built Vulcan Foundry, 1943.

A modern example of a broad gauge light tank engine for local passenger traffic is illustrated. Built for the Indian Railways, its up-to-date features include superheater, multiple valve regulator in the header, and Caprotti valve gear.

75. Victorian Government Railways (Plate 93C, page 369).

Class R. 4–6–4. Built North British Loco., 1951–52.

In view of a possible change to standard gauge these engines have been built for easy conversion. The bar frames are 5 in. thick, and distance pieces are fitted between them and the cylinders, motion plates and other outside attachments. The cylinders are of cast steel with cast-iron liners and the wheel centres of SCOA-P type. Mechanical stokers are provided and the engines handle 500-ton trains at speeds up to 70 m.p.h.

76. South Australian Government Railways (Plate 93D, page 369).

Class 520. 4–8–4. Built Islington, 1943–44.

These engines were also designed for possible change-over to standard gauge and the only alterations necessary will be new wheel centres and brake cross beams. Recent tests have proved that the 520 class, designed to operate 500-ton trains at up to 70 m.p.h., are capable of handling loads increased to 675 tons, 2,600 I.H.P. being developed at 70 m.p.h. They were the first engines in Australia to have roller bearings on all axles and great saving of weight has been accomplished by welded construction including the cylinders. By reducing the coal and water carried they can run over 60 lb. rails and the tablet exchanging apparatus operates at up to 50 m.p.h. Maximum grade encountered is 1 in 45 with uncompensated curves of 660 ft radius.

77. U.S.S.R. Railway System (Plate 94A, page 370).

No. 6998* (no class). 4–6–4. Built Voroshilovgrad, 1938.

Since 1937 a few streamlined express passenger engines have been built in the U.S.S.R. with a view to ascertaining the possibilities of high-speed services which should be feasible on the long clear stretches of line available. No. 6998 is very similar to the three engines of Class V: both have 7 ft 2½ in. coupled wheels and are designed for a speed of 112 m.p.h. On test this engine hauled 900 tons up a grade of 1 in 167, developing 3,400 H.P., the speed falling from 68 to 47 m.p.h. With the increase of diesel and electric traction further developments on these lines are unlikely.

78. U.S.S.R. Railway System (Plate 94B, page 370).

Class P 36. 4–8–4. Built Kolomna, 1950.

The prototype of this class was built in 1950 and, in 1954, the building of a considerable number began. They are familiar to many visitors from the West since taking over a number of principal passenger services, including Leningrad-Moscow and Brest-Litovsk to Moscow. With tenders holding 23 tons of coal they are well adapted to long runs. The very generous Russian loading gauge is at once noticeable in their height of 16 ft 4·46 in. This is by no means the maximum permissible, as engines with a height of 17 ft 4·74 in. have been built.

79. U.S.S.R. Railway System (Plate 94C, page 370).

Class LV. 2–10–2. Built Voroshilovgrad, 1954.

This is a larger-boilered version of the numerous Class 2–10–0's, and it has a similar route availability. It was, however, desired to enable these engines to run tender first with equal facility and so an extra trailing axle was provided. After experimenting with the rebuilt 2–10–2 No. CO-18 in 1952, the prototype was built in 1954 and quantity production started the following year. The adhesive weight can be increased from 91 to 96 tons if desired by altering the fulcrums of the truck compensating beams. Maximum thermal efficiency obtained on test was 9·27 per cent.

* Works number used for its official designation.

Part VIII. Cape gauge: 3ft 6 in.

80. South African Railways and Harbours (Plate 95A, page 371).

Class 25NC. 4–8–4. Built North British Loco., 1953.

The South African Railways, with nearly 12,600 route-miles, is the world's largest 3 ft 6 in. gauge railway system. Another characteristic is the very large number (over 1,300) of "Mountain", or 4–8–2 type, engines used on all classes of traffic, and far more extensively than on any other railway. First introduced in 1904, this type has been developed to its maximum capacity and only recently eclipsed by the enormous 4–8–4's of Class 25. The NC (non-condensing) engines number fifty, whilst ninety others have condensing tenders for working over long stretches without water supply (*see* Chapter 9, Part IV, page 472). The engine frames are a single steel casting weighing 18 tons. Mechanical stokers with a capacity of over 5 tons per hour are fitted. Larger and heavier than most European engines on the standard gauge, they are a fine illustration of what can be achieved on a narrow gauge.

81. South African Railways and Harbours (Plate 95B, page 371).
Class GMAM. 4–8–2+2–8–4 Garratt. Built Beyer Peacock, 1956.

The other notable feature of S.A.R. motive power is the extensive use of Beyer-Garratt locomotives, since their introduction in 1921. By 1929 these had been developed to the 214-ton GL class, with a tractive effort of 90,000 lb. Class GMAM are however lighter, for service on 60 lb. rails, and in order to make the maximum use of the weight available, a separate water tank wagon (in effect a tender) is coupled to the engine. The rear engine unit carries coal only, in place of the usual coal and water as on the otherwise similar GMA class. The engine frames are one-piece steel castings. Mechanical stokers and mechanically-operated rocking grates are fitted.

82. South African Railways and Harbours (Plate 95C, page 371).
Class FD. 2–6–2+2–6–2 modified Fairlie. Built North British Loco., 1926.

The "Fairlie" was amongst the earliest successful forms of articulated locomotive and the principle is adaptable to a variety of forms. The type illustrated superficially resembles the Beyer-Garratt, but both tanks and boiler are carried on a single girder frame pivoted at the centres of the coupled wheelbases of the engine units. The tanks are tapered inwards at their outer ends to reduce the overhang on sharp curves.

83. Rhodesia Railways (Plate 96A, page 372).
Class 15. 4–6–4+4–6–4 Garratt. Built Beyer Peacock, 1940.

These engines were introduced for operation on long runs of nearly 500 miles without intervening locomotive depots, crews being changed *en route*. With a ruling grade of 1 in 80, they are rated to take 500-ton passenger trains with a maximum speed of 50 m.p.h. and, if required, 900-ton freight loads. An unusual feature for Beyer-Garratts lies in their being intended to run chimney-first only, and the front tank has been shaped both to give a clear view ahead and act as a smoke deflector. The valve gears are also arranged so that the die blocks are in the same position in the links of both engine units, instead of the usual reversed position. (This is accomplished by setting the return cranks on the rear engine at 180° relative to the front unit.)

84. New Zealand Government Railways (Plate 97A, page 373).
Class K. 4–8–4. Built N.Z.G.R., 1932–36.

For many years the bulk of the N.Z.G.R. main line traffic was handled by the Class Ab 4-cylinder compound "Pacifics", with over 140 in service, but in 1930 a 50 per cent increase in power was decided upon. The resulting K Class, thirty of which were built at the Hutt Works in 1932–36, are a remarkable achievement in compact design. A tractive effort of 30,815 lb. was obtained within the restricted limits of 11 ft 6 in. height, 8 ft 6 in. width and 14-ton axle load, while the grate area of 47·7 sq. ft was considered the limit for hand firing, although the engines were subsequently converted to oil-burners. Capable of handling 1,000-ton freight trains on the level at 30 m.p.h., or 500-ton passenger trains at 50 m.p.h., speeds up to 69 m.p.h. have been attained. Thirty-five improved Ka class followed in 1939–50, with roller bearings on all axles and frames of high-tensile steel. On the 1 in 50 uncompensated Raurimu spiral a Class Ka locomotive accelerated from 23 to 26 m.p.h. with 215 tons. Six Class Kb locomotives, built at Hillside in 1939, have siphons and boosters, which bring the T.E. up to 36,815 lb. These can take 560 tons freight over the 1 in 50 grades of the Arthur Pass. Both Ka and Kb classes have had their streamlining removed.

85. New Zealand Government Railways (Plate 97B, page 373).
Class Ja. 4–8–2. Built North British Loco., 1939.

Whilst the "K's" were built for 70 lb. rails, increased power was also required for stretches laid with 53–56 lb. track, limiting the axle load to 11½ tons. Forty J class 4–8–2's were built by the North British Locomotive Co. in 1939, followed by thirty-five Ja class from the Hillside shops and, finally, sixteen oil-burners from N.B.L.Co. in 1951. Designed for 800-ton freight or 400-ton passenger trains, they have put up many remarkable performances, with recorded speeds up to 67 m.p.h. The "J's" have 4-in. bar frames, whilst the frames of the "K's" are of 1¼-in. plate.

86. Indonesian State Railway (Plate 98A, page 374).
Class D52. 2–8–2. Built Krupp, 1952.

Before World War II the highly developed railway system of Java included a variety of modern and specialized locomotive classes. After the ravages of those years and the impossibility of obtaining spares from former suppliers, the principal requirement for rehabilitation

PART VIII. CAPE GAUGE 3ft. 6in.

EXAMPLE NO.		80 (Plate 95A, page 371)	81 (Plate 95B, page 371)	82 (Plate 95C, page 371)	83 (Plate 96A, page 372)	84 (Plate 97A, page 373)	85 (Plate 97B, page 373)	86 (Plate 98A, page 374)	87 (Plate 98B, page 374)	88 (Plate 98C, page 374)
RAILWAY		SAR	SAR	SAR	RR	NZGR	NZGR	ISR	JNR	NR
Class		25NC	GMAM	FD	15	K	Ja	D52	E10	51
Type		4-8-4	4-8-2+2-8-4	2-6-2+2-6-2	4-6-4+4-6-4	4-8-4	4-8-2	2-8-2	2-10-4T	0-8-0 T.-T.
Cylinders (no.) dia. × stroke	in.	(2) 24×28	(4) 20·5×26	(4) 15×24	(4) 17·5×26	(2) 20×26	(2) 18×26	(2) 19·7×23·6	(2) 21·65×26	(2) 18×23
Coupled wheel dia.	in.	60	54	45·5	57	54	54	59·17	49·25	42·75
Wheelbase, coupled	ft in.	15' 9"	14' 5"	8' 6"	10' 6"	14' 3"	14' 3"	15' 9"	19' 0·3"	13' 3·75"
Wheelbase, total	ft in.	81' 4·7"	86' 4"	58' 7"	84' 3"	61' 10·5"	58' 0"	57' 5"	38' 0·7"	36' 9"
Tractive force (85%)	lb.	51,400	68,800	36,312	42,750	30,815	24,960	29,970	47,850	26,670
Boiler pressure	p.s.i.	225	200	180	180	200	200	227·6	227·6	180
Grate area	sq. ft	70	63·2	40·87	49·5	47·7	39	37·66	35·2	13·35
Heating surface	sq. ft	3,390	3,197(a)	1,745	2,336(a)	1,922	1,462	1,840(f)	1,568	872
Superheating surface	sq. ft	630	747	362	494	392	271	646	833	—
Weight full, engine	lb.	260,848	424,580	255,584	402,080	194,208	154,784	153,282	229,600	105,504
tender	lb.	237,216	—	—	—	119,672	93,408	111,113	—	84,000
Coal capacity	lb.	40,320	31,360	11,200	22,400	—	—	11,000[3]	8,960	8,960
Oil capacity	gal.	—	—	—	—	1,600	1,400	1,225[3]	—	4,800[3]
Water capacity	gal.	10,500	2,100[1]	3,800	7,000	5,000	4,000	5,500	1,760	28,000
Max. axle load	lb.	41,440	34,496	27,664	29,680	30,968	25,760	28,396	33,488	—
Over-all length	ft in.	92' 0·6"	94' 4"	67' 8·25"	92' 6"*	69' 8"	66' 11·4"	64' 5·6"	47' 4·9"	53' 7·25"
Over-all height	ft in.	13' 0"	13' 0"	12' 9"	13' 1"	11' 6"	11' 6"	12' 2·5"	13' 0·75"	12' 4"
Over-all width	ft in.	9' 11"	10' 0"	9' 0"	10' 0"	8' 6"	8' 6"	10' 1·25"	9' 7·6"	9' 5·75"
Normal max. speed	m.p.h.	55	(45)	—	50	35/55	35/55	56	37	—
Min. track radius at slow speed	ft	275	275	300	275	330	330	460	328	280

[1] A separate water tank wagon is attached when necessary. [2] 1,000 gallons in side tanks, 3,800 gallons in tender. [3] Coal or oil as alternatives.
Dimensions marked * have been scaled from drawings and diagrams or computed from data.

was a large class of simple, general-purpose locomotives. As in France, the 2–8–2 tender type was chosen, with provision for both coal and oil burning. One hundred of this type were supplied to Indonesia by the Krupp Locomotive Works in 1951–52. They can haul 600 tons at 56 m.p.h. on the level or at 25 m.p.h. on a grade of 1 in 100.

87. Japanese National Railways (Plate 98B, page 374).

Class E 10. 2–10–4T. Built Kisha Seizo Kaisha Ltd., 1948.

Since World War II, Japanese locomotive practice has followed that of North America and 3-cylinder locomotives have been retired or rebuilt. The heavy freight locomotive illustrated is a fine example of recent design, the wheel arrangement being unique for tank engines.

88. Nigerian Railway (Plate 98C, page 374).

Class 51. 0–8–0T–Tender. Built Hunslet, 1955.

Under certain conditions locomotives require additional water supply. In the early days of overseas railways, tank engines adequate for the original sections were supplied with tenders, to enable them to work over greatly increased distances as new sections were opened up. In desert areas, where water supply points are far apart, tender engines run coupled to one or more tank wagons. (The Garratt engines described (No. 81) afford another example.) The Nigerian engines illustrated are principally used for shunting operations around Port Harcourt, but can also work main-line freight. Similar engines are employed in Ghana for the manganese ore traffic to Takoradi Harbour.

Part IX. Metre gauge: 3 ft $3\frac{3}{8}$ in.

89. East African Railways and Harbours (Plate 96B, page 372).

Class 59. 4–8–2+2–8–4 Garratt. Built Beyer Peacock, 1955.

These are the most powerful metre gauge locomotives in the world. The railway climbs to over 9,000 ft altitude, but the heaviest traffic is on the 329-mile stretch between the port of Mombasa and Nairobi at 5,600 ft, where an axle load of 21 tons is now permitted. The thirty-four Class 59 engines haul 1,200 tons on this stretch, with a ruling grade of 1 in 66, a 70 per cent increase over the previous maximum load.

90. East African Railways and Harbours (Plate 99A, page 375).

Class 31. 2–8–4. Built Vulcan Foundry, 1955.

Like all recent metre gauge locomotives of the E.A.R., these engines can be readily converted to the 3 ft 6 in. gauge with the minimum of trouble, by fitting wider tyres. The design is based on the Nigerian "River" class, and a lighter version (Class 30) is used in Tanganyika. Designed for a wide range of branch-line duties and heavy shunting, they will replace the older 4–8–0's. The replacement of 4–8–0's is the reverse process to that of the Argentine, where the broad gauge allows sufficient width for the firebox.

91. Brazilian National Railways (D.N.E.F.) (Plate 99B, page 375).

Class 242.N. 4–8–4. Built Locomotives-Batignolles-Chatillon (Nantes), 1951.

Designed under the direction of M. Chapelon, this and the following class well illustrate how large and powerful machines can be skilfully contrived for narrow-gauge lines laid with light rails. In spite of these severe restrictions, 4-in. bar frames and a large boiler are included, with an axle load of under 13 tons. These engines are intended for heavy passenger and fast freight work, burning low-grade Brazilian coal. For this they are required to maintain an output of 1,785 H.P.

92. Brazilian National Railways (D.N.E.F.) (Plate 99C, page 375).

Class 142.N. 2–8–4. Built Fives-Lille, etc. 1951.

In addition to the 4–8–4 engines described, sixty-six lighter engines were constructed at the three Works, Creusot, Cail and Fives-Lille, with an axle load of only 10 tons. Three forms of tender were supplied, one heavy and two light, for work in different areas:

(1) Six-wheel bogies, 26,450 lb. coal, 3,970 gallons of water.

(2) Four-wheel bogies, 15,430 lb. coal, 3,085 gallons of water.

(3) Four-wheel bogies, 15,430 lb. firewood, 2,865 gallons of water.

93. Sorocabana Railway (Brazil) (E.F.S.) (Plate 100A, page 376).

No. 902. 2–10–2 (3-cylinder). Built Henschel, 1930.

The design of metre-gauge locomotives is much more restricted than for the 3 ft 6 in. gauge, and the difference, although only 2·6 in., is critical. Four-cylinder engines have been built for the latter gauge, but not more than three cylinders have been used for non-articulated types on the metre gauge. Of these, some powerful examples can be found, such as the Brazilian 2–10–2's illustrated.

94. Central Railway of Brazil (E.F.C.) (Plate 100B, page 376).

No. 1307. 2–8–8–4. Built Henschel, 1937.

The sudden development of heavy traffic, particularly

Mavag

PLATE 89A. No. 64: Hungarian State Railways (M.A.V.) (*see page* 357).

Henschel & Sohn

PLATE 89B. No. 65: Bulgarian State Railways (B.D.Z.) (*see page* 359).

H.M. Le Fleming

PLATE 90. No. 66: Bulgarian State Railways (B.D.Z.) (*see* page 359). From an oil painting by H.M. Le Fleming.

Henschel & Sohn

PLATE 91A. No. 67: Turkish State Railways (T.C.D.D.) (*see* page 359).

Henschel & Sohn

PLATE 91B. No. 68: Turkish State Railways (T.C.D.D.) (*see* page 359).

P. Ransome-Wallis

PLATE 91C. No. 69: Hellenic State Railways (C.E.H.) (*see* page 359).

La Maquinista

PLATE 92A. No. 70: Spanish National Railways (R.E.N.F.E.) (*see* page 359).

La Maquinista

PLATE 92B. No. 71: Spanish National Railways (R.E.N.F.E.) (*see* page 359).

Vulcan Foundry Ltd

PLATE 92C. No. 72: Argentina (F.N. General Roca) (*see* page 359).

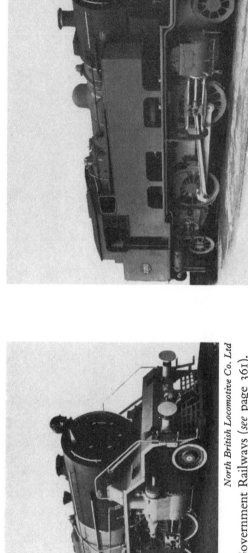

North British Locomotive Co. Ltd

PLATE 93A. No. 73: Indian Government Railways (*see* page 361).

South Australian Government Railways

PLATE 93D. No. 76: South Australian Government Railways (*see* page 361).

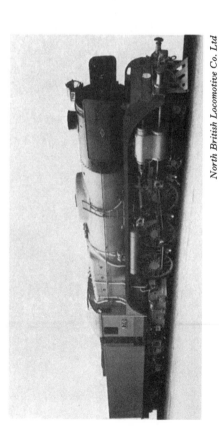

North British Locomotive Co. Ltd

PLATE 93C. No. 75: Victorian Government Railways (*see* page 361).

Vulcan Foundry Ltd

PLATE 93B. No. 74: Indian Government Railways (*see* page 361).

Soviet Ministry of Railways

PLATE 94A. No. 77: USSR Railway System (*see* page 361).

Josef Otto Slezak

PLATE 94B. No. 78: USSR Railway System (*see* page 361).

Soviet Ministry of Railways

PLATE 94C. No. 79: USSR Railway System (*see* page 361).

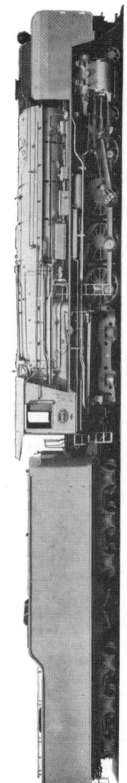

North British Locomotive Co. Ltd

PLATE 95A. No. 80: South African Railways and Harbours (*see page* 361).

Beyer, Peacock and Co. Ltd

PLATE 95B. No. 81: South African Railways and Harbours (*see page* 362).

North British Locomotive Co. Ltd

PLATE 95C. No. 82: South African Railways and Harbours (*see page* 362).

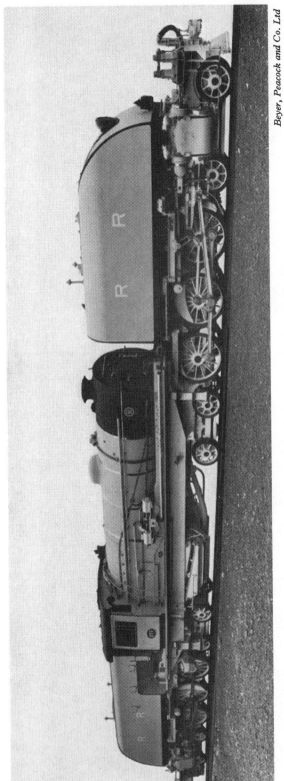

Beyer, Peacock and Co. Ltd

PLATE 96A. No. 83: Rhodesia Railways (*see* page 362).

Beyer, Peacock and Co. Ltd

PLATE 96B. No. 89: East African Railways and Harbours (*see* page 364).

New Zealand Government Railways

PLATE 97A. No. 84: New Zealand Government Railways (*see* page 362)

New Zealand Government Railways

PLATE 97B. No. 85: New Zealand Government Railways (*see* page 362).

Fried Krupp

PLATE 98A. No. 86: Indonesian State Railway (*see* page 362).

Kisha Seizo Kaisha

PLATE 98B. No. 87: Japanese National Railways (*see* page 364).

The Hunslet Engine Co. Ltd

PLATE 98C. No. 88: Nigerian Railway (*see* page 364).

Vulcan Foundry Ltd

PLATE 99A. No. 90: East African Railways and Harbours (*see* page 364).

Soc. Anon. Batignolles-Chatillon

PLATE 99B. No. 91: Brazilian National Railways (D.N.E.F.) (*see* page 364).

Soc. Anon. Batignolles-Chatillon

PLATE 99C. No. 92: Brazilian National Railways (D.N.E.F.) (*see* page 364).

Henschel & Sohn

PLATE 100A. No. 93: Sorocabana Railway (Brazil) (E.F.S.) (*see* page 364).

Henschel & Sohn

PLATE 100B. No. 94: Central Railway of Brazil (E.F.C.) (*see* page 364).

North British Locomotive Co. Ltd

PLATE 100C. No. 95: Malayan Railway (*see* page 384).

R. Stephenson and Hawthorns Ltd

PLATE 101A. No. 96: Indian Government Railways (*see* page 384).

Mfr. Esslingen

PLATE 101B. No. 97: Argentina (F.N. General Belgrano) (*see* page 384).

PLATE 102A. No. 98: Colombian National Railways (Giradot–Tolima–Huila) (*see* page 384).

R. Stephenson and Hawthorns Ltd

PLATE 102B. No. 99: Sierra Leone Government Railways (*see* page 384).

Beyer, Peacock and Co. Ltd

PLATE 102C. No. 100: South African Railways and Harbours (*see* page 384).

Société Anglo-Franco-Belge

FERROCARRILES NACIONALES.

British Railways

PLATE 103A. Stationary Testing Plants: Rugby: Class 7, Pacific No. 70005, *John Milton*, under test.

British Railways

PLATE 103B. Stationary Testing Plants: Swindon: Class 9, 2–10–0 No. 92178 with double chimney and blast-pipe, under test.

Kenneth Leech

PLATE 104A. Controlled Road Testing: near Swindon: British Railways 3-cylinder 4–6–2 No. 71000, *Duke of Gloucester*, with indicator shelter, hauling a train of twenty coaches and a dynamometer car.

Kenneth Leech

PLATE 104B. Controlled Road Testing: near Hullavington: Ex L.N.E.R. 3-cylinder 2–6–2 class V–2, No. 60845 with indicator shelter, hauling a train of twenty coaches and a dynamometer car.

P. Ransome-Wallis

PLATE XI. British Railways (Southern Region): a winter-time train from London to the Kent coast in charge of No. 30793, "Sir Ontzlake", Class N–15. These engines are representative of many hundreds of simple 2–cylinder 4–6–0 locomotives which have efficiently handled passenger and fast freight traffic in Britain during the past twenty-five years.

PART IX. METRE GAUGE 3ft 3⅜in.

EXAMPLE NO.		89	90	9	92	93	94	95	96	97
		(Plate 96B, page 372)	(Plate 99A, page 375)	(Plate 99B, page 375)	(Plate 99C, page 375)	(Plate 100A, page 376)	(Plate 100B, page 376)	(Plate 100C, page 376)	(Plate 101A, page 377)	(Plate 101B, page 377)
RAILWAY		EAR	EAR	DNEF	DNEF	EFS	EFC	MR	IGR	FN
Class		59	31	242N	142N	(No. 902)	(No. 1307)	56·4	YL	24
Type		4-8-2+2-8-4	2-8-4	4-8-4	2-8-4	2-10-2	2-8-8-4	4-6-2	2-6-2	0-12-2T
Cylinders (no.) dia. × stroke	in.	(4) 20·5 × 28	(2) 17 × 26	(2) 17 × 25·2	(2) 17 × 22	(3) 20·9 × 22	(4) 17 × 22	(3) 13 × 24	(2) 12·25 × 22	(2) 15·74 × 15·74 (f) 15·74 (r) (2) 26·77 × 19·7
Coupled wheel dia.	in.	54	48	59	50	48	42	54	43	37
Wheelbase, coupled	ft in.	15' 0"	13' 3"	15' 9"	13' 5·8"	17' 0·7" (k)	11' 9·75"	10' 0"	8' 4"	19' 11"
Wheelbase total	ft in.	92' 6·5"	52' 2"	70' 3·7"	60' 6·4"	56' 3·9"	71' 11·6"	51' 7"	39' 6·5"	25' 11"
Tractive force (85%)	lb.	83,350	26,600	35,056	30,085	50,864	53,978	23,940	13,700	55,145 (r)
Boiler pressure	p.s.i.	225	200	284	213	199	210	250	210	199
Grate area	sq. ft	72	30	57·35	43	53·28	75·35	27	17·75	37·66
Heating surface	sq. ft	3,560	1,652	1,805 (s)	1,315 (s)	3,165 (a)	2,346 (fs)	1,109 (a)	525 (a)	2,292
Superheating surface	sq. ft	747	328	732	484	1,016	915	218	130	932
Weight full, engine	lb.	563,752	157,164	205,030	154,324	206,080	259,043	133,280	86,324	241,857
tender	lb.	—	100,709	160,386	132,277	94,528	115,743	99,232	55,776	—
Coal capacity	lb.	—	—	39,683	26,455	15,680	26,455	22,400	8,960	—
Oil capacity	gal.	2,700	1,680	—	—	—	—	—	—	906
Water capacity	gal.	8,660	4,108	4,850	3,967	3,740	3,960	2,510	2,000	2,336
Max. axle load	lb.	47,040	25,200	28,660	22,266	35,168	26,455	28,448	17,920	38,140
Over-all length	ft in.	104' 1·5"	62' 6"	80' 11·25"	71' 1·3"	66' 6·8"*	81' 2·2"	62' 0"	49' 7·75"	43' 3"
Over-all height	ft in.	13' 5·5"	12' 6"	12' 10·3"	12' 2"	12' 10·8"*	12' 9·5"	12' 9"	11' 3"	13' 3·4"
Over-all width	ft in.	10' 0"	9' 4·25"	8' 10·3"	8' 9"*	9' 10·2"*	8' 10·9"*	9' 2"*	8' 6"	10' 3·6"
Normal max. speed	m.p.h.	45	45	50	37	37	31	50	(45)	25 (r)
Min. track radius at slow speed	ft	290	290	262	262	230	230	330	358	394 (r)

(r) On rack sections speed is limited to 7·5 m.p.h. and the minimum track curvature 820 ft radius. Dimensions marked * have been scaled from drawings and diagrams or computed from data.

in ores in areas served by light, narrow-gauge track, presents the designer with difficult problems. In this instance, power equivalent to a heavy European 2–10–0 was required on a 12-ton axle load and with an ability to traverse curves of only 230-ft radius with just under 1 in. gauge widening. Four such engines were built, and on the curvature quoted, the smokebox front is 19 in. off centre.

95. Malayan Railway (Plate 100C, page 376).

Class 56.4. 4–6–2 (3-cylinder). Built North British Loco., 1938–47.

The Malayan Railway was amongst the earlier overseas railways to adopt the "Pacific" type, sixty engines of the H class being built between 1907–14. With a small volume of highly rated freight traffic it has been possible to adopt standard engines for both passenger and freight services. As a result of experience gained with subsequent classes of "Pacific" the class illustrated was designed and sixty-eight engines eventually built. They have bar frames, steel fireboxes and the three cylinders are provided with rotary cam poppet valves, the camshaft being divided into two parts, independently driven from each side of the engine. This avoids complete immobilization in case of breakdown on a long stretch of single track. These engines have now all been converted to burn oil fuel.

96. Indian Government Railways (Plate 101A, page 377).

Class YL. 2–6–2. Built R. Stephenson & Hawthorns, 1952.

In India two gauges, the broad (5 ft 6 in.) and the metre (3 ft 3 3/8 in.), were adopted at the outset and standard engine classes suitable for the many lines of both gauges were introduced at an early stage. The present standard series were started in 1925 and the engine illustrated is a typical example of those for the metre gauge.

97. Argentina (F.N. General Belgrano) (Plate 101B, page 377).

Class 24. 0–12–2T (rack and adhesion). Built Esslingen, 1955.

This remarkable engine was built for the severe conditions near Jujuy in North-western Argentina. It is required to operate a 400-ton train on a grade of 1 in 16 with 984-ft radius S-curve, at a speed of 7½ m.p.h. Built on 3·15-in. bar frames, the gear ratio of the four-coupled rack engine is 1 : 2·44. When operating over the rack (Abt system), compound working is used, with the rack cylinders taking low-pressure steam from the adhesion engine (*see* Chapter 9, page 473).

Part X. Narrow gauge: 3 ft 0 in.–2 ft 0 in.

98. Colombian National Railways (Giradot-Tolima-Huila) (Plate 102A, page 378).

Kitson-Meyer. 2–8–8–2T. Built R. Stephenson, 1935.

The Kitson-Meyer type engine illustrated has a rear tank under the bunker in addition to side tanks. Thus the usual tank engine arrangement is mounted on a girder frame, to which the bar-frame engine units are pivoted. The locomotive was designed to haul 330-ton loads at 9–10 m.p.h. up the 1 in 22 grades between Giradot and Bogota.

99. Sierra Leone Government Railways (Plate 102B page 378).

Class 63. 4–8–2+2–8–4 Beyer Garratt. Built Beyer Peacock, 1956.

Beyer Garratt locomotives of the 2–6–2+2–6–2 type were introduced into Sierra Leone in 1926. These larger locomotives provide an increase in power of 35 per cent to deal with increasing traffic over a railway which has a maximum permissible axle load of 5 tons and a ruling gradient of 1 in 50.

100. South African Railways and Harbours (Plate 102C, page 378).

Class NG 15. 2–8–2. Built Anglo-Franco-Belge, 1951.

On the highly-developed 2-ft gauge lines of South Africa, which total some 800 miles, single trains of up to 600 tons are operated, and altitudes of over 4,000 ft reached. The modern "Mikado" illustrated is superheated and, like most designs for very narrow gauges, has outside frames, cylinders and balanced cranks, the firebox being above the framing.

PART X. NARROW GAUGE 3ft 0 in.–2ft 0 in.

EXAMPLE NO.		98 (Plate 102A, page 378)	99 (Plate 102B, page 378)	100 (Plate 102C, page 378)
RAILWAY		G.T.H.	S.L.G.R.	SAR
Class		—	63	NG15
Type		2–8–8–2T.	4–8–2+2–8–4	2–8–2
Gauge	ft in.	3′ 0″	2′ 6″	2′ 0″
Cylinders (no.) dia. × stroke	in.	(4) 17·75×20	(4) 12·63×16	(2) 15·75×17·75
Coupled wheel dia.	in.	37·5	33	33·9
Wheelbase, coupled	ft in.	10′ 2·5″	9′ 4·5″	9′ 7″
Wheelbase, total	ft in.	53′ 11″	56′ 10″	44′ 7·75″
Tractive force (85%)	lb.	58,564	22,990	18,825
Boiler pressure	p.s.i.	205	175	171
Grate area	sq. ft	51	22·5	16·7
Heating surface	sq. ft	2,567	924	796
Superheating surface	sq. ft	640	178	180
Weight full, engine	lb.	291,760	148,568	82,208
Weight full, tender	lb.	—	—	70,000
Coal capacity	lb.	—	8,960	12,320
Oil capacity	gal.	1,100	—	—
Water capacity	gal.	4,000	1,600	2,860
Max. axle load	lb.	32,480	11,200	15,148
Over-all length	ft in.	66′ 4·75″	64′ 7·5″	54′ 3·9″
Over-all height	ft in.	12′ 4·25″	10′ 5″	10′ 5″
Over-all width	ft in.	10′ 0″	7′ 6″	7′ 6·5″
Normal max. speed	m.p.h.	—	20	30
Minimum track radius at slow speed	ft	360	230	—

CHAPTER 6

The Testing of Locomotives

by S. O. ELL

Part I. Steam locomotive theory and data

ACTION OF THE LOCOMOTIVE

The steam locomotive comes within the general classification of Heat Engines, because it converts the heat
energy that is liberated by the combustion of fuel into
mechanical energy or work. The method by which this
is done, or the action of the locomotive, falls into two
divisions:

(i) The automatic supply of the working medium
(steam) to the cylinders.

(ii) The conversion of the working medium into
tractive force and displacement.

Automatic supply of the working medium. The steam
demanded by the cylinders is generated in the boiler from
the combustion of fuel, and because the requirements of
the cylinders vary very widely, it is essential that the rates
of steam demand and combustion be automatically
linked together. Combustion requires air, and the rate of
combustion depends on the rate of air supply to the
furnace. By making the exhaust steam expel the products
of combustion and thereby induce air to enter the furnace
at a rate proportional to the rate of combustion, the
essential link is made.

The sequence of events is known as the Stephenson
Cycle. Originally incorporated in the *Rocket*, it was responsible for this locomotive being the first with an automatic action. With rare exceptions, all later steam locomotives have depended on this cycle for their successful
operation and for this reason they are known as
Stephensonian engines.

The action is illustrated by Fig. 80, which is based on a
form of diagram due to Lawford H. Fry. It represents the
results of a series of trials on an engine of a well-known
British locomotive class.

Fry aptly calls the action a linked triangle of the three
elements: water, air and fire, and the diagram is, therefore, threefold. The mean relation between the rate of
combustion and the rate of steam production and
exhaustion is represented by the curve *A*, that between
the rate of steam production and exhaustion and the rates
of gas ejection and air supply by curves *B* and *C* res-

pectively; that between the rate of air supply and the
rate of combustion is shown by curve *D*. Thus for a
steam rate of 20,000 lb./hr, the gas ejected is 33,000 lb./
hr, and the air supplied 30,000 lb./hr, as indicated
by the dotted lines. Going anti-clock-wise to the
adjacent relation, the dotted lines, continued, show
that for a rate of air supply of 30,000 lb./hr, the rate
of combustion is 2,200 lb. of coal per hour. For this
rate, by following the dotted line into the next section,
the original steam rate of 20,000 lb./hr is reached. Going
back clockwise to the previous section, the straight line
E is drawn to represent the air theoretically required,
which, for the coal concerned, is 10 lb. per pound of
coal fully burnt, so that 22,000 lb. of air per hour is
theoretically necessary in the case of the example, as
shown by the lowest dotted projections to the air axis.
The excess air is, therefore, 8,000÷22,000 of the theoretical air, or 36½ per cent. Air in excess of that theoretically
necessary is always required to complete combustion
under practical conditions.

If the curves *D* and *E* be extended, they would eventually intersect at some point beyond the limit of the
diagram, this point being the theoretical limit of the

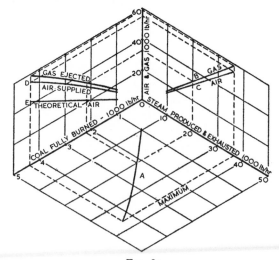

FIG. 80.

action. But, since air 20–30 per cent in excess of theoretical requirements is at least necessary to sustain combustion, the practical limit corresponds to a much lower rate than this. Owing to the losses attendant on the combustion and heat transfer processes, there is a limit also to the rate of steam production, this limit being reached when the sum of the losses in the two processes reaches 50 per cent of the heat released. In the boiler concerned, the limiting rate of steam production is 36,000 lb./hr, the conditions then obtaining being indicated by the outer set of dotted co-ordinates in the figure. The aim should be to provide the engine with draughting arrangements effective enough to supply air 20–30 per cent in excess of theoretical requirements at the corresponding limiting rate of combustion.

The steam rate links the action of the boiler with the action of the cylinders and the former is only dependent on speed, cut-off and throttling, in so far as these in combination affect the steam rate.

Conversion of the working medium into tractive force and displacement. It may be said that the railway age was engendered in the invention of the Stephenson cycle for the steam engine, which, in other respects, was pre-eminently suitable for traction purposes, especially having regard to the techniques of the times. The availability of pressure energy for starting and a speed which is both of a low order and wide range, made direct transmission of power from cylinders to wheel treads both feasible and possible. Stephenson and his contemporaries soon solved the problem of expansive working with reversibility, which this form of transmission calls for. Their link motions were not only in character, but are still notable engineering achievements, because they form one of the few examples where efficiency of performance is almost unimpaired by wear. Whilst the drawbacks were, and are, unevenness of turning moment, production of hammer blow and liability to produce oscillations of various kinds, their effects have all been minimized as the locomotive has evolved. So for a hundred years and more, the natural "aptitude" of steam for traction purposes kept it unchallenged in this field until the evolution of costly and elaborate transmissions, made possible by the techniques of our own times, could harness the efficient, but relatively inflexible, diesel to traction.

In a given engine, the steam rate forms the parameter of both the speed–tractive force relation and the speed–cut-off relation. (A parameter is the quantity which remains constant when others vary, but varies with different conditions.)

When the steam rate is held constant (Fig. 81) whilst the speed varies, the speed–tractive force relation is given by co-ordinates of the single curve *A* whilst, at the same time, the speed–cut-off relation is given by the co-ordinates of the single curve *B*, both *A* and *B* being identified with the steam rate. Thus, at 70 m.p.h., with a

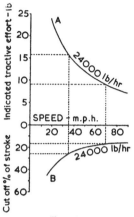

FIG. 81

steam rate of 24,000 lb./hr, the total cylinder tractive force (which is proportional to the total mean effective pressure on the pistons) is 9,040 lb. and the cut-off is 17 per cent of the stroke, as shown by following the dotted lines running through the given speed to the two curves, and then vertically downwards to the horizontal axes. Similarly, at 35 m.p.h., the steam rate remaining unchanged, the cylinder tractive force is 15,800 lb., and the cut-off is 26 per cent of the stroke.

From Figure 80 may be found the corresponding boiler conditions.

The steam rate is the heat energy stream produced by the boiler and fed to the cylinders for conversion into power. Constant power would be produced with a constant steam rate if the efficiency of conversion and quality of the steam over the speed range also remained constant. Thus, the product of tractive force and speed, which is a definition of power, would be constant for all speeds. In practice this is never achieved because the efficiency varies with speed and so curve *A* is never a true hyperbola, although the quality of the steam remains unchanged.

Taking equal increments of steam rate, the complete diagram for the engine appears as in Figure 82.

BOILER PERFORMANCE

To obtain the important relation between the steam produced and coal consumed, and to separate and quantify the losses associated with the combustion and heat transfer processes, the evaporative range of the boiler must be covered by a series of carefully conducted tests. An essential condition for each test is the mainten-

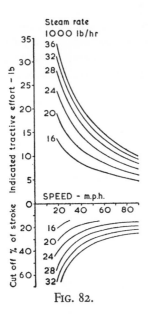

FIG. 82.

water consumption) plotted against the heat per hour taken up by the boiler and appearing in the steam. To avoid using very large figures, this is usually divided by 970 and called the *equivalent evaporation from and at 212° F.*, 970 B.T.U., being the heat required to evaporate one pound of water into steam at this temperature, or the latent heat of steam at 212° F.

(ii) *Heat in steam produced and heat released in the firebox.* This, being the second of the principal relations, is given by curve *B* (Fig. 83) of the section of the diagram above and to the left of the origin. The vertical axis concerns the rate of combustion, or the rate at which coal is fully burnt, in pounds of coal per hour of a certain calorific value in B.T.U. per lb. The vertical scale is, therefore, one of hourly rate of heat release in the firebox in denominations of the calorific value per pound. For the boiler concerned the latter is 14,260 B.T.U./lb.

The rate of combustion is generally computed by Fry's Method, thus: From the smoke box gas analysis, the weight of gases produced by the combustion of each pound of coal can be determined. Knowing the weight and temperature of the smoke box gases, the amount of heat carried out of the smoke box for each pound of coal burnt can be found. This deducted from the calorific value

ance of a constant rate of steam production. The corresponding rate of coal consumption may then be measured by the average rate at which coal is fired.

The data required necessitates the determination of the heating or calorific value of the coal, and its proximate and ultimate analyses. The smoke box gas must be analysed periodically and its mean temperature as discharged from the tube system ascertained. The mean pressure of the steam produced must be observed, together with its mean temperature if a superheater is fitted. The feed-water quantity and temperature must be known.

The principal relationships. The four principal relationships of the boiler are given in Fig. 83 (which refers to the particular boiler concerned in Fig. 80).

(i) The relation between the hourly rate of heat production in the steam and the weight of steam produced in the same time is shown in the section below and to the left of the origin by the curve *A*.

(ii) The relation between the hourly rate of heat produced in the steam and the hourly rate of heat release by combustion in the firebox is represented by the curve *B* in the section above and to the left of the origin.

(iii) The hourly rate of heat release by combustion related to the hourly rate of heat fed to the firebed in the form of coal fired, is shown in the section above and to the right of the origin by curve *C*.

(iv) The relation between the coal consumed and fired per hour, and the weight of steam produced per hour, is represented by the remaining curve *D*.

(i) *Heat and weight of steam produced.* Considering the first relation in detail, curve *A* (Fig. 83) is the weight of steam produced per hour (which is equal to the hourly

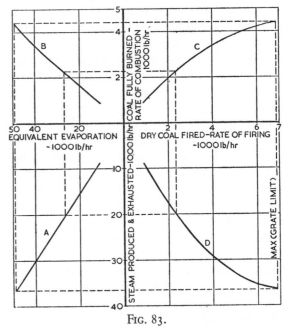

FIG. 83.

of the coal per pound gives the heat transferred to the boiler per pound of coal actually burnt. Dividing this into the total heat taken up and appearing in the steam produced (as defined in the previous section), gives the rate of combustion, assuming no loss by external radiation, to include which the rate thus found may be increased by 1 per cent.

388

Curve *B* is the mean of this relatively stable relation for the particular boiler.

For each point on curve *B*, the heat taken up and produced in the steam, divided by the heat released in the firebox (the co-ordinates of the point), is the transmission efficiency, usually expressed as a percentage. The value is about 84 per cent at low rates of working, falling slightly and virtually in a straight line as the rate increases (Fig. 84).

(iii) *Heat liberated by combustion and heat in coal consumed.* Curve *C* in the third section of the diagram (Fig. 83) represents this relation. The heat input, as for the coal fully burnt, is shown in denominations of the "dry" calorific value per lb.

The characteristic shape of *C* indicates very clearly how the difference between the heat liberated and heat in the coal consumed increases as the rates rise. It is a loss, and a most serious loss at high working rates, due to rejection of unburnt fuel, mostly lifted in small pieces from the

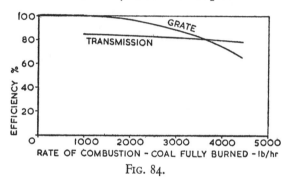

FIG. 84.

firebed and rejected through the tube system before combustion has been completed.

From each point on curve *C*, the heat released from the coal as a percentage of the heat in the coal fired (obtained from the co-ordinates of the point) is the grate efficiency. Its value at low rates of consumption is asymptotic to 100 per cent but then falls off sharply and uniformly as the rate of consumption rises, as shown in Fig. 84, where it is shown plotted on a base of rate of combustion, as for the transmission efficiency.

(iv) *The steam – coal relation.* This is generated from the other curves, as shown by curve *D* (Fig. 83), which is seen to reach a maximum, known as the "grate limit".

The principal quantities for any given condition may be found from Fig. 83. The dotted lines show that when 20,000 lb. of steam is produced per hour, the equivalent evaporation from and at 212° F. is 26,000 lb. per hour, the rate of combustion being 2,200 lb. and the rate of consumption 2,300 lb. of coal per hour. The maximum rate of evaporation of the boiler on the given coal is 36,500 pounds per hour.

The boiler thermal efficiency is usually given on a basis of coal consumed per hour, or of coal consumed per square foot of grate per hour. It is the product of the grate efficiency and the net transmission efficiency.

Calculated in this way, but plotted on the base of coal consumption, the boiler efficiency is shown in Fig. 85. Its linear form on this base is characteristic of the locomotive boiler over most of its working range.

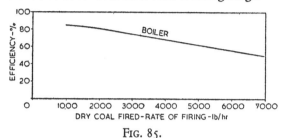

FIG. 85.

COALS

(*see* Chapter 4, Part II, Fuels, page 315, *and* Chapter 7, Effect of fuel, page 411).

Origin and nature. Coals occur in rocks in the earth's crust produced by the decay of plant materials accumulated millions of years ago. During the decay of plant life, the rather inert carbon enters but slowly into other combinations, whereas the more active hydrogen and oxygen react more quickly and form apparently gaseous combinations which leave the mass. As a result, the remaining products of the decaying process grow richer in carbon but lower in oxygen, hydrogen, and incidentally, nitrogen. In general, the higher the geologic age of the formations the more nearly does the fuel approach the state of the mineral graphite, which is 100 per cent carbon.

Bituminous coal. For locomotive purposes the most important member of the coal series is bituminous coal, the boundaries of which lie between 50 per cent volatile matter, 50 per cent fixed carbon on the one hand and 25 per cent volatile matter, 75 per cent fixed carbon on the other, these proportions being on a dry ash-free basis. The heating or calorific value of coal increases with its carbon content.

All bituminous coals show a banded structure of bright and dull bands parallel to the bedding planes. These are of four types, according to the Stopes classification – vitrain, clarain, durain, and fusian. Vitrain and clarain produce what are known as "Brights" whilst durain produces the hard and dull coal known as "Hards". Hards have a higher ash content generally than Brights from the same seam but the ash fuses at a higher temperature, which makes them much favoured for locomotive firing, as for example, the Hards of the East Midlands coal field.

The volatile matter content and caking properties of the South Wales coals are different from those of coals mined in other parts of the country. These coals lie in the high carbon – low volatile part of the bituminous range, with which is associated a relatively high calorific value. Again a high fusion ash temperature of 1,300° C. makes coals such as those from the Tredegar group much favoured for locomotive firing. In contrast to the Hards, the South Wales locomotive coals are all soft and friable but possess marked caking properties and swelling power. These coals soften to a certain extent during the combustion process and become plastic. In a certain temperature range, fluid material is extruded on to the particles and grains, cementing them together to form a light crust which offers high resistance to the lift of particles from the firebed.

Coals of relatively high volatile matter content, burn with long flames; low volatile, e.g., South Wales coals, burn with short flames. The first show to advantage where sudden increases in load are demanded as in a locomotive but are apt to be smoky. The less rapid release of the volatiles of the second class permits them to burn with hotter flames, a better regulated supply of air and a high firebed temperature. This allows the coke to burn to carbon dioxide, so that little carbon monoxide is formed in the fuel bed. On the other hand, the high carbon coals require a relatively greater draught, a greater proportion of air as primary air, or air which passes through the firebed, and a relatively thicker firebed.

Proximate or engineering analysis. The proportion of moisture, ash, volatile matter and fixed carbon give a great deal of information as to the kind of fuel. A statement of these proportions by weight is called the proximate or engineering analysis. With the calorific value, a very valuable description is obtained.

The proximate analysis is always given on the "as received" or "as used" basis for obvious reasons. But when a comparison between two fuels is required this should be on a moisture-free basis because of the accidental nature of the hydroscopic or extraneous moisture.

Moisture varies within fairly wide limits and it is usual to distinguish between "free" or surface moisture and the "inherent" moisture which is in equilibrium with the atmosphere.

Ash which is intimately associated with the coal is called "inherent" ash as distinct from "adventitious" ash derived from shale, clay, pyrites and dirt in the coal seams. The fusion temperature of the ash is high in a good locomotive coal.

Volatile matter consists of water from the decomposition of the coal (not from the inherent moisture), gas and tar.

Fixed carbon is the solid residue or "coke" left after the volatile matter is driven off, minus the ash, because the "coke" contains all the inorganic constituents present in the original coal which go to form the ash.

Calorific value. The calorific value is the number of heat units developed by the complete combustion of unit weight of the fuel. In this country, the heating or calorific value becomes the number of British Thermal Units (B.T.U.) developed when one pound of the given fuel is completely burned.

Ultimate or chemical analysis. The ultimate or chemical analysis is required when it is necessary to establish a detailed and accurate heat balance, as in locomotive testing. The essential difference between the ultimate and proximate analyses is that the first gives the component parts of the volatile matter item in the second, which it sub-divides into hydrogen, sulphur, oxygen and nitrogen, and by determining the total carbon, makes possible the determination of volatile carbon from the total carbon of the proximate analysis by difference.

Grading for locomotive purposes. Locomotive coals are graded on the basis of mileage with clean fires. British Railways class coal suitable for journeys of 400 miles as Grade 1, for 150 miles as Grade 2, and less than 150 miles as Grade 3. Within these grades, qualities are denoted by the letters A and B.

Properties of representative coals. To illustrate the foregoing descriptive properties of coals the table on page 391 has been drawn up for several well-known British locomotive coals.

COMBUSTION

Definition. Combustion is the union of an element or compound with oxygen (O_2) in which heat is evolved. Of these elements, three only need be considered for the present purposes:

carbon (C), hydrogen (H_2) and sulphur (S) these being the elementary combustibles. Sulphur is, however, of comparatively negligible importance.

Hydrogen and carbon are chemically combined in the coal in various forms, but on being heated in the presence of oxygen, they separate and combine with the oxygen independently of one another.

The chemistry of combustion. In the simplest chemical language the changes are expressed by the equations:

$$C + O_2 = CO_2$$
$$2C + O_2 = 2CO$$
$$2H_2 + O_2 = 2H_2O$$

The first two indicate that the "oxydization" of carbon can take place in two ways, depending on the relative amounts of carbon and oxygen present and on the conditions under which they are mixed together. The oxydization of hydrogen, as expressed by the third, can only take place in one way.

The first equation means that, since the atomic weight

DESCRIPTION OF REPRESENTATIVE LOCOMOTIVE COALS

Colliery N.C.B. Division Area B.R. Grade Appearance	South Kirby North Eastern No. 4 1A Hard, dull, large, no dust		Markham South Wales No. 6 1A Soft, friable, large, to dust		Blidworth East Midland No. 3 2B Hard, dull, Cobbles, no dust		Bedwas South Wales 2A Fairly soft, friable large, to dust		Ireland East Midland No. 1 3A Dull grey, some bright friable		Celynen South Wales No. 6 3A Bright, friable, 2-in. to dust	
Gross calorific value B.T.U.-lb.	As received. 13,850	Dry 14,140	As received. 14,310	Dry 14,460	As received. 12,790	Dry 14,030	As received. 14,130	Dry 14,260	As received. 12,660	Dry 13,210	As received. 12,940	Dry 13,470
Proximate or engineering analysis % by weight												
Moisture	2·1	–	1·1	–	8·8	–	0·9	–	4·2	–	3·9	–
Vol. matter less moisture	38·0	38·8	18·6	18·9	33·9	37·1	26·4	26·7	38·4	40·0	28·7	29·9
Fixed carbon	56·2	57·4	72·7	73·5	53·4	58·6	66·2	66·8	52·7	55·0	57·4	59·7
Ash	3·7	3·8	7·6	7·6	3·9	4·3	6·5	6·5	4·7	5·0	10·0	10·4
Sulphur	1·77	1·81	0·74	0·74	0·75	0·83	0·75	0·75	2·84	2·97	0·63	0·65
Iron in ash	28·0		6·6		6·3		8·0		9·1		4·1	
Incombustible volatile matter	40·3		20·45		38·8		28·55		42·1		33·4	
Fixed carbon	59·7		79·55		61·2		71·45		57·9		66·6	
Ultimate or chemical analysis % by weight												
Carbon	74·65		76·90		74·55		76·07		69·90		69·80	
Hydrogen	4·98		3·86		4·85		4·25		4·63		4·04	
Oxygen	13·51		9·65		14·22		11·18		16·25		13·86	
Nitrogen	1·25		1·25		1·25		1·25		1·25		1·25	
Sulphur	1·81		0·74		0·83		0·75		2·97		0·65	
Ash	3·80		7·60		4·30		6·50		5·00		10·40	

of carbon is 12 and that of oxygen 16, 12 parts by weight of carbon combine with 32 parts of oxygen to form 44 parts by weight of carbon dioxide (CO_2).

The second means that 12 parts by weight of carbon combine with 16 parts of oxygen to form 28 parts by weight of carbon monoxide (CO).

The third indicates that, since the atomic weight of hydrogen is 1, 4 parts by weight of hydrogen combine with 32 parts of oxygen to form 36 parts by weight of water (H_2O).

The oxygen is supplied to the fire-grate in the air admitted to it directly, or with the fuel – the "primary" air. Air may be considered as 79·1 per cent nitrogen and 20·9 per cent oxygen by volume, the nitrogen playing no part in combustion, although it has to be heated up to the temperature of the other gases by expenditure of heat evolved in the process.

The air brings the oxygen into contact with the heated carbon in the lowest layers of the firebed, combustion taking place in those surfaces in contact with the oxygen, and producing CO_2. This, rising through the hot upper layers of the firebed, takes up more carbon, forming carbon monoxide, the chemical equation being

$$CO_2 + C = 2CO$$

In the presence of sufficient oxygen, and providing the temperature is above 1,210° F., all, or the greater part of the carbon burns to CO_2, the equation being:

$$2CO + O_2 = 2CO_2$$

the net result is the same as if the carbon had been burnt directly to CO_2, as previously given.

Whilst this combustion of the solid carbon is taking place, the volatile matter is being distilled off and is burning in the firebox where it is burnt with the "secondary" air, or air admitted above the fire.

Combustion is, therefore, brought about by the use of both primary and secondary air and efficient combustion depends on the correct proportioning of the two, which again depends on the description of the coal. Secondary air must be so controlled as to complete combustion within the firebox and with as little excess air as possible.

For each pound of carbon in the CO_2 the heat produced is 14,540 B.T.U., while for each pound in the CO the heat produced is only 4,380 B.T.U. Clearly 70 per cent of the heat in the carbon is lost when CO is allowed to escape unburnt.

In practice, complete combustion cannot be obtained unless more than the theoretical amount of air is admitted.

This is due partly to the difficulty of mixing air and combustible volatiles above the firebed and partly to the necessity of completing combustion within the firebox. In any case, the speed of combustion depends on the rate at which oxygen can reach the fuel and the "scrubbing" action of the gases on the burning fuel promotes the interchange of O_2 and CO_2 and accelerates the combustion.

The relative ease with which secondary air can be drawn into the firebox directs attention to the importance of controlling the excess. The proportion needed to give complete combustion depends on the firebox gas temperature. As a general rule, with anything over 14 per cent CO_2, carbon monoxide and hydrogen will be present, which will cause a greater loss of efficiency than anything gained by a higher percentage of CO_2.

The presence of 1 per cent of CO in the waste gases causes a loss of 4 to 5 per cent of the heat in the coal burned, and 0·1 per cent of CO causes a loss of 0·5 per cent. The quantity of excess air is between 50 per cent and 30 per cent.

The physical complement of combustion (*see* Chapter 4, Part I, Primary Air and Secondary Air). To burn the incandescent fixed carbon on the grate, it must be vigorously scrubbed by the air, because combustion takes place on the surfaces. The air stream has the tendency to detach small particles. When in suspension these burn completely. When the boiler is being forced, however, the volume of air which must pass through the firebed is so considerable that detached pieces are carried by the gas stream through the tube system and rejected before they are fully burned. A good locomotive coal must therefore possess, amongst other things, a high resistance to the detachment of particles. The friable coals of South Wales possess this by reason of their caking qualities. The hard, non-caking durains also possess it (*see* Coals). But at high rates of combustion even the best coals tend to divide along the lamination planes and to fracture in other planes at right angles. It is when the pieces thus formed are discharged incompletely burned that the loss is the most serious.

The mechanics of this physical action stem from the Bernouilli Theorem. The lifting force is the product of a constant, depending on the shape and smoothness of the piece, its cross-sectional area, the density of the fluid (which is the airstream in this case) and the square of its velocity. When the other factors remain sensibly constant, the lift of unburnt coal from the firebed is proportional to the square of the air velocity, but because the loss is serious only when the pieces are rejected through the tube system, the velocity at parts other than the firebed is also involved.

CYLINDER PERFORMANCE
(*see* Chapter 4, Part I, Diagram factor, Pressure, mean effective; Steam)

There are four inter-related factors in the operation of the locomotive engine: steam supply, tractive force, speed and cut-off. At any speed there is a considerable range of tractive force dependent on steam rate and cut-off; any given tractive force can be produced over a considerable range of speed by varying cut-off and steam rate. Any two of the four conditions may be selected as conditions or requirements, the other two follow.

Tractive effort/speed curves (as Fig. 82) are shown as a series of continuous curves, each curve corresponding to a constant rate of steam supply and implying a definite cut-off variation. Indicated tractive effort is the effort exerted at the wheel treads due to the mean effective pressure, Pm, on the pistons, assuming no intervening losses by friction. It is expressed by the formula:

$$Ti = \frac{D^2 \times S \times Pm \times n}{2W} \text{ lb.}$$

where $Ti =$ indicated tractive effort in pounds.
$D =$ diameter of the cylinders in inches.
$S =$ stroke of the cylinders in inches.
$N =$ number of cylinders.
$W =$ diameter of the driving wheels in inches.
$Pm =$ mean effective pressure in p.s.i.

(*See* Chapter 4, Part I, Tractive Effort.)

Maximum tractive force is realized only under full boiler pressure with the latest cut-off. Latest cut-off can be maintained without impairing pressure or exhausting the steam supply only at low speeds up to about 10 m.p.h., the exact speed depending on boiler capacity in relation to cylinder volume. The maximum mean effective pressure on the pistons at starting is about 90 per cent of the boiler pressure, but the actual force at the wheel rims or drawbar is less than this owing to the intervening friction. The value of the starting effort at the drawbar can safely be assumed for modern locomotives as the value given when Pm is reduced to 85 per cent of the boiler pressure P, leading to the expression for the Rated Tractive Effort:

$$Tr = \frac{85 \times P \times D \times S \times N}{100 \times 2W} \text{ lb.}$$

Above 10–20 m.p.h. maximum tractive effort is limited by steam supply and speed, so that cut-off must be varied to keep as much within boiler capacity as the duties require.

Work is the product of effort and distance, its unit being the foot-pound. Horsepower is the rate of performing work, one H.P. corresponding to 33,000 foot-pounds per minute. The connection between H.P. and tractive effort for locomotives is:

$$\text{H.P.} = \frac{\text{Tractive effort} \times \text{speed in m.p.h.}}{375}$$

This is a general expression. If the tractive effort is the indicated effort, the corresponding H.P. is the I.H.P., if the effort is at the drawbar the H.P. is drawbar H.P., etc.

In testing procedure, the characteristics are obtained by measuring the mean effective pressure on the pistons from indicator diagrams (*see* Chapter 4, Part I, Diagram Factor).

The Indicator Diagram is produced by a pencil actuated by a small piston under the influence of the steam pressure, pressing against a spring having a deflection proportional to pressure. The pencil movement is vertical. The diagram paper is moved horizontally by the crosshead at a reduced scale and the resulting figure is a pressure-volume diagram. Its area is proportional to the work done per stroke and its mean height is proportional to the mean effective pressure.

Fig. 86 shows both a typical indicator diagram, *a b c d e f*, and the idealized or hypothetical diagram *g h j k l*. The valve events and the parts of the steam cycle are clearly worked. The base line of the hypothetical diagram is zero absolute pressure, 14·7 p.s.i., below the atmospheric.

From the properties of the hypothetical diagram the behaviour of the actual engine can be predicted and explained. To the locomotive engineer, one of the most interesting examples is the effect of full and partial regulator working on steam consumption.

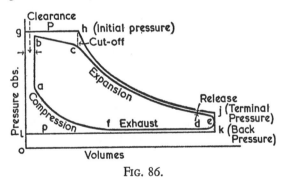

Fig. 86.

Tests by Willans on stationary engines showed that throttling was less economical than early cut-offs with full throttle, as predicted from the hypothetical diagrams. Tests made for the first time at Swindon in 1947 confirmed the applicability of Willans' results to the locomotive. Generally speaking, in the locomotive, throttling is obligatory, especially at higher speeds, when working rates are below two-thirds of boiler capacity. Above this,

throttling is an alternative to full regulator working, but is less economical. Most economical working rate is between one quarter and two thirds of maximum.

Steam per I.H.P. hour for the modern passenger locomotive with superheat up to 300°–350° F. at maximum rates of evaporation, is $13\frac{1}{2}$ to 14 lb. in the most economical band of hourly steam consumption, and at 300–375 revolutions per minute.

$$r.p.m. = \frac{336\,V}{W} \text{ where}$$

$V =$ speed in m.p.h.
$W =$ diameter of driving wheels in inches.

The above figures apply to piston valves and associated gear. A rate as low as $12\frac{1}{4}$ lb. per I.H.P. per hour was achieved on tests of the 3-cylinder B.R. Class 8 locomotive with British Caprotti poppet valve gear at the Swindon Locomotive Testing Plant. French and German compound locomotives with high superheat have given even lower rates, down to 11·1 lb.

Steam consumption per I.H.P. hour is not, however, an exact basis of comparison because of variations in total heat in the steam supply to the cylinders and in the temperature conditions of the boiler feed.

There are actually three efficiencies which are of interest:

(i) The absolute thermal efficiency found experimentally.

(ii) The absolute thermal efficiency of the ideal engine working between the same limits of pressure and temperature.

(iii) The ratio of the two efficiencies.

(i) The absolute efficiency is found experimentally from the work done as measured by the indicator and the heat supplied above the feed-water temperature. Representative values for modern locomotives are $13\frac{1}{2}$ per cent over the range specified for minimum steam per I.H.P. hour.

(ii) The ideal engine is considered to work on the Rankine cycle.

(iii) The efficiency ratio for modern simple steam locomotive works out at 68 per cent over the range specified, for minimum steam per I.H.P. hour.

Minimum steam consumption does not occur with earliest cut-off. This is because steam per I.H.P. hour is theoretically equal to 2,545 divided by the heat drop per pound, the number 2,545 being the equivalent of the H.P./ hour in heat units. At a given speed the heat drop is maximal when the cut-off is about 20 per cent of the stroke. In later cut-offs, though admission temperature is higher, exhaust temperature is also higher and there is a greater amount of heat exchange between steam and cylinder walls, the heat drop being less for this reason.

Superheat is mainly justified by its influence on cylinder condensation and by removing the principal loss of heat by the cylinder walls. Condensation is practically eliminated when steam is dry at cut-off, this requiring at least 150° F. of superheat. At average working pressures this represents a steam temperature of at least 550° F. It is the general practice to secure a temperature 50°–100° F. more than this.

(*See* Chapter 4, Part I, Steam and Superheater.)

DRAUGHTING
(*see* Chapter 4, Part II, Blast-pipes)

Definition. The draughting arrangements in the smoke-box must be regarded as an apparatus for doing work. It receives its energy from the exhaust steam from the cylinders. It accomplishes work by ejecting the smoke-box gases and ash into the atmosphere. To do so a pressure less than atmosphere must be maintained in the smokebox so that there is a constant flow of air from the atmosphere through and over the fire, and of the gases of combustion from the firebox through the tube system into the smokebox and from the smokebox back to atmosphere again.

In order that the energy of the exhaust jet may be sufficient, the cylinders must exhaust against a back pressure, which tends to lower cylinder efficiency.

The pressure in the gas circuit less than the atmosphere is called the draught and this varies in different parts of the circuit. The smokebox draught is equal to the total resistance offered to the air and gas stream which, beside the resistance through the firebed and tubes, includes the additional resistance of smokebox deflectors and spark arrestors when fitted.

Draught is easily measured by manometers containing water and it is usual to express it in inches of water.

Gas flow is of the "turbulent" type so that the resistance to gas flow varies as the square of the flow.

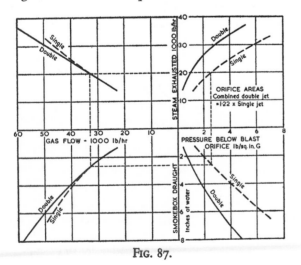

FIG. 87.

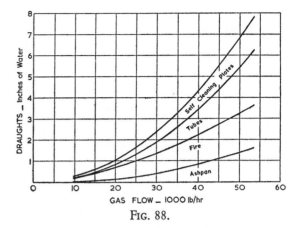

FIG. 88.

Operation. The action of the draughting arrangements is shown by the four-fold diagram of Fig. 87, which is related to Fig. 80. The top left quadrant shows the gas–steam relation to be sensibly linear. This is the overall performance of the arrangement, $1\frac{3}{4}$ lb. of gas being ejected for each pound of steam exhausted, which will satisfy most boilers. The bottom left quadrant shows the relation between the gas ejected and the smokebox draught, or the resistance of the spark arrestor if fitted. The top right quadrant shows the relation between the steam supply and its mean pressure just below the blast orifice. It is obviously desirable to keep the pressure as low as possible for a given rate of exhaustion in order to retain for use in the cylinder as much of the usable energy in the steam as possible. The lower right quadrant shows the relation between the mean steam pressure just below the orifice and the smokebox draught required. If the steam were of constant enthalpy, the slope of this line would represent the efficiency of the draughting system. Although this is not quite the case, it is a most useful "yardstick".

Fig. 88 shows the loss of draught in the various parts of the gas circuit over the range of operation of a typical boiler fitted with "self-cleaning" gear. Should the self-cleaning gear be removed it will not generally be advisable to allow the gas–steam ratio to increase as this would probably result in increase in the loss by unburnt fuel. Instead, the blast orifice should be increased to reduce the energy supply to the system.

The ejector action. At one time it was commonly believed that the behaviour of the exhaust jet was similar to that of a pump, the separate exhausts each filling the chimney with a ball of steam which, by its momentum, pushes before it a certain volume of smokebox gas until both steam ball and gas emerge from the top of the chimney.

But in 1902, Professor W.F.M. Goss published the results of his investigations into the action, carried out on the Purdue Testing Plant, which clearly established that the action was really very different. It is a two-fold action, one effect being to induce motion in the gas immediately surrounding the jet, the other being an enfolding and entraining effect by which the gases are made to mingle with the substance of the jet.

Goss was the first to show:

(i) that the action continued in the chimney to a point nearly at the top;

(ii) that the jet largely accommodates itself to the chimney; and

(iii) that the entraining action almost wholly occurred in the peripheral steam, which has a much lower velocity than the central core of the jet.

Plain double and multiple jets are now common and much favoured (*see* Chapter 4, Part I, Blast Nozzle).

Figs. 89 and 90 show recommended proportions for plain single and double jet designs, which are the same for

velocity to the gases immediately surrounding it, this being the principal action in the part between the base of the petticoat and the blast pipe top, though some entraining also takes place. The gases are thus induced to enter the base of the chimney where they attain the same velocity as that of the steam. From here onwards the final stage of entraining the gases in the peripheral steam is performed.

Since the action owes so much of its effectiveness to the peripheral steam, the over-all effectiveness and efficiency can be greatly enhanced by dividing the jet. The improvement in efficiency is indicated by the fact that the orifice area can be increased by 20 per cent when a single jet is replaced by a double jet (*see* Fig. 87), without increasing the loss by unburnt fuel. The effectiveness is invariably extended over a wider range of the boiler's evaporative range. The performance of the double jet

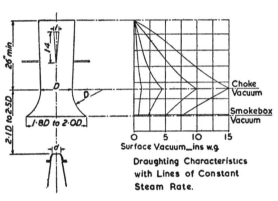

Limiting Diameter of Blast Orifice.

$$d = 1.128 \sqrt{9.5 + .0059 S} \text{ ins.}$$

Where S = Evap. Heating Surface In Sq. Ft

Diameter of Choke.

$$D = 2.85d \text{ to } 2.95d \text{ ins.}$$

FIG. 89.

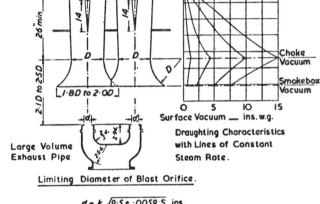

$$d = k \sqrt{9.5 + .0059 S} \text{ ins.}$$

Where S = Evap. Heating Surface In Sq. Ft.

$$k = .8365 \text{ to } .8850$$

Diameter of Choke

$$D = 2.85d \text{ to } 2.95d \text{ ins.}$$

As a first approximation work to the lower limit of k and upper limit of D

FIG. 90.

the two cases. These proportions have been extensively applied in recent years to British locomotives of all types, ranging from locomotives with grate areas of from 14 to 42 square feet, in the case of the single jet, and up to 48½ square feet in the case of the double jet. Manometers placed in various places in the chimney from the petticoat to the top readily give the vacuum characteristics of the design, as shown in the diagrams.

The action between the orifice and bottom of the petticoat is principally that of inducing velocity in the gases immediately surrounding it.

The jet has a wave form at its periphery due to the higher velocity of its core. The peripheral steam imparts

design owes a good deal to the relatively long chimney or entraining tube.

HEAT TRANSFER

The heat evolved by combustion (*see* Combustion) must be transmitted as efficiently as possible to the heating surfaces. Some of the heat generated is lost in the waste gases, and, of the heat absorbed, some is lost in external radiation. Heat is transferred in three ways — by *conduction* through a boiler plate, and the quantity depends on the temperature difference on the two sides. Assuming uniform plate thickness, the larger the surface the greater the quantity conducted in a given time. Moving particles in liquids and gases transmit heat from one to the other by *convection* if they differ in temperature, and they transmit heat by contact when brought against the surface of a solid. Heat can be transferred from a hot body to a cooler body separated by a suitable medium by *radiation* as in a locomotive firebox. Transfer by conduction and convection both depend on temperature difference and little on temperature level, in contrast to the transfer by radiation which does depend on temperature level.

The locomotive firebed and firebox are almost perfect radiators and absorbers of heat by radiation for two reasons:

(i) because the firebox is an enclosed chamber so that the radiation not at first absorbed is repeatedly reflected until it is finally absorbed;

(ii) because of the presence of incandescent carbon particles, carbon being almost a perfect radiator and absorber.

A close estimation of how heat is absorbed in the parts of the boiler by the modes of heat transfer can be obtained from the results of boiler tests if the gas temperatures in the firebox have been taken. For the locomotive boiler concerned in Figs. 80 and 83 much of the necessary information has been given in the preceding sections and this is completed in Fig. 91.

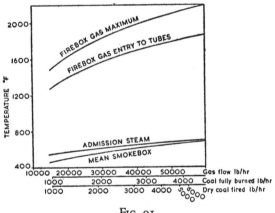

FIG. 91.

Some of this heat goes to superheating the steam. Knowing the quantity passing in unit time and the heat taken up per pound from the pressure and temperatures, this may be readily calculated and deducted from the total delivered to the tube system. The remainder is delivered to the evaporative surfaces.

The absorption as a percentage of the heat released in the firebox, or as a percentage of the heat in the coal fully burnt, is shown in the diagrammatic Heat Balance in Fig. 92 and as a percentage of the heat in the coal fired in Fig. 93.

Of the heat in the coal fully burnt which is not absorbed, part is as sensible heat in the mixed smokebox gases, part as latent heat in the water vapour produced by the combustion of hydrogen, and part in the carbon monoxide produced.

Of the heat absorbed some is lost by external radiation – about 1 per cent of the heat in the coal fully burnt. This is shown in the Heat Balances as the overlap of the parts representing heat absorbed and heat rejected.

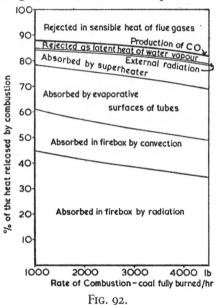

FIG. 92.

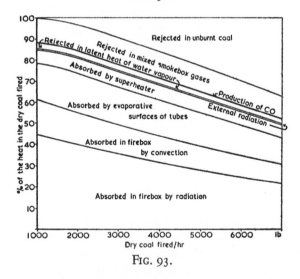

FIG. 93.

THE MEASUREMENT OF COAL AND WATER CONSUMPTION

The measurement of these quantities in performance and efficiency tests requires the exercise of much care and skill if uniform accuracy is to be achieved. Of the two, the measurement of coal consumption is the more difficult.

The first essential of all such tests is the maintenance of a uniformly constant rate of steam production throughout the test period, this having the desired effect of maintaining a uniformly constant rate at which heat is being taken from the boiler, since the enthalpy or total heat per pound under these conditions remains constant.

Under the Stephenson cycle, the rate of coal consumption is a direct function of the rate of steam production and exhaustion. Hence, if the rate of steam production and exhaustion is constant, there corresponds some constant rate of coal consumption, which, however, can be determined only when there is virtual equilibrium between rate of consumption and rate of firing. Herein lies the difficulty, because steam may be produced temporarily at a rate higher or lower than it is, in fact, being generated, and the firebed may vary in content because the actual rate of consumption may, for the time being, be higher or lower than the rate at which coal is being fired. The mean rate is subject to an error due to the variations in the rate.

Since the mere measurement of coal fired over the test period is obviously not enough, arrangements are made to measure intermediate rates during the test. This is conveniently done by weighing the coal into bags of one hundredweight. The fireman takes his coal from a "scuttle" which is replenished when empty from the bagged coal. The number of bags discharged at any one replenishment is such that five to seven minutes elapse between replenishments. This weight of coal, one, two or three bags, as the case may be, is called the increment and remains unaltered throughout the test.

Fig. 94 gives a good idea of the efforts of a fireman to balance coal supply with the constant demand. It concerns a portion only of a test (Fig. 95) in order to show, on a large scale, the actual pattern of firing. The base is one of time elapsed, and the vertical scale is one of weight of coal fired. Replenishments are marked by points – 2, 4, 6, 8, 10 – and between these are the carefully recorded times of firing each shovelful, that is, each "step" is a shovelful of coal fired. The pattern that emerges is typical – a few shovelfuls fired in succession, followed by

a pause to note the response of the boiler, indicated by a tendency for the pressure to rise, subsequently followed by a tendency to fall, when the firing is repeated. The mean rate over these three increments is apparently represented by the best straight line that can be drawn through them (Fig. 95).

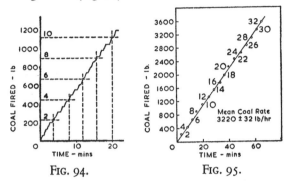

FIG. 94. FIG. 95.

Taking the test as a whole, the mean rate is best taken as the mean of the intermediate rates, and the probable error calculated from their variance by the method of least squares.

The test commences after a preliminary period under test conditions, which period is not only for the purpose of stabilizing temperatures, but also to enable the firing rate to become stabilized as far as can be judged. At a convenient replenishment of the coal increment, the test proper then commences. It finishes also on a convenient replenishment after the necessary equilibrium has been established according to the Code. All points between must be considered of equal value or "weight", and on no account must any portion of the test be discarded, or the commencement and termination be altered. Significant error in the rate at the commencement, or any apparent lapse during the test, lengthens the duration of the test. In the extreme cases the duration may be unduly prolonged which virtually means that it is invalid; the minimum period is one hour.

Part II. Stationary testing plants

OBJECTS AND ORIGINS

From the earliest times in railway history, engineers have realized the need to measure the performance and efficiency of locomotives to technical standards comparable with those used in stationary power plant testing. But the measurement of the performances and efficiency of the steam locomotive is one of especial difficulty. The interdependence of boiler, cylinders and draughting (*see* Action of the Locomotive) makes it necessary to test it as an entity. The large thermal capacities of the boiler

require, as a primary condition for testing, a sensibly constant rate of heat output maintained for a relatively long period. In these circumstances, a constant rate of heat input can be assumed.

But one outstanding feature distinguishes steam locomotive testing. This is that all the thermal rates and quantities can be measured only indirectly. The rate of heat output is a function of the steam quantity rate which can only be measured indirectly by the feed-water rate. The rate of heat input can be evaluated only by the mean

397

rate at which coal of known heating value is fired, an indirect method of measuring consumption. Hence the necessity of a means by which the action of the loco-motive can be controlled for the relatively long periods required to establish indirectly the mean rates which form the basis of efficiency and performance. The stationary testing plant provides these facilities. The incidental resistances applying to track conditions are absent. A dynamometer measures the work done, not on the trailing load but at the wheel rim. Hence much data of importance to railway operation is not obtainable. But the locomotive is not affected by changes in external con-ditions and every part of it is accessible so that the stationary tests are particularly suited to development and research, for comparing thermal efficiencies of different types of locomotives and for studying the effects of changes in design. A constant rate of heat output from the boiler (known as "constant conditions") is readily ob-tained by running at constant speed, fixed cut-off and sensibly constant boiler pressure and water level, so that the plant has to have efficient means of absorbing power and controlling speed. The braked rollers must be capable of accommodating various wheel arrangements and varying wheelbases. There must be accurate means of measuring water and weighing coal, of "indicating" the cylinders, of analysing the smokebox gases, of measuring pressures, vacuum and temperatures. There must be facilities for the analysis of the fuel and the determination of its heating value. In short, the stationary locomotive testing plant is a locomotive laboratory.

The first successful series of locomotive trials on a stationary plant were made by Professor W.F.M.Goss on a plant he had built at Purdue University in 1891. Although it appears that Borodin in 1886 had succeeded in running a locomotive for periods at constant speed on a small plant at Kiev, it is to Goss that the honour of attain-ing results to the desired standard unquestionably belongs. His work encouraged the building of the other plants in the United States of America, one by the Chicago and North Western Railway in 1895 followed four years later by one at Columbia University.

For the St. Louis Exposition of 1904 the Pennsylvania Railroad Company built a plant which was afterwards removed to the company's shops at Altoona. After Purdue, this plant and one built later at Illinois Univer-sity, are the three most important testing plants in the United States.

The year of origin of the Altoona Plant was also that of the first successful testing plant of Europe. This was designed by G.J.Churchward, chief Mechanical Engineer of the Great Western Railway Company of England and built in the Company's workshops at Swindon. After considerable initial difficulties, this plant was eventually brought into the successful state in which it remains today.

In 1936 the London Midland and Scottish Railway and the London and North Eastern Railway decided to pro-ceed jointly with the construction of a testing plant at Rugby for their own use. Work was stopped on the project by the outbreak of war in 1939 and the plant was not completed until 1948 after the nationalization of the railways.

In France, a fine plant at Vitry, near Paris, was opened in 1934.

BRIEF DESCRIPTION OF THE BRITISH STATIONARY PLANTS

Swindon. Though remaining unaltered in essentials since it was first built, the plant was considerably modi-fied in detail in 1936 at which time the brake control system was redesigned to enable the plant to cope with the increased power of locomotives. Improvements to the main bearings were also carried out. The success of these efforts enabled much progress to be made in the developing of testing technique, and design and pro-vision of equipment of the special nature that locomotive testing requires.

There are five rollers, of which four are actually braked, with provision for braking the fifth when neces-sary. Power is absorbed by eight hydraulically operated friction brakes of the differential type running totally immersed in tanks having a continuous flow of water. The control is by means of adjustment of pressure in the brake cylinders through the momentary operation of valves on the control table. To each nominally constant testing condition corresponds a practically stable brake pressure which may be from 20 to 120 p.s.i., according to the nominal torque. In operation, when this pressure has been found, adjustment is necessary only to counteract the effect of small variations in the rate of cylinder steam flow which would otherwise produce a variation in speed. Speed is kept constant with the aid of a variation in steam flow meter.

The brakes can absorb any power that can be trans-mitted to the rollers by the adhesion of steam locomotive wheels carrying the maximum loads permitted in this country.

Adhesion conditions are kept to a very high standard by the use of a dry medium which immediately removes oil and moisture on the wheel treads and rollers and counteracts the burnishing of their surfaces which tends to appear soon after the commencement of a test series.

The plant is well equipped as regards instrumentation, which includes means by which any physical variable

PLATE XII. British Railways (London Midland Region): 4-cylinder 4-6-2 locomotive, No. 46247, "City of Liverpool".

that can be converted into electrical quantity may be photographically recorded on strip chart. Facilities for fuel analysis and the determination of calorific value are available.

A present-day view of the plant showing a test in progress is given in Plate 103B, page 379.

Rugby. The testing plant itself consists of seven pairs of rollers, which support the locomotive, and up to five of which are driven by the coupled wheels of the locomotive. Each of these five pairs of rollers is coupled to a Froude hydraulic brake or dynamometer capable of absorbing up to 1,200 H.P., and designed for a maximum speed of up to 130 m.p.h.

The characteristics of a Froude dynamometer are such that, for any one setting of the controls of the locomotive and of the brake, the combination of locomotive and brake is stable for any minor change of power output, only very small change of speed resulting. The fundamental design of the Froude dynamometer is such that torque increases and decreases as a function of speed, and this provides the dynamometer with valuable self-governing properties which assist in the maintenance of steady speed, irrespective of any adjustment of the control.

In measuring the power at the wheel rim, the Froude dynamometers fulfil an additional function. Arms transmit the torque load to a spring and the deflection of the spring is transmitted electrically to the control room. Each dynamometer can be controlled separately, or all can be operated together by a master controller. This enables the load on all driving axles to be increased or decreased together, or the separate loads on each driving axle may be altered separately.

Measurement of the power of the locomotive is finally made at the drawbar by means of an Amsler dynamometer. The pull exerted by the locomotive is applied to the piston of a hydraulic cylinder, the pressure of which is both shown by a gauge calibrated in pounds pull and also recorded by means of a small hydraulic cylinder and piston connected to an accurately calibrated spring and to a pen.

The Amsler recording table not only records the pull but also has integrating devices to obtain and record the work done and the power developed as well as the actual speed.

Controls and recording equipment are housed in a soundproof control room which contains most of the instrumentation. A laboratory and drawing office are attached to the plant. A general view of the Plant in operation, is shown in Plate 103A, page 379.

Part III. Diesel locomotives
(*see* Diesel Railway Traction, Part VIII, Testing)

Power is the product of tractive effort and speed or, in an electrical transmission, the product of current and voltage. Therefore the ideal characteristic curve which fully utilizes the available output of the diesel engine is a rectangular hyperbola representing constant power (*see* Fig. 96). If a generator with this characteristic could readily be built no further apparatus would be required, and the available output of the engine would be available at all train speeds. The design of generator with series,

shunt and externally excited windings, provides the best approach to the ideal characteristics, but the form of the curve is similar to *B* and *C*. *B* utilizes the full output at *x* only and at one train speed. At all other speeds, power is less than the maximum available but the engine is not overloaded. *C* utilizes full output at *y* and *z* only. Between these the engine is overloaded, and beyond them it is underloaded.

A torque regulator is, therefore, incorporated which fully loads the engine between *y* and *z* and to avoid overloading. Available output at low speeds is not required as this produces very high starting efforts so that up to *z* the natural characteristic is followed. The natural characteristic is followed at high speeds beyond *y* and arrangements are made to unload the engine in this range.

It is, of course, necessary to provide for outputs below the maximum engine rating and the control system has to be arranged to select any one of a number of curves within the band of output of the engine, at the choice of the driver.

The diesel engine efficiency curve is relatively flat, being a maximum at one-third full power and two-thirds maximum engine speed when it attains a value of

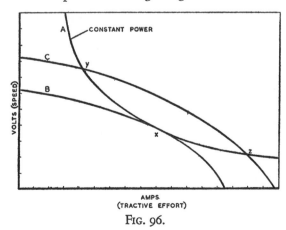

FIG. 96.

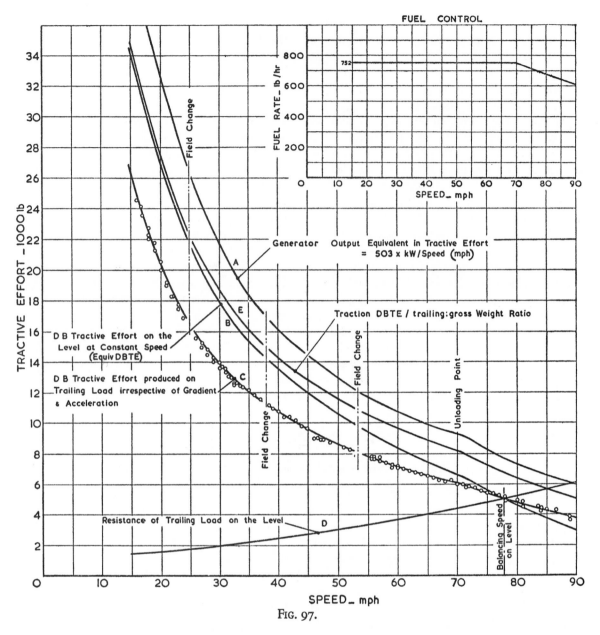

FIG. 97.

about 36 per cent. The inclusive efficiency of the electric transmission is nearly 83 per cent when output is from 40 to 100 per cent of full. Auxiliaries require about 5 per cent of the available engine power.

Figure 97 shows by curve C the measured pull of a 2,000 H.P. locomotive on a train of 392 tons at full power. D is the resistance on the level of the load, B the derived equivalent tractive effort on the level at constant speed,

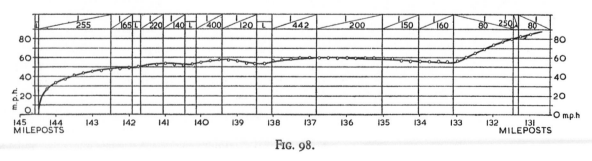

FIG. 98.

A the measured generator output, *E* is the curve from which the full power pull on any other load may be calculated by dividing the values by the gross weight and multiplying by the trailing weight. Fuel control, showing unloading at high speed, is given in the inset diagram.

Calculating the speed–distance curve from *C* and *D* for the given load in the route, Fig. 98, produces the full line. The circles represent the actual speeds attained on the trial, the agreement verifying curves *C* and *D*. The testing system is the controlled road system.

Part IV. Road testing

DYNAMOMETER CARS

General description. A dynamometer car is, as its name implies, a vehicle equipped for measuring force – the tractive force or pull of locomotives. All dynamometer cars are equipped for measuring and recording much more than this, in fact they are mobile traction laboratories. Primarily they measure and record the function of force, time and distance.

(i) The product of force and distance is work done.

(ii) Work done in a given time is horsepower.

(iii) Speed is the distance moved in a given time.

(iv) Acceleration is the rate of change of speed.

The cars are therefore equipped for mechanically or electrically calculating these quantities from the primary measurements:

(i) Pull from a dynamometer.

(ii) Distance from the revolutions of a calibrated rolling wheel.

(iii) Time from a master clock or from special constant-speed electric motors.

On large-scale recorders, in which the charts move proportionally to distance and to time (separately or alternatively), all these functions are drawn or indicated. Mileposts and miscellaneous events are marked, also the instant of making all observations, so that the chart when fully annotated forms a master record of a test.

Dynamometer cars form the measuring and recording units of road tests on controlled conditions and are valuable as observation units in revenue earning services.

Origin and development. The origin of the dynamometer car can be traced back to within a decade of the Rainhill trials of Stephenson's *Rocket* in 1820. The railway age was engendered in these trials and the rapid growth of railways which followed made ever-increasing demands for more and more power, improved performance and higher efficiency. Trial and experiment were necessary and, at the instance of the Directors of the Great Western Railway Company of England, Charles Babbage built instruments for recording tractive force and time into a vehicle provided by I.K.Brunel. In his autobiography *Passages from the life of a Philosopher*, Babbage describes how he experimented with it in 1839 on the line from Maidenhead to Drayton, the experiments lasting five months and costing £300. Besides tractive force and time, the vertical, lateral and end "shake of the carriage" at its centre, and at its end, were recorded on rolled sheets of paper, each 1,000 feet long.

Though the details of what must be regarded as the world's first dynamometer car are fragmentary, its value was evident to Daniel Gooch who built for himself a car when he took over from Brunel the burden of providing locomotive power for the Great Western Railway. This was called a "Measuring Van", and it commenced its career by carrying out a series of resistance trials on freight vehicles in 1856, the records of which are still extant.

In 1899, G.J.Churchward, who was shortly to follow William Dean as fourth Locomotive Superintendent of the Great Western Railway, designed and built a dynamometer car to replace Gooch's, which had then been forty-three years in service. The new car commenced its career in 1901 and is still in service with its original dynamometer and recorder. The instrumentation associated with it is modern, up to date and specialized, as required by modern locomotives of all kinds.

The Great Western dynamometer car was loaned to the North Eastern Railway in 1903, and the North Eastern afterwards built a car, modelled on the Great Western car, which went into service in 1906.

For the Lancashire & Yorkshire Railway, George Hughes designed and built a dynamometer car in 1913. This became No. 1 Car of the L.M.S. Railway at the grouping of the railways. The L.N.W.R. also had a dynamometer car which was adapted by the L.M.S.R. for other purposes when the L.N.W.R. was incorporated in that system.

In all the above cars the dynamometer is, or was, a spring, the pull of the locomotive being taken through the spring, so that its deflection from the no-load position is proportional to the pull, this deflection being recorded.

In modern dynamometer cars, of which there are two on British Railways, the dynamometer is of the hydraulic type; this type avoids extension of the drawbar under pull and is capable of withstanding larger maximum pulls.

A hydraulic dynamometer of the Amsler type consists of two coaxial oil-filled cylinders, back to back, with the pistons connected together, and to the drawgear, so that pressure of oil in one cylinder is a measure of pulling force, and in the other is a measure of buffing force. The inevitable leakages of oil from these cylinders are made good by an electric pump, so that the pistons are always maintained approximately in the same position. Excessive shocks to the dynamometer are prevented by spring friction draught gear which is incorporated in the drawgear. The pressure in the drawbar cylinders is transmitted to three smaller cylinders of alternative sizes, the piston of the selected cylinder pressing against a finely made, calibrated helical spring. The deflection of this spring is, therefore, a measure of the drawbar force to a scale determined by the size of the selected cylinder. The spring deflection moves a pen on the recorder, driven alternatively on a distance or time base, so that a continuous record of pull is recorded. Other pens, actuated from the principal measurements through various arrangements of spherical integrators, record speed, drawbar horsepower, and work done on the drawgear. The latest cars are equipped with a rotary accelerometer for recording work done against acceleration, this displacing the heavy pendulum formerly fitted.

Arrangements for recording location and observations are generally similar in all dynamometer cars.

METHODS AND SYSTEMS OF ROAD TESTING

Discussion. The first attempts to obtain traction data took place when the first dynamometer car was built in 1839 (*see* Dynamometer Cars). A dynamometer car can measure the pull of a locomotive and integrate it over the distance run. It can therefore provide the total work done on the trailing load. Total fuel and water may be measured and thus the mean efficiency on the work done of the drawbar can be calculated. Apparently, therefore, with the aid of observation of engine working and general behaviour, an assessment of performance and efficiency is possible. This kind of test was very general in this country following the building of several dynamometer cars in the early years of this century. The practice cumulated in an extensive series of tests in 1948, following nationalization, known as the Locomotive Interchange Trials.

But such results virtually defy analysis, because of the influence and efficiency of so many factors which may vary from one route to another, and from one train to another on the same route. Efficiency, relative to rate of working, is very far from being linear in a steam locomotive (this being a feature which distinguishes it from the diesel) and it is quite impossible to expect operating conditions to permit the same programme of working rates to apply to each and every train on each route.

Even the working rates themselves are unknown. The efficiency also contains a factor which is quite independent of thermal efficiency and customarily ignored – the ratio of trailing load to gross load. There is also the marked influence on the results, of error caused by the coal fired not being the true consumption by reason of the thermal capacity of the firebox.

These objections have been recognized for many years. They were the reasons which brought into being the stationary locomotive testing plant, and also directed the thoughts of engineers to devise road testing techniques which would enable essential factors to be controlled. America favoured the stationary plant. Whilst eminently suited to development and research, and to studying the effects of changes in design, the stationary plant cannot produce much of the data that is needed in traction problems; the total locomotive resistance, for instance, as influenced by speed, wind and curvature. The minds of many were exercised by the so-called artificial conditions of the stationary plant and the doubt whether the results from tests on the stationary plant are reproducible in service.

Chiefly, however, road tests under controlled conditions came into existence, partly owing to the dearth of expensive stationary plants, and partly because it is necessary to know a good deal about the resistance of vehicles of all kinds in various circumstances, before operating calculations can be made (Plates 104A, 104B, page 380). Indeed, Lomonosoff is renowned more for his contribution to the economics of railway operation, and the application of railway mechanics, than for his pioneer work in developing the first successful road-testing methods.

Origin of road testing under controlled conditions. The first really successful locomotive tests were carried out by Lomonosoff in Russia on the Karkov–Nikolaiev Railway between 1898 and 1900. Lomonosoff made use of certain very long inclines of almost constant gradient that exist in South Russia. On such lines the locomotive reaches a constant speed (dependent on the weight of the train) when cut-off and boiler pressure are kept constant. The conditions of the stationary testing plant were, therefore, simulated, and its methods for the determination of performance and efficiency were applicable. But, in addition to the locomotive data, Lomonosoff's method gave the resistance of the trailing load. This system became the standard method of testing steam locomotives in Russia from 1905.

Next in historical order came the method of Czeczott, appearing on the Polish State Railways in 1921. Lacking inclines of constant gradient sufficiently long to attain constant speed as in Lomonosoff's method, the Polish method uses an approximately straight and level track, ninety miles long. Behind the dynamometer car is a "regulating" locomotive and an appropriate number of vehicles. The purpose of the "regulating" locomotive is to keep the speed constant, as it could either assist the locomotive under test or apply its own brakes and those of the vehicle behind it, as required. This system differs from Lomonosoff's in that, as the load is artificial, data on vehicle resistance is not produced.

Following Czeczott, came that of Günther and Nordmann in 1925 on the German State Railways. Also using an approximately straight and level track (between Grünwald and Magdeburg) the power of the test locomotive in the German method is absorbed by one or more "brake" locomotives, attached to the rear of the dynamometer car. The braking is effected by air (with water from the boiler for cooling) being compressed in the cylinders of the "brake" locomotive and discharged under pressure. This brake system is known as the Riggenbach system. By the German method, braking may be continued indefinitely, without overheating, though several "brake" locomotives may be needed to provide sufficient adhesion for the transmission of the braking force to the rail.

Use of "regulating" locomotives has been made in France and Russia and of "brake" locomotives and units in France, Holland and England.

All these systems are constant speed systems, and do not differ in essentials, but only in the means used to obtain constant speed. Save in producing locomotive resistance, they achieve no more than the stationary plant.

The next system evolved was a radical departure from these. This was the controlled road testing system originated on the Great Western Railway of England by Ell in 1947. Its initial object was to seek traction data, of the kind unobtainable from the stationary testing plant, for use in the technical train timing work and the studies in the economics of traction which that company had pioneered in England since 1930. A second object was to reconcile results from the company's stationary plant with normal service performance.

Out of the work on the stationary plant had grown the concept of the steam rate parameters of force and speed, and of speed and cut-off, whilst the rate also governs consumption in accordance with the Stephenson cycle. By finding a means of keeping constant the cylinder feed through the normal locomotive controls, the somewhat artificial device of constant speed could be dispensed with. The test train could be a normal train operating at variable speed and subject, therefore, to Newton's second law of motion, in exactly the same way as a normal train. Tractive force on the mass of the trailing

load, and its resistance, uniquely determine the speed–distance and speed–time relations of the mass moving over a given route. If all but resistance are recorded, resistance itself, as a function of speed, emerges as a mathematical solution to the equation of motion. Besides this unique feature, tractive effort recordings are self-adjusting for gradient and acceleration.

A simple means of indicating variation in flow was evolved in 1946. Experience in control by the meter was gained with actual trains in 1947. In the year following nationalization, the first attempts at a complete series of tests were made on the South Wales line with a prototype standard locomotive which had just completed draughting trials on the Swindon stationary plant. These were consistent and successful. Under the name of the *Controlled Road Testing System*, many series of trials have taken place with steam locomotives of a variety of power classifications. The results of these and current trials are periodically published by the British Transport Commission, in Bulletins. In 1955, the system was extended to the testing of diesel locomotives, merely by replacing the steam rate by the fuel rate in the scheme of things.

Following the lead of Grünther and Nordmann in Germany on constant speed trials with "brake" locomotives, the London, Midland & Scottish Railway Company, in 1936, considered the possibility of conducting similar tests, with the "brake" locomotive replaced by braking units capable of maintaining a constant speed. As built, the braking units are, in effect, powerful motor coaches equipped for rheostatic braking only, and each carried on two four-wheeled bogies, having a generator driven from each axle. Three units are available. The power generated in opposing the drawbar pull of the locomotive is dissipated by banks of metallic resisters heating two columns of air drawn by fans from beneath the coach and expelled through the roof.

The auxiliary power for the fans, for the blowing and excitation of the generators, is provided by a 100-H.P. diesel-generator set with stand-by battery. The main generators of the three units are electrically similar and are continuously rated at 375 H.P. They differ in gear ratio, the three units having, respectively, maximum speeds of 50, 90 and 120 m.p.h. Below 50 m.p.h., all three units are in operation, providing maximum retarding force and adhesion. Above 50 m.p.h., one unit only is in use.

A dynamometer car is used in association with the braking units, being formed between the test locomotive and the braking units. Besides its normal functions, this car functions also as the control vehicle.

Results of the first series of locomotive tests made with

the braking units or mobile testing plant were published in 1953 by the British Transport Commission, and regularly since that date. In order to economize in testing time, recent tests of steam locomotives have been made in steps of constant speed, the locomotive controls being adjusted, in accordance with an Ell variation in-flow indicator. The steps of constant speed are thus a reproduction on the line of what had been done on the Swindon stationary plant in the development of this instrument.

Results of the first tests of diesel-electric locomotives by the mobile testing unit were published by the British Transport Commission in 1958.

Comparative observational tests. To reconcile the performance and efficiency of a locomotive under service conditions, with the performance and efficiency characteristics obtained by one of the analytical road testing systems, is practically essential and theoretically desirable. Without this, test results are inclined to appear academic. There are transitory periods in the action of the steam locomotive in service for which no parallel can be devised in the analytical systems; yet these transitory performances are a legitimate feature, and a most valuable practical asset, of this type of motive power. The economics of traction are greatly affected by wastages due to run-down condition, sub-standard firing and driving, fuel for auxiliaries, standby losses, faults in the schedule and delays of one sort or another.

Whilst the old type of dynamometer car testing in service trains had very limited value, the situation is completely changed when the efficiency and performance characteristics of the locomotive from analytical road testing are available. A method of making such comparative observational tests was devised on the Western Region of British Railways by Ell and applied with consistent results to express and cross-country steam-hauled trains, from 1957.

In the case of steam trains, the method consists of assigning to each small increment of time, the fuel and water equivalent of pull and speed, according to the characteristics. These are summed for the total traction consumptions; these and the actual, are reconciled by regarding the difference in water as being evaporated with a boiler efficiency of 84 per cent. If the non-traction consumptions are reasonable (about 6 per cent for long-distance express trains, 15 per cent for cross-country trains with standby periods), the probable efficiency and consumption at all points on the route can be reliably shown against the measured average.

In the case of diesel-hauled trains, the method is much simpler, because consumption is directly measured by meter, and because of the approximately linear character

of the relation of consumption with work done, either at the rail or on the trailing load. In the latter case, if efficiency is raised in the proportion of gross weight to trailing weight, a fairly constant value of 27 per cent is obtained for passenger trains, irrespective of the type of duty. For main line diesels of 2,000 H.P. to 3,000 H.P. this figure very clearly approximates to the efficiency at the rail, because the specific resistance of locomotive and trailing load are approximately equal.

RESISTANCE OF LOCOMOTIVES

The resistance of a locomotive is of three parts:

(i) *Machinery resistance*, which is the friction of pistons, guides, pins and all moving parts as far as, but not including, the axle journals.

(ii) *Inherent resistance*, which is resistance due to journal and flange friction, the action of the wheel treads on the rail and the resistance due to motion in still air; and

(iii) *Incidental resistance*, which includes resistance due to gradient, curvature, acceleration and wind.

If resistance is expressed for straight level track, the incidental resistance, except wind resistance, are excluded. That is to say, the resistance is then the sum of the machinery, inherent and wind resistance, for steam locomotives and of inherent and wind resistance for diesel and electric locomotives.

Figure 99 shows total resistance expressed in this way for a representative Class 8 express steam locomotive of 150 tons and a 2,000 H.P. diesel of 133 tons, both curves having been obtained from actual tests. A contrary wind of 7½ m.p.h. natural velocity striking at an angle of 45° to the direction of travel may be assumed, the wind speed being about the average for the country and the direction being that which produces maximum resistance for the wind speed.

The resistance of other locomotives may be estimated from whichever curve in Fig. 99 is appropriate, the resistance being considered as proportional to total weight.

This assumption can only be accepted within small

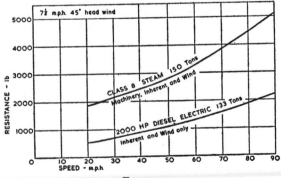

FIG. 99.

margins of difference in weight because many of the component resistances are either not a function of weight at all, or are only indirectly connected with weight.

Nevertheless, until recently resistance was commonly expressed as resistance per ton weight (or "specific" resistance) in relation to train speed. The relationship was usually considered as having the mathematical form:

$$r_e = A + BV + CV^2$$

where

r_e is the specific resistance in pounds per ton.

V is the train speed in m.p.h.

A is a co-efficient which expresses the so-called constant resistances – mainly machinery and journal friction.

B is a co-efficient expressing tread and miscellaneous resistances which vary with train speed.

C is the wind resistance.

The curves of Fig. 99 have this form, but it is doubtful if they can be satisfactorily reduced to such simple terms and a limited extension to other locomotives on a weight basis must be understood.

RESISTANCE OF VEHICLES

Coaching Stock. The controlled road testing system has produced a considerable amount of data on the resistance of representative British coaching stock under a variety of wind conditions. Co-ordination has enabled the principal factors to be sorted out. These concern the train speed, the relative wind velocity, and the yaw, or the angle of the resultant wind with the direction of motion of the train.

In operating calculations the resistance of the train must not be underestimated and it is usual to assume a wind of 10 m.p.h. at 45°. For these conditions the diagram shown in Fig. 100 gives the resistance–speed relation. It also shows the relation for still air.

Multiple-unit main-line stock. Resistance can be assumed, as for coaching stock, with the addition of head-on resistance per train as follows:

Speed	Additional Resistance lb.
10	150
20	166
30	210
40	300
50	450
60	640
70	860
80	1,100

Freight vehicles. Resistance of freight vehicles depends very largely on the type of vehicle and the load. These vary greatly. But since it is impossible to particularize in operating calculations, a resistance per ton is needed

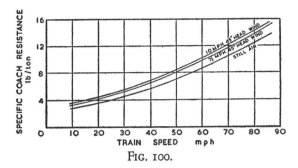

FIG. 100.

which can be generally applied. From controlled road tests of 500-ton trains, each made up of similarly classified and similarly loaded vehicles, the total mean resistances shown in Fig. 101 were obtained, the conditions being average. These may be broadly divided into empty and loaded vehicle resistances. The top curve covers the empty vehicles and the middle heavy curve is an envelope

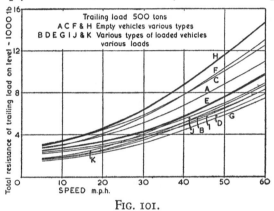

FIG. 101.

curve for the loaded vehicles. Taking these curves, the following table emerges as practical values of the resistance *per ton*:

	Loaded		Empty
Speed	Res. per ton (lb.)	Speed	Res. per ton (lb.)
0	20·00	0	20·00
10	5·29	10	6·70
20	6·54	20	9·10
30	8·17	30	12·76
40	11·07	40	17·50
50	14·52	50	23·20
60	18·64	60	29·40

Brake vans should be considered as loaded vehicles.

Most operating calculations concern loaded vehicles. When empty vehicles have to be accounted for, a recommended rule is to consider an empty vehicle as equivalent to one and one-third times its tare weight for the purpose of estimating the load of a train in terms of loaded vehicles. It is clear from the foregoing table that there can be no real loading equivalent, because resistance on the level is not the only consideration.

On gradients the actual weight is the deciding factor. For practical purposes, therefore, a compromise must be made between resistance and weight, and the above rule is simple and as suitable as any.

TRACTION RELATIONS

The equivalent drawbar tractive effort is a force available at the drawbar, at constant speed on the level.

The rail tractive effort is a force available at the rail, at a given speed, irrespective of whether the locomotive is on a gradient or on the level and whether accelerating, at constant speed or decelerating.

The actual drawbar tractive effort is a force exerted on a trailing load.

The drawbar tractive effort at constant speed on the level, or equivalent D.B.T.E., is the most common relation produced. It is most useful for comparing one locomotive with another. But it is not immediately useful in estimating the actual performance with a given load over a given route, because this requires a knowledge of resistance, the speed is not constant and the gradients vary. The equivalent D.B.T.E. is only the actual drawbar pull when the speed is constant and the line is level; at lower speeds the actual pull is less, at higher speeds it is more, whether on the level or on a gradient. The difference between the indicated tractive effort of the cylinders of a steam locomotive (*see* Cylinder Performance) and the equivalent D.B.T.E. is the resistance on the level, this including machinery, wind and inherent resistance. The difference between the rail tractive effort of a diesel and the equivalent D.B.T.E. is the resistance on the level, including only journal friction of the axle-boxes, the wind resistance and the inherent resistance.

The actual work done at the drawbar on the trailing load (as measured by a dynamometer car) is that due to the exertion of the actual drawbar tractive effort in overcoming resistance, and in overcoming the gradient component of the weight on rising gradients (on falling gradients effort is reduced by the gradient component and in providing acceleration).

With a given load and at a given steam rate, or fuel rate in a diesel, the actual D.B.T.E. does not have to be adjusted for gradient and acceleration but, unlike the others, it depends on the ratio of trailing load to gross load, and on the resistance of the trailing load. But when the specific resistance of the locomotive is equal to the specific resistance of the trailing load (as in many diesels hauling passenger trains), the actual drawbar tractive effort is equal to rail tractive effort when multiplied by the gross load and divided by the trailing load.

These and other less simple relations are shown below. Constant speed testing systems readily pro-duce the equivalent D.B.T.E. at constant speed on the level, because the drawbar pull measurements need only a correction for the gradient. The variable speed controlled road testing system records the actual D.B.T.E. (which is proportional to rail tractive effort when specific locomotive and train resistances are equal), and needs no correction for gradient and acceleration. The resistance of the trailing load on the level is the mathematical complement of the actual D.B.T.E., this enabling the equivalent D.B.T.E. curve to be constructed from the actual.

W_e = weight of locomotive.
W_c = weight of trailing load.

$$n = \frac{W_c}{W_e}$$

$$k = \frac{W_c}{W_e + W_c} = \text{trailing : gross weight ratio}$$

F = fuel rate

V_0, V_1, V_2, V_3, etc. = train speed

r_c = specific resistance of trailing load on the level.

r_e = specific wind and track resistance of locomotive on the level.

$R = r_c \times W_c$ = total resistance of trailing load on the level.

$C = r_c \times W_e$

P = equivalent drawbar tractive effort or drawbar tractive effort on the level at constant speed = rail tractive effort—$r_e W_e$.

ρ = tractive effort on trailing load or traction drawbar tractive effort (*trac. D.B.T.E.*)

Residual forces bc, $b'c'$, etc.

On level:

ab, $a'b'$, etc., accel. force on trailing load.
ac, $a'c'$, etc., accel. force on locomotive.
Locomotive and trailing load have same acceleration.

$$\therefore \frac{ab}{ac} = \frac{a'b'}{a'c'} = \frac{W_c}{W_e} = n$$

On gradient:

cd, $c'd'$, grad. component of locomotive.
eb, $e'b'$, grad. component of trailing load.

$$\therefore \frac{cd}{eb} = \frac{c'd'}{e'b'} = \frac{W_e}{W_c} = \frac{1}{n}$$

ad, $a'd'$, accel. force on locomotive.
ae, $a'e'$, accel. force on trailing load.
Locomotive and trailing load have same acceleration.

$$\therefore \frac{ae}{ad} = \frac{a'e'}{a'd'} = \frac{W_c}{W_e} = n$$

Thus ρ is not qualified by gradient and acceleration.

Consider a given fuel rate F and speed V_3 as:

$$W_c = nW_e, R = nC \text{ and } k = \frac{n}{n+1}$$

$$\rho = P - (P - nC)\frac{W_e}{W_e + W_c}$$

$$= P - P\frac{W_e}{W_e + W_c} + nC\frac{W_e}{W_e + W_c}$$

$$= P\left[1 - \frac{W_e}{W_e + W_c}\right] + nC\frac{W_e}{W_e + W_c}$$

But

$$1 - \frac{W_e}{W_e + W_c} = k \text{ and } W_c = nW_e$$

$$\therefore \rho = Pk + Ck$$

$$\therefore \rho = k(P + C) \text{———————} ①$$

Note:

(i) This result is independent of n and therefore holds for all values of n.

(ii) Thus for a given fuel rate and speed:
P and C are constant, $\rho \propto k$ from ①

$$\therefore \frac{\rho^1}{k^1} = \frac{\rho^2}{k^2} = \frac{\rho^3}{k^3} = \ldots \ldots \ldots \ldots$$

(iii) This may be extended to traction d.b.h.p. and fuel/traction d.b.h.p.hr. as follows:

traction d.b.h.p. $\propto \rho \propto k$

fuel/traction d.b.h.p.hr. $\propto \dfrac{1}{\rho} \propto \dfrac{1}{k}$

(iv) When $r_e = r_c$, $C = W_c r_c = W_e r_e$
$$\rho = k(P + C) = k(P + W_e r_e)$$
$$= k \times \text{rail T.E.}$$
When $r_e > r_c$
$$\rho = k[\text{rail T.E.} - W_e(r_e - r_c)]$$

The systematic relationships between the traction relations are explained in Fig. 102. In steam locomotives, the cylinder tractive effort curve lies above that marked p/k, and so does the traction equivalent of generator output in diesel-electric locomotives, but in the latter, the rail tractive effort is almost identical with

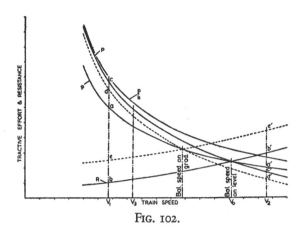

FIG. 102.

this curve, and is co-incident when specific locomotive and load resistances are equal.

The criterion of the cost of energy of movement is the fuel-per-ton-mile, which is equal to the fuel-per-hour divided by the product of speed in m.p.h., and the tons hauled. This is the same as stating that the fuel-per-ton-mile is proportional to the power/weight ratio and the fuel-per-horsepower-hour, and inversely proportional to the average speed.

The power/weight ratio and the average speed are functions of the duties assigned to the locomotive and the fuel-per-H.P.-hour is a function of the built-in thermal efficiency. For a given duty, the running that makes fuel-per-ton-mile a minimum is the most economic in fuel consumption. The factors of the fuel-per-ton-mile are, however, interdependent variables.

With the diesel, fuel-per-rail-H.P.-hour is practically a constant, and the fuel-per-ton-mile is, therefore, proportional to the H.P./ton (the power/weight ratio) and inversely proportional to the speed.

With steam locomotives, fuel-per-rail-H.P.-hour rises with the rate of working, above two-thirds of the maximum continuous rate of evaporation, so that, though the approximation given above for diesels is roughly applicable to steam locomotives at nominal rates of working, the economy of this type of power is greatly reduced when the locomotive is forced.

Part V. Performance and cost of energy
(*see* Chapter 7, Technical Train Timing)

There are three principal elements in rail transport – the track, the rolling stock and traffic control (or the time table). These are inseparable, but the predominating element is traffic control. It is impossible to plan efficiently the working of a railway, to design efficiently

the technical elements or to determine properly the economics of a traffic pattern, without familiarity with the basic laws of motion of a train along the permanent way and to be able to apply these laws to practice. Without the application of railway mechanics it is impossible to calculate, even approximately, how any novelty, such as change of form of motive power, will affect the cost of movement.

It is this important sphere (railway mechanics) which contains the end-products of locomotive testing. Formerly, whilst the railways had a monopoly of transport, whilst the form of motive power was predominantly of one type and whilst the traffic patterns remained unchanged, little use was made of scientific methods.

For all forms of motive power it is now the practice to plan for all classes of traffic on technical train timings, prepared for a wide range of power and load. In these, all technical factors are taken into account including the power margins necessary to maintain the time-table in the face of certain contingencies that are always likely to affect it. Fuel consumption is included, as this, for any form of motive power is greatly affected by the perfor-

mance and duties required of power units – factors outside their "built-in" efficiencies. A form of presentation of this information extensively used in this country is the Ell performance and cost of energy diagram. This gives the connection between power, load and overall time for the route, with the corresponding point-to-point timings, with power margins. Fuel consumptions are shown by iso-fuel lines. Such diagrams can be applied to operating projects and fuel costing by non-engineering personnel.

The criterion of the working costs is not dependent on any one factor, but on the cost of the net ton mile which is dependent upon a large number of factors both economic and technical. These are expressed in the well-known equation of Lomonosoff which has been specially arranged in the form given in Fig. 103. Fuel consumption and cost are represented by the last terms but are an important item in the whole.

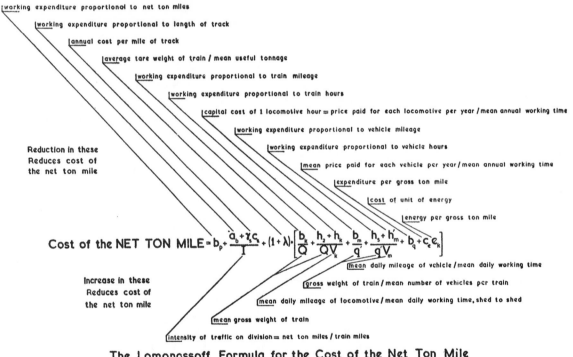

The Lomonossoff Formula for the Cost of the Net Ton Mile

FIG. 103.

CHAPTER 7

The Steam Locomotive in Traffic

by O. S. NOCK

Part I. Conditions of service

INTRODUCTORY

The working of steam locomotives in traffic is normally beset by such a variety of factors that train schedules have in the past usually been decided, not from a scientific assessment of locomotive capacity, but from an accumulation of running experiences gained over many years. The results came to include many anomalies, and it is only in recent years that a more precise approach to the problem has been made, by appreciation and use of the fundamental relations linking the basic performance of the locomotive with the mechanics of the train in motion. Modern methods of testing have shown that if the rate of evaporation in the boiler be kept constant, or nearly so, the performance of a locomotive with any given load can be predicted by reference to the physical characteristics of the route. Put in another way, if it is desired to operate a service at a certain average speed, the same method of analysis will determine with precision the load that can be hauled on that schedule by the various classes of locomotive available. This method of analysis, as evolved by the Western Region, British Railways, is known as "Technical Train Timing". There are, however, many factors to be taken into account, other than the optimum capacity of the locomotive, and before discussing technical train timing other aspects of engine performance have to be considered (*see* Chapter 6, Part V, Performance and Cost of Energy).

FIRING RATES

With large express passenger and mixed traffic locomotives that are hand fired it is the capacity of the fireman that determines the maximum sustained output in traffic. On British Railways, on continuous runs of three to four hours' duration the maximum firing rate has been agreed at 3,000 lb. of coal per hour. This corresponds to a steaming rate of approximately 25,000 lb. per hour, on a good modern type of locomotive having a nominal tractive effort of 35,000 to 40,000 lb. at 85 per cent boiler pressure. Such locomotives have been shown capable of maximum sustained steaming rates of 32,000 to 40,000

lb. per hour, and the difference between these latter values and the maximum service demand of 25,000 lb. per hour provides a margin that is considered desirable in British practice. There are times when individual engine crews, in recovering lost time or coping with an exceptional load, may steam a locomotive at considerably more than 25,000 lb. per hour, for short periods, but such reserve capacity is not taken into account in the planning of train schedules.

On the large steam locomotives used in the U.S.A., prior to dieselization mechanical stokers were a necessity, and with such equipment, no less on oil-fired steam locomotives, it was customary to work more nearly at the maximum capacity of the boiler than is usual in Great Britain. On long runs over the main lines of the Eastern, London Midland, and Western Regions, the aim has been to provide locomotives that will meet the traffic demands while steaming at economical rates of evaporation and admission in the cylinders. The differences between British and North American practice in this respect are brought out in more detail in the chapter commencing on page 319.

EFFECT OF FUEL

The quality of fuel available is an important factor to be considered in the planning of train services likely to make severe demands upon locomotive power. In Great Britain the numerous varieties of bituminous coal are graded according to calorific value and their suitability or otherwise for long runs at high rates of evaporation. Calorific value is not the only factor to be taken into account. The technique of firing is quite different, to take extreme examples, with a hard, quick-burning coal from South Yorkshire or Nottingham from that necessary with the slower-burning soft Welsh or Kentish grades. Sometimes two varieties are encountered in a single round trip, as with the Continental Boat Train workings of British Railways, Southern Region. Locomotives starting from London are provided with hard coal, whereas their tenders are topped up with a soft Kentish grade at Dover for the return journey. If the firing technique is suited to the grade of coal there is no reason why the rate of evaporation should not be in proportion to the calorific value of the coal; but in the present circumstances fuel of the second quality is often supplied broken small, and including much fine dust, and then it needs a very expert fireman to secure an adequate output of steam. In many locomotives built since World War II design features such as multiple-jet blastpipes have been introduced to assist in securing good steaming with the poorest quality of coal. Such devices often result in an increase in basic fuel consumption, due to the fact that much of the really small content is thrown unburnt through the chimney. In such circumstances, however, reliable steaming is more important than the maintenance of a low coal consumption in relation to the actual work performed, as represented by the average horsepower sustained at the drawbar per hour. The total work done on any one trip is measured as so many drawbar horsepower hours (*see* Chapter 4, Part II Fuels, page 315 and Chapter 6, Coals, page 389).

INFLUENCE OF GRADIENTS

The sustained speed at which a locomotive will haul a given load up a gradient can be readily determined from the first principles of mechanics if the power output capacity is known; but instances where the speed is sustained for any length of time are rare, and the conditions under which an ascent is commenced often outweigh the capacity of a locomotive to sustain that speed. On a road of undulating character, if the gradients do not exceed a maximum steepness of about 1 in 270 (0·37 per cent) it is possible for a locomotive to be worked at an approximately constant rate of evaporation, whether running uphill or down, and the resulting average speed will be little, if at all, different from that which could be maintained on level track. Speed restrictions rather than gradients have a greater influence in limiting the over-all averages. Providing the alignment of the road is good enough to permit of really high downhill speeds, a fast over-all average can still be maintained, despite heavy intermediate gradients. The main line of British Railways, Southern Region, between Salisbury and Exeter, eighty-eight miles, provides an excellent example of this characteristic. This route has many stretches inclined at 1 in 80 (1·25 per cent), and yet it was found possible to run trains loading up to 450 tons tare with relatively small 4–6–0 locomotives of 25,320 lb. nominal tractive effort, at an over-all average speed of 55 m.p.h. Impetus played a big part in the working of such trains. The Seaton bank – six and a half miles at 1 in 100–80 (1·0 to 1·25 per cent) – was often commenced at 85 m.p.h., but it will be appreciated that a reduction of speed, either for signal or

engineering work, at the foot of one of these steep gradients would cause serious loss of time.

The Scottish Region main line between Edinburgh and Dundee, covering the Forth and Tay bridges, provides an example of the opposite kind. Although the gradients are neither so long nor so severe as those between Salisbury and Exeter, the permanent speed restrictions are many, and are located at the foot of some of the worst gradients. In consequence, the highest average speed booked in recent years, with large "Pacific" engines of 34,000 lb. tractive effort, and loads of 450 to 480 tons, was no more than 45 m.p.h. No opportunity existed to run fast downhill, and the gradients had to be climbed from initial speeds of 25 to 30 m.p.h. It is over this kind of road that a skilful engine crew can steam a locomotive intermittently at a higher rate of evaporation than could be sustained for any length of time. The sections of adverse grading are relatively short, and the succeeding descents are run with steam shut off, or on very light steaming, thus enabling the boiler to recover for the next severe uphill spell. Such working is not recommended as a regular feature, but it is a useful attribute to have when there is lost time to be recovered, or when an exceptional load is to be conveyed.

CIVIL ENGINEERING RESTRICTIONS

Present-day difficulties and restrictions in the operation of traffic are often due to circumstances existing at the time the railways in question were constructed. All railways in Great Britain suffer from the handicap of the early loading gauge adopted. As a consequence there are, for example, many locomotives running on the 3 ft 6 in. gauge railways of South Africa that are taller and wider than the largest in Great Britain. On many railways, economy in first constructional costs was a dominating factor; earth-works were kept to a minimum, expensive bridges and tunnelling were avoided, and routes built through difficult country included much severe curvature, heavy gradients, lightly constructed bridges, and light section of permanent way. Such features were adequate at the time of construction, but with the growth of traffic they have proved a most serious handicap. On many railways in dominion and colonial territories, development of the country has taxed operating facilities to the utmost. On single-tracked main lines the only solution has been to run freight services on a relatively few trains, but each carrying very heavy loads. If double- or triple-heading is to be avoided, special consideration requires to be given to the design of locomotives for such routes. Use of standard designs is rarely possible. The locomotive must be tailored to suit the particular duty.

In Great Britain at one time there was a tendency to

build special locomotives for special work. The West Highland section of the former London and North Eastern Railway was a case in question. Gradients include long stretches of 1 in 60 (1·67 per cent), for the most part with severe curvature; the speed limit is 40 m.p.h. throughout, thus giving no chance to climb gradients with the aid of impetus gained by fast running on the approaches. Until the year 1937, the largest and most powerful locomotives permitted to be used on the route were limited to a maximum load of 220 tons, and in consequence much double-heading was necessary. In 1937 a special design of 2–6–0 was worked out, to enable loads of 300 tons to be taken without assistance, while keeping within the limitations of axle-loading laid down. In obtaining the high nominal tractive effort for a 2–6–0, of 36,600 lb., coupled wheels of no more than 5 ft 2 in. diameter had been used, and while the locomotives of this class did admirable work on the mountain section, the complete run from Glasgow to Fort William included the fast and level stretch alongside the Clyde estuary, not subject to the 40 m.p.h. speed restriction. The special 2–6–0 locomotives found difficulty in keeping time on this fast stretch, and the hard working involved, with high piston speeds, set up mechanical troubles. Ultimately they were taken off passenger work, and replaced by a standard type of 4–6–0; these latter could not take such heavy loads as the special 2–6–0's, and required assistance more frequently, but the work was done with reliability.

ROSTERING OF LOCOMOTIVES

The aim of all locomotive diagramming is to operate the traffic allocated to each depot with the minimum of locomotives, by securing maximum mileage from each. As the majority of depots are responsible for a variety of duties the ideal is to have locomotives capable of as wide a range of workings as possible, so that the allocation of special locomotives to special duties is the smallest proportion practicable. It is also desirable to have the fewest different classes of locomotive, as this not only simplifies diagramming, but makes possible the stocking of less variety of spare parts, special tools and heavier appliances. The general principles may be followed by reference to working at a large depot where the main line locomotives consist of two classes of 4–6–0, an express passenger type capable of the fastest work scheduled in Great Britain today, and a two-cylinder general utility type. This shed is mostly concerned with passenger traffic, though a limited number of freight workings are also rostered.

Both varieties of 4–6–0 are grouped in links according to the severity of the duties concerned. One express passenger locomotive in first-class condition is selected and set aside for the exclusive working of an extra high speed service (average speed 67½ m.p.h. start to stop). This engine is given special attention and maintenance and is kept on this one duty for months at a time. The mileage per month is not high, but in view of the importance of reliability in working a much-advertised, and well-patronized businessman's train, the procedure is justified. For the next group of express passenger services worked by this depot, having average speeds of 56 – 65 m.p.h., a group of engines having relatively low mileage since last overhaul are set aside. These engines take other duties as well, in order to secure good monthly mileages, and all are manned on a common user system, no crews having particular engines allocated to them regularly. As mileage since last overhaul mounts up, engines are relegated to less exacting workings, to stopping trains, spare duties, "special", and even to goods. Even though a locomotive may be unfit for fast express duty, it can continue to do useful work until the time comes for another visit to works for overhaul. On returning to duty it may be selected for the 67½ m.p.h. turn. In such a cycle of working a locomotive will run 90,000 to 100,000 miles between successive shoppings in a period of about eighteen months. A similar cycle, though on less spectacular turns, is worked by the mixed traffic 4–6–0 class. When newly returned from works, engines of this type are occasionally requisitioned for first-class express work, and their duties gradually descend to stopping passenger and intermediate goods trains.

With large express passenger locomotives of the "Pacific" type an attempt is made, by careful servicing, to keep them in heavy express work almost up to the time for going to works for overhaul. Prior to World War II this was successfully done with the Gresley streamlined locomotives of the former L.N.E.R., even though some of the turns worked included such a train as the Coronation, booked at an average speed of 72 m.p.h. between London and York.

Cyclic workings. In order to obtain the best possible mileages from locomotives it is sometimes arranged for circular tours of multiple duty to be worked, in which an engine is manned by successive crews from several different depots, and may be away from its home station for two days or more. An example from the practice of the London Midland, and Scottish Regions of British Railways will illustrate the extent to which this practice is sometimes developed. The duty described on the next page was at one time regularly worked by engines of the "Royal Scot" class stationed at Polmadie (Glasgow).

Each engine on this cycle was away from its home shed for approximately forty-eight hours, after which it was on shed for twenty-four hours before going out on the next

Duty	Mileage	Duration hr	Duration min.	Average speed m.p.h.
11.20 p.m. Glasgow (Central) to Crewe	244	5	43	42·7
Turn round at Crewe	—	4	22	—
9.25 a.m. Crewe to Perth	292	8	00	36·4
Turn round at Perth	—	3	55	—
9.20 p.m. Perth to Carlisle	151	5	25	26·1*
Turn round at Carlisle	—	4	15	—
7.0 a.m. Carlisle to Glasgow (St Enoch)	115½	3	02	37·7
Turn round at St Enoch	—	2	48	—
12.50 p.m. St Enoch to Carlisle	115½	2	38	43·8
Turn round at Carlisle	—	4	08	—
7.36 p.m. Carlisle to St Enoch	115½	2	30	46·2

* Express fish train.

cycle. The mileage covered on each round was 1,087, in the course of which six different trains were worked, and the engine itself was handled by ten different engine crews, apart from relief men. Out of each forty-eight hours on duty, thirty-three were spent in traffic, and engines thus engaged were averaging more than 2,500 miles per week.

The above table serves to emphasize the difficulty in Great Britain of securing a high utilization from locomotive stock, owing to the turn-round times necessitated by waiting for suitable return, or continuation workings. To improve upon this, long through-turns between London and Glasgow were instituted with the Stanier "Pacific" engines, and turn-round times cut, in some cases, to three and a half hours. Thus a locomotive going out with the 11.45 p.m. from Euston was in London again in less than twenty-two hours, having covered 802 miles in the meantime. The combination of long through-workings and short turn-round times probably reached its furthest extent in the U.S.A. on the Santa Fé Railroad, where locomotives of the 4–8–4 type worked through over the 1,234 miles from La Junta to Los Angeles, and on some duties from Kansas City to Los Angeles, 1,788 miles. Normal turn-round times on locomotives working *The Fast Mail* and *The Chief* were approximately eleven hours, after a twenty-six-hour through-working; but the turn-round was often reduced in practice to seven hours. These locomotives were oil fired, and in the course of the La Junta–Los Angeles run were handled by nine different crews.

ROUTE AVAILABILITY
(THE EFFECT OF HAMMER-BLOW)

On routes where weight restrictions are imposed the effect of hammer-blow (or dynamic augment) on underline bridges is more important than dead weight. To secure good riding it was at one time conventional practice to balance two-thirds of the reciprocating parts in two-cylinder locomotives. This involved the use of heavy weights cast integrally with the coupled wheels, and while the locomotive rode well, a severe hammer-blow effect was imparted to rails and underline structures. The use of three or four cylinders enabled locomotives to have a better inherent balance, and the weight added to couled wheels was very much less. Consequently axle-loadings could be increased without risk of setting up undue stresses in bridges and so on. At the same time, however, the two-cylinder machine is cheaper to build and maintain, and in recent years a departure has been made from former conventions by balancing only 30 per cent of the reciprocating parts. Locomotives used on the Great Eastern line of British Railways provide an interesting example:

Type class	Year introduced	Cylinders	Nom. T.E. lb.	Adhesion w. tons
4–6–0 B.12	1911	2-inside	21,969	44
4–6–0 B.17	1928	1-inside & 2-outside	25,000	54
4–6–0 B.1	1942	2-outside	26,878	52½

As a result of this change in design practice the "B1" 4–6–0 with only 30 per cent of the reciprocating weight balanced, and a low hammer-blow value, has a very high route availability (*see* Chapter 4, Part I, Masses).

LOCOMOTIVE FOR SPECIAL SERVICE

In Great Britain, instances of roads requiring special consideration are relatively few, and in recent years the policy has been to use standard types wherever possible. In countries of the British Commonwealth, particularly where rail gauges are less than the standard 4 ft 8½ in., running conditions which would be exceptional in Britain, predominate and provide the conditions for which the standard passenger and freight locomotives of the railways concerned must be designed. The Mombasa–Nairobi–Kampala main line of the East African Railways is a good example. Long gradients of 1 in 65 (1·55 per cent) have to be climbed; the line is laid with rails weighing only 50 lb. to the yard, and axle loads in consequence are limited to a maximum of twelve tons. Locomotives of the Beyer-Garratt type are extensively used. Despite the structural limitations, machines having a nominal tractive effort in excess of 45,000 lb. have been successfully introduced. These are of the 4–8–4+4–8–4 type, and convey loads on the various gradients as follows:

Gradient	Load (tons)
1 in 50 (2%)	500
1 in 65 (1·55%)	650
1 in 85 (1·2%)	860

Train working includes round trips of 1,100 miles, from Nairobi to Kampala and back. The caboose system is worked, with two sets of enginemen, each pair working for a period of eight hours.

TECHNICAL TRAIN TIMING

In applying the basic "force : mass" data for any locomotive to the calculation of train schedules, certain definite principles have to be observed:

Ruling rate.

(i) What is termed the "ruling rate" is the rate of steam production in the boiler that will produce the necessary over-all average speed for the train. In practice, however, some variations from the ruling rate may be necessary in order to avoid too great a variation from the average speed. The variations will largely depend upon the severity, or otherwise, of the gradients. As an example, it would be practicable with service trains to run at a constant steam rate from Euston to Rugby, London Midland Region, but quite impracticable to do the same between Ayr and Stranraer, Scottish Region.

(ii) It is appreciated that variations from the ruling rate, to avoid substantial fluctuations in speed, can become a discontinuity in working, and reduce the over-all efficiency of the locomotive in operating the service.

(iii) On duties where the demands are considerably below the maximum capacity of the locomotive it is usual to find variations predominating over the ruling rate, whereas in heavy duty the working approximates much more closely to a steadily maintained uniform rate, very near to the ruling rate.

(iv) Where the kinetic energy of a train has to be reduced, in preparation for a speed restriction or a stop, as much as practicable should be absorbed by resistance, without braking.

From the principal performance relations obtained from a locomotive on test, the accelerating "force : mass" ratio can be derived, and a series of time–distance graphs drawn for different steaming rates, for different loads, and for different gradients. The appropriate graphs are applied, successively, in accordance with the varying gradients along the line, and a composite time-distance graph for the whole journey produced. Thus the effect of different loads can be studied.

Theoretical diagrams. From such studies a diagram such as Fig. 104 can be produced, which shows the traction coal per ton mile against variations of average booked speed and load. The lower diagonal line across the diagram shows the basic maximum performance, approximately 25,000 lb. of steam per hour, while the hatched area indicates the range in which the locomotive is working at maximum efficiency. This diagram relates to

a particular route approximately 100 miles long, including moderate gradients and one lengthy permanent speed restriction at approximately one-third of the total distance. From the diagram it will be seen that for an average speed of 60 m.p.h. the maximum load that can be conveyed within the range of maximum locomotive efficiency is 375 tons, but if the locomotive was steamed up to the maximum agreed limit for hand firing, a load of 480 tons could be conveyed at the same average speed.

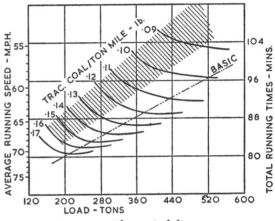

FIG. 104. Theoretical diagram.

Recovery time. In Great Britain it is customary to include in any train schedule making maximum demands upon a locomotive, a proportion of what is termed recovery time; this is designed to provide a margin against time lost by temporary speed restrictions due to engineering work on the line. When a schedule has been planned to require, for example, a steam rate of 25,000 lb. per hour for three hours on end, a common practice is to insert an extra four minutes for every 100 miles. The recovery allowance is not usually spread uniformly over the entire journey but added as "recovery time" between specific timing points. With the maximum firing rate agreed by all parties concerned, and recovery margins agreed, the point-to-point schedule times can be worked out on a scientific basis from performance data obtained on test from the locomotive class concerned.

For a locomotive booked to run in a schedule emanating from the hatched area on the diagram, recovery to a certain extent is relatively easy. Extra physical effort produces corresponding results, and a good driver, by using extra energy where it is profitable to do so, can effect the recovery with a scarcely appreciable rise in fuel consumption.

Recovery to a considerable degree beyond this can be accomplished by a responsive locomotive. The driver makes high power demands of the locomotive where time can be most profitably recovered – rates often higher than could be physically met continuously by the fireman. Advantage is legitimately taken, not only of the steam reserve but of the reserve of heat energy in the firebed. The actual limit to recovery as far as power supply is concerned is set by the highest continuous steam rate that the boiler can supply or the fireman satisfy. It is always wasteful in fuel, being well removed from the rates of the economical band. From the mechanics of the train, each successive minute regained is more difficult than its predecessor. This applies to both diesel and steam traction, being worse in the latter on account of the decrease in thermal efficiency and the physical effort entailed. It is seen, therefore, that the best booked timings for steam locomotives almost settle themselves, and the close agreement often found between empirical timings and those from the hatched areas in the diagram is no mere coincidence. Power classification is automatically performed by the hatched bands in the diagram.

Part II. Standardization of locomotive designs

INTRODUCTORY

Under the heading Civil engineering restrictions (page 412), reference was made to the practice of designing special locomotives for particular duties. The subject is controversial, and both sides of the argument need consideration. On most railways a considerable diversity of locomotive duties is to be found, taking into account all classes of traffic operated. These range from high-speed passenger to slow, heavy freight, and it would be difficult to imagine a single type of locomotive that could deal competently and efficiently with the entire range. Express passenger work requires a machine that will run freely at high speed. To reduce piston speed to a minimum, and the frequency with which steam must enter and be exhausted from the cylinders, a large diameter driving wheel is desirable (at least 6 ft 6 in.) and the boiler must be designed to provide a high sustained output of steam. On the contrary, the working of a heavy mineral train usually makes no more than intermittent demands on the boiler. High tractive power is needed for starting a train from rest, and in surmounting steep gradients, but for the major part of a run the locomotive is being worked under easy steam.

Between the two extremes of high-speed passenger and heavy mineral, there are many categories of service, some due to the traffic operated, some due to the physical characteristics of a particular route. In Great Britain in particular, prior to grouping of the railways in 1923, the old individual companies had many varieties of locomotive intended for specific duties, or groups of duties. On some, indeed, the specialization went to the extent of reserving locomotives for a single roster. The former Caledonian Railway allocated certain of its large 4–6–0 engines to regular duties, with always the same engine crew, and those turns were worked to the exclusion of all others, except when the locomotive was stopped for major overhaul. Such methods achieved a regularity and reliability in service that was of benefit to the owning company and to the travelling public; but the monthly mileages run by such locomotives were small, and a serious situation was immediately created if, for some reason, the special locomotive was at any time not available for the duty.

GENERAL UTILITY LOCOMOTIVES

Need for general utility types. It is desirable to obtain the highest possible utilization from each motive power unit, so as to operate the traffic with the fewest locomotives, and secure the best return for the capital represented by those locomotives. On the other hand utilization can be carried to the extent that insufficient time is available between turns for proper servicing, and a degree of unreliability arises in consequence. One of the most frequent causes of failure on locomotives programmed to run long monthly mileages is blocked boiler tubes, so that a balance must be struck between the advantages of the limited mileage, closely regulated duty, or group of duties, and the alternative of putting the engine into a "link" or "pool" and using it to the utmost, with the accepted hazard of an occasional failure. Today the tendency is towards high utilization, and with such a policy laid down by the top management of many railways, the locomotive design and operating staffs have had to develop their existing practices so as to produce locomotives suitable for very diverse duties. In earlier years the one factor of driving wheel diameter seemed an unsurmountable obstacle in the way of producing a general utility locomotive. Customary dimensions in this country lay between 7 ft and 8 ft diameter for express passenger work, and between 4 ft 6 in. and 5 ft 6 in. for freight locomotives.

British 4–6–0 general utility types (Plates 106A, 106B, 106C, page 428). Today, the combination of modern valve gears and good steaming boilers has enabled general utility locomotives of the 4–6–0 type to be used, originating from the Great Western "Hall" class, introduced in 1928. From this Great Western development has come perhaps the most generally useful and versatile

P. Ransome-Wallis

PLATE XIII. British Railways (London Midland Region): the last express steam locomotive to be built for British Railways (1954). The 3–cylinder 4–6–2, No. 71000, "Duke of Gloucester" with rotary cam poppet valves.

stud of locomotives to operate in the steam era on any railway. On British Railways today there are four distinct groups of these engines, and of that originating on the Great Western there are two varieties.

4–6–0 GENERAL UTILITY LOCOMOTIVES

Original company	Chief mechanical engineer	Coupled wheel dia. ft in.		Nominal tractive effort at 85% B.P. lb.	Total number in service
G.W.R.	C.B.Collett	6	0	27,275	330
G.W.R.	C.B.Collett	5	8½	28,875	80
L.M.S.R.	Sir W.A.Stanier	6	0	25,455	842
L.N.E.R.	E.Thompson	6	2	26,878	300
B.R.	R.A.Riddles	6	2	26,120	172

The ex-Great Western engines have the layout of Stephenson link motion as developed by Churchward in 1903, while the L.M.S.R., L.N.E.R., and British Railways varieties have the Walschaerts radial valve gear. The extent to which a well-designed modern mixed traffic 4–6–0 fulfils the role of "common-user" can be shown by reference to present duties of the Stanier class, as built in such large numbers by the former L.M.S.R.

Stanier class "5" 4–6–0 workings. On the L.M.S.R. these locomotives undertook, with complete success, the following:

Section	Duties
L. & N.W.R.	Any work previously undertaken by ex-L. & N.W.R. express, mixed traffic and freight types.
Midland	All present-day passenger duties. All freight except heaviest mineral.
L. & Y.R.	Any present-day duties.
Caledonian	All present-day duties except heaviest Anglo-Scottish expresses.
Glasgow & South Western	All present-day duties.
Highland	All present-day duties.

Since nationalization they have been used on the West Highland line (former L.N.E.R.), between Edinburgh and Perth (former L.N.E.R.), and on certain Western Region workings between Shrewsbury and Pontypool Road. The extent to which one standard type can replace a great variety of designs is further emphasized by the table in the next column.

To the above forty pre-grouping types could also be added the Midland Compound 4–4–0's built by the L.M.S.R., the 3-cylinder 4–6–0's of the "Patriot" class

LOCO. TYPES REPLACED OR SUPERSEDED BY STANIER CLASS 5. 4–6–0

Railway	4–4–0	4–4–2	4–6–0	2–6–0	0–6–0	0–8–0	Total
L. & N.W.R.	1		3			1	5
Midland	3				1		4
L. & Y.R.	1	1	1		1	1	5
Furness	2				1		3
Caledonian	2		10	1	1	1	15
G. & S.W.R.	1		1	1	1		4
Highland	1		4				5

and the Horwich-built 2–6–0's. Perhaps the most interesting individual case is that of the former Caledonian Railway, where the Stanier locomotives were found capable of working, with complete success, duties formerly split up between no fewer than ten different classes of 4–6–0 ranging from the special, small-wheeled types for the Oban line to main-line freight, and heavy express passenger engines. The only difference is that the Stanier engines perform these diverse duties at higher speeds, with heavier loads, and with far greater efficiency than did the older engines designed specifically for the jobs.

RANGE OF STANDARD DESIGNS
Great Western practice. On the Great Western Railway, in the period 1903–11, the 2-cylinder engine design, using 18 in. × 30 in. cylinders, was applied to an extensive range of types for various duties, as follows:

Type	Coupled wheel dia. ft in.		Duties
4–6–0	6	8½	Heavy express passenger.
4–4–0	6	8½	Intermediate express passenger.
4–4–2T	6	8½	Fast suburban passenger.
2–6–0	5	8½	Mixed traffic.
2–6–2T	5	8½	Mixed tank engine duties.
2–8–0	4	8½	Heavy freight.
2–8–0T	4	8½	Short haul, heavy freight.

At that stage in the development there was no use for a general utility type for heavy work. A 4–6–0 with 5 ft 8½ in. wheels had been envisaged, but no example was actually built. The long-lap, long-travel valves made all types in this range free running and the 2–6–0 tender engine, in particular, proved capable of running up to 75 m.p.h., with comfort and ease. The success of these relatively small mixed traffic engines led to designs of 2–6–0 for similar work being prepared and built on many British railways, though limitation in boiler capacity precluded their use on the heaviest passenger duties. The natural development in this respect came on the Great Western Railway in 1924, with the production of a prototype 4–6–0 with 6 ft coupled wheels. This formed the forerunner of the "Hall" class, referred to earlier.

Six-coupled suburban tank locomotives (Plate 107, page 429). Another type of locomotive that has come into prominence, by standardization, is the six-coupled suburban tank, of the 2–6–2, or 2–6–4 type. The tractive effort is high, in relation to the loads hauled; but a prime necessity in this class of duty is rapid acceleration from station stops. At the same time the modern long-lap long-travel valves provide a free-running engine, capable of speeds up to 75 or 80 m.p.h., with 5 ft 9 in. coupled wheels. Such locomotives can be used successfully on outer suburban, or fast residential trains having non-stop runs of twenty miles or more. The use of the Stanier Class 4, 2–6–4T type on the former L.M.S.R. enabled existing types to be replaced as follows:

TANK ENGINE TYPES REPLACED BY STANIER 2–6–4 TYPE

Railway	4–6–4	4–6–2	0–6–4	4–4–2	0–6–2	2–4–2	0–4–0
L. & N.W.R.		1		1	1		
Midland			1				1
Midland (L.T. & S)	1			2	1		
L. & Y.R.	1					1	
Furness	1			1	1		
Caledonian		1					1
G. & S.W.R.	1				1		
Highland			1				

VALVE DESIGN: ITS IMPORTANCE

Before the end of the nineteenth century, certain types of four-coupled passenger locomotives with driving wheels having a diameter of less than 7 ft, were displaying a capacity for speed hitherto thought possible only with single-driver machines, with 7 ft 6 in., or larger, wheels. In particular, the London and North Western 2–4–0 "Precedent" class with 6 ft 9 in. wheels, and the Caledonian "66" Class 4–4–0's with 6 ft 6 in. wheels, were outstanding. Speeds up to nearly 90 m.p.h. were being attained by these classes in the 1890's. In both classes very careful attention had been given to the valve design. The layout of the ports and passages provided for a very free flow of both live and exhaust steam, and in both classes the back pressure was low. It was thus possible to attain high piston speeds without any choking effect. Both these nineteenth-century designs of cylinders and valves were, however, applied to small locomotives, with inside cylinders. As enlargement became necessary, to meet the

increasing demands of traffic, it was not possible within the limits of the British loading gauge to develop the same compact design. Valves had to be placed either above or below the cylinders instead of between them; steam passages became longer and less direct, and although the tractive capacity of locomotives was enormously increased, their freedom in running declined. Particularly was this the case on the Caledonian Railway, which company built, between 1900 and 1914, no fewer than six different varieties of the 4–6–0 type alone, each for a specific duty.

Contribution to standardization. Since it was unlikely that locomotives would remain small, other means had to be sought to secure freedom in running at high speeds. The piston valve was developed on the North Eastern Railway, and elsewhere, while various devices were invented for reducing the friction of slide valves. At the same time little if any thought was then being given to the idea of a general purpose locomotive capable of true, common-user service. The trend towards standardization took the form of establishing cylinder sizes that would be common to passenger, freight, and local tank types, uniformity in fittings, interchangeability of boilers, and so on. The development of cylinder and valve design, in which the Great Western Railway played a leading part, came from a determined drive by G.J. Churchward to secure a high output of power from his 4–6–0 passenger engines at high speed. Hitherto 60 m.p.h. had been considered the standard of express speed with a passenger train, on level tracks; speeds of 75, 80 or even 90 m.p.h. had been attained by free-running locomotives in favourable conditions, when working under a relatively light head of steam. In heavier working the valves would not clear the volume of steam required, and this restriction dictated the maximum speed of the locomotive when steaming hard. It was Churchward's development that paved the way for the modern general utility locomotive.

Long-lap, long-travel valves (see Chapter 4, Part I, Travel, Valve *and* Valves for Steam Distribution). For economical running at maximum power it is necessary to work at a short cut-off with the regulator wide, if not fully, open. Thermodynamic losses due to the throttling at admission of steam to the cylinders are thus reduced to a minimum. With conventional layouts of valve gear, however, the port opening to steam was gradually reduced as the motion was linked up, so that even if a driver attempted to run in the most economical manner the entry for the steam to the cylinders would be so restricted as to nullify other theoretical advantages. It was thus common practice to work express locomotives with the valves cutting off at about 25 or 30 per cent of the piston stroke, so as to get a good port opening and reasonable freedom in running. Churchward set out to produce a valve gear design that would ensure a full port opening when working at the earliest cut-off needed for yielding the high power he desired at 70 to 75 m.p.h., namely 15

with ever-increasing loads and speeds. Little could be achieved towards the establishment of standard designs.

Comparison of main-line express passenger power, in its development, between Great Britain and the U.S.A., may be epitomized in the following table:

Country	Railway	Year	Type	Class	Wt. engine only tons	Nomin: T.E. lb.
England	Great Western	1914	4–6–0	"Star"	75	27,800
England	L.M.S.R.	1938	4–6–2	"Duchess"	105	40,000
U.S.A.	Pennsylvania	1915	4–6–2	"K4"	138	44,000
U.S.A.	New York Central	1938	4–6–4	"5445"	163	43,400
U.S.A.	Santa Fé	1938	4–8–4	"3765"	228	66,000

per cent. He did this by increasing the length of the valve laps to almost double the then conventional size. This meant increasing the travel of the valve, from 4 in. in full gear, to something over 6 in., and this many engineers of the day were disinclined to do. It was feared that maintenance costs on the valves and gear would be increased to an inordinate amount. But the use of long lap, long travel valves gave the required answer so far as high-speed passenger working was concerned, and Great Western 4-cylinder 4–6–0 locomotives of the "Star" class having a nominal tractive effort of 12·4 tons at 85 per cent boiler pressure, have given sustained drawbar pulls of three tons at 70 m.p.h.

OVERSEAS PRACTICE: A COMPARISON

In Great Britain, the intensity of service over a very small area means that a high proportion of the traffic consists of short hauls, made up into moderate loads. In consequence the general utility engines, such as the Stanier 4–6–0 and 2–6–4T designs, represent a high proportion of the total stock. In France and the United States of America – two countries where the steam locomotive was brought to a high state of development – traffic requirements since the end of World War I eliminated the need for locomotives of medium power, of a general utility kind. Locomotives were built to provide maximum power for express passenger and heavy freight duties, comprising long through hauls,

Development in the U.S.A. was confined to relatively small classes of locomotives. On the Santa Fé, which could claim some of the most powerful and hard-working express passenger types in the North American continent, the 4–8–4's of the "3765" class numbered only twenty-one machines (see Chapter 5, No. 3). There was not the need for standardization of types. Extreme power and reliability was needed for a few exceptional duties and locomotives were developed for the purpose. Although practically all locomotives on the Class One railroads in the U.S.A. were built by one or other of the three firms, Alco, Baldwin, and Lima, and had fittings in standard use by those manufacturers, the railways themselves had individual requirements in detail. But although the numbers of locomotives on different railways may have been low, it might in the cause of unification have been possible to use one design of large 4–8–4 of about 65,000 lb. tractive effort on many different roads. The New York Central, Norfolk and Western, Union Pacific, Denver and Rio Grande Western, and others, in addition to the Santa Fe, all used large express passenger locomotives of the 4–8–4 type, and the manner in which standardization between them all might have been accomplished has been shown by their subsequent acceptance of standard types of diesel-electric locomotives for duties formerly carried out by the previous individual varieties of steam power (see Chapter 2, Part V, The Operating of Diesel Locomotives).

Part III. Human factors in locomotive running

INTRODUCTORY: THE TRAINING AND SELECTION OF ENGINEMEN

Conditions existing in Great Britain today differ considerably from those prevailing in France, and in the hey-day of steam in the U.S.A.: this is to a great extent due to the different methods of selection and training of

enginemen in the countries concerned. In Great Britain the most senior and arduous duties are attained by a gradual process of promotion, on seniority in service, through the various groups of duties at each depot; progression is from pilot and spare turns, to slow goods, stopping passenger, main-line goods, and finally express passenger. A driver is usually in his early "fifties" before he reaches top link work, and his suitability is that derived from long experience, a good safety record, and a proved ability to handle locomotives. The top links are thus bound to contain men of widely differing outlook and temperament.

In France conditions are quite different. Prior to World War II, although each of the former railways merged into the S.N.C.F. had their own designs of express passenger locomotives, nearly all were 4-cylinder compounds, and many were, and are, complicated machines to handle. They require special skill in driving if the best results are to be obtained. The many controls provide, for example, for independent linking up of high- and low-pressure valve gears, and for the admission of a limited amount of live steam direct into the low-pressure steam chest when a high output of power is needed for a short period. In consequence special measures are adopted for the training of men destined for top link duties. In the first place they are selected; then they are given both theoretical and practical training as mechanical engineers, in addition to the necessary training in the operation of a railway. They reach top link duties at a much earlier age than drivers do in Great Britain, and retire at the age of fifty. In America, while drivers do not receive any special training, generally speaking the admission of men to the highest classes of work is much more selective than in Great Britain, and once attained the position is held only on record of performance. The attitude towards running in America could be summed up in the words of a superintendent, speaking on the vexed question of recovering lost time. He said: "If an engineer don't make up time he goes off the job."

THE BRITISH POSITION

The variations arising from the British system of promotion can be broadly summarized thus:

(i) Keen locomotivemen who do their best to maintain schedule, no matter how difficult or frustrating the circumstances.

(ii) Average men, who keep point to point times, recover delays when circumstances are favourable.

(iii) Erratic individuals, whose work seems to reflect their varying moods.

To get the best work, collectively, out of links of such diverse character much depends on leadership of district officers, shedmasters and locomotive running inspectors. The schedules must be realistic, to which end the practice of technical train timing (see page 415) is now contributing; but it is important also that the tasks should be sufficiently arduous to provide a challenge to the skill and enginemanship of the men concerned. Slack

schedules are in themselves conducive to slack and slovenly methods. The shed staff has an important part to play. The footplate men cannot give of their best if locomotives are sent out incompletely serviced, or if attention has not been given to minor defects reported on previous trips. Discouragement of this kind hardly creates enthusiasm to run the trains to time, especially when out-of-course speed reductions or signal delays are experienced (see Chapter 8, The Motive Power Department).

ALLOCATION OF ENGINES TO CREWS

The way in which locomotives are rostered to duties has already been referred to (see page 413). The impact of the varying methods upon engine crews must now be noticed. In Great Britain today, men working in the highest class of main-line service may be allocated locomotives in one of four ways:

(i) An engine reserved for the one individual duty.

(ii) Any one of a group of low-mileage engines generally confined to first-class duty.

(iii) An engine from a common-user pool.

(iv) Their "own" regular engine, which they man in partnership with one, or at the most two other crews, and which they always have on whatever train they are working. (The French "machine titulaire".)

Without any doubt (iii) gives the least satisfactory results, and from the viewpoint of train running, engine maintenance costs and the general contentment and well-being of all personnel involved (iv) is by far the best. In the majority of districts, however, there are today many factors which make it almost impossible for men to "keep their own engines". Difficulties arise mostly with the rostering of enginemen's duties. At some depots where passenger duties extend over many different routes, it would seem practicable to organize links specializing in those routes, to which a regular allocation of locomotives could be made. At those same depots, however, seasonal traffic fluctuation may entail the running of a considerable number of special trains on one particular route. These are operated by the first crews available. To enable them to be run without the inconvenience of using road pilots, the express passenger links must be sufficiently large to have drivers who know the road, available for the special train workings. The links are thus arranged to cover all routes radiating from the depot, and the men work over each, frequently enough to retain the requisite road familiarity. With such a diversity, including varieties of loading, gradients, and schedules, it is not possible to use the same class of engine on all duties, let alone confine engines to regular crews.

The common-user principle, particularly where cyclic

workings are in force, is the most difficult to maintain from the engine servicing viewpoint, and the least rewarding for the crews. For example, a driver and fireman are booked to make a sixty-mile trip on a locomotive rostered for a 300-mile through working: they take over at a point eighty miles from the start of the duty after the locomotive has already been handled by two different crews, and at the end of their turn they hand over to a fourth crew. The locomotive is stationed at a depot, not merely out of their district but in another Region. They take over their charge in the course of a station stop lasting five minutes at the most, and the time is occupied in stowing kit and rations, hearing details of the load, and so on; there is no time to learn any peculiarities or minor defects of the engine. All too frequently the only advice received is unwelcome, such as: "She won't steam, mate." The crew cannot be blamed for lack of interest, seeing that they will be handing over to someone else in little more than an hour. The position is not so bad with engines from a common-user pool, which a single crew will work from end to end of the run, especially if the duties of the driver and fireman include preparation of the engine on shed. But on some duties, preparation is done by relief crews, whose work includes bringing the engine actually on to the train, ten minutes, or less, before starting time. With any system of common-user rostering or cyclic working of locomotives, everything depends on good teamwork; the slipshod carrying out of one operation, either in servicing or running, can easily prejudice the rest of the diagram, and with it the punctual working not only of the rostered trains, but of others that my be following, or making connection at junctions.

The allocation of engines to individual crews invariably results in improved performance. Drivers and firemen instinctively take pride in a machine that is "their own". Cab fittings begin to shine, little gadgets for the men's comfort are fitted, and having the same engine month in, month out, drivers are in a much better position to detect and diagnose troubles, and to get things put right. That sense of pride in their craft, which is inherent in all locomotive enginemen, to a greater or less degree, comes to the surface, and almost invariably there grows a spirit of healthy rivalry between men in the same link that is good for the service and the public.

Interesting present-day examples of locomotives confined to regular enginemen, in teams of two and three, are seen:

(i) In the heavy main line "Pacific" duties worked from Haymarket Shed, Edinburgh. These engines are double-manned, and drivers and firemen take the same locomotive whether bound for Glasgow, Perth, Dundee or Newcastle.

(ii) Before the recent advent of extensive diesel operation, the "Britannia" class "Pacifics" allocated to Stratford for working between London and Norwich, were triple manned, while those allocated to Norwich were double manned.

PSYCHOLOGY IN RUNNING

The variations in attitude and temperament are reflected in work on the road. Among the keenest men, for example, there are those who seem to take a delight in running ahead of time in any circumstances. Sometimes such practice results from a flair for showmanship, sometimes from a spirit of *joie de vivre*, and in one recorded instance because the driver concerned did not possess a watch and merely knew he had to go hard! Such men are not necessarily heavy on coal, but running ahead of time can often be an embarrassment. Much more acceptable from all points of view is the man who times his train with accuracy on a normal run, rarely being half a minute early or late at intermediate timing points, but who will run very much harder than normal when necessary to recover lost time. Some men in their anxiety to do their best in such circumstances are inclined to "flog" their engines, while a lesser proportion, who are real artists in the craft of enginemanship, seem to be able to secure the extra effort with little extra in the way of coal consumption, and wear and tear.

In the middle strata of enginemen are many who, while handling their locomotives with care, skill and economy, seem unwilling to go very far beyond what the timetable demands on a clear run, so that when delays do occur, the train is correspondingly late at its destination. Sometimes, if running conditions are exceptionally favourable, they will make up time, but not often. Finally there are the erratic men. There is no point in attempting to analyse their vagaries, as they are legion; but the men themselves are, at most depots, numerically few. Examples of the more common forms of erratic behaviour in enginemen may be mentioned:

(i) The "coal-dodger", who makes poor time on the open road, and attempts to regain a little by running hard over stretches where speed is restricted.

(ii) The man who takes any opportunity for losing time, and when checked is so slow in recovery that the actual loss in time is far greater than the amount strictly attributable to the check.

(iii) The man who on rare occasions runs extremely hard, not in the interests of good operating, but for some quite personal reason of his own or of his fireman.

SIGNALLING AND AUTOMATIC TRAIN CONTROL

Much discussion has taken place, ever since the establishment of railways, as to the best way of presenting information on the state of the road ahead to the driver. Signal engineers in all countries have been divided among themselves; the schools of thought and of practice have been diverse. But among enginemen an indifference to the system of signalling would often be detected.

Signal sighting. Drivers, as a race, are extremely quick in learning the varying signs and symbols displayed; it is all taken as part of their prized craft – road knowledge. Universally, however, the plea comes from the footplate "do make it so that we can see the signals". Many factors do arise to render sighting of signals difficult, and for that reason the use of visual, or audible cab signals in conjunction with automatic control of the train brakes is becoming an increasingly important factor in train operation.

Today, the term automatic train control is used loosely, and incorrectly in the majority of cases, so that it is important to appreciate the various methods of warning a driver that are included under this general term.

Four systems may be noticed in particular:

(i) Automatic train stop, in conjunction with the stop signals (London Transport, and certain other suburban electric lines).

(ii) Intermittent warning system, at distant signals, with automatic brake application only in event of the driver failing to acknowledge warning indication (British Railways, main lines).

(iii) Continuous cab signalling inductively controlled by coded track circuits; no brake application (Pennsylvania Railroad, and other American lines).

(iv) Continuous cab signalling, as above, but with automatic speed control according to signal indications (American railways).

British systems (*see* Chapter 4, Part I, Automatic Train Control). In the automatic train stop system (i) the driver is powerless once a signal at danger has been passed. Control is taken out of his hands and the train stopped by an emergency brake application. The brakes can be released only after the driver has climbed down from his cab and re-set the trip cock. It is applicable only on railways where speed is relatively slow, where the brake power of all trains is the same, and where it is possible to standardize the clearance margins ahead of signals that are available in the case of an over-run.

The system (ii) adopted for main line use on British Railways has two forms, though both give a similar set of indications displayed by the distant signals. Accidents due to the over-running of signals almost invariably have their origin in an obscured, or tardy sight of the distant signal when running fast. The British warning system is, therefore, designed to "alert" a driver as he is approaching

the distant, by giving an audible indication in the engine cab of the aspect displayed by that signal:

a siren or horn sounds if the signal is at "caution";
a bell rings if the signal is at "all-clear".

The driver is given the opportunity of acknowledging a "warning" indication; the horn or siren sounds until he operates an acknowledging lever. If he does not acknowledge, the warning continues to sound, the brakes are applied and the train is stopped.

The basic principle of all British practice where automatic train control is concerned is to provide apparatus to assist the driver in observing and acting upon the indications displayed by the wayside signals. The audible cab signals must never be regarded as a substitute. Only in the extreme case of a driver failing to acknowledge a warning is control of the train taken out of his hands. The earliest form of this warning system is that standardized on the former Great Western Railway, in which the link between the engine and the wayside signals is made through a contact ramp. The more recent form, now standardized for general service on British Railways, has no physical contact. The link is provided through the action of magnetic and electro-magnetic inductors in the track.

The criticism is sometimes made of this system: "Why have an automatic brake application feature if the driver can cancel it by just moving the acknowledging switch?" Quite apart from the bad effect on morale of taking control out of a driver's hand, it is only in something like one in a million cases that a driver is inadvertently taking a risk, and in countless cases every day drivers use the distant signal warning indication with great skill, checking the speed of their trains on first sighting it, and then continuing at a reduced speed ready to stop dead at the home signal, if necessary. In many instances their judgement and vigilance are rewarded by the home signal clearing before they have had to stop. With an irrevocable brake application at an adverse distant signal such trains would suffer delay from being stopped and having to restart (Plate 108, page 430).

American practice. The American attitude to cab signalling is notably different from the British. The installation of cab signals that are continuously controlled and illuminated in the full view of the driver tends to distract attention from the wayside signals. In the U.S.A. this is not considered to be detrimental, in fact on one section of the Pennsylvania in the Pittsburgh district the wayside signals have been removed altogether, and complete reliance placed upon the continuously-controlled cab signals. On the Pennsylvania and other lines, such store is set upon the efficiency of the cab signals that no brake application feature is now included at all. A most important difference between British and American practice lies in the fact that in the U.S.A. (and in some recent installations in France), the cab indication is continuously displayed. Not only this, but the aspect shown, controlled according to the occupancy of the track ahead, is changed before the driver sights the next wayside signal. Where trains are following each other closely, it is a great advantage in running to know in advance from the cab signal indication that the next wayside signal ahead has cleared to a more favourable indication. Running is improved, and the confidence of enginemen

increased. In some of the earlier American installations, notably one on the Atchison, Topeka and Santa Fé Railroad, the cab signals displayed the letter H (high speed), M (medium) and L (low), and speed-controlled brake equipment was fitted to the locomotives concerned to constrain the speed to predetermined maximum values corresponding to the three indications, for example, 75 m.p.h. for "high", 45 m.p.h. for "medium" and 15 m.p.h. for "low". Generally speaking, the continuously-controlled cab signal, without brake application, is now favoured in the U.S.A.

Part IV. Locomotive performance: an analysis of some severe passenger duties

BRITISH, FRENCH AND AMERICAN WORK COMPARED

Interesting comparisons may be made between a number of duties on regularly-booked express passenger trains in different parts of the world, as showing the degree to which steam locomotives were extended. The runs chosen for analysis and tabulated below are actual trips, on which work something above average was performed.

TABLE A. GREAT BRITAIN

Region	Route	Distance, miles	Average speed m.p.h.	Load tons
1 Western	Bristol–Paddington	117·6	72·7	265
2 Western	Paddington–Exeter	173·5	66·8	480/420*
3 L.M.	Euston–Rugby	82·6	63·5	510
4 L.N.E.R.†	Kings Cross–York	188·7	72·5	325

* Reduced from 480/420 tons by slip coach, 95 miles from Paddington.

† Pre-nationalization schedule.

Taking one run from each country for first comparison, the efforts involved can be related to the weights of the locomotives concerned. All these runs were made

TABLE B. FRANCE

Railway	Route	Distance, miles	Average speed m.p.h.	Load tons
1 P.O.Midi	Poitiers–Angoulême	70·1	70·8	630
2 P.O.Midi	Poitiers–St Pierre	62·8	68·0	530
3 Nord	Paris–Arras	119·4	64·3	505
4 Nord	Paris–Brussels	193·1	64·7	305

TABLE C. U.S.A.

Railway	Route	Distance, miles	Average speed m.p.h.	Load tons
1 Milwaukee	Chicago–Milwaukee	85·0	73·5	365
2 Milwaukee	New Lisbon–La Crosse	59·8	73·3	365
3 Milwaukee	Sparta–Portage	78·3	82·0	465
4 N.Y.C.	Toledo–Elkhart	133·0	71·0	1025
5 Pennsylvania	Englewood–Fort Wayne	141·0	71·6	630

over good roads where high sustained efforts can be prolonged without many interruptions on account of speed restrictions (Plates 109A, 109B, 110B, 111A, pages 431, 432 and 433).

From Table No.	Country	Run	Load tons	Speed m.p.h.	Wt. of engine only tons	Wt. of tender tons
(A.1.)	Great Britain	Bristol–Paddington	265	72·7	80	46½
(B.2.)	France	Poitiers–St Pierre	530	68·0	98·0	60
(C.3.)	U.S.A.	Sparta–Portage	465	82·0	185¼	167½

The ratio of engine and tender weight to weight of train is thus 1 : 2·1 for the British locomotive, 1 : 3·3 for the French, and 1 : 1·32 for the Milwaukee. The average speed maintained was highest in the American case, though on one special occasion a British 4–6–0 locomotive of the same class (G.W.R. "Castle") maintained a start to stop average speed of 81·6 m.p.h., with a train of 195 tons, or with a ratio of 1 : 1·54, engine to train. In the Milwaukee case, the booked average speed was 80·8 m.p.h., and came in the course of a through run on which start-to-stop journeys booked at 68·0, 73·7, 76·8, 64·7, and 68·0 m.p.h., were included. The locomotives were specially designed for this duty, the *Hiawatha* express, and worked through over the 400 miles between Chicago and St Paul. By contrast, the Western Region locomotive making the Bristol–Paddington run covers only 236 miles in the day's work, made up of two separate runs of 118 miles each. The figures for the French "Pacific" serve once again to emphasize the almost incredible work performed by these compound engines.

York Central "Hudsons" of class "J–3a" were large and heavy machines for their type; both had only two cylinders, and the tractive effort is low in relation to the work they were required regularly to perform. The N.Y.C. 4–6–4's were fitted with boosters, that increased tractive effort by a further 12,100 lb. at low speeds. But on runs such as that analysed, when an average speed of 78 m.p.h. was maintained for 116 miles on end, with an average of 85 m.p.h. for thirty-one miles and a maximum of 94 m.p.h., the booster would not be in action at all, except in starting from rest. In view of the high ratios of load to tractive effort, and load to total heating surface, the locomotive had to be steamed extremely hard. The regulator was kept fully open all the time, and the cut-off kept as late as 42 per cent up to 75 m.p.h., and at 35 per cent up to the maximum of 94 m.p.h. Speed was sustained at over 80 m.p.h. on dead level road, and the maximum was attained down no steeper gradient than 1 in 1,000 (0·1 per cent).

The power output of the locomotive at high speed

| Country | Railway | Average speed m.p.h. | Trailing load tons | Locomotive | | | | | | Ratios | | |
				Type	Weight eng. only tons	Weight tender only tons	Nom.: T.E. lb.	Total H.S. sq. ft	Load: eng. wt.	Load: T.E.	Load: H.S.
Great Britain	Western	72·7	265	4–6–0	80	46·7	31,625	2283	2·1	18·8	0·116
,, ,,	L.N.E.R.	72·5	325	4–6–2	103	62·4	35,455	3325	1·95	20·5	0·101
U.S.A.	Pennsylvania	71·6	630	4–6–2	141	95·5	44,000	5020	2·67	32·1	0·125
,,	N.Y.C.	71·0	1025	4–6–4	160·7	139·4	43,440	5932	3·45	52·8	0·173

The London–York run of the *Coronation* express (Table A, No. 4), came in the course of a through locomotive working of 393 miles; but although the train load was heavier the ratio of engine and tender weight to train weight, 1 : 1·95, was much the same as that of the Western Region's *Bristolian*, shown in the same table (A.1). The British schedules were planned so that the locomotives concerned could work regularly at economical rates of admission. The "Castle" class engine making the 72·7 m.p.h. run with the *Bristolian*, was working at less than 20 per cent cut-off for practically the entire journey, and on the *Coronation*, the Gresley "A.4" Pacifics were run in 15 per cent on the level and slightly adverse grades, with an advance to 25 per cent, at the most, on long stretches of 1 in 200 (0·5 per cent) ascent required to be climbed at about 75 m.p.h (Plates 109A, 109C, page 431). The most arresting performance in the table on page 425 is that on the New York Central system, where a load of 1,025 tons was worked at a start-to-stop average speed of 71·0 m.p.h. Comparison in more detail may now be made of the locomotives concerned.

The Pennsylvania "K4" class "Pacific", and the New

makes a most interesting comparison with certain maximum British efforts, as showing the demands that were habitually made on American locomotives in the concluding years of steam haulage. The New York Central 4–6–4 sustained, at 80 m.p.h., an indicated horsepower of 3,650. (Table C, No. 4.) (Plate 110C, page 432.) It should be emphasized that none of the British performances quoted were made in ordinary service. They were made in special test conditions, for the most part with good coal, and with the locomotives steamed almost to the maximum limit of the boiler and front end. In relation to the sizes of the boilers concerned, the American locomotive does not show as high a horsepower as might be expected; but clearly the limitation lay in the cylinder volume. The power output per pound of nominal tractive effort is much the highest in the American case, though it is the former Great Western 4–6–0 of the "King" class that shows the highest output in relation to its boiler and firebox. The tests on which the data tabulated on page 435 are based were carried out before the more recent modification to these engines, fitting a twin-orifice blastpipe and double chimney (*see* Chapter 5, No. 29).

M.W. Earley

PLATE 105. British Railways (Southern Region) rebuilt 3-cylinder 4–6–2 "Merchant Navy" class, No. 35020, *Bibby Line*, on a dynamometer car test with a down express from Waterloo to Exeter. The train is leaving Basingstoke.

P. Ransome-Wallis

PLATE 106A. Ex G.W.R. "Hall" class, No. 5964, *Wolseley Hall*, on the up *Cornish Riviera* express about to leave Penzance.

P. Ransome-Wallis

PLATE 106B. Ex L.M.S.R. class 5, No. 45462 working a train of empty coal wagons. The train is descending Beattock Bank.

P. Ransome-Wallis

PLATE 106C. Ex L.N.E.R. class B–1, No. 61015 on loan to the Southern Region and working an up Kent Coast express, leaving Herne Bay.

PLATE 106. British General Utility 4–6–0 Mixed Traffic Locomotives

P. Ransome-Wallis

PLATE 107. British General Utility Tank Locomotive: Ex L.M.S.R. 2–6–4T, No. 42108 working a fast train between Bradford and Penistone. The later British Railways standard 2–6–4T locomotives are based on this Fairburn–Stanier design.

British Railways

PLATE 108. Great Western Railway of England: system of automatic train control, showing energized ramp placed between the running rails; the shoe on the locomotive is in contact with the ramp. The bell and the cancelling handle can be seen in the right-hand corner of the cab, above and to the right of the reversing gear. The locomotive is a 4-cylinder 4–6–0 of the "Castle" class.

P. Ransome-Wallis

PLATE 109A. The pre-war Coronation Express of the London and North Eastern Railway, Gresley 3-cylinder 4–6–2 class A–4 No. 4489, *Dominion of Canada*.

M.W. Earley

PLATE 109B. British Railways (Western Region) up Bristolian Express passing Tilehurst. The locomotive is ex Great Western 4-cylinder 4–6–0 "Castle" class No. 7034, *Ince Castle*.

PLATE 109. British High-Speed Express Trains

P. Ransome-Wallis

PLATE 110A. Great Western 4-cylinder "King" class 4–6–0 No. 6021, *King Richard II*
as built.

P. Ransome-Wallis

PLATE 110B. Ex L.M.S.R. 4-cylinder "Duchess" class 4–6–2 No. 46231, *Duchess
of Atholl* with a down express climbing Shap.

New York Central System

PLATE 110C. New York Central 4–6–4 class J–3a in its final, streamlined form. Engine
No. 5429 was one of a long series of "Hudson" type locomotives built for the N.Y.C.
system. They were some of the finest machines ever put on rails. The three locomotives
above, each of different type and size, have, none the less, the same nominal tractive effort.

P. Ransome-Wallis

PLATE 111A. S.N.C.F. Chapelon 4-cylinder compound 4–6–2 No. 231. E.6 working The Blue Train between Calais and Paris on the Nord Region.

P. Ransome-Wallis

PLATE 111B. Pennsylvania Railroad 4–6–2 locomotive of class K–4s, No. 3740.

American Locomotive Co.

PLATE 111C. Chicago, Milwaukee, St Paul and Pacific Railroad 4–6–4, class F–Y, No. 100. High-speed locomotive with 7 ft diameter driving wheels, built for working the famous Hiawatha expresses in 1938.

433

Beyer, Peacock and Co. Ltd

PLATE 112A. 4–6–4+4–6–4 class 15 locomotive of the Rhodesian Railways, 3 ft 6 in. gauge.

East African Railways and Harbours

PLATE 112C. 4–8–2+2–8–4 locomotive of the East African Railways, Class 60, working between Nairobi and Mombasa. These are metre-gauge locomotives which can be converted to run on the 3 ft 6 in. gauge.

Beyer, Peacock and Co. Ltd

PLATE 112B. 4–6–2+2–6–4 express locomotive for the Algerian State Railways, at Calais while running trials. Thirty of these locomotives were built between 1932 and 1941 for service between Algiers, Oran and Constantine, 4 ft 8½ in. gauge.

PLATE 112. Beyer-Garratt Locomotives for service in Africa

POWER OUTPUT AT 80 M.P.H.

Railway	Loco. Type	Loco. Class	I.H.P.	Total Heating Surface sq. ft	Grate area sq. ft	Nominal T.E. lb.	Ratios : I.H.P. per		
							sq. ft of H.S.	sq. ft of grate area	per lb. of T.E.
N.Y.C.	4–6–4	J3A	3650	5932	82·0	43,440	0·615	44·5	0·084
B.R. (W)	4–6–0	"King"	2100	2514	34·3	40,300	0·835	61·3	0·052
B.R. (E)	2–6–2	"V2"	1970	3111	41·25	33,730	0·584	47·7	0·059
B.R. (Std)	4–6–2	"Britannia"	2100	3192	42·0	32,150	0·657	50·0	0·065
B.R. (Std)	4–6–2	71000	2600	3181	48·6	39,080	0·815	53·5	0·066

In France the existence for so many years of the legal maximum of 120 Km. per hour (74½ m.p.h.) led to the development of locomotive designs that would enable high power to be sustained at speeds of 60 to 70 m.p.h., so that a high average speed could be held throughout the entire journey. The climax in this respect was demonstrated on a run with a Chapelon 4–6–2 of the P.O.-Midi, which on a special test ran from Bordeaux to St Pierre-des-Corps, 216·7 miles at an average speed of 72·2 m.p.h. from start to stop, without exceeding the legal maximum at any point. Part of this run is shown in Table B, No. 2. The load was 400 tons (Plate 111A, page 433).

One of these locomotives on a further test, with a load of 630 tons, developed an indicated horsepower of 3,700 in the following circumstances:

Speed	74 m.p.h.
Cut-off H.P. cylinders	59 per cent.
Cut-off L.P. cylinders	54 per cent.
Steam chest pressure p.s.i. (H.P.)	228
Steam chest pressure p.s.i. (L.P.)	84

This effort came at the conclusion of an uphill section of 24·0 miles; the summit point was passed in 22¼ minutes from the dead start. It took place during the run from Poitiers to Angoulême (Table B, No. 1), when this distance of 70·1 miles was covered at an average speed of 70·8 m.p.h. start to stop. In this particular table the French runs are at a disadvantage in comparison with the British and American, in that speeds up to and including 100 m.p.h. are contained in the latter two instances, while all the French journeys are limited to a maximum of 74½ m.p.h.

Analysis of the maximum effort of the Chapelon compound 4–6–2 taken from the dynamometer car records, gives some striking results:

Total heating surface	2,771 sq. ft
Grate area	49 sq. ft
Nominal tractive effort	34,000 lb.
Maximum sustained I.H.P.	3,700
Ratios:	
I.H.P. per sq. ft of heating surface	1·33
I.H.P. per sq. ft of grate area	75·6
I.H.P. per lb. of nominal T.E.	0·109

These figures are much in advance of any British or American performances, but although the water level was steadily maintained in the boiler it is only fair to say that this effort lasted less than twenty minutes and was succeeded by a long spell of running on favourable gradients. The locomotives were not capable of sustaining such standards of work over any lengthy period. They were rebuilds of old and far less powerful machines, and in their rebuilt form the chassis was not capable of standing the stresses involved in the output of such power. The run from Paris to Arras (Table B, No. 3) is more characteristic of the best quality work by these locomotives in ordinary service.

In Great Britain the 4–6–0 is the most popular type for all-round service. In the heaviest duties it is sometimes thought to have limitations due to the necessity of using a narrow firebox, and a relatively small grate area. Some dynamometer car records are now tabulated concerning test runs with Great Western "King" class locomotives on both special and service trains (Plate 110A, page 432).

Special train. Test at constant rate of evaporation: flying average over test length of approximately 70 miles.

Load of train (excluding engine and tender)	781 tons
Average speed (pass to pass)	61·7 m.p.h.
Steam to cylinders lb. per hr	28,700
Coal consumption lb. per hr	4,150
lb. per mile	67·2
lb. per sq. ft of grate per hour	121
lb. per d.b.h.p.hr	3·33
Sustained I.H.P. (at 80 m.p.h.)	1,980
Range of cut-offs (57 to 81 m.p.h.)	22–27 per cent
I.H.P. per sq. ft of heating surface	0·787
I.H.P. per sq. ft of grate area	57·7
I.H.P. per lb. of nominal T.E.	0·049

Service train; Cornish Riviera Express with dynamometer car attached:

Length of run (London–Newton Abbot)	193·6 miles.
Load of train:	
to Heywood Road (slip coach)	480 tons
to Newton Abbot	420 tons
Average speed (start to stop)	62·5 m.p.h.

Steam to cylinders lb. per hr	26,950
Coal consumption lb. per hr	3,245
lb. per mile	51·9
lb. per sq. ft of grate area per hour	94·5
lb. per d.b.h.p.hr (ex. of auxiliaries)	3·21

Average drawbar horsepower:

| under power | 978 |
| over-all for trip | 828 |

Values of indicated horsepower were not recorded, but at the average steam rate of 27,000 lb. per hr at speeds between 60 and 85 m.p.h., the recorded value of the I.H.P. on the Swindon stationary testing plant is almost constant at 1,870. Thus in service conditions the ratios are:

I.H.P. per sq. ft of heating surface	0·745
I.H.P. per sq. ft of grate area	54·6
I.H.P. per lb. of nominal T.E.	0·0465

The first two ratios are higher than the corresponding values for the New York Central 4–6–4; but the cylinder volume of the ex-Great Western locomotive is very large in relation to the size of the boiler, and the performance was achieved with cut-offs in the cylinders of about 20 per cent except on the steepest gradients. Performance of this quality was necessary daily with the Cornish Riviera Express between the years 1928 and 1939, but today, although the scheduled speed is the same, a re-arrangement of train services, with more frequent departures from London, has enabled the load of the Cornish Riviera Express to be lightened.

A dynamometer car record taken with a very typical express train of today, the Atlantic Coast Express of the Southern Region, gives a good impression of the demands made in ordinary service upon modern British steam locomotives.

Exeter–Salisbury.

| Load | 11 coaches |
| Weight (gross) excluding engine and tender | 385 tons |

Timing:

Exeter–Sidmouth Junction:	
12·2 miles	booked 18 min.
	actual 20 min.
Sidmouth Junction–Salisbury:	
75·8 miles	booked 79 min.
	actual 74½ min.
Maximum speed	82 m.p.h.

Maximum drawbar horsepower:

Exeter–Sidmouth Junction	1,050
Sidmouth Junction–Salisbury	1,350
Average drawbar horsepower	715

On the non-stop run from Salisbury to Waterloo, 83·8 miles booked in 85 minutes start to stop, with a load of 450 tons, a signal check occurred near Basingstoke resulting in a dead stop. The two halves of the journey can be analysed thus:

Salisbury–Winklebury signals.

Distance	35·1 miles
Time	40 min.
Maximum d.b.h.p.	1,200
Average d.b.h.p.	830
Maximum speed	81 m.p.h.

Winklebury–Waterloo.

Distance	48·7 miles
Time	47 min.
Maximum d.b.h.p.	900
Average d.b.h.p.	543
Maximum speed	80 m.p.h.

The stretch from Salisbury to Winklebury is adverse, and that onwards to Waterloo almost entirely favourable. The average speed over the 30·2 miles from Hook to Surbiton was 77·5 m.p.h., and yet the drawbar horse-power required in hauling 450 tons never exceeded 900. The locomotive was one of the rebuilt "Merchant Navy" class "Pacifics," No. 35020 (Plate 105, page 427).

On the express train service of the Santa Fé, particularly between Chicago and Los Angeles, very heavy work was performed by the 4–8–4 oil-fired locomotives of the "3765" class. A typical piece of work on *The Fast Mail* involved the running of 134·5 miles from Dalies to Gallup, start to stop, in 124 minutes with a load of 1,590 tons. The ruling gradient over this section is 1 in 220 (0·45 per cent), and although there are both adverse and favourable stretches the tendency is adverse. In making the over-all average of 65·2 m.p.h. speed varied between 59 and 90 m.p.h., and on the climbing sections the indicated horsepower was 5,500 continuously. Analysis of this performance gives the following:

Total heating surface	7,679 sq. ft
Grate area	108 sq. ft
Nominal tractive effort	66,000 lb.
Maximum sustained I.H.P.	5,500

Ratios:

I.H.P. per sq. ft of heating surface	0·72
I.H.P. per sq. ft of grate area	50·9
I.H.P. per lb. of nominal T.E.	0·083

The power in relation to tractive effort is almost the same as that of the N.Y.C. 4–6–4, again revealing a very high power output per cylinder volume, resulting from the use of late cut-offs.

Fuel consumption on American steam locomotives was very heavy. The Pennsylvania run from Englewood to Fort Wayne (Table C, No. 5), can be taken as typical of coal-burning types. Details of this run are next analysed:

Pennsylvania R.R. Class "K4" 4–6–2.

Length of run	141 miles
Average speed, start to stop	71·6 m.p.h.
Load of train	630 tons
Flying average over 121·4 miles	74·7 m.p.h.
Locomotive: heating surface	5,020 sq. ft
grate area	70 sq. ft
nominal T.E.	44,400 lb.
Indicated horsepower (80 m.p.h.)	2,300

Ratios:

I.H.P. per sq. ft of heating surface	0·0518
I.H.P. per sq. ft of grate area	32·9
I.H.P. per lb. of nominal T.E.	0·052
Water consumption, per mile	78 gal.
Coal consumption: per mile	135 lb.
per hr	9,670 lb.
per I.H.P. hr	4·2 lb.

By comparison the British Railways standard Class 8 4–6–2, No. 71000, hauling a similar load of 590 tons, sustained an average speed of all but 80 m.p.h. on level road, an I.H.P. of 2,500, while the coal rate was 4,850 lb. per hr, and the water rate 30,000 lb. per hr. Related to distance covered, these British figures are 65 lb. per mile, coal, at 75 m.p.h., and 40 gal. per mile, water. Allowance must be made for the poorer quality of locomotive fuel in the U.S.A., and also that the mechanical stokers are inclined to be wasteful. The British figure of 2·6 lb. of coal per I.H.P. hr, at 80 m.p.h., compares closely with approximately 2·5 lb. per I.H.P. hr attained on the P.O.-Midi rebuilt compound "Pacific" locomotives on the heavy test run referred to on page 425 (Plate IIIB, page 433).

STEAM VERSUS DIESEL TRIALS: N.Y.C. SYSTEM

Perhaps the most severe endurance tests set to steam locomotives occurred during trials between the "Niagara" 4–8–4 locomotives of the New York Central System (*see* Chapter 5, No. 4), and diesel-electric locomotives, prior to the widespread introduction of the latter. The following average figures were obtained for six "Niagara" engines on a basis of annual performance:

Total hours	8,760
Hours for shopping and inspection	672
Hours available to running dept.	8,088
Hours unavailable due to servicing, etc.	1,435
Hours available but not used	573
Hours used	6,080
Per cent availability	75·9
Per cent utilization	69·4
Total mileage run	314,694
Average miles per month	26,226
Average miles per day	862

All the above service was performed in the haulage of very fast and heavy trains, mostly carrying a trailing load of more than 1,000 tons, and requiring sustained speeds of 80 m.p.h. or more on level track. It is interesting to observe that with careful maintenance, and equal service conditions, the "Niagara" locomotives showed an availability equal to that of the competing diesels, though the latter showed a slightly higher mileage per day (*see also* Chapter 2, Part II, The Diesel at War's End).

MILEAGE OF "HIAWATHA" TYPE 4–6–4 HIGH-SPEED LOCOMOTIVES CHICAGO, MILWAUKEE, ST. PAUL AND PACIFIC RAILROAD

Engine No.	Date put into traffic	Mileage to 1 June, 1939	Approximate average miles per month
100	19.8.38	108,521	11,750
101	24.8.38	101,179	11,115
102	30.8.38	103,138	11,450
103	7.9.38	95,180	10,850
104	15.9.38	97,904	11,550
105	24.9.38	95,430	11,500

These six locomotives were then engaged in working the high-speed "Hiawatha" expresses, running through over the 400 miles between Chicago and St Paul, with a trailing load of 465 tons at average point to point speeds of 65 to 80 m.p.h. (*see* page 425). The above mileages gave a good indication of the degree of reliability, combined with high utilization achieved in these locomotives. The average daily mileage of the whole stud over the period quoted is 375 (Plate IIIC, page 433).

WORKING OF BEYER-GARRATT LOCOMOTIVES IN AFRICA

In contrast to the runs on which a high sustained output of power is required at maximum speed, as on duties in Great Britain, France, and the U.S.A., locomotive duties that require a high standard of all-round performance, and the utmost reliability, are to be found on the substandard gauge trunk lines in East and South Africa. The East African main line between Mombasa and Nairobi, certain routes of the South African Railways and almost the entire network of the Rhodesia Railways are operated by Beyer-Garratt locomotives, which, despite the relatively low average speeds, work regularly high monthly mileages. On the Rhodesia Railways in particular, with end to end average speeds of only 30 m.p.h., on duties approaching 500 miles in length, Beyer-Garratt locomotives are making monthly mileages up to and exceeding 10,000. With express passenger trains the run of 484 miles from Mafeking to Bulawayo is performed by one locomotive in a continuous trip of seventeen and a half hours' duration. There is no running shed at any intermediate point of the journey. Moreover there is no appreciable lay-over after the completion of this trip. The locomotive works back after a short turn-round time making a round trip of 968 miles as a single turn of duty. This round trip is done with both passenger and freight trains.

The largest locomotives of the Beyer-Garratt type at present working on the Rhodesia Railways are of the 4–8–2+2–8–4 type, weigh 225 tons in working order, and have a nominal traction effort of 69,330 lb. Their rostered loads on various sections of the railway system are:

Section	Gradient (compensated)	Load	
		Tons (short)	Tons Imperial
Bulawayo–Gwelo	1 in 80	1,650	1,470
Wankie–Bulawayo	1 in 130	1,950	1,745
Kafue–Broken Hill	1 in 64·5	1,400	1,255

Again the loads are more, due to the maximum number of axles convenient for operating rather than representing the maximum capacity of the locomotives. The ability of the Beyer-Garratt locomotives to haul very heavy loads has not only avoided the wasteful expedient of double-heading, but has enabled the railway to handle a vastly increased traffic. Practically the entire main-line mileage is single-tracked, and unless very expensive additions to track facilities were made, the only way to cope with increased tonnage offered was to convey it in much heavier trains. The success with which this has been done is reflected in the astonishing fact that between the years 1948 and 1955 the total tonnage hauled on the Rhodesia Railways has doubled. To date no fewer than 250 locomotives of the Beyer-Garratt type have been ordered for working in Rhodesia. Engines of the 15th class, having the 4–6–4+4–6–4 wheel arrangement, have averaged nearly a quarter of a million miles between general repairs, and 100,000 miles between tyre turnings. (See Chapter 4, Part I, Articulated Locomotive.) (Plates 112A, 112B, 114C, page 434.)

CHAPTER 8

The Organization of a Steam Motive Power Depot

by G. FREEMAN ALLEN (*Parts* I, II *and* III) *and* P. RANSOME-WALLIS (*Part* IV)

Part I. The motive power department

On the larger railway systems of the world it is usual practice to place the allocation and management of locomotives in traffic in the hands of an organization distinct from that responsible for the design and construction of motive power. Those charged with the running of locomotives must work in close conjunction, both with the railway's mechanical engineering department and with the traffic operating department. It is, however, better that they be, to some extent, independent of both. Too close a relationship with the mechanical engineer's department may result in faulty appreciation of everyday operating requirements, and, since traffic operators are rarely erudite locomotive engineers, an operating department cannot exercise effective technical control of a locomotive running organization, particularly where steam is the preponderant means of traction. The performance of a diesel or electric locomotive may be predictable within precise limits and sufficiently independent of human and temporal factors to need no separate organization to run it day by day. This is not the case with steam power, whose performance is varied by circumstances, many of which can be controlled by an on-the-spot management skilled both in matters mechanical and in staff relations.

The need for the motive power, or locomotive running department of a steam-operated railway to be largely independent is the general view of those professionally involved, but it is by no means the rule. There are even divergences within a unified system like British Railways, where four regions have made their motive power departments responsible to the regional operating organizations, so far as the daily provision of traction is concerned, while in the remainder the motive power organizations remain independent. Whatever the arrangement from the traffic viewpoint, the motive power department must remain answerable to the mechanical engineer on matters of design, maintenance and repair procedure.

ALLOCATION OF LOCOMOTIVES

The headquarters of the motive power department orders the disposition of the railway's locomotives. This is based on the traffic offering in each area of the railway and the abilities of the various types of locomotive owned by the railway to meet the traffic requirements. It is essential, therefore, for the headquarters to be possessed of complete data on the performance and condition of its locomotives, so that it can calculate maximum loadings and optimum timings for the routes it covers. It must work with the railway's operating department to ensure that, so far as is compatible with public need, the working timetable is framed to match the capabilities of the locomotives, and to permit the arrangement of duties that will make the most economic use of the available motive power. The aim in view is to evolve a series of working rounds – diagrams or rosters as they are usually known – for the railway's locomotives which, without overtaxing manpower or maintenance resources, will keep them in revenue-earning service for as much of each twenty-four hours as possible. Finally, since the motive power department is in the closest day-by-day touch with the railway's locomotives, its headquarters is the mechanical engineer's best source of information on design defects and causes of proneness to failure. From the motive power department may also come suggestions for the modification of existing types in the light of service experience, or for the addition of new designs to the railway's stock to meet new traffic requirements.

When a new timetable has been prepared, the locomotive diagrams to operate it are agreed, and given to the motive power depots most conveniently placed to fulfil them. The motive power headquarters will then allocate its locomotive stock to depots accordingly. These depots provide accommodation and major servicing for locomotives out of traffic, and degrees of repair facilities according to the importance of the depot. They are established:

(i) At traffic centres where other operating requirements make it convenient to allow time for changes of locomotives on through traffic.

(ii) Where there is a high proportion of, or completely terminal working.

In addition, sub-depots are set up at locations such as freight yards, where it is desirable to have a small establishment at which locomotives can be turned, refuelled and berthed, but at which full servicing facilities are not provided. Such sub-depots provide a convenient point at which enginemen can sign off and on between outward and home workings.

The allocation of locomotives must grant to each depot a reserve over and above requirements to meet the diagrams. Locomotives must be withdrawn from service periodically for major servicing and repairs; special workings must also be provided for. The allocations cannot be inflexible, for, with freight especially, there will be seasonal ebbs and flows of traffic which must be covered by switches of locomotives from depot to depot.

DISTRICT ORGANIZATION

On a large railway system the motive power depots will be grouped into divisions, or districts, the boundaries of which will be determined by a combination of geographical circumstances and operating convenience. The aim of this is to avoid the evils inherent in over-centralization in any organization. Thus the district motive power superintendent will have deputed to him many of the administrative functions of his headquarters. So far as possible his district will be made self-supporting. The allocation of locomotives to his district will be made so that, to some extent, the district superintendent should be able to meet the special occasional needs of his depots by moves of locomotives from depot to depot within his district. At his headquarters depot there will be equipment for heavier repairs than elsewhere in the district, and also the district's heavy breakdown train.

LOCAL ORGANIZATION – THE SHEDMASTER AND HIS STAFF

The officer in charge of a motive power depot, commonly known as a shedmaster, must be both a skilled technician and an expert in human relations. His task is to have the maximum number of his locomotives fit and available for work at any given moment; and since the steam locomotive is so dependent on the human element in maintenance as well as operation, to infuse his staff with an *esprit de corps* as well as ensuring their technical proficiency.

The staff of a motive power depot can be divided into five groups:

Clerks. Clerical work at a big depot is very extensive, and covers a wide range of statistics, returns, wages and tax calculations, and correspondence. For example, the depot clerical office:

(*a*) Maintains detailed records of the mechanical history of every locomotive on the allocation.

(*b*) Prepares reports, required by headquarters, on such matters as locomotive failures, periodic examinations, fuel consumption and locomotive performance. These items are then related to the specific diagrams of each particular locomotive.

(*c*) Deals with all matters concerning personnel, such as accidents, sickness, leave, welfare and so on.

(*d*) Works out the wages calculations. These are often very involved, as enginemen's pay may include bonuses for mileage worked, for time-keeping or the regaining of lost time, and for economy in fuel and oil consumption.

(*e*) Makes out demands for stores and reports on plant and machinery.

Enginemen. Enginemen are grouped into sets or "links" according to their experience. No engineman is allowed to work over a route until he has learned it, and each link will, therefore, cover a series of engine and crew diagrams requiring knowledge of a defined route area of the railway served by the depot concerned. The senior links will comprise men who have qualified by footplate experience and route knowledge to operate the most important diagrams allocated to the depot, and the links below them will work diagrams on a declining scale of importance. This organization can best be understood by explaining a particular case.

At the important Holbeck motive power depot at Leeds, England, which serves the main line from London (St Pancras) to Leeds, Carlisle and Glasgow, and its branches in the West Riding area, the enginemen are grouped in twelve links, each consisting of eight to twelve crews. The first three links are equal in seniority, and provide for the working of passenger trains in different directions from Leeds; in No. 1 Link, the men must be qualified to work to Bradford and London, in No. 2 Link to Glasgow, but no farther south than Derby, and in No. 3 to London and Morecambe. The duties of these three links include what, in Britain, are known as "lodging turns" – that is, runs of such lengthy duration that the men must be given time for sleep between out and home working (in Britain eight hours is the accepted length of one footplate shift). At depots to which lodging turns are operated it is necessary to provide an enginemen's hostel in which visiting crews can be fed and given sleeping accommodation.

The next three links, Nos. 4, 5 and 6, cover almost as much territory as the first three, but the services worked by their crews are not in the same flight and include freight as well as passenger trains. Below No. 6 Link, the route knowledge required is progressively more limited. Nos. 7 and 7A deal with local passenger trains and freight, while the men in No. 8 Link have to know no more than the short distance from the motive power depot to Leeds

City Station; this enables them to relieve at the latter point, main-line crews who have completed an eight-hour shift on arrival, and to take their locomotives away to Holbeck depot. Nos. 8A and 9 consist of men who, by reason of deteriorating health or age, have had to be relegated to freight work in the Leeds area and movements within the depot precincts. Finally, the most junior link is made up of men who have just graduated to footplate work and who are confined solely to preparation and disposal jobs and to moving locomotives as required by the mechanical staff within the depot.

The preparation of men for footplate work varies from country to country. In France, for example, a thorough mechanical training in workshops is a pre-requisite, whereas in Britain the would-be engineman is taken on as a locomotive cleaner, and his early knowledge of the mechanics of the steam locomotives is gained in that job. It should be added, however, that most British motive power depots run voluntary but well attended "Mutual Improvement Classes", at which expert instructors lecture regularly on mechanical detail and driving method; moreover, every B.R. engineman now receives free some useful text-books on the working of locomotives. It is usual for the footplateman to commence as a fireman in the lowest link, progressing to the most senior grade, and then to revert to the lowest link when he commences his driving, so that he reaches a senior driving job at the prime of life. He will be required to pass an examination before he can become a fireman, and another before beginning his driving career; there may be an interval between taking these examinations and his posting to a regular link, during which, if he has satisfied the examiner, he will be regarded as a "passed cleaner" or "passed fireman", able to turn out for a firing or driving job respectively if the depot is short of a regular man for a turn of duty.

The diagrams for the locomotives and enginemen of a motive power depot are compiled by the railway's central train diagramming office, which, like the motive power department headquarters in its allocation of locomotives, views the matter on an all-line basis. It is left to the individual depot to arrange the diagrams allocated to it into cycles that will form suitable working weeks for the locomotives and enginemen available. Such diagrams are kept constantly under review and if, in the light of experience, it is thought that more economical working could be achieved by detail alteration, headquarters is so informed. In addition, every depot makes a weekly return which shows the working of each locomotive in relation to its diagram.

The arrangement of crew diagrams within the links should ensure that each set of men works over all the routes covered by a link at regular intervals, so that their route knowledge is kept up to date. No man is allowed to work over a route unless he has signed a declaration to the effect that he has familiarized himself with it, and has done so within a specified period before the day he is required to operate over it. Moreover, the enginemen rosters should be ordered so that crews alternate between day and night shift work. It is generally considered best to allocate passed firemen and drivers to tasks within the depot or its immediate neighbourhood, so that they are readily available to deputise if the regular men for any reason fail to report for duty.

Running foreman and locomotive inspectors. The day-to-day nomination of individual locomotives for the diagrams of a depot, subject to the provision of a locomotive of the power class specified in the diagram, is the responsibility of the running foreman. It is also his duty to see that a proper crew is provided for each locomotive. He must, therefore, be possessed of up-to-date records of the availability or otherwise of every locomotive on his depot, whether they are part of his own allocation or locomotives being serviced between out-and-home workings of a "foreign" depot. He must hold footplate records of all the depot's enginemen, so that he knows the capabilities of each and the routes for which they have signed; he is also responsible for the signing on and off duty at their appointed times for work.

The running staff of a district headquarters depot will also include locomotive inspectors who will be selected men from the ranks of the senior drivers. Their principal occupation is with locomotives on the road – training and advising enginemen in the more efficient working of their locomotives, or in the use of a new item of equipment. They investigate enginemen's complaints of the behaviour of a particular locomotive or piece of apparatus. Other matters concerning running, such as the siting of signals, also come within their province, and they examine cleaners and firemen for graduation to firing and driving respectively.

Shed grades. There are many duties of widely different types which are necessary for the efficient conduct of a locomotive depot. These duties vary with the size of individual depots and with the facilities available at each depot. In smaller sheds, the grades may overlap, i.e. the men of one grade also doing the duties of another one or more grades. The titles of the various grades are mostly self-explanatory, and they fall into two main sections:

(i) Those duties which involve work on the depot.

(ii) Those duties which are concerned with the preparation and maintenance of locomotives on the shed.

In the first section are such duties as:

Ashfillers, who load ashes from ash-pits into wagons, or manipulate mechanical ash-handling plant.

Coalmen, who unload coal from wagons and issue it, by various means, to engine tenders. Where a mechanical coaling plant is provided, they are responsible for the operation of the plant.

Lampmen, who examine, clean and trim all engine lamps.

Messroom attendants, who keep the enginemen's messroom clean and attend fires and cooking ranges.

Sandmen, who unload sand from wagons to sand-drying plant and issue it to locomotives as required.

Shed labourers, who are engaged in the general maintenance and cleanliness of the depot. They keep the pits, turntables, drains, etc., clean.

Storekeepers, who are responsible for the maintenance and control of stores, including lubricating oils, waste, enginemen's cloths, etc. In large depots they supervise the issue of stores by store-issuers.

Telephone attendants.

Toolmen, who should see that each engine has a full set of tools before leaving the depot.

In the second section are such duties as:

Barmen, whose duties are the examination and renewal when necessary of firebars. They examine and keep clean brick arches and inside firebox plates.

Boiler-washers, who are responsible for the washing out of locomotive boilers. They also blow the boiler down and refill with water.

Fire-droppers, who are responsible for emptying the smokebox of ash and for cleaning the fire. If the engine is not required for service again soon, the fire is dropped or thrown out.

Fire-lighters and steam-raisers, who light fires with special fire-lighters and are responsible for seeing that sufficient steam is available in the boiler at the time the engine is needed.

Tube-cleaners, who are responsible for cleaning boiler tubes by any of the accepted methods available.

The movement of locomotives on the shed and their berthing in correct sequence for their next turn of duty, is the work of enginemen in the lower links.

Tradesmen. This group includes the many skilled men who are responsible for the mechanical upkeep and repair of locomotives and tenders. Blacksmiths, coppersmiths, fitters, turners, welders and others are required in a large depot. They are controlled by the mechanical foreman, who is also responsible for organizing the periodical and mileage examinations of his locomotives at the intervals prescribed by the mechanical engineer, and for maintaining their serviceability by such repairs as are within the limits of his equipment. If a locomotive is beyond his resources to repair, it must be proposed for attention at the district headquarters depot or at workshops. It is the aim of the mechanical foreman always to have the minimum number of engines stopped at any one time.

Part II. The planning and layout of a running shed

The use of the phrase "engine shed" to describe a motive power depot can be misleading, for covered accommodation is only a part of the layout. With steam locomotives, it is desirable that repairs and some servicing operations be carried out under cover, but if a locomotive is proceeding to a depot merely for refuelling and minor requirements between workings, there is no need for it to be housed; moreover, some servicing operations require apparatus of a size that cannot be brought under cover. To save expense, therefore, a shed for berthing locomotives need be no larger than is adequate to house about half of the depot's allocation; at any given moment a good proportion of the remainder will be out at work, while others, and locomotives visiting from "foreign" depots. will need minor attention that can be carried out in the open air if they come on to the depot.

The ideal in planning a motive power depot is to devise a layout that, first of all, will enable locomotives coming in to pass continuously forward through a

sequence of operations necessary as a preliminary to disposal after duty, and, in part, to servicing between stages of a duty. Since it would be needless expense to provide two sets of apparatus for these operations, one for locomotives to be berthed, the other for those returning to duty, the layout should make provision for the latter to by-pass stages of attention they do not require, so that they are not held up by locomotives taking the longer service. These operations completed, the locomotives should then be able to reach berthing points in the shed, or stabling points in the yard, with the minimum of shunting, and without recrossing their paths through the disposal sequence. The berthing and stabling roads should be laid out to enable locomotives to be extracted for duty with the minimum of disturbance to others on the depot, and the exit to the main line should be separate from the path of incoming locomotives. There must also be berths in which locomotives can be placed without fear of having to be moved, so that repairs can be carried

PLATE XIV. Indian Government Railways: "The Frontier Mail" (Amritsar to Bombay Central) leaving New Delhi. The locomotive is a standard W.P. Class Pacific, No. 7544.

out. The space available for a depot is never limitless, and often considerable ingenuity is needed to devise a suitable layout.

There are two types of locomotive shed, *the round-house* and *the parallel-road shed*. The former was used in the United States in the days of steam almost universally. Indeed, the very name "roundhouse" in America is used to mean a locomotive shed (Plates 113B, 114, pages 453, 454). In Britain and the West of Europe, both types of structure are used; the type adopted depends largely upon available space at the site. Most new sheds, however, are of the parallel-road type.

The roundhouse has a centrally placed turntable from which berths radiate like the spokes of a wheel, with the advantage that any berth in the shed can be reached without disturbance of locomotives already on the depot; moreover, it makes no difference which way round an engine is stabled, since it can be turned to the required direction on its way out. On the other hand, the roundhouse has several disadvantages. It tends to be wasteful of space since each berth must be designed to accommodate the longest type of locomotive expected to work into the depot, with the result that when it is occupied by a smaller engine there is unused trackage; in addition, a greater multiplication of pipe lines and servicing equipment is necessary than with the parallel-road shed, and the arrangement of smoke troughs in the roof, which is itself more expensive to construct, becomes exceedingly complicated. A major difficulty is that if the turntable should be incapacitated for any reason, the whole shed is put out of commission; the limited number of inlet and exit roads is a further source of impaired efficiency (Plate 29, page 121).

The parallel-road shed suffers from the disadvantage that to berth a locomotive in the shed may require the remarshalling of several other engines already in the shed. If, however, the shed is of the "through" type much

of this trouble is done away with, though sheds of this type require much more space and yard trackage. On the other hand, this expense is offset by the cheaper costs of constructing the shed itself, compared with a roundhouse. In a small depot, the limited number of engine movements will probably permit the use of a single-ended (dead-end) parallel-road shed without operating inconvenience. Decision in favour of a parallel-road layout will certainly require the construction of a separate shed to deal with major servicing operations and repairs, which need the locomotive to be berthed for a considerable period without being moved. In a roundhouse, where each berth is self-contained, repairs not involving the use of such bulky apparatus as a wheel-drop, can be carried out wherever the locomotive is berthed, without fear of disturbance.

One of the most recently built steam motive power depots in the world is the modern installation of British Railways at Ipswich, Suffolk (Fig. 105 and plate 113A page 453), which may be taken as an example of contemporary thought on layout, In discussing this, it will be useful to outline the provision for disposal and preparation of locomotives, and the means for ensuring their appropriate examinations and repairs.

The Ipswich depot is laid out in rectangular form. Locomotives coming on to the depot take a path down one side of the rectangle, in the course of which they pass through such disposal stages as turning, examination, watering and coaling and then reverse on to ashpits, over which the fireboxes and smokeboxes can be cleaned. In the centre of the layout is a double-ended parallel-road shed, and from the ashpits a locomotive can be run direct to any of its roads; or, if the locomotive is not to be berthed, it can be run round the shed to a position in the yard at the other side of the shed, whence there is immediate access to the main line.

The first operation in the disposal of a locomotive on

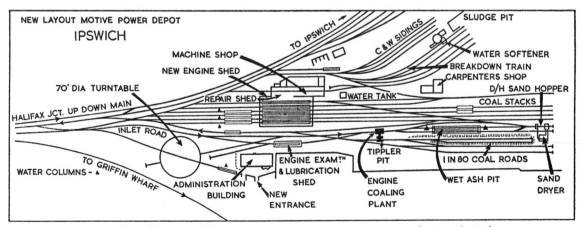

FIG. 105. Plan of layout of the new Motive Power Depot at Ipswich, British Railways.

Ipswich depot is turning. (If space permits, the expense of a turntable can be avoided by incorporating a triangle, or wye, in the depot layout.) In modern practice, turntables are powered either by an independent electric motor, or by vacuum or compressed air from the locomotive's brake system according to which method is standard on the railway concerned. At Ipswich, the turntable is not inserted in the disposal path, but to one side of it, with connecting leads, so that locomotives coming to the depot which do not require turning for their next duty can by-pass ones that do.

The next stage in the disposal sequence for locomotives proceeding to berth, is the daily examination which is carried out in a separate shed built over the disposal road and equipped with fluorescent lighting in the walls and in the inspection pit between the rails. In Britain, it is the rule that all locomotives used on passenger and brake-fitted freight trains must be examined daily by a fitter at the disposal of the running foreman, though other locomotives are examined but once a week. A loop in the disposal road, therefore, by-passes the inspection shed, to enable locomotives visiting the depot for fuel only to proceed direct to the next stage. In the inspection shed the fitter will examine wheel-tyres, coupling rods, valve gear, big- and little-ends, slide-bar bolts, glands, axles and axle-boxes, springs, brake and sanding gear, and the firebox, reporting any defects on the locomotive's repair card, of which more will be said later.

From examination, the locomotive takes one of two tracks past the water points, where its tanks are refilled. In districts where the water supply exceeds a certain level of hardness, and where the depot concerned is of a size to require more than approximately ten million gallons a year, stationary plant for water treatment should be provided. The purpose of this is to remove from the water matter that will otherwise corrode the boiler and cause the formation of scale, thus impairing the locomotive's efficiency and making stoppages for boiler maintenance undesirably frequent and costly.

The locomotive next stops under the coaling plant (Plate 115, page 455). The simplest method of coaling a locomotive, of course, is by hand from wagons on an adjoining track, or from a coal stage, but it is also the most costly in terms of time and labour. For a depot with an allocation of more than thirty main-line locomotives, therefore, the expense of mechanical coaling plant is undoubtedly justified. For a small depot, an early type of mechanical coaler still in use involves an element of manual labour; the coal is unloaded manually from wagons into narrow-gauge tub wagons which are then traversed by hand to a powered hoist that elevates them

and tipples their contents down an overhead chute into the locomotive tender. For larger depots it is usual to provide a far more massive structure incorporating storage bunkers, each of 100 tons or more capacity, above the tracks; separate bunkers in the one structure can be reserved for different grades of fuel, with a separate track beneath each bunker, so that each locomotive can obtain the type of coal suited to its work and one or more locomotives can be coaled simultaneously from the same plant. Early installations of this kind were designed so that wagons could be hoisted bodily from ground level up to the top of the bunkers and there rotated to empty their contents, but modern practice favours a ground-level tippler to empty the wagons into a large sunken hopper, from which a skip raises the coal to the lip of the bunkers.

At the Ipswich depot under discussion, the final operation before the locomotive proceeds to the ashpits is to refill its sanders. A Kelbus sand drier is installed in a building at the side of the track, and from this the sand is piped, with the aid of compressed air, to a gantry spanning the disposal track, from which there are pipe connections which can be attached to the sandboxes of the locomotive.

The locomotive has now reached the limit of the disposal track and from the sanding point it reverses on to the ashpits. If it is to be berthed in the depot, the dropping of the fire is left until the end of the disposal sequence, in order to avoid the harm caused by a sharp drop in boiler temperature. If a locomotive is merely visiting the depot between stages of a diagram, its visit to the ashpits will be for smokebox cleaning only, after which it will by-pass the shed and run to a position in the yard ready for its next duty.

The ashpits are of the "wet" type, as opposed to "dry", that is to say, the pit is filled with water to keep down dust and especially to prevent dust and ash from penetrating the bearings of the locomotive being serviced. Two adjacent tracks are built over a single wide pit with inclined sides, so that the deposited ash or clinker will accumulate in a sump between the tracks. The sump is covered with easily removable gratings that make a firm surface for the staff during servicing. Above runs the track of a wheeled grab-crane; periodically, the gratings are removed and the grab crane is brought up to remove the ash from the pit and empty it into wagons for disposal. Except for those which are equipped with self-cleaning smokebox arrangements, a visit to the ashpits will figure in the routine of every locomotive coming on to the depot. From the ashpits the layout affords immediate access to any road in the shed.

At the largest motive power depots more elaborate

ash-handling plant is provided. Narrow-gauge tracks are laid in the ashpits and on these run tubs into which the ash and clinker from a locomotive's firebox and grate are shovelled. The tubs, when full, are propelled to the centrally located handling plant, which consists of a capacious pit below rail level and an overhead ash bunker. The tubs are tipped to empty their contents into the pit, which is V-shaped, with a skip at the apex of the V; as soon as the skip is full, a weight-ometer switch operates, closing the chutes admitting to it, and setting in motion the electric hoisting gear; this lifts the skip up to the overhead bunker and tipples its contents therein, whereafter the skip returns to the bottom of the pit and the chutes are reopened for a fresh load (Plates 116A, 116C, page 456).

With the exception of the offices of the mechanical foreman and his staff, which are incorporated in the repair shop building, all the necessary offices, staff accommodation and stores of the Ipswich motive power depot are housed in one building conveniently sited alongside the layout and close to the running and repair sheds. On the ground floor a spacious entrance lobby provides ample room for the display of staff notices, as well as details of locomotive and locomotive-men's duties and special operating notices. Adjoining the lobby are the offices of the running foreman and the timekeeper, at which enginemen sign on and off duty. These offices are located so that the running foreman can readily command the disposal of locomotives after duty and their dispatch from the depot for work. At the other end of the ground floor, the well laid out stores are placed so that they are only a few paces from the shed.

The staff accommodation includes mess-rooms, a lecture room and a spacious changing-room equipped with handbasins and shower-baths. Every man has his own locker for his belongings.

Part III. The routine of a large running shed

CLEANING OF ENGINES

The routine attention given to a locomotive not stopped for repairs, between disposal and preparation for its next duty, principally involves cleaning, both inside and out. Apart from the fact that a high standard of external cleanliness has a great public relations value (the converse is also only too true), it has considerable importance from the viewpoint of mechanical efficiency, for the entry of dirt into bearings can cause failure; moreover, an accumulation of grime can sometimes obscure a defect from the eyes of examining staff. It is highly desirable, therefore, that a depot should be possessed of an adequate cleaning staff to ensure that every one of its locomotives is cleaned at least once a week, and preferably more often in the case of the more important types on the allocation. A keen engine-crew, particularly where it is possible to maintain double-manning of locomotives, will usually appropriate a great deal of the responsibility for the cleanliness of their locomotive to themselves.

The cleaning can be done by hand, using paraffin-type cleaning oil for the greasiest parts, such as the motion, and composition cleaner or petroleum jelly elsewhere. In many depots cleaning is done with the aid of a pressurized jet through which a suitable solution can be projected to wash the engine. Most modern depots, however, are equipped with a form of steam jenny that can be powered from an engine in steam or from the depot's stationary boiler, and connected by piping to easily portable guns; this produces a mixture of oil and water at high temperature and at high pressure – up to 400 p.s.i. – and is both highly effective in removing oily deposits of dirt and immensely time-saving, by comparison with manual cleaning.

Internal cleaning is concerned mainly with the firebox and tubes. It is desirable, but not always practicable, to carry out once a week, cleaning of the inner firebox to remove the deposits left by combustion. At this time also, opportunity is taken to replace broken firebars and to make good any defects in the brick-arch. When a concrete arch is fitted, it is cleaned of deposits and inspected for signs of cracking.

Deposits of soot must be removed from the flue tubes. This may be done manually by a "pull through" or by a rod having a mop of flax or hemp at one end. However, except for superheater tubes which are always manually cleaned, compressed air blast is used for tube cleaning at most modern depots, such depots being usually well equipped with compressed air points for the operation of grease guns and other equipment.

PREPARATION OF ENGINES

The preparation of a locomotive for its next duty begins with steam raising. Although it may on occasions be necessary to steam an engine in a hurry, by forcing the draught, it should not be the normal practice, for rapid rises in temperature are harmful to the boiler and cause tubes and stays to leak. Steam should be raised with cold water in the boiler and without forcing the fire. The usual method is to lay in the grate a thin bed of coal which is ignited with highly combustible firelighters (wood is not recommended, as it tends to form unwanted clinker). The lighting-up should be timed so that there will be 50–100 p.s.i. in the boiler by the hour at which the engine crew are scheduled to sign on for duty; in the case of small engines this will mean that fire-lighting must begin about three hours before the enginemen book on, but over five hours may be needed to raise the requisite pressure in the largest main-line locomotive.

The first task of the enginemen after signing on for duty is to study the notice boards in the lobby for details of any new restrictions or other special developments affecting the route they are booked to take. Adjacent to each other in the lobby should be the lists of the depot's engines and enginemen diagrams, with their code numbers, and the running foreman's daily list of the locomotives he has selected for each diagram. In a large depot, information is available for the enginemen concerned, as to the berth on which they will find their engine.

The crew's next call is at the store to take out the regulation set of footplatemen's tools and also supplies of the different kinds of oil they will need – lubricating, cylinder, burning and paraffin. To avoid wastage, the storekeeper will be in possession of an issuing scale showing the correct allowances for every diagram worked by the depot; these allowances will be based on the mileage worked on the diagram and the type of locomotive rostered to it.

Reaching their engine, the crew proceed to the final stages of preparation for a new duty. The driver satisfies himself that booked repairs have been effected or reasonably deferred, and then the crew proceed to an examination of the whole engine. The driver walks round the locomotive, looking over the valve-gear, oiling crossheads, big-ends and side-rods, satisfying himself that the mechanical or displacement lubricators are adequately supplied with oil and all trimmings in order, and making sure that nuts and cotters are secure. Meanwhile, the fireman will begin to accelerate the raising of steam pressure by work on the fire and use of the blower. It is also the fireman's job to test footplate equipment such as gauge-cocks, injectors and sanding gear, to ensure their working order. Climbing on to the tender he trims the coal forward to his liking, or he will satisfy himself as to the working of the mechanical stoker, if one is fitted. From three-quarters of an hour to an hour after signing on (the diagrams for enginemen must allow for this preparation time) the crew and engine will be ready to move off the depot for their duty.

REPAIRS AND THE X-DAY SCHEME

The question of repairs depends first of all on the locomotive's crew. One of the tasks of a crew coming off duty, in addition to reporting to the running foreman details of the turn they have just completed, is to submit an account of any defects they have noticed in running. On some railway systems, depots maintain repair books or repair sheets on which these details are entered, but although this makes for some ease of reference, it does entail a considerable amount of recopying when instructions are issued to fitters to deal with the defects. In many ways, the system used on British Railways, whereby each driver coming off duty has to fill up a repair card for his locomotive, is to be commended; even if his locomotive proved entirely trouble-free, the driver must return a "no repair" card, so that there is an up-to-date record available in the depot of the mechanical condition of all its locomotives.

The repair cards are handed in to the running foreman. He has at his disposal a fitter for the routine daily examination of locomotives coming off duty, and the fitter will add to the card the details of any defect his examination reveals to require attention. At regular periods in the day a fitter from the mechanical foreman's staff will collect the repair cards from the running foreman.

It is obviously desirable that the stoppages of locomotives be reduced to the minimum. Since a steam locomotive must perforce be withdrawn from service at frequent intervals for its boiler to be washed out, a proceeding which immobilizes it for several hours, it is as well to take advantage of this necessity by using the time to carry out as much essential servicing and repair work as possible, rather than stop the locomotive for a further period. British Railways, indeed, have adopted as standard a scheme introduced by the L.M.S.R. in 1928, whereby the whole procedure of periodical and mileage examinations, and as much repair work as possible, is bound up with the appointed days for a locomotive's boiler washout – or "X-Days" as they are termed. The period between X-Days differs from type to type, and depends partly on construction characteristics and partly on the work being performed. It is based upon experience. The various examinations of parts of the locomotive are timed to occur at multiples of the specified interval between washouts, and as far as possible, repair work is deferred until the locomotive's next X-Day.

The adoption of this scheme has played a large part in the increase of British locomotive availability achieved since World War II. Nowadays, many British depots are improving upon the target of 85 per cent availability at any given moment, which allows for a margin of 4 per cent of the allocation to be in or awaiting workshops, 2 per cent to be on shed awaiting materials to effect repairs, 5 per cent to be out of traffic at sheds under repair or examination, and 4 per cent to be completing repair or examination and available for work on the next shift.

To provide a record of the work to be carried out on a locomotive on its X-Day a depot keeps an X-Day card for each engine on its allocation. On this card are listed all the various non-moving parts, such as the safety valves and brake gear, which have to be examined on a periodical basis, and the moving parts, such as the valve gear, which are examined after specified mileages; in each case the stipulated period between examinations is indicated, and spaces are provided for the entry of the last dates and mileages at which each examination was carried out. (It should be added that additional cards are maintained for each locomotive providing more detailed records in respect of items such as wheels and tyres.)

At the depot, when a driver has booked defects against a locomotive, the mechanical foreman must decide what is to be done. If the safety of the locomotive is jeopardized, or the defect makes it unworkable, it must be stopped from traffic until the trouble is remedied. If, however, it is still workable and safe, a fitter will be detailed to attend only to so much of the repair work as can be carried out before the engine's next duty. Whenever possible, in order to maintain the engine's availability, it will be returned to traffic and the defects will be entered on the X-card for attention during its next boiler washout.

Since this matter of maximum availability is all-important, it follows that the mechanical foreman must plan his washouts carefully, so that the number of his locomotives stopped for X-Day is roughly constant at any given moment. For a large depot it would be reasonable to expect no more than 5 or 6 per cent of the allocation to be stopped for this purpose on any one day, and the work should be staggered throughout the whole twenty-four hours to reduce simultaneous withdrawals from service to the minimum. It goes without saying that the mechanical foreman will keep the running foreman closely advised of his proposals for stoppages.

Where a motive power district is reasonably close-knit, and the adoption of such a principle does not incur considerable light-engine running, there is a good case for concentrating all but minor running repairs at the district headquarters depot. Not only does this encourage the provision of modern machine-tool equipment that could not be economically entertained for every depot on the system, but it also makes possible a concentration of highly skilled labour. Similarly, it is generally not economical for district headquarters to be so well equipped that they can usurp the functions of the system's locomotive works. On most railways, therefore, there is a grading of repair equipment necessary at the various levels of depot, headquarters depot and works, and limits are laid down as to repair jobs the first two can undertake, beyond which the task must be passed on to the next echelon.

On the other hand, depots must be restrained from submitting locomotives for shopping when they are themselves perfectly capable of providing at least some interim attention that will keep the locomotive in service until it is next due for works. For the works, too, need a steady flow of work that must be programmed as closely as possible with regard both to the capacity of the shops and to the traffic needs of the railway. It is, therefore, customary for a railway to schedule the shopping of its locomotives for general overhaul on a mileage basis, but leaving a margin of workshop capacity for unforeseen needs.

The modern Ipswich depot of British Railways, previously referred to, has a fluorescently-lit repair shop containing two roads, and with adjacent blacksmiths' and electricians' shops; its major equipment consists of a small, electrically-powered, overhead travelling crane and an electrically-operated wheel-drop, while there are compressed-air supply points for the operation of pneumatic tools and also electric welding equipment. The depot can thus quickly remove engine wheels and axles for refitting of boxes or for attention to wheels and tyres, and can tackle most repairs that crop up between general overhauls.

On British Railways, locomotives are called forward for periodical overhaul at workshops by the Shopping Bureau of each region, which bases its action on the regular reports of an engine's condition and the mileage it is putting in. These reports are furnished by the depots. The largest passenger types will be called forward every 12–18 months, or 70,000–100,000 miles, and at the other end of the scale a small freight locomotive may run for over $2\frac{1}{2}$ years between general overhauls.

X-DAYS – BOILER WASHOUTS AND PERIODICAL EXAMINATIONS

The interval at which boiler washouts take place will vary according to the work performed by the type of locomotive, and may be from ten days in the case of large engines putting up high daily mileages, to a month in the case of small classes performing local short trips. It is preferable that the washing out of boilers be done with hot water because:

(a) Wide changes of temperature in a boiler should be avoided.

(b) It is possible to use for washing out the hot water

blown down from the boiler when the engine first comes on to the plant. Such water, however, must be filtered before re-use in this manner.

(*c*) The time taken for washing out is reduced.

(*d*) At the conclusion of the operation, the boiler is left hot and so steam-raising is expedited.

On the other hand, cold water is more efficacious in removing certain types of scale, and where this occurs, cold water washouts are introduced into the maintenance schedule.

Plants are manufactured for motive power depot installation which, with the aid of steam power from a stationary source, carry through the whole cycle of filtering water extracted from the locomotive's boiler and re-supplying it for the washout at a pressure of some 60 p.s.i. For a large depot, hot water washout plants are available of a size to cater for six locomotives simultaneously. The engines are grouped in pairs to undergo the three different sequences of the operation – two having their boilers emptied, two being supplied with filtered hot water for washout, and the remaining two having their boilers refilled after washout.

Locomotives due for washout should come off the ashpits with 80–100 p.s.i. pressure, to enable the boiler to be blown down, after connection has been made between the engine's blow-down cock and the hot water plant. If there is to be a cold water washout, an interval of some hours must elapse to enable the boiler to cool down after it has been drained, but with a hot water treatment the job can be commenced as soon as the boiler has been emptied. The plugs and mudhole doors are then removed and a connection is led from the washout line of the hot water plant, first to the plug holes in the firebox and then to the front tube plate. At the same time rods are inserted through plug holes, close to that through which the water is being injected, and worked about to dislodge scale, which the water will carry away (hence the need for the latter to be admitted under pressure). After the washout has been completed, the boiler and firebox are examined thoroughly, the plugs and mudhole doors replaced, and the blow-down cock connected to the filling line of the hot water plant. The boiler is then refilled (Plate 116B, page 456).

Apart from using the stoppage for washout to effect any requisite repairs, the opportunity can also be taken to carry out necessary periodical and mileage examinations of the locomotive. As previously observed, the X-Day scheme of British Railways has been devised to time these examinations at multiples of the day intervals at which boiler washouts are desirable for individual classes of locomotive. Moreover, the specified time and mileage intervals roughly coincide; thus, mileage examinations necessary at 500–600, 10,000–12,000, 20,000–24,000 and 30,000–36,000 miles can be carried out at the same time as periodical examinations of non-moving parts stipulated respectively for 3–5 weeks, 7–9 weeks, 11–15 weeks and 5–6 months.

Below is given a table showing some of the intervals at which various components are sheduled for examination.

Part to be examined	Remarks	Mileage or period
Superheater flues, elements	Express passenger locos	6–8 days
Firebox water spaces, tubes, etc.	With boiler empty on locos over 200 p.s.i.w.p. With boiler empty on locos below 200 p.s.i.w.p.	12–16 days 24–32 days
Ejectors	Holes in ejector and blower ring cleaned and examined	3–5 weeks
Gauge glasses	To be changed on engines over 200 p.s.i.w.p. To be changed on engines below 200 p.s.i.w.p.	3–5 weeks 7–9 weeks
Fusible plugs	To be renewed	7–9 weeks
Live steam injector, wheels and tyres	Examined, gauged and condition recorded	5–6,000 miles
Big and little ends	Except shunting engines and those with bushed ends Bush type	10–12,000 miles 20–24,000 miles
Crank axles	Except shunting engines with solid cranks	10–12,000 miles
Pistons, rods, cylinders, ports, blast pipe, etc.	Except engines solely on shunting work	30–36,000 miles
Piston valves	Except engines solely on shunting work	30–36,000 miles

Part IV. Steam engine terminals in the United States

The steam locomotive in the United States attained greater size and power than anywhere else in the world. On many Class One railroads, steam locomotives also attained higher mileage and availability figures than those achieved in any other country. There is no doubt that such intensive utilization was, in the later years of steam power, largely the result of the increasing competition offered by the diesel locomotive. Be that as it may, in order to achieve this high utilization, engine terminal facilities were developed to such a degree that it was possible for a large simple articulated locomotive to come off a train, be completely serviced, inspected, coaled, watered and turned, and be ready to move off the terminal, all in the space of some forty minutes.

Engine terminals (steam locomotive sheds or motive power depots) were almost invariably of the roundhouse type, with ten or more stalls radiating from a central turntable (Plates 29 and 114, pages 121 and 454).

On the railways of America, in earlier days, inspection, running repairs, maintenance and servicing were carried out, as they still are in many other countries, in the roundhouse. It is a sobering thought that only in recent years, and when other forms of motive power are displacing it, has any real effort been made to provide modern terminal facilities for the steam locomotive. All over the world, many engine sheds, engine terminals, *depots des machines* – call them what you will – have remained dark, dingy, dirty and draughty places. It was obvious, therefore, that if greatly improved turn-round times were to be achieved, new and more modern methods must be employed.

The achievements of rapid servicing and maintenance in the United States were made possible mainly by the building at engine terminals of special *servicing sheds*. This enabled the roundhouse to be kept for stabling and storage and for boiler washouts and repairs, especially those necessitating fixed plant such as lifting gear and wheel drops. It also kept the roundhouse free from the smoke, dirt and dust which are inevitable with servicing.

The servicing shed was a straight-through structure, typically containing two tracks, smoke extraction ventilators in the roof, and inspection pits between the tracks. The brick and concrete structure was centrally heated, and lined with glazed tiles which could easily be kept clean. Four main doors, electrically operated, could be opened for the ingress and egress of locomotives, but otherwise kept closed. Fluorescent lighting gave good general illumination in the shed, and frequently-spaced lighting fixtures in the pits and along the side walls enabled the whole locomotive to be accurately inspected.

The lubrication of a large American locomotive was an extensive and complicated procedure, no fewer than six different types of oil and grease being required. In the servicing shed all lubrication was carried out from sets of hoses. These sets were arranged, four on each side of each track, sixteen in all, spaced at intervals, so that the two units of a simple Mallet locomotive could be dealt with simultaneously. Each set comprised six hoses, each hose having a different nipple to correspond with the correct fitting on the locomotive. The lubricants supplied in this manner were:

(i) Valve oil for valves, cylinders, etc. (through mechanical lubricators).

(ii) Engine oil for driving-box shoes and wedges, crosshead guides and many other locations (also through mechanical lubricators).

(iii) Soft grease used mostly for valve gears.

(iv) Semi-fluid grease for roller bearings.

(v) Hard grease for side-rod lubricating cups; these cups are filled by air-operated grease guns, and the hose in this instance was an air-line.

(vi) Extreme-pressure oil for side-rod roller bearings, etc.

Supplies of lubricants were steam heated and pumped to the servicing shed from an oil distribution centre nearby. Normally ten minutes in the servicing shed sufficed for lubrication, inspection and minor adjustments to be efficiently carried out.

Outside the servicing shed were the *coaling and sanding plants*, the *washing platform, ash-disposal plant and stand pipes* (water columns). The locomotive was coaled from overhead hoppers and the sand boxes were filled in like manner and at the same time. Five minutes was the maximum time required.

At the ash-handling plant, the locomotive fires were cleaned and ashpans and smokeboxes were emptied direct into concrete hoppers between the rails. Streams of water kept dust to a minimum, and, mixing with the ash, carried it to a sump from which it was pumped up into a settling tank and the water drained off. The dry ash was then dumped into special vehicles for conveyance away from the site. While on the ash plant the locomotive was being washed with hot water pressurized sprays, and the tender tanks were filled from the stand pipes. These procedures occupied fifteen minutes.

After leaving the servicing plant the engine was turned and was then ready for its return working.

The whole of these servicing facilities could be used in either direction, continuously and progressively, according to the direction of working of the locomotive. Thus, a locomotive could go straight from its train, through the servicing shed, thence to the washing platform, ash disposal, water, sand and coaling plants, finally being turned; or the procedure could be reversed.

Engine terminals in the United States were of two classes: *turnaround terminals* and *maintenance terminals*. In the former were found the servicing facilities described, and sufficient machinery and plant to deal with such matters as the remetalling of bushes and axle-boxes and, in general, emergency running repairs.

The maintenance terminal, while having the facilities already listed, also had well-equipped machine shops and facilities for re-tubing a boiler or even re-wheeling a locomotive completely. The degree of maintenance and repairs carried out at maintenance terminals depended upon the policy adopted by each railway concerned.

Much saving of time out of service was made by the careful allocation of spare parts, so that renewable components could be fitted at *turnaround terminals* during, say, a monthly boiler washout, in the roundhouse.

Hot water boiler washout plant utilized the blow-off water from the locomotive, washed the boiler out, and re-filled the boiler with hot water, all of which procedures were operated by push-button controls. A boiler could be emptied, washed and refilled in a period of eight hours. Many roundhouses were equipped with facilities for the *direct steaming of boilers* – steam being piped from a stationary boiler, direct to the locomotive boiler. Thus the boiler was heated evenly before the fire was lit and many of the stresses set up by unequal expansion in the boiler plates and tubes were avoided.

In order to save time on repairs to the brick arch, or for small repairs to the inner firebox and flues, *asbestos suits* were provided to enable men to work in the firebox almost immediately after the fire had been dropped.

By the adoption of these and other modern servicing methods, many steam locomotives in the United States were running up to 25,000 miles a month, with out-of-traffic time reduced to as little as forty-eight hours, and sometimes less. Such utilization has never been equalled by steam locomotives anywhere else in the world, and is never likely to be so.

Despite it all, the steam locomotive in the United States is now little more than a nostalgic memory.

British Railways

PLATE 113A. Running shed of the straight-through type: British Railways motor power depot at Ipswich.

P. Ransome-Wallis

PLATE 113B. Round-house type of shed at Boulogne, S.N.C.F.

Norfolk and Western Railway

PLATE 114. Norfolk and Western round-house at Shaffers Crossing, Roanoke. The long buildings on each side of the picture are "Lubratoria" where high-speed lubrication and servicing are carried out. The round-house proper is in the background. The Norfolk and Western brought the servicing and maintenance of steam locomotives to the highest pitch of speed and efficiency of any of the world's railways.

P. Ransome-Wallis

PLATE 115. British Railways (Southern Region): coaling plant.

C.R.L. Coles

PLATE 116c. Manual removal of ash from smokebox. The ash is shovelled from the smokebox into the small hoppers in the pit. The hose keeps the dust damped down.

The late S.C. Townroe

PLATE 116a. Emptying a hopper ash-pan.

The late G.F. Burtt

PLATE 116b. Washing out a locomotive boiler.

PLATE 116. Work at a Motive Power Depot

P. Ransome-Wallis

PLATE 117A. Sentinel 150 H.P. railcar for the L.N.E.R.

P. J. Bawcutt

PLATE 117C. Sentinel 2-2-2-2 passenger locomotive for Egyptian State Railways. The two engine units are independently geared to the second and third axles.

P. Ransome-Wallis

PLATE 117B. Sentinel 100 H.P. shunting (switching) locomotive for the L.N.E.R.

Western Maryland Railway

PLATE 118A. Shay locomotive – engine side.

Western Maryland Railway

PLATE 118B. Shay locomotive – non-powered side.

British Railways

PLATE 119A. Midland Railway: Paget multi-cylinder locomotive.

C.C.B. Herbert

PLATE 119B. Southern Railway: "Leader" class multi-cylinder locomotive, No. 36001.

Coras Iompair Eireann

PLATE 119C. Coras Iompair Eireann: peat-burning multi-cylinder locomotive, No. CC.1.

S.N.C.F.

(B)

PLATE 120A. Southern Railway 4–4–2 No. 2039 *Hartland Point* experimentally fitted with sleeve-valve cylinders before such cylinders were fitted to the new "Leader" class locomotives (*see* plate 119B).

PLATE 120B. S.N.C.F. Chapelon 6-cylinder compound locomotive No. 160, A-1. This locomotive had two high-pressure and two low-pressure cylinders between the frames and two low-pressure cylinders outside. It was equipped with two superheaters.

PLATE 120C. The Kitson-Still locomotive.

P. Ransome-Wallis

(A)

Locomotive Publishing Co.

(C)

CHAPTER 9

Unconventional Forms of Motive Power

by P. RANSOME-WALLIS

In this section an attempt is made to review briefly some of the many different forms of railway locomotives which have been invented in an attempt to improve upon the fundamental Stephenson concept (*see* Chapter 6, Part I). This relates, of course, to steam locomotives, but also included in this section are locomotive experiments in other fields which have not, anyway as yet, become standard forms of motive power. The gas turbine is the largest and most important competitor to come under this heading and, indeed, it bids fair soon to become a well-established form of railway locomotive and no longer to be classed as "unconventional".

To cover anything like the whole subject of unconventional motive power would require a large volume in itself and the writer is conscious of many important omissions. This is particularly true of the many interesting and unusual locomotives which have been built for, and used successfully on, industrial and agricultural railways. Space, and space alone, prevents their inclusion not only in this section but indeed in this volume.

Experimental devices used on conventional locomotives (e.g. special forms of feed water heaters) are not dealt with here as they do not constitute an unconventional form of motive power in the exact sense.

It is a tribute to the simplicity and soundness of the original Stephenson concept that few, if any, of the many designs of unconventional steam locomotives have proved lastingly successful. Their inventors have usually fixed their eyes firmly on the low overall thermal efficiency of the Stephenson engine and in trying to save a ton of coal have neglected the far more important factors of simplicity of design, reliability in traffic, and ease and economy of maintenance and repair. For this reason, many of the locomotives here described are mainly of historic interest. Their inclusion in a review of current motive power was considered justified, and indeed necessary, in order to present as complete a picture as possible of why the locomotive of today is what it is.

Part I. Multi-cylinder steam locomotives

Conventional locomotives have, in the past, been built with up to six cylinders (Triplex Mallet type, and S.N.C.F. 160 A.1) (Plate 120B, page 460), while three- and four-cylinder locomotives are in common use. In an effort to provide more even torque and tractive effort characteristics three and more cylinders have been used in conjunction with some form of gear drive, while in other designs six or eight small cylinders have been used in conjunction with direct drive. While gear-driven steam locomotives have still a limited application, multi-cylinder direct-drive locomotives had faded from the railway scene until the introduction in 1958 of the six-cylinder Bulleid peat-burning locomotive for Coras Iompair Eireann.

Some of the more important locomotives in both categories are described briefly.

RECIPROCATING STEAM LOCOMOTIVES WITH GEAR DRIVE

The use of gears has never been accepted in locomotive practice with equanimity. Nevertheless some very successful and useful geared locomotives are in service to this day.

Geared locomotives are generally of the multi-cylinder type, as in this combination excellent torque characteristics can be achieved by the use of reduction gearing and the locomotives are specially suited for shunting and for duties where a high tractive effort with even torque is required at very low speeds.

Geared locomotives fall into two main classes:

(i) Those in which spur gears are employed, drive being through a jackshaft and side rods, or by gear drive direct on to the axle, or by chain drive. Many different types of such engines have been built, especially in the

United States. Probably the best and most efficient of all such locomotives is the Sentinel Patent Locomotive in which the final drive may be through gear wheels or through chains.

(ii) Those in which the engine(s) are mounted at right angles to the axle and drive is taken through shafts and bevel gearing to the axles. By far the most important type in this category is the Shay.

The Sentinel Patent locomotive. The Sentinel Wagon Co., Ltd, of Shrewsbury, England, have for many years produced a wide range of highly successful geared steam locomotives for shunting (switching) duties and for railcars. They have been used extensively for industrial duties and also on the main line railways of the world. Main line locomotives having six two-cylinder engine units, and capable of hauling 200-ton trains, have also been built.

Sentinel locomotives vary in size and gear ratios according to the duties they are required to perform, and the modern unit is of the 0-4-0T or 0-6-0T wheel arrangement.

A vertical water-tube rapid-steaming boiler is provided which **may** be coal fired, but preferably is fired automatically by oil fuel. Steam can be raised from cold in forty minutes, and working pressure is 275 p.s.i.

The engine consists of two vertical two-cylinder double-acting units with poppet valves; normal maximum speed is 500 r.p.m., and the cylinders are $6\frac{3}{4}$ in. diameter by 9 in. stroke.

The drive is through reduction gearing to a lay shaft from which chains and sprockets transmit the power to the axles. Sufficient fuel and water is carried for up to sixteen hours steaming without replenishment. Power/weight ratio compares very favourably with diesel locomotives of similar size (Plates 117A, 117B, and 117C, page 457).

The Shay locomotive. A geared articulated locomotive having vertical cylinders (usually three in number) mounted on the frames on one side only and necessitating the boiler's being offset. The drive is taken through a horizontal driving shaft having flexible joints and being geared individually to the ends of the axles by reduction gearing of the bevel type. Boiler cab, tanks and bunker may be all on one frame, or a tender may be used, in which case the drive is extended to the tender axles.

The Shay locomotive was invented by Ephraim Shay, a lumberman, in 1880, the first engine of the type having a vertical boiler, two cylinders and two bogies. Shay locomotives have the merit of being able to travel over very rough tracks and to negotiate very sharp curves. They are used in lumber haulage very successfully.

The Western Maryland engine illustrated (Plates 118A, 118B, page 458), was used to push loads of 200 tons at 5 m.p.h. up a short colliery branch having a ruling gradient of 1 in 11 and curves of 310-ft radius. It could go anywhere that a standard freight wagon (car) could go. The gear drive gave great holding power and acceleration on the steep gradients. The colliery has now been closed and the engine is preserved.

It has the following particulars:

Driving wheel diameter	4 ft 0 in.
Cylinders (3)	17 in. × 28 in.
Working pressure	200 p.s.i.
Tractive effort (85 per cent W.P.)	59,740 lb.

MULTI-CYLINDER STEAM LOCOMOTIVES WITH DIRECT DRIVE

The Paget locomotive (1908). This locomotive was conceived by Mr (later Sir) Cecil Paget, when Works Manager at Derby, Midland Railway. It was built in the company's works during 1907–8. This engine, a 2-6-2, had eight single-acting trunk piston cylinders placed in two blocks of four cylinders each between the frames, and between the first and second, and second and third coupled axles, respectively. The outer two cylinders of the first block drove on to the leading coupled axle. The inside two cylinders of the first block and the outer two of the second block drove the middle coupled axle, while the inner two cylinders of the second block drove the third coupled axle. Thus the first axle had two lateral cranks, the middle axle four cranks, and the third axle two medially-placed cranks. The frames were outside the driving wheels in order to allow room between the frames for the cylinders. Steam distribution was by rotary valves and exhaust occurred through ports in the cylinder liners. Drive for the valves was through a longitudinal rotating shaft, driven from a gearbox which also contained gear mechanism for reversing the engine. The gearbox drive was from a transverse shaft driven by a crank and a backwards extension of the outside coupling rods.

The boiler was of the firetube type but was unconventional, in that in order to avoid the high cost of a stayed copper firebox, the dry-back type was adopted. There were no water-legs, back and sides of the firebox being of firebrick. Two fireholes were provided for ease of firing a grate of fifty-five square feet. Working pressure was 180 p.s.i.

This locomotive came very near to success and proved itself well able to travel up to 80 m.p.h. (with but 5 ft 4 in. wheels), with a full load. Main difficulties were steam leakage past the pistons, and trouble with the rotating valves. The sleeves of the valves were of phosphor bronze, rotating in a cast-iron tunnel. The differential expansion between the two metals caused much trouble with the valves seizing up. The boiler, although an excellent steamer, gave trouble with firebricks cracking and breaking – an experience which has been shared by all who have designed dry-back boilers for locomotive work (Plate 119A, page 459).

The engine was finally broken up during World War I.

The Henschel 1–Do–1 locomotive (1941). During the early years of World War II the firm of Henschel found time to complete a large streamlined experimental locomotive with individual axle drive. The four driving axles had inside axle-boxes and situated outside one wheel on each axle was a V-type two-cylinder steam engine. The engines were staggered – two units on each side of the locomotive. They each had two cylinders of an equal diameter and stroke of 11·85 in. Valves were of the piston type, the eccentrics being gear driven from the main axle. There was no link, the engines being reversed by means of bevel gearing operated from the cab. The drive to the axle was through linkage of a type similar to that used for electric locomotives, the crankshaft of the engine being in line with the axle it drove.

The boiler was standard with that of the Class 44, 2–10–0 with a working pressure of 285 p.s.i.

The results of trials with this locomotive are not known. After the war it was seized by the Americans and taken to the United States. Its later history and fate are unknown.

The Southern Railway "Leader" class (1948). Following upon his unusual "Pacific" designs for the Southern Railway ("Merchant Navy" and "West Country" classes), Mr O.V.Bulleid designed a double-ended locomotive which was intended for fast main-line service. The cab and controls were duplicated. The locomotive followed the design of the Paget locomotive in that it had a dry-back boiler and was of the multi-cylinder type.

The locomotive consisted of a single unit, the boiler, coal bunker and water tanks being carried on a welded girder frame structure supported by two six-wheel bogies. The bogies had no centre pivot or bolster and were similar to those used on electric locomotives.

Three cylinders, $12\frac{1}{4}$ in. by 15 in., formed an enclosed engine unit on each bogie, there being thus six cylinders in all. Steam distribution in each cylinder was by means of a sleeve valve which not only reciprocated but was partly rotated (through 30°) during its travel. The sleeve was arranged so that it covered and uncovered steam ports in the cylinder wall, the design being very reminiscent of the Uniflow. Steam was admitted when ports in the sleeve coincided with steam ports in the cylinder, and exhausted when similar exhaust ports were uncovered. The valve gear was a radial one, similar in computation to Walschaerts, but being driven through chains.

Each three-cylinder engine formed a totally enclosed unit, the drive to the axles being by chains, which also coupled the three axles in each unit. Thus the whole weight of the locomotive, 131 tons, was available for adhesion. The driving wheels were 5 ft 1 in. diameter and the tractive effort was 26,300 lb., at 85 per cent working pressure. T.I.A. water softening treatment was provided.

The boiler was of the fire-tube type but without water legs, steel plates lined with firebrick forming the back and sides of the firebox. The crown was, however, covered by water, and the firebox contained four thermic syphons. The firehole was offset in the back-plate and the centre-line of the boiler was offset from the centre-line of the locomotive, in order to give more room to the fireman. Working pressure was 280 p.s.i. The boiler was of all-welded construction, no rivets at all being used.

This locomotive was given many trials from Brighton Works, where it was built. Many troubles developed. The first was a human one, the small compartment in the middle of the engine allocated to the fireman proved far too small for the normal swing of the shovel, and became far too hot for comfort. As on previous occasions when they have been used, the firebricks lining the firebox cracked, broke and fell down.

Trouble developed in the cylinders, for steam leaked past the sleeves and lubrication of the sleeves proved difficult and they seized up in the cylinders. The chains of the valve gear stretched, and valve events were uncertain.

As a result, the "Leader" was withdrawn from service in 1951 and four sister engines, two of which had been laid down at Brighton, were not proceeded with.

It is of interest to note that the "guinea pig" locomotive for the "Leader" was 4–4–2 Class H-1, No. 32039, a Marsh Atlantic of 1905. She ran with sleeve valve cylinders and a multiple jet blast pipe, from 1947 until being broken up (Plates 119B and 120A, pages 459 and 460).

Coras Iompair Eireann peat-burning locomotive. (1958). On leaving the Southern Railway after nationalization, Mr O.V.Bulleid became Chief Mechanical Engineer of the railways of Eire.

Coal in Ireland is expensive and has to be imported, and the railways have a comparatively small amount of traffic. Utilization of the country's large supplies of peat as a locomotive fuel is, therefore, a very attractive proposition, and for many years experiments with peat-burning boilers have been carried out at Inchicore, both in stationary plant and in locomotives. The experiments came to fruition in 1958, when just before he retired, Mr Bulleid's peat-burning locomotive was completed at Inchicore.

A full description of this interesting locomotive is not available at this time, but it has many resemblances to the Southern Railway "Leader" class. The boiler, two peat bunkers and water tanks are mounted on a girder frame supported on two six-wheel bogies. The boiler is of the firetube type with a large dry-backed square firebox having no water legs. Two mechanical stokers are adapted to feed peat into the firebox from the two bunkers. The working pressure is 250 p.s.i.

The combustion characteristics of peat make the intermittent blast of the reciprocating locomotive unsuitable, and two turbine-driven fans – one at each end of the loco-motive – provide draught for the boiler. The hot gases from the boiler are led through pre-heaters to the water tanks (again one at each end of the locomotive), before exhausting to atmosphere.

Two totally-enclosed, three-cylinder, piston-valve, high-speed steam engines are mounted one on each bogie, and drive the trailing axle of each bogie through single reduction gearing, a jack-shaft and chains. The axles of each bogie are also coupled by chains. Driving controls are duplicated in two cabs, one at each end of the middle section of the locomotive (Plate 119C, page 459).

Part II. Steam locomotives using very high pressures

It has long been appreciated that, following marine and stationary practice, very high steam pressures can produce both economies in fuel and water consumption, allow smaller cylinders to be used, and also permit compounding and even triple expansion to become a real advantage.

Nonetheless, locomotives using very high pressures must be classed as unconventional in that the Stephenson form of boiler is not used, though in most cases the automatic action of the Stephenson locomotive has been preserved.

When pressures above 300 p.s.i. are used, the conventional Stephenson type locomotive boiler becomes unsuitable and various other types are adopted. Over the years many forms of boiler have been tried, most of them involving the use of water tubes, though it would be entirely wrong to suppose that the use of water tubes has been confined only to very high pressure boilers.

The use of such boilers has been exploited by the railways of several countries, and locomotives of this type have been built, notably in Germany, Great Britain, France, Switzerland, Canada and the United States.

It is convenient to divide very high pressure boilers into four main classes:

(i) Those in which a purely water-tube boiler is employed. Instance of this was the L.N.E. 4–6–4 compound locomotive No. 10000 (1929). Working pressure was 450 p.s.i., and the boiler was of the five-drum Yarrow type (Plate 121B, page 479).

(ii) Those in which a water-tube firebox is combined with a fire-tube boiler. Instance of this was the Delaware and Hudson four-cylinder, triple expansion 4–8–0, No. 1403, "L.F.Loree" (1933), with a working pressure of 500 p.s.i. (Plate 121A, page 479).

(iii) Those in which three separate pressure systems are used:

(a) a primary closed circuit of water tubes and drums in which steam is generated, at 1,400–1,600 p.s.i.

(b) a high-pressure drum in which steam is raised to 800–900 p.s.i. by heat transfer from the tubes of the primary circuit. This boiler supplies steam to the high-pressure cylinder(s) of the locomotive.

(c) a low-pressure fire-tube boiler, supplying steam at 200–250 p.s.i., to the low-pressure cylinders of the locomotive, such steam being usually mixed in a receiver with exhaust steam from the high-pressure cylinders.

This is the Schmidt-Henschel principle and boilers of this type were fitted to such locomotives as the L.M.S. three-cylinder compound 4–6–0 No. 6399 "Fury" (1929), Canadian Pacific 2–10–4 No. 5905 (1930), and the P.L.M. four-cylinder compound 4–8–2 No. 241. B.1 (1930), though in this instance steam was used in the H.P. cylinders at the reduced pressure of 207 p.s.i.

(iv) Those in which steam is raised in a series of tube circuits, there being no boiler in the accepted sense. Of such a type was the Löffler system and it was applied in 1930 to a 4–6–2 locomotive of the German State Railways. This locomotive had two high-pressure cylinders and one low-pressure.

In 1936 a P.L.M. locomotive was fitted with a Velox boiler, but as with all high-pressure systems the additional complication involved proved entirely unsuitable for locomotive work. Draughting has also always been a difficulty in these experimental boilers, particularly was this so in the case of the Yarrow-boilered L.N.E.R locomotive in which great difficulty was experienced in keeping the combustion spaces airtight.

No locomotives using very high pressures are at work today.

Part III. Steam turbine driven locomotives

The reciprocating steam locomotive, rugged, flexible and reliable though it is, suffers from three major disadvantages, inherent in its design and conception:

(i) It has a low over-all thermal efficiency, this being of the order of 10 to 12 per cent in well-designed engines (*see* Chapter 5, Part I, Thermal Efficiency).

(ii) It imparts to the rail a severe dynamic augment, or hammer blow, caused by the balancing of the reciprocating masses (*see* Chapter 5, Part I, Masses). This factor, probably more than any other, limits the size and power of the reciprocating railway locomotive.

(iii) The torque developed at the wheel tread is not even, varying from zero at the end of the stroke to a maximum at mid-stroke.

(iv) It requires a supply of good water to be obtainable at convenient points along the route over which it works. This presents little difficulty in most European countries, but in countries where long hauls have to be made over desert and arid lands or areas where, in winter, water supplies become frozen, it becomes a major problem, and may involve the inclusion in the train of extra water-carrying vehicles. This may mean that some 10 per cent of the train tonnage is taken up with a non-paying load.

The success of the diesel in railway motive power has provided the answer to all these problems, but has in some few cases raised a new one. Many railways have been built to serve areas rich in coal, and against the advantages of the diesel must be offset the increased cost of oil as fuel, where coal is available and cheap.

In such areas recently, and on many railways before the advent of the diesel, the steam turbine as a means of motive power held undeniable attractions as an answer to one or more of the major disadvantages of the reciprocating locomotive. Many experimental engines have been built, several achieved a large measure of success. None are, as far as is known, still in operation on main-line railways.

Such locomotives fall into four main groups:

(i) Condensing turbines with electrical transmission.
(ii) Condensing turbines with mechanical transmission.
(iii) Non-condensing turbines with electrical transmission.
(iv) Non-condensing turbines with mechanical transmission.

CONDENSING TURBINE LOCOMOTIVES WITH ELECTRICAL TRANSMISSION

The idea of obtaining the undoubted advantages of electric traction without the high capital expenditure of fixed plant and power lines, is one which has for long appealed to locomotive designers. The conception of a mobile power plant with a steam engine as the power generator first came to fruition in 1894 when a small unit using an opposed piston reciprocating engine was built by Brown, Boveri of Baden. Two years later a much larger locomotive was built by the same firm, and this may be regarded as the first practical locomotive using electrical transmission. It was designed by a French engineer, Robert Heilmann, and power was derived from a high-speed, six-cylinder vertical steam engine developing 1,350 H.P. at 400 r.p.m. This engine was direct coupled to two D.C. electric generators – one at either end of the crankshaft, and each of 450 kW capacity. A separate two-cylinder steam engine drove a dynamo for auxiliaries and for excitation of the main generators. A locomotive-type boiler provided steam at 185 p.s.i.

The Heilmann locomotive ran on two eight-wheeler bogies, each axle of which was driven by an electric motor of 125 H.P. through the medium of compression rings.

The next venture in steam electric propulsion came in 1910 and was a condensing turbo-electric locomotive.

The Reid-Ramsay turbine-electric locomotive (1910). This locomotive was built by the North British Locomotive Co., and was of unusual wheel arrangement, being carried on two eight-wheel bogies. The bogies each comprised two carrying axles, themselves mounted as a bogie, and two driving axles. Four D.C. series-wound traction motors were constructed with their armatures built on the four driving axles. Current at 600 v. was supplied by a generator driven by an impulse turbine running at 3,000 r.p.m. Steam was generated by a locomotive-type boiler with superheater, and the exhaust steam was led to an "ejector condenser" for which cooling water was supplied by the water in the tanks of the locomotive. A turbine-driven fan provided forced draught for the boiler. There is little information available about the results achieved with this locomotive (Plate 123A, page 481).

The Ramsay turbine-electric locomotive (1920). This was the second British enterprise and consisted of a two-framed locomotive having the wheel formation 2–6–0+0–6–2. A locomotive-type boiler supplied steam at 200 p.s.i., and 300° superheat, to a multi-stage impulse turbine. The steam was then led to a condenser, the tubes of which were rotated slowly through cooling water, and further cooled by a fan.

The electric generator was flexibly coupled to the turbine, and three-phase A.C. was supplied to four A.C. traction motors, each of 275 H.P. The motors were grouped in pairs, one pair driving each set of coupled wheels through reduction gearing on to a jackshaft, which drove the wheels through coupling rods.

This locomotive developed, at the rail, under 1,000 H.P. – far too little for operating on British main-line railways; furthermore, with a gross weight of 155 tons, the power/weight ratio was abnormally high. At the date at which it was built, A.C. traction motors were far from ideal for railway motive power. The condenser leaked badly and much water was lost, and the boiler was a poor steamer. The engine was scrapped in 1924 (Plate 122C, page 480).

Union Pacific R.R. 4–6–0+0–6–4 units (1938). In many ways this locomotive anticipated much that was soon to become standard in diesel practice in the United States. Built by the General Electric Company, this locomotive consisted of two identical units, each of the 4–6–0+ 0–6–4 type, each developing 2,500 H.P., and being independent in itself. Thus the locomotive could be run as two separate units, or coupled together to form a single locomotive of 5,000 H.P., operated by a form of multiple unit control.

In each unit, steam was generated in a semi-flash type of water-tube boiler, the capacity of which was only 100 gallons water. The working pressure was 1,500 p.s.i., and automatic oil firing was used, the rate of combustion being controlled by the steam demand. Power was generated by a cross-compound turbine, the H.P. and L.P. turbines meshing individually with a single helical gear wheel; the gear ratio was approximately 10 : 1. The turbine speed was 12,500 r.p.m., and the D.C. electric generator speed 1,200 r.p.m. The boiler and generating plant were carried in the first part of each unit and the second part of each unit carried a vertical air-cooled condenser having four large fans.

The generator supplied current at 700 v. to six traction motors on each unit, suspended one on each of six driving axles and geared to them by single reduction gearing.

The complete two-unit locomotive weighed 550 tons of which 340 tons was available for adhesion. Starting tractive effort was 162,000 lb., and 13,000 lb. at the maximum speed of 110 m.p.h. The locomotive ran trials on the Union Pacific, Great Northern, and New York Central Railroads and showed considerable promise. Further development and trials were arrested by World War II, and later the wholesale conversion to diesel-electric traction spelled the doom of a most interesting and justifiable experiment.

CONDENSING TURBINE LOCOMOTIVES WITH MECHANICAL TRANSMISSION

Before World War II considerable research was expended on this type of motive power and a number of locomotives were built in Britain, Germany and Switzerland. The later engines achieved some measure of success, although the gain in thermal efficiency was slight, the saving in fuel often negative, but the saving in water was as high as 80 per cent over the reciprocating locomotive. Increased first cost and high maintenance more than offset any other advantages. Failures were mostly due to the condensers being unable to maintain sufficient vacuum, and to the boilers being poor steamers, doubtless due to draughting problems.

Some of the more important engines are briefly reviewed here.

The Zolly turbine locomotive (1921). This engine was built by the Swiss Locomotive Works and used a Zolly impulse turbine located in front of the smoke box, a separate turbine being used for reverse running. The main turbine developed 1,200 H.P., and drove the coupled wheels through reduction gearing, a jackshaft and coupling rods. Two surface condensers placed one on either side of the boiler, used cooling water from the tender. The boiler, of the locomotive type, had a working pressure of 170 p.s.i.

The Krupp turbine locomotive (1922). This locomotive was developed contemporaneously with the previously-mentioned Swiss locomotive and was a development of it. It differed mainly in the condenser which, located in the tender, took the form of a "locomotive tower" and employed split tubes giving a very large evaporative surface. It used three auxiliary turbines in addition to the main turbine and the reverse turbine. These drove a condenser fan, a draught fan, and the third drove a dynamo for lighting, and other auxiliaries. The reverse turbine remained in mesh when the engine was running forwards and absorbed 410 H.P. – a severe handicap. The draught fan used exhaust steam and its output therefore was proportional to the demand for steam in the main turbine.

A somewhat similar but more successful turbine locomotive was built by Maffei in 1921 and ran successfully for many years on the German Railways in express train service (Plate 123B, page 481).

The Ramsay Macleod turbine locomotive (1924). This locomotive was a reconstruction by the North British Locomotive Company of the earlier Reid-Ramsay Turbo-Electric Locomotive (q.v.). It was a single-unit machine having no tender, the boiler, condenser water tanks and bunker being mounted on a single frame carried on two bogies. Each bogie had a two-axled guiding truck and two driving axles. The wheel arrangement was 4–4–0+0–4–4.

The drive was by two 500 H.P. turbines, one mounted on each bogie and driving through double reduction gearing on to the driving axles. The two-stage (or compound) principle was employed, the H.P. turbine being on the trailing bogie, and the L.P. on the leading bogie. The boiler was of the locomotive type with a working pressure of 175 p.s.i., using draught from a turbine-driven fan.

The condenser, placed at the leading end of the locomotive, was of the air-cooled evaporative type with a turbine-driven fan supplementing the induced draught due to the progress of the locomotive. The air was cooled by a water spray, which also sprayed on to the condenser tubes. This locomotive developed only 1,000 H.P., and this was insufficient for normal traffic requirements. It was also a very low figure when the high weight, 135 tons, of the locomotive was considered (Plate 122B, page 480).

The Ljungstrom turbine locomotives (1924–28). These four locomotives were forerunners of the non-condensing turbine locomotives designed by the same firm in later years. They had reaction turbines developing 1,800, 2,000 (2) and 2,200 H.P., in their respective engines. The single turbine drove three driving axles through double reduction gearing. Reversing was effected by the insertion of an idler gear in the main gear train.

The boilers in all four engines worked at the, for then, high pressures of 280 to 300 p.s.i., which partly contributed to the compact design of the turbines. The boiler was carried on a separate vehicle from the turbine.

The Ljungstrom condenser was of the air-cooled type carried on and occupying most of the space on the same vehicle as the turbines. It employed three turbine-driven fans and the steam was condensed in flattened copper tubes which formed a "roof" over the whole condenser vehicle.

The four locomotives were supplied as follows:

(i) To the Swedish Railways in 1921. It was rebuilt in 1922 with a larger turbine and withdrawn in 1924.

(ii) To the Argentine State Railways (metre gauge) in 1925. This engine was oil-fired. It was required to work in an area of scarce and bad water and used only 4 per cent of the water used by an ordinary locomotive. It was, however, too complicated for operation by the local enginemen, and costly in maintenance. It was withdrawn in 1931 being replaced by a Henschel reciprocating 2–8–2 locomotive with a condensing tender in which no vacuum was used.

(iii) For the L.M.S. Railway in Britain in 1926. This locomotive ran trials during two years on the Midland main line. Apart from a large economy in water, it showed no advantage over a Hughes 2–6–0 of Class 4 against which it was tried (Plate 122D, page 480).

(iv) For the Swedish State Railways in 1927. This engine was similar to that supplied to the L.M.S. but it embodied many detail modifications and was in service until after 1939.

NON-CONDENSING TURBINE LOCOMOTIVES WITH ELECTRICAL TRANSMISSION

The earlier attempts in locomotives to use turbines with electrical transmission had, in common with those using mechanical transmission, always incorporated condensers. It is well known that the efficiency and power of a turbine is only fully realized when the exhaust steam is led to an efficient condenser. Condensers, however, have seldom proved able to stand up to the hard usage, fluctuating demands, and severe vibration experienced in steam locomotive practice. Used purely as a means of saving water and not essentially as a means of increasing the efficiency of the steam engine, they have had some success when carried in a separate tender and used in conjunction with reciprocating locomotives, notably in Russia and South Africa (q.v.). In turbine locomotive practice, they have usually proved to be the Achilles heel of the design and the more recent experiments with turbines in this field have made use of non-condensing turbines.

The two outstanding modern examples of non-condensing turbo-eletric locomotives are both American.

Chesapeake and Ohio R.R. No. 500 (1947). The Baldwin Locomotive Works supplied to the C. & O. R.R., the world's largest passenger locomotive, and two more of similar design followed early in 1948. A coal-burning locomotive type boiler with mechanical stoker supplied steam at 310 p.s.i., to an impulse turbine developing 6,000 H.P. at 6,000 r.p.m. This turbine was coupled, through single reduction gearing, to two generators of 2,000 kW. capacity, generated current being at 580 v. pressure. The two main generators were each in two units of 1,000 kW. capacity and each of the four units supplied current to two 620 H.P. traction motors connected in parallel. These eight motors were axle hung and drove the axles through single reduction gearing.

All axles ran in roller bearings, and the whole of the lubrication for the engine and chassis was automatic. Coal was carried in a bunker in the front end of the locomotive, and water in a tender at the rear end. The boiler occupied the middle section and turbine and generators the after part of the locomotive – all mounted on a cast steel engine bed.

This huge machine weighed 594 tons of which five-eighths was available for adhesion. The wheel arrangement was 4–8–0+4–8–4, or perhaps more correctly 4–8–2+4–6–2+0–4–0.

Despite the obvious desire of this railroad to use the large supplies of cheap coal available in the area which it serves, the purchase of three enormous and expensive machines of completely untried design gives rise to a certain wonderment. In the event, these engines were failures, expensive to run and expensive to maintain, and they were scrapped three years after being put into service.

Norfolk and Western R.R. No. 2300 (1951). Four years after the ill-starred C. & O. engines had gone to the scrap heap, a single non-condensing turbine-electric locomotive was built by Baldwin-Lima-Hamilton in conjunction with the Westinghouse Corporation and Babcock & Wilcox. The reason for this locomotive was much the same as that which had decided the C. & O. to experiment with this form of power seven years earlier. No. 2300 was, however, designed for freight service.

The N. & W. locomotive came nearer to power-station practice than perhaps any other locomotive in that the Babcock & Wilcox boiler was of water-tube type and equipped with a chain grate for the first time in locomotive history. Working pressure was 600 p.s.i. Coal was fed to the grate by a Standard mechanical stoker; the rate of firing, the rate of air flow and the feed water supply were controlled automatically by the demands for steam made upon the boiler. Draughting was by air from a turbo-blower so that the turbine exhaust was not impeded by blast arrangements and back pressure was nil.

The power plant consisted of an impulse turbine developing 4,500 H.P. at 8,000 r.p.m. A single generator was driven through reduction gearing of ratio 8·9 : 1. Dynamic braking equipment was also driven off the generator shaft. There were twelve D.C. traction motors, one to each axle, so that the whole of the 586 tons weight of locomotive was available for adhesion. Tractive effort was 144,000 lb. at 9 m.p.h., and driving wheels were 3 ft 6 in. diameter. Timken roller bearings were fitted to all axles, and the traction arrangement and equipment followed standard diesel-electric practice.

Coal was carried in a bunker in the front part of the locomotive and water in a tender. Water softening was provided and continuous blow-down of the boiler was arranged.

This locomotive, which was the largest single-unit locomotive in the world, was in revenue-earning service for four years before being withdrawn for scrap in 1958. It showed a marked economy in coal consumption, some 30 per cent, over comparable reciprocating steam

power, but the automatic controls referred to developed faults and the company stated that "electrical, turbine and feed-water components" gave much trouble and caused excessive delays and heavy maintenance costs. The fact that the engine was an individual example and that no similar engines were built for other companies no doubt played a large part in keeping initial costs high and preventing further development. This, against the standardized reliability of diesel-electric traction, must have prejudiced any chances of success which this locomotive might have had.

It is probable that No. 2300 was the last of the great experiments in motive power using steam as a prime mover. Vast sums of money have been spent in an effort to improve upon the reciprocating steam locomotive. Some have come very near to success, but all have in the end been abandoned (Plate 124C, page 482).

NON-CONDENSING TURBINE LOCOMOTIVES WITH MECHANICAL TRANSMISSION

Belluzzo's engine (1907). A small experimental 0–2–2–0T locomotive was built by Signor Belluzzo in Italy in 1907. This locomotive was the first to be constructed to use steam turbines. (A locomotive built in 1834 in America used rotating cylinders and was not a true turbine.)

Four turbines were fitted to this Italian locomotive, two on each side. Steam passed through all four turbines in turn before being exhausted to atmosphere up the chimney. The turbines drove the four wheels independently through gear wheels (Plate 122A, page 480).

Grangesberg-Oxelosund (Sweden) 2–8–0 locomotive (1932). The Ljungstrom Company designed, and Nydquist & Holm built a successful freight locomotive in which the 2,000 H.P. turbine was placed ahead of the smokebox and drove the coupled wheels through triple reduction gearing, a jack shaft and coupling rods. Reversing was carried out by a gear change operated by a screw and crank.

The boiler had a working pressure of 185 p.s.i., and was standard with that of three-cylinder simple reciprocating locomotives of the same type. Tractive force was 47,400 lb., at 85 per cent of the working pressure.

This locomotive was simple and robust, and a saving of fuel of 10 per cent over comparable reciprocating locomotives was claimed. Maintenance costs were low.

Stanier 4–6–2 turbomotive for the L.M.S.R. (1935). The success of the Swedish locomotive described above led to further consideration of a non-condensing turbine express locomotive for main-line duties on the L.M.S.R. in Britain. In 1933 the first Stanier four-cylinder "Pacifics" were put into service. An initial order for three such

Swiss Federal Railways

PLATE XV. Brienz–Rothorn Railway (Switzerland): 0–4–2T locomotive, 800 mm gauge operating on the Abt rack system. This mountain railway is 4.7 miles long and has a maximum gradient of 1 in 4 (25 per cent).

locomotives was amended, two were built, and the frames, boiler and other details of the third engine were used in the construction of a strictly comparable turbine locomotive No. 6202. The Ljungstrom Company and Metropolitan Vickers collaborated in the design, the latter company supplying the turbines and gearing.

Unlike the Swedish engine, there were two turbines, a main one of 2,600 H.P. for forward running and a smaller turbine for reverse running. The main turbine was permanently geared through double helical triple reduction gearing to the leading coupled axle. The reverse turbine drove through an additional reduction gear which was out of mesh except when backwards running of the locomotive was required. Turbines and gearing were lubricated by a forced feed closed circuit actuated by three oil pumps.

The boiler was of dimensions standard with that of the "Princess Royal" class locomotives with which it was interchangeable. The working pressure was 250 p.s.i. It had initially, however, a thirty-two-element superheater and, later, one with forty elements. A double blast-pipe and chimney were fitted, and a tubular feed water heater was provided in series with an exhaust steam injector.

Roller bearings were fitted to all carrying and driving wheels of engine and tender.

Despite the originality of the design No. 6202 gave excellent main-line service for fifteen years after its introduction in 1935. It had of course many failures, especially with the small reverse turbine. This was too small to push a train back out of Euston Station, London, up Camden Bank to the carriage sidings, and even after a larger reverse turbine was fitted, a powerful shunting engine was provided to deal with trains worked into the terminus by No. 6202.

Throughout the life of the locomotive the main turbine failed only twice, though in both cases these failures were severe and costly. In one case the main turbine spindle fractured while the engine was travelling at high speed. This resulted in the rotor fouling the stator diaphragm and required a complete reconstruction of the turbine. Such mishaps kept the engine out of traffic for long and expensive periods, a state of affairs which could, of course, have been greatly alleviated had spares been immediately available, as would be the case with a standard locomotive.

One of the great difficulties with single experimental locomotives when they are put into main-line service is the fact that in the link to which they are allocated they may be driven by any one of, say, forty sets of men. Ideally, three or four sets of men specially instructed in driving methods applicable to the new machine would be the fairest way of assessing its performance and main-

tenance requirements. From the operating angle this would be very difficult; from the trade union angle it is impossible. For much of her life No. 6202 ran with a trained C.M.E. fitter as third man on the footplate. Later on, and especially after nationalization, this practice was dropped, failures increased, and the engine was withdrawn from service in 1951.

During its career No. 6202 ran many trials against locomotives of the "Princess Royal" and "Coronation" classes. The turbomotive showed little or no gain in thermal efficiency over the modern well-designed reciprocating locomotive. There was, however, some small economy in the consumption of coal and water, but this was obviously insufficient to offset the higher initial cost of the locomotive. It is perhaps unfair to try to assess maintenance costs, as these of course would have been greatly reduced if more locomotives of the type had been built.

No. 6202 was replaced in 1952 (officially rebuilt) by a four-cylinder 4–6–2 reciprocating locomotive carrying number 46202 and named *Princess Anne*. This locomotive was so badly damaged in the Harrow disaster of October the same year, that it was scrapped (Plate 124A, page 482).

Pennsylvania R.R. 6–8–6 class S-2 (1946). The successful results achieved by the British turbomotive largely influenced the Pennsylvania R.R., in conjunction with Baldwin Locomotive Co., and Westinghouse Electric Corporation, to consider in 1940 the production of a similar, though much larger machine. Designs for a 4–8–4 locomotive developing 6,900 H.P. were completed, but there the matter stopped until the end of World War II. When it was reconsidered, in 1946, special alloy steels were in very short supply and in order to preserve the intended power output, extra weight had to be accepted and the four-wheel trucks were altered to six-wheel trucks, the design becoming 6–8–6.

Apart from size, the Pennsylvania engine differed from the L.M.S. in several important particulars. The main turbine transmitted its power to two driving axles – the second and third – through double reduction helical gearing having quill drive on the main pinions. There was rod drive to the first and fourth pairs of driving wheels. The reverse turbine developed only 1,500 H.P., and was clutched into gear as required, the gearing being quadruple reduction. These arrangements gave a tractive effort of 70,000 lb. going forward, and 61,000 lb. in reverse.

The boiler had a mechanical stoker and a working pressure of 310 p.s.i. It was capable of producing 95,000 lb. of steam per hour in normal service. Timken roller bearings were fitted to all axles.

In fast main-line service the engine handled 1,000-ton

trains on schedules up to 75 m.p.h., during a daily round trip of 580 miles. On test it handled a train of 980 tons at an average speed of 105 m.p.h. on thirty miles of level track.

Like all steam locomotives in the United States the S-2 Turbine fell a victim to dieselization. Heavy maintenance costs and excessive delays when failures occurred due to the absence of standardized spares showed up badly against diesel-electric over-all availability (Plate 124B, page 482).

OTHER APPLICATIONS OF THE STEAM TURBINE TO THE RAILWAY LOCOMOTIVE

These mostly consist of various attempts to utilize the exhaust steam from the reciprocating engine in an exhaust steam turbine. One example may be quoted.

In 1928 the German Railways fitted a turbine tender behind a standard 4–6–0 locomotive of Class P-8. A three-stage Zolly turbine was used for forward running and a reverse turbine was also fitted. The turbine drove through triple reduction gearing and coupling rods on to two driving axles. Steam from the normal cylinders was exhausted into a receiver from whence it was fed to the turbine. A turbine-driven fan provided draught for the boiler.

An additional 300 H.P. was claimed – a low figure for such a complicated and expensive addition.

Part IV. Condensing tenders for reciprocating locomotives

Tenders of conventional locomotives carrying condensing apparatus are used to save water in districts where this is scarce and/or of bad quality. They have been used for many years in Russia, the Argentine, and more recently in South Africa. They are not primarily intended to increase the efficiency of the locomotive, but mainly to supply pure water for the boiler. Thus the maintenance of a high vacuum or indeed of any vacuum at all may be considered unnecessary.

South African 4–8–4 locomotives. Class 25. Sixty of these engines have Henschel condensing tenders, and this has resulted in considerable alteration of the exhaust arrangements. On leaving the cylinders, the exhaust steam is led to a receiver from which it passes through a turbine driving a fan to provide draught for the boiler as normal blast pipe arrangements are not available. The steam then passes through a long pipe, in which is incorporated a grease separator, to the front of the tender where it drives a low-pressure turbine before finally passing into the condenser tubes which are ranged along the sides of the tender. The low-pressure turbine drives a horizontal shaft on which are bevel gears which drive five condenser cooling fans, revolving in filtered air intakes, vertically disposed along the top of the tender.

The condensate from the condenser tubes is collected in a hot well in the bottom of the tender from whence it is fed back to the boiler by means of two turbine-driven rotary pumps, as it is at too high a temperature to be fed efficiently by injectors.

Provision is made for live steam at reduced pressure to enter the exhaust steam system so that draughting may be maintained when the engine is not steaming.

These tenders show a saving in water of 88 per cent over the conventional locomotives (Plate 123C, page 481).

Part V. Rack and similar locomotives

RACK LOCOMOTIVES: STEAM

When trains have to be worked over steep gradients, it may be that the locomotive employing ordinary "smooth wheel to smooth rail" adhesion is ineffective. On such lines other methods need to be employed to enable the locomotive to haul its train up the incline, and to prevent its running away down the incline. In early days, and indeed in some isolated examples up to the present day, some form of cable haulage has proved adequate assistance where short steep gradients are encountered on freight lines. Where gradients are long and steep, and particularly if passenger traffic is involved, some form of rack is usually employed. Here a steel 'ladder" or a toothed rail is fixed between the running rails and pinions or cogged wheels on the locomotive engage in the serrations. The cogged wheels are driven by steam cylinders and pistons through the medium of suitable rods, cranks and, in some systems, gearing.

The gradient at which normal adhesion working becomes impracticable depends largely upon the length of the gradient, the curvature of the line and the weight of train to be operated. Generally, it may be said that long gradients more severe than 1 in 20 (5 per cent) are impracticable for adhesion working only.

Rack locomotives fall broadly into two categories.

(i) *Rack and adhesion:* these locomotives work on lines where part of the route is so heavily graded that rack working is necessary, but where other parts are less steep, or even level, and can be worked by adhesion only. Such locomotives have two complete engines supplied with steam from the same boiler. The outside cylinders and motion drive coupled wheels in the normal manner, while inside cylinders drive cogged wheels or pinions, usually through reduction gear wheels. There are two such pinions and they are synchronized by coupling rods. They are supported in their own frames which are, in turn, fixed to the locomotive main frames.

Other designs of rack and adhesion locomotives employ compounding, the rack engine cylinders using the exhaust steam from the adhesion engine. In such designs, the rack engine cylinders may be placed above the cylinders of the adhesion engine and a common piston valve incorporating a selector valve used for each pair of cylinders on either side of the locomotive. Although using low pressure steam, the rack engine cylinders are usually made the same diameter as those of the adhesion engine. When working on the rack both engines are used. When working adhesion only, the exhaust from the cylinders is turned to atmosphere and the rack engine isolated.

In rack and adhesion locomotives having two high-pressure engines, two separate regulators and reversers are fitted and each engine is a separately controlled unit.

At the commencement of each rack section there is a sprung "lead-in" section to the rack-rail or ladder, enabling the cogged-wheels on the locomotive to engage smoothly. In some cases, the pinions on the locomotive are sprung. Rack systems most used on the rack and adhesion railways throughout the world are either of the Riggenbach or Abt systems (q.v.).

Modern rack and adhesion locomotives are illustrated and further described in Part 5, No. 48, page 346 and No. 97, page 384.

(ii) *Rack only*. Pure rack locomotives differ from rack and adhesion locomotives in that their tractive effort is transmitted only through the cogged-wheels to the rack rail and there is no drive to the flanged carrying wheels. Such wheels are used for carrying and guiding the locomotive, but may also be used as adhesive wheels for braking purposes. Such locomotives are used principally for mountain railways, and typically, the freight wagons or passenger car(s) are pushed up the incline, and therefore are kept by gravity in close contact with the locomotive while ascending or descending. For this reason they are seldom coupled to it, and this also ensures that should the locomotive run away or become derailed, it will not drag the passenger car with it. The car is fitted with an independent emergency braking system (q.v.).

Rack locomotives in the past were fitted with vertical boilers to avoid the effect of gradient on coverage of the firebox plates. All the more modern types have conventional locomotive boilers. The boiler is, however, inclined on the frames downwards towards the front end so that the boiler top is roughly horizontal when the engine is on the incline. It is customary for rack locomotives, and indeed all locomotives working on severe gradients, to work chimney first up the hill. This ensures that the firebox crown will normally be kept well covered with water (Plates 125A, 125B and 125C, page 483 and Colour Plate XV; page 469).

RACK LOCOMOTIVES: DIESEL

Diesel power is not used to any extent on rack railways. The best-known example is on the Manitou and Pikes Peak Railway in Colorado, U.S.A. This railway operates on the Abt system and for many years was operated by 0–4–2T steam locomotives. A petrol-electric motor coach was then tried and this was followed by three diesel-electric motor coaches. These units propel a trailer-coach, thus making a two-coach train. The trailer-coaches have independent braking and are not coupled to the motor coach, which is equipped with dynamic braking (*see* Chapter 1, Part VIII, page 104).

Railcars for the Monte Generoso Railway (1958). The latest diesel-powered mountain railcars have recently been supplied to this 800 mm. gauge Swiss railway. They are of lightweight welded steel construction with two six-cylinder Bussing diesel engines each of 180 H.P., placed horizontally under the floor.

Transmission of power to the cogged wheels is through a five-speed gearbox having electric pneumatic control. The cogged wheels are mounted on the outer axles of each of the two bogies, and drive is by cardan shaft. The Abt rack system with twin racks is used on the Monte Generoso Railway which is 9 Km. (5 5/8 miles) long and has a ruling gradient of 1 in 4·5. A maximum speed of 18 Km/h. ($11\frac{1}{4}$ m.p.h.) is allowed on the ascent and 12 Km/h. ($7\frac{1}{2}$ m.p.h.) on the descent. Oerlikon compressed air brakes, automatic spring brakes and exhaust brakes are fitted to the vehicles (Plate 126B, page 484).

RACK LOCOMOTIVES: ELECTRIC

Electric traction is used for many rack and rack and adhesion railways throughout the world, notably in Switzerland (Plate 126A, page 484). A full description of typical electric rack locomotives is given in Chapter 3, page 211.

SOME SYSTEMS OF RACK WORKING

The first rack railway was devised by Blenkinsop in 1811 for the Middleton Colliery, near Leeds, England. At that time while the advantages of the railway were being more and more appreciated, the dictum that "smooth wheels would grip smooth rails" was not understood. It was natural, therefore, that when this steam-operated colliery line was planned, some form of rack and pinion system was considered necessary (Fig. 106).

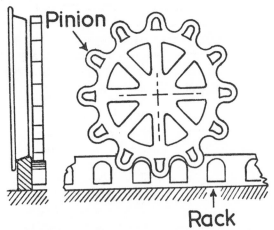

FIG. 106. Blenkinsop's rack and adhesion system (1811).

A single rack line was laid, and attached to the outer side of one of the running rails. The locomotive had four carrying wheels, and the power from the two cylinders was transmitted to the rack pinion through rods, cranks and reduction gear wheels. The cylinders were placed vertically and partly immersed in the top of the boiler. In the earliest engine they were single acting, but in later engines double-acting cylinders were fitted. These engines, which were actually designed by Mathew Murray and built by Messrs. Fenton, Murray & Wood, achieved a fair measure of success and one was able to haul a load of 140 tons up an incline of 1 in 440 at 3 m.p.h.

Since these early times, many different systems of rack working have been devised. Those which have survived to remain in general use are due to Riggenbach, Abt, and Locher.

The Riggenbach system (FIG. 107). This consists essentially of a steel "ladder" laid between the carrying rails, and a cogged wheel or pinion on the locomotive

which engages in the "rungs" of the ladder. The "rungs" typically five inches in width, are riveted into channel irons – one either side – which form the sides of the "ladder". The drive to the pinion is from steam cylinders through pistons, rods, cranks, and gear wheels. The Riggenbach system was invented by Nicholas Riggenbach, a Swiss locomotive engineer, and was first used in 1874 on the Kahlenberg mountain railway in Austria (Plate 125C, page 483).

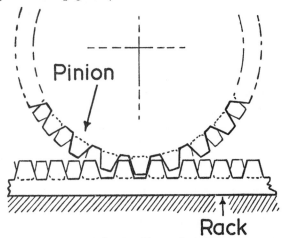

FIG. 108. Abt Double Rack System.

The Abt system (Fig. 108). This is the most widely used rack system. It was developed by Dr. Roman Abt in 1882 and first used on the Blankenberg Railway in Germany in 1884. It employs the same principle as the Riggenbach system, but instead of a ladder, two toothed rack rails are used, and are fixed, in close apposition the one to the other, between the running rails. The two racks are so arranged that the teeth of one are in line with the indents of the other and the cogged wheels or pinions of the locomotives are also arranged that they mate exactly with the rack rails. The teeth of both rack and pinion are slightly tapered (Plates 125A and 125B, page 483).

The Locher system (FIG. 109). This system was devised

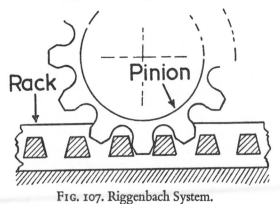

FIG. 107. Riggenbach System.

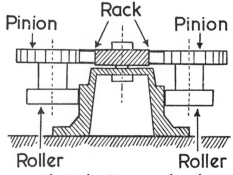

FIG. 109. The Locher System used on the Mount Pilatus Railway.

by a Swiss engineer, Dr Edouard Locher, for use on the Mount Pilatus Railway, on which there are gradients as steep as 1 in 2 (50 per cent). Vertical rack systems were not considered safe on such a steep incline, and Locher therefore used horizontal rack-rails with horizontal pinions on the locomotive to engage in them. The racks are cut in both sides of a single rail placed on its side between the running rails. Beneath the rack rail on both sides are smooth rails along which run guiding wheels placed below, but on the same shaft as the horizontal pinions of the locomotive. The carrying wheels are flangeless and serve no other purpose, propulsion being entirely through the cogged wheels or pinions, and the engine is guided entirely by the horizontal guiding wheels described.

The original Pilatus locomotives were steam, having a boiler athwart the frames and steam cylinders driving the vertical shafts. Present-day operation is by electric railcars of special type – having the seats arranged in tiers across the vehicle. Two sets of rack and guiding wheels are provided.

The line is operated on 1,550 v. D.C., with overhead power lines. Braking is similar to that described in Chapter 3, page 211 (Plate 126A, page 484).

BRAKING SYSTEMS

Braking of locomotives and trains used on mountain railways is of great importance and several systems of braking are provided on each engine. Those in common use are:

(i) Hand-operated band brakes which act on drums attached to the carrying wheels and to the cogged wheels.

(ii) Counter-pressure braking in which air is drawn into the steam cylinders by way of the exhaust ports, having first passed through filters. The air is compressed in the cylinders before being released through a graduated valve under the control of the driver. Overheating of the cylinders and valves by the compression of the air is avoided by injecting into the cylinder, water from the boiler. A silencer may be provided to quieten the explosive noise of the air suddenly released from the cylinders.

(iii) Automatic steam braking which is applied if the locomotive should exceed a pre-determined speed. This may be actuated by a centrifugal governor or by other special device. The steam brake usually acts on the band brakes described above in (i).

(iv) Supplementary brakes may be fitted to one or more of the axles of the passenger coaches on mountain railways. These are usually operated only if the train should exceed the pre-determined speed, and are actuated by a centrifugal governor. If the speed rises to too high a level, the governor actuates a clutch which applies band

brakes fixed to a brake cogged wheel engaged with the rack.

The coaches are also fitted with hand brakes operating through the same bands. The braking of rack and adhesion locomotives may employ any of the above systems for use when the locomotive is working on rack sections. In addition, the normal type of steam, vacuum, hand and air brakes acting through brake blocks on to the adhesion wheels are used, those fitted being whichever type is applicable to the standard practice of the railway. It should not be forgotten that many rack and adhesion locomotives may spend much of their working life being used as ordinary locomotives.

Braking systems used for electrically operated rack railways are described in Chapter 3, page 211.

THE FELL SYSTEM

For working over gradients of moderate severity (of the order of 1 in 10 (10 per cent) to 1 in 20 (5 per cent)) a British railway engineer, John Barraclough Fell (1815–1901) devised a centre-rail system in which no rack was used. A large double-headed rail was laid on its side and secured to the sleepers (ties) between the running rails. Horizontally opposed driving wheels on the locomotive were brought to bear, by powerful springs, on either side of this rail. The pressure at which the springs acted on the wheels was under the control of the engineman, and could be varied from zero to about thirty tons according

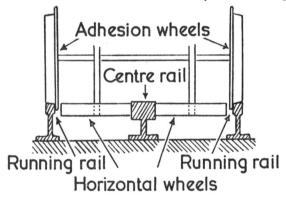
FIG. 110. Fell System.

to the gradient and the load being hauled. Two inside cylinders drove the horizontal wheels and these were independent of two outside cylinders which drove the normal flanged adhesion wheels in the usual way. The locomotives could, therefore, be used as ordinary motive power units in much the same way as can a rack and adhesion locomotive. The horizontal driving wheels provided additional braking power when the train was descending. Brake shoes on the brake vans of the train were also so arranged that they bore directly on the centre rail and were under the control of a brakeman. Normal air brakes were also provided on both locomotive and train (Fig. 110).

The first Fell-operated railway was built over the Mont Cenis pass in Switzerland and was opened in 1868. The ruling gradient was 1 in 12. It was closed a few years later as the Mont Cenis tunnel was opened in 1871. It was worked by steam locomotives.

The most famous line operated on the Fell system was in the North Island of New Zealand. Six 0–4–2T locomotives built, four in 1875 and two in 1886, worked on the three miles of the Rimutaka incline from the day in 1878 when it was opened, until 3 November, 1955, when the line was closed. A new and more easily graded route through the mountains had, at long last, been built.

The Fell tank engines handled sixty-five tons each on the 1 in 15 gradient of the Rimutaka, and four locomotives equally spaced throughout the train were required for the 260-ton trains which were the heaviest allowed on the bank.

The last remaining Fell-operated railway is the 3 ft 6 in. gauge Snaefell Railway in the Isle of Man, on which the ruling gradient is 1 in 12. It is operated by single–deck tramcar-type vehicles. These each have two bogies with each of the four axles motorized, giving a total 100 H.P. for each car. Current is 550 v. D.C., with overhead power lines.

The Fell centre rail is used only for braking purposes on the descent and there are no traction motors to the horizontal wheels which are grooved and "clasp" the centre rail on each side.

Part VI. Miscellaneous unconventional motive power

DUAL POWERED LOCOMOTIVES
Diesel and electric locomotives. In some routes on world railways it is necessary for diesel locomotives to run over electrified tracks during part of their regular schedules. It is, however, seldom that it has been thought worth while to add complication to the locomotive to enable it to work as a straight electric over such routes and as a diesel over non-electrified routes. Such locomotives can be classed as unconventional.

The outstanding examples of such dual power locomotives are those of the New York, New Haven and Hartford R.R., in the United States. Diesel locomotives are not permitted to work inside the New York City boundary, and the electric power into Grand Central Terminal is 600 v. D.C., with third-rail conduction.

Two locomotives built for light-weight high-speed trains by Baldwin-Lima-Hamilton in 1956 have Maybach diesels driving through Mekydro transmission. Also coupled to the transmission are two D.C. electric motors which derive their current through a retractable shoe from the third rail.

Provision is made for changeover from diesel to electric operation, and vice versa, to be made at speed. The electric motors are of considerably less horsepower than are the diesel motors, as high-speed running is not permitted, nor possible, in the New York area.

Diesel electric-electric locomotives. This combination of dual power is also to be found on the N.Y., N.H. and H. R.R., and for the same reasons as already given. Thirty 3,500 H.P. twin-unit locomotives were built by General Motors (E.M. Division) late in 1957. There are eight traction motors and provision is made to change over from third rail supply to diesel-generator supply and vice versa, without any reduction of speed.

ELECTRICALLY HEATED STEAM LOCOMOTIVES
During World War II, the Swiss Federal Railways were extremely short of coal to fire their numerous steam shunting locomotives. In order to overcome this difficulty Messrs. Brown, Boveri supplied two immersion type electric heating units which were fitted to two standard 0–6–0 tank locomotives. Current was collected from the overhead conductors through a pantograph mounted on the cab of the locomotive. Current consumption was, as expected, exceedingly high, and no further locomotives were equipped in this manner.

THE KITSON–STILL LOCOMOTIVE, 1927
One of the greatest assets of the Stephenson locomotive is its flexibility, and it is doubtful if, even today, diesel and electric traction can equal the steam locomotive in this respect. Certainly in the earlier days of diesel rail locomotives the steam locomotive was a much more flexible machine.

In an endeavour to combine the flexibility of the steam locomotive with the increased thermal efficiency of the diesel, Messrs. Kitson & Co., Ltd, of Leeds, built, in 1927, the Kitson-Still locomotive. This locomotive, a 2–6–2 tank, had eight cylinders, horizontally opposed, each pair of opposed cylinders driving on to a common crankpin, one connecting rod being forked. The cylinders were those of a solid injection diesel working on a four-stroke cycle, and were double acting. Steam was used on one side of the pistons when starting, accelerating or in other conditions of overload. The steam was generated in a completely cylindrical fire-tube boiler, with a corrugated firebox and two sets of fire-tubes. It was fired by an oil burner when steam was being used, the steam exhaust passing through a blast pipe in the smokebox and thence through the chimney in the normal manner. The first set of fire-tubes were thus used as in a conventional locomotive boiler. The second set of boiler tubes received the exhaust gases of the diesel side of the engine, which imparted much of their heat to the water in the boiler before escaping to atmosphere.

The drive from the engine crankshaft was through gears to a jack shaft, having a large central gear wheel incorporating a spring drive. Drive from the jack shaft to the six driving wheels was by cranks and coupling rods. Reversing of the engine involved the use of two reversers—the diesel side of the engine being reversed by cams, while the steam side was reversed by altering the position of the vertically disposed piston valves through the medium of a valve gear of the Hackworth type.

The Kitson-Still locomotive ran trials on the L.N.E.R. Trouble was experienced by the excessive leakage of steam and water through the expansion joints between cylinder liners and water jackets. This was later cured. The steam boiler proved inferior to the conventional type of locomotive boiler. Dynamometer car tests showed that the thermal efficiency of the engine using steam and internal combustion was about 16 per cent, and this rose to about 24 per cent when internal combustion was in use.

Generally, the tests showed that this might well be a form of motive power worth developing. Unfortunately,

during the depression of 1930–32, Kitsons were forced to close their works and no further work on the Kitson-Still locomotive was possible and it was scrapped (Plate 120c, page 460).

PROPELLER-DRIVEN RAILCARS

Experiments in propeller propulsion were carried out in Germany over a number of years.

In 1903–5 two motor coaches were built by Siemens and A.E.G., in which electrically-powered propellers were used. These coaches ran on the Zossen-Marienfelde experimental track and reached speeds of up to 130 m.p.h. The current, three-phase A.C., was collected from overhead conductors.

In 1919–20 an ordinary railway bogie-coach ran between Berlin and Hamburg at speeds of over 100 m.p.h. It was driven by two propellers powered by old petrol aero-engines, one unit at each end of the coach.

No further experiments were carried out until 1931 when Dr Friederich Knuckenberg designed his "Rail Zeppelin". This vehicle was driven by a single petrol aero-type engine with a four-bladed propeller. Engine and propeller were mounted in the rear of the coach and so acted as a pusher. The coach was aerodynamically streamlined and had an aluminium body which could seat forty passengers.

It was in experimental service between Berlin and Hamburg and attained a top speed of 120 m.p.h., and an average speed between the two cities of 99.5 m.p.h. The "Rail Zeppelin" never went into regular operational service, but formed the pattern for the early diesel-electric "Flying Hamburger" units.

CHAPTER 10

The Gas Turbine in Railway Service

by P. RANSOME-WALLIS

The gas turbine as a prime mover is a comparative newcomer to railway service. The pioneer work in this field was carried out by the Götaverken Company in Sweden who built, in 1933, a locomotive using a turbine supplied with power gas from a free-piston diesel compressor. This principle had been evolved by the firm in 1924 and stationary and marine plants had been constructed. The system later became known as the turbo-diesel (Part II, page 492).

Gas turbines with axial flow compressors driven from the turbine shaft have, with one exception, so far been used only with electric transmission. This exception is a recent 3,200 H.P. C–C type locomotive built by Skoda in Czechoslovakia. Mechanical transmission with a two-speed gearbox is used in this machine, and provision is made for the fitting of alternative sets of gears suited for high speed (passenger) or low speed (freight) operation. The turbine rotates at 6,000 r.p.m. at full power.

Electric transmission is used in all other gas turbine locomotives which are in revenue-earning service at the present time (1958). The first successful locomotive was put into service in 1943, and since then, gas turbine-electric locomotives have been built for several of the world's railways, but only in the United States, on the Union Pacific Railroad, have such locomotives been constructed in any number. Elsewhere the gas turbine has remained largely experimental and has certainly offered as yet no serious competition to other forms of motive power.

From the survey which follows it will be seen that, for railway work other than for long hauls, such as those on the Union Pacific, the gas turbine has certain disadvantages which will probably result in its being classified as unconventional motive power for some time to come.

Part I. Gas turbine-electric locomotives

OUTLINE OF THE BASIC PRINCIPLES OF THE WORKING OF A GAS TURBINE-ELECTRIC LOCOMOTIVE

Air is compressed to 40–50 p.s.i. in an axial-flow compressor driven by the turbine shaft. The compressed air is heated by being passed through a heat exchanger, the heat being derived from exhaust gases from the turbine. Part of this hot compressed air enters a combustion chamber where it comes into contact with, and ignites, pre-heated and vaporized fuel issuing from a burner. The greater part of the air, however, by-passes the combustion chamber and is ducted to meet and to mix directly with very hot gas which emerges from the combustion chamber at high velocity, and which consists of the products of combustion of the fuel. The air has a cooling effect on this very hot gas, and the consequent mixture is normally at temperatures of 1,000°–1,250° F., when it enters the turbine and comes in contact with the blading. Within the turbine, the hot, rapidly flowing gases expand through the blading, giving up heat in the form of energy and causing the turbine to rotate.

On leaving the turbine, the still hot gases are led through the heat exchanger where more heat is given up to the air entering the combustion chamber, and the cycle is completed, the gases finally escaping to atmosphere through a duct in the roof of the locomotive.

The one or more electrical generators are normally of the D.C. type and are driven by the main shaft of the turbine through reduction gearing. The power is fed to D.C. traction motors, the general layout and drive being similar to that of diesel-electric and electric locomotives.

The functioning of the gas turbine involves a complete cycle, and this cycle may be started in various ways. The first ignition of the explosive mixture in the combustion chamber is usually effected electrically, and continued by the heat of compression after the turbine has attained sufficient speed for the compressor to function. The initial rotation is achieved by motoring the main generator, current for this being obtained either from storage batteries, or, more usually, from an auxiliary diesel-driven generator. This diesel generator supplies power for other auxiliaries – blowers, fans, lighting, heating, fuel pumps, circulating pumps, air or vacuum brake pumps, and so on, and it may provide power to propel the locomotive, light, about engine terminals and yards, without the

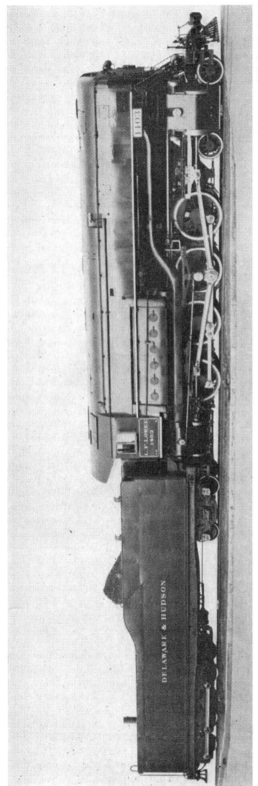

Delaware and Hudson Railroad

PLATE 121A. The last of the four high-pressure locomotives for the Delaware and Hudson Railroad – No. 1403, *L.F. Loree*. This engine had a combined water-tube and fire-tube boiler, steam space being in the drum of the water-tube part. The fire-tube part was filled with water. Working pressure was 500 p.s.i. There were four cylinders working on the triple-expansion principle with opposed drive. H.P. cylinder was under the cab on the right-hand side, I.P. cylinder under the cab on the left-hand side, and the two L.P. cylinders were under the smokebox. R.C. poppet valves were used. A six-coupled booster-unit formed the rear bogie of the tender and used boiler steam at 500 p.s.i.

P. Ransome-Wallis

PLATE 121B. L.N.E.R. 4-cylinder compound 4–6–4 No. 10000 with Yarrow-type water-tube boiler; working pressure was 450 p.s.i. Here seen south of Edinburgh with an up East Coast express.

Institution of Mechanical Engineers

PLATE 122B. Reid-Macleod condensing turbine–mechanical loco–motive (1924) showing water jets for cooling of condenser tubes.

Locomotive Publishing Co.

PLATE 122D. Ljungström condensing turbine–mechanical locomo–tive for the L.M.S.R. (1926).

Institution of Mechanical Engineers

PLATE 122A. Belluzzo's non-condensing turbine–mechanical locomotive (1907).

Locomotive Publishing Co.

PLATE 122C. Ramsay condensing turbine–electric locomotive (1920).

Institution of Mechanical Engineers

PLATE 123A. The Reid-Ramsay condensing turbine-electric locomotive (1910).

Institution of Mechanical Engineers

PLATE 123B. The Krupp condensing turbine-mechanical locomotive (1922).

Locomotive Publishing Co.

PLATE 123C. South African Railways 4–8–4, Class 25 locomotive with Henschel-type condensing tender.

P. Ransome-Wallis

PLATE 124A. Non-condensing steam turbine locomotive: L.M.S.R. 4–6–2 No. 6202.

Pennsylvania Railroad

PLATE 124B. Non-condensing steam turbine locomotive: Pennsylvania Railroad
6–8–6 No. 6200, Class S–2.

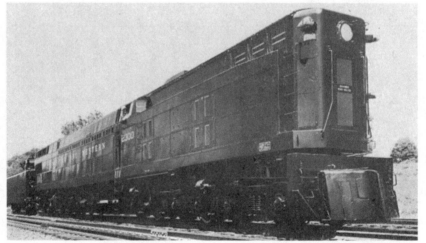

Norfolk and Western Railway

PLATE 124C. Non-condensing steam turbine-electric locomotive: Norfolk
and Western 6–6–6–6 No. 2300.

P. Ransome-Wallis

(B)

PLATE 125A. Snowdon Mountain Railway (Wales) 0–4–2T, No. 5, *Moel Siabod*. Abt rack system.

PLATE 125B. Schneeberg Railway (Austrian State Railways) 0–4–2T, No. 999.01 with Giesl ejector. Abt rack system.

PLATE 125C. Madeira Railway 0–4–0T, No. L–6. Riggenbach rack system.

PLATE 125. Steam Rack Locomotives

P. Ransome-Wallis

(A)

P. Ransome-Wallis

(C)

483

P. Ransome-Wallis

PLATE 126A. Electric Rack Motor Coaches: Pilatus railway (Switzerland). The summit of the line which is worked on the Locher rack system.

Swiss Industrial Co.

PLATE 126B. Diesel Rack Railcar: Monte Generoso Railway (Switzerland). Abt rack system.

P. *Ransome-Wallis*

PLATE 127A. Brown, Boveri gas turbine–electric locomotive until recently in service on the Swiss Federal Railways.

P. *Ransome-Wallis*

PLATE 127B. Brown, Boveri gas turbine–electric locomotive in service on the Western Region of British Railways.

General Electric Corporation

PLATE 128. General Electric 8,500 H.P. gas turbine–electric locomotive for the Union Pacific Railroad.

necessity for starting the turbine. High-grade fuel such as diesel oil is used in the turbine until the cycle is fully established, when a change is made to the lower-grade fuel normally used.

The over-all thermal efficiency of the gas turbine-electric locomotive is greater than that of the steam locomotive, but less than that of the diesel-electric, comparative figures being of the order of 10 per cent for steam, 14 per cent for the gas turbine and 30 per cent for the diesel-electric. The efficiency is greatest when the turbine is at full power, and the lowest when the turbine is idling or running at very low power output.

The main advantages claimed for the gas turbine as a prime mover in railway service may be summarized as follows:

(i) First cost is less than that of a diesel-electric of equal power, though much more than that of a steam locomotive.

(ii) It requires less maintenance than either diesel or steam as a prime mover, as it has no reciprocating parts.

(iii) It is able to burn very low-grade fuel, but this is partly offset by the higher fuel consumption of the turbine, especially at low speeds.

(iv) It shows considerable economy in lubricating oil over both diesel and steam.

(v) It has better starting and torque characteristics than has the diesel, and is far more flexible. Its characteristics approach the "automatic action" of the steam locomotive, but as, so far, the gas turbine in locomotives has mostly been used with electric transmission, this advantage has largely been nullified.

(vi) When used with electric transmission, it is well adapted to the use of dynamic braking.

(vii) It is claimed that on a weight-per-horsepower basis, the gas turbine scores heavily over the diesel. This claim however, is hardly justified when the weight of compressors, gearing and auxiliaries are taken into consideration, and any advantage which the turbine may have in this respect is certainly not a very great one.

(viii) As the gas turbine requires no water either as fuel or coolant, it is particularly well suited for long hauls through water-starved terrain.

(ix) Recent experiments show that it is possible to use propane gas as fuel. This will effect economies in fuel cost in areas where this gas is available. Trials are also being made in the use of pulverized fuel, and an experimental 1A1A–A1A1 locomotive is now under construction in Britain at the North British Locomotive Company's Works in Glasgow. In the United States, similar experiments are being pursued under the auspices of the Locomotive Development Committee.

The main disadvantages of the gas turbine for railway service may be summarized as follows:

(i) It attains greatest efficiency only when running at full power and is very inefficient when idling and when it has frequently to be stopped and started. Fuel consumption in these circumstances is very high.

(ii) A great deal of the power output of the gas turbine is absorbed in driving its own compressor. This is of the order of from 65 to 75 per cent of the total power output.

(iii) The gas turbine revolves at a very high rate, 4,000 –5,000 r.p.m. or more. Such rates of rotation are much beyond that which can be accepted for normal electric generators, and so reduction gearing must be used between the turbine and the generator to bring the drive speed of the generator down to 1,000–1,500 r.p.m. The introduction of such gearing increases first cost, maintenance and weight-per-horsepower.

(iv) The efficiency and power output of the gas turbine are very dependent upon the temperature and pressure of the atmosphere. Its power output may be of the order of 25 per cent greater at sea level at 50° F. than at 2,500 feet at 90° F. ambient.

(v) The noise level of the turbine, especially when starting up, is very high and would preclude their use in large numbers in passenger service. The noise from, say, ten such locomotives in the confined spaces of a station like King's Cross would be prohibitive for both travellers and near-by residents.

BRIEF DESCRIPTION OF SOME GAS TURBINE-ELECTRIC LOCOMOTIVES

The first gas turbine-electric locomotive (1943).[1] Gas turbines were built in Russia by General Kuzminske in 1887. In 1922 an attempt was made to install such a turbine in a locomotive, but the attempt was not successful, largely due to the metal of the turbine blading failing to stand up to the high temperatures of the propellant gases. It was thus left to the famous firm of Brown, Boveri of Baden, Switzerland, to produce in 1943 the first locomotive using a gas turbine as a prime mover. This firm had pioneered gas turbines for stationary use since 1910, but it was in the face of wartime scarcities and difficulties that Swiss Federal Railways 1–Do–1 locomotive No. 1101 of Series Am 4/6 first ran track trials in 1943. Even with their at-that-time unrivalled experience in this field, the firm took over four years in the design and construction of the locomotive, the order for a 2,000 H.P. gas turbine-electric locomotive having been placed by the Swiss Federal Railways early in 1939. The locomotive

[1] This locomotive has now been reconstructed (Feb. 1959) as an experimental electric locomotive capable of operating on $1\frac{1}{2}$ kV. direct current and on alternating current of 15 kV. $16\frac{2}{3}$ cycles and 25 kV. 50 cycles.

was intended primarily to work light, fast, passenger trains on routes which normally handle insufficient traffic to justify electrification. No. 1101 has consistently handled loads of up to 300 tons in regular service (Plate 127A, page 485).

At the time of its completion in 1943, the approximate cost per horsepower of the three principal prime movers in locomotive service were:

Steam   £7 per horsepower.
Diesel electric   £17 10s per horsepower.
Gas turbine-electric   £13 per horsepower,

and the relative over-all efficiency was considered to be 6 to 8 per cent for steam, 26 to 28 per cent for diesel electric, and 15 to 16 per cent for gas turbine-electric. The advantages and disadvantages of this form of prime mover were considered to be those already stated (page 487).

The decision to use electric transmission rather than any form of mechanical or hydraulic transmission was made because at that time electric transmission was by far the most reliable, and was widely used in diesel locomotives. Furthermore there was a wealth of experience already accumulated from electric locomotives using D.C. traction motors. To introduce any form of transmission other than electric would, therefore, be to add complication to an already untried and experimental form of traction.

The gas turbine operates on the principles already described, a heat exchanger being fitted and reduction gearing reducing the turbine speed of 5,200 r.p.m., at full power, to 812 r.p.m. at the generator shaft. The turbine develops 8,000 H.P., of which 6,000 H.P. is absorbed by the compressor. The whole unit is built on a bed-plate and lowered into the vehicle from above. It rests in the vehicle on a three-point elastic suspension. A 100 H.P. diesel generator provides power for motoring the main generator and for auxiliaries. It also may be connected directly to the four traction motors supplying enough power to move the engine light, at up to 10 m.p.h., for shunting and hostling purposes.

Starting the main turbine is semi-automatic. The driver first starts the auxiliary diesel and this in turn motors the main generator and rotates the turbine and its compressor. When the speed of rotation has reached about 1,000 r.p.m., the driver makes a switch which starts the fuel pump and also ignites the fuel by means of an electric igniter. Diesel oil is used for starting, and the change-over to low-grade fuel is made when the turbine has warmed up. When further acceleration has occurred, the driver cuts out the auxiliary diesel, or he may divert the current from it to the traction motors, so being able to move the locomotive without using power from the main generator while the turbine is warming up. Time taken from starting the auxiliary diesel until being able to move the locomotive is about four minutes.

Operation of the locomotive is effected by a single control handwheel, subject to the overriding control of a dead-man's pedal. The regulation of power and its co-ordination between turbine, generator and traction motors is effected by a servo-field regulator operated by the oil pressure governing system so widely used in diesel-electric practice (see Chapter 1, Part III, page 49).

The locomotive develops a starting tractive effort of 29,000 lb., and a continuous tractive effort at 45 m.p.h. of 11,000 lb. Maximum speed is 70 m.p.h.

Gas turbine-electric locomotives for British Railways. Up to the present time, two gas turbine-electric locomotives have been operating, somewhat sporadically, on British Railways. Both are allocated to the Western Region, and indeed it is to the erstwhile Great Western Railway that credit is due for their introduction.

The first locomotive, No. 18000, was introduced in 1949, and was built by Messrs Brown, Boveri of Baden, Switzerland. This locomotive is a slightly more powerful version of the Swiss Federal Railways' locomotive already described. No. 18000 develops 2,500 H.P. at the turbine coupling and develops 31,500 lb., of tractive effort at 21 m.p.h. It is of the A1A–A1A wheel arrangement with a driving cab at either end. Controls are made as simple as possible and, for example, the various operations involved in starting up the main turbine are fully automatic once they have been initiated by manipulation of the starting switch.

The locomotive weighs 119 tons, of which 81 tons is available for adhesion. On this locomotive and on the second British Railways machine, steam heating boilers with automatic oil fuel and feed water controls are fitted (Plate 127B, page 485).

The second locomotive, No. 18100, was placed in service in 1951 and was built by Metropolitan Vickers and British Railways. It is of the Co–Co type, with a driving cab at each end.

No. 18100 is more powerful than the Brown, Boveri engine, and can develop 3,500 H.P. at the turbine coupling; with the turbine blading at present fitted, however, the output is limited to 3,000 H.P. A starting tractive effort of 60,000 lb. is provided with a total weight of 130 tons, all available for adhesion and divided between six axles grouped in two three-axle trucks (bogies).

The cycle of events in this gas turbine is somewhat different from that of the Brown, Boveri machine. There is no heat exchanger; compression heating and expansion of the air occurs in a compressor, combustion chamber and three-stage turbine, all arranged in line as a straight-through single unit. The turbine has two output shafts, one of which drives two of the three main generators through reduction gearing; the other shaft drives the third main generator, the auxiliary generator and the exciter, again through reduction gearing. These auxiliaries absorb some 150 H.P. of the total output of the turbine. The turbine rotates at the high speed of 7,000 r.p.m., but the drive to the generators is reduced to 1,600 r.p.m. by the reduction gearing.

Each main generator supplies current to two of the six nose-suspended D.C. traction motors which in turn drive the six carrying axles through single reduction gearing.

No auxiliary diesel generator is provided in the Metropolitan Vickers design; the turbine is started in the usual way by motoring the main generators, but the power for this is obtained from storage batteries which are kept charged by an auxiliary generator driven off the second turbine shaft. The turbine is started by a single starting-button control, the subsequent events until full power is available are entirely automatic. When the starting button is depressed, an electrically driven sequence controller determines the events which follow in the following order:

(i) auxiliary feed pump starts;

(ii) main fuel pumps and lubrication pumps start;

(iii) igniters are switched on;

(iv) automatic starting valve moves, and the turbine begins to rotate;

(v) when the rotation of the turbine reaches 1,000 r.p.m., the starting valve commences fuel delivery through the idling jets in the combustion chamber;

(vi) when rotational speed has reached 2,500 r.p.m., the battery automatically disconnects from the main generators and the turbine runs under its own power;

(vii) the turbine accelerates to 4,000 r.p.m., at which idling speed it is allowed to warm up for ten minutes before being opened up to give full power.

In service, turbine and generator are controlled electrically and are so arranged that their speed and output match the requirements of the traction motors as determined by the driver through a master controller.

As this goes to press it is learned that this locomotive has been converted to an A.C. electric locomotive with the A1A–A1A wheel arrangement. The primary transformer winding is so arranged that it can be used with either 25 kV. or 6·25 kV. current. The reconstructed locomotive is now No. E 1000 and is to be used for crew training for the Manchester–Crewe electrification scheme.

Gas turbine-electric locomotives for the Union Pacific Railroad of America. The Union Pacific Railroad is alone among the railways of the world in its employment of gas turbine-electric locomotives to handle a significant percentage of its freight traffic.

In 1948 the American Locomotive Company, with the General Electric Company of America, built the first gas turbine-electric locomotive in America. After preliminary trials this locomotive went into service on the Union Pacific Railroad in 1949. Its performance was very satisfactory and during its first twenty-one months on the Union Pacific it ran more than 106,000 miles in main-line service.

The compressor, combustion chamber and two-stage turbine are constructed in line and there is no heat exchanger. The turbine delivers 4,800 H.P. at 6,700 r.p.m., at an altitude of 1,500 feet and at an ambient temperature of 80° F.

The turbine is connected through reduction gearing having a ratio of 65 : 18 to four main generators, which provide 4,500 H.P. to the eight D.C. traction motors. The locomotive is carried on four two-axle trucks (bogies) each axle being motored. A diesel generator supplies power for auxiliaries and for motoring one of the main generators to start the turbine. It also permits dynamic braking to be used when the locomotive may be travelling down long gradients with the turbine shut down. Starting is effected in much the same sequence as has already been described for the Swiss and British gas turbine-electric locomotives. The rate of rotation at which the fuel is ignited is, however, lower – being 500 r.p.m., and the changeover from diesel oil to "Bunker C" fuel is made at 4,700 r.p.m.

Ten locomotives of this design were placed in service in 1952, and give every satisfaction in freight service between Ogden (Utah), Cheyenne and Omaha, a route which includes two summits, at Aspen Tunnel, 7,230 feet above sea level and at Sherman Summit, 8,013 feet. The ruling gradient is 1 in 88.

In 1954 a further fifteen locomotives of similar design and power went into service. Externally, these later locomotives are distinguishable from the earlier engines in that they have external walk ways or aisles. Both series have a driving cab at one end only. In the later engines, the combustion chambers are twelve inches longer and, as later practice has been to filter the "Bunker C" fuel oil before supplying it to the locomotive, the oil filtering apparatus on the locomotive has been omitted (Colour Plate XVI, page 496).

In order to increase the range of these locomotives between fuelling stops, they now have tenders which carry 24,000 gallons of fuel oil, in addition to the 7,200 gallons carried in the locomotive. In fact, the fuel tanks of the locomotive are, as far as possible, kept filled in order to

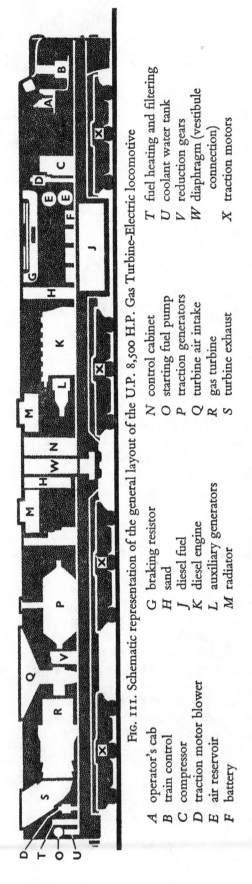

FIG. III. Schematic representation of the general layout of the U.P. 8,500 H.P. Gas Turbine-Electric locomotive

A operator's cab
B train control
C compressor
D traction motor blower
E air reservoir
F battery

G braking resistor
H sand
J diesel fuel
K diesel engine
L auxiliary generators
M radiator

N control cabinet
O starting fuel pump
P traction generators
Q turbine air intake
R gas turbine
S turbine exhaust

T fuel heating and filtering
U coolant water tank
V reduction gears
W diaphragm (vestibule connection)
X traction motors

stabilize the adhesion weight. The tenders have been re-built from former steam locomotive tenders, and are carried on twelve wheels. They are fitted with 100 kW. electric heaters powered from the auxiliary generator, in order to reduce the viscosity of the oil during the severe cold weather experienced in this region. As a result of this practice the gas turbine-electric locomotives can operate 1,500 miles without refuelling.

In order to retire the large articulated steam locomotives of the 4–6–6–4 and 4–8–8–4 types which have for the last eighteen years worked heavy freight over the Ogden–Cheyenne–Omaha routes, more power is needed than is available from one 4,500 H.P. gas turbine-electric unit. Accordingly, these locomotives have been run in tandem, the two units being placed back to back with the tender between them. Multiple unit control has been arranged, though this involves the use of long cables owing to the position of the tender between the two units.

The success, over eight years, of the twenty-five 4,500 H.P. locomotives has induced the Union Pacific Railroad to claim that on its system, this form of motive power is no longer experimental or unconventional. Over this period, in fact, the availability of these locomotives has exceeded 80 per cent. Although fuel costs are higher than those of comparable diesel-electric locomotives, the gas turbine-electric is capable of higher running speeds in service and the maintenance costs are less than those of the diesel-electric. The turbine blading has stood up well, requiring replacement only after some 10,000 to 12,000 hours service, and more than half the reported failures of locomotives in service have been due to the electrical rather than to the turbine side of the locomotives.

As a result of the success achieved with the 4,500 H.P. locomotives, it is not surprising that the Union Pacific should order further and more powerful gas turbine-electric locomotives from the same builder. Fifteen 8,500 H.P. two-unit locomotives are now (1958) being delivered. These locomotives are intended to replace the 4–8–8–4 ("Big Boy") articulated steam locomotives. The two units of the new locomotives are permanently coupled, each unit being carried on two three-axle trucks (bogies), and fuel carried in a separate tender of 23,000 gallons capacity.

The first unit contains the driving cab and the auxiliary machinery. This consists of a six-cylinder 1,000 H.P. diesel engine, direct coupled to a D.C. generator and also, driving through a gearbox, a smaller generator for battery charging. Brake compressors (two), traction motor blowers and cooling fans are also housed in this unit and 2,500 gallons of diesel oil is carried in a tank under the platform. Auxiliary control apparatus is housed behind the driving cab, and propulsion control for the unit at the rear end of the vehicle.

The second unit carries the main power plant – the compressor, combustion chambers and turbine and all its associated equipment. The main control apparatus is housed in the forward end of the unit. The turbine is of a simple-cycle type without heat exchanger. The compressor is of a sixteen-stage axial type and supplies air to ten combustion chambers arranged radially around the compressor. The compressed air must therefore change direction 180° after leaving the compressor to reach the combustion chambers. Having passed through these chambers it must again reverse direction to reach the two-stage turbine. This arrangement allows the whole power unit to be much shorter and allows the use of only two main turbine shaft bearings. The turbine driving shaft emerges from the compressor end of the turbine and not from the exhaust end as in previous practice.

The turbine develops 8,500 H.P. at 4,860 r.p.m., at 6,000 feet altitude and at an ambient temperature of 90° F. It is interesting to note that the rating of the 4,500 H.P. locomotives was taken at 1,500 feet altitude and 80° F. ambient. The horsepower rating of the turbines of the new locomotives increases to 10,700 H.P. at this datum, and thus a fair amount of reserve power is available in the new machines, although the electrical gear must be the limiting factor, as it is designed to deal with only 8,500 H.P. between 18 and 40 m.p.h.

Starting the turbine is fully automatic and follows the usual sequence of events commencing with the motoring of one of the main generators with power from the auxiliary diesel generator. The starting events are, however, controlled by an air-driven controller with cam-operated mechanical interlocking in place of the usual electrical time-relays, which have in practice been found to get out of adjustment.

The electric propulsion equipment follows standard diesel-electric practice and design. Four D.C. generators are driven off the main turbine shaft, through reduction gearing, at 1,050 r.p.m. The generators are mounted back to back in two sets installed side by side across the locomotive. Great care has been taken to provide clean blower air for the cooling of the main generators.

There are twelve standard G.E.-752 type traction motors – one to each carrying axle, and all permanently connected in parallel. Each of the four main generators supplies current to the motors of one truck (bogie). The motors drive the axles, each through a 74 : 18 reduction

gearing. The locomotive is geared for a maximum speed of 66 m.p.h., with driving wheels of 3 ft 4 in. diameter. A starting traction effort of 240,000 lb. is available, with a continuous tractive effort of 145,000 lb. at 18 m.p.h.

Careful consideration has been given to the riding qualities of the locomotive, and the four carrying trucks (bogies) are of the swing bolster type, each bolster being supported on four pads of rubber and steel bonded together. Stability of the truck is maintained by the use of hydraulic shock absorbers which are fitted between the equalizer and the side frame, and between the bolster and the side frame.

The driving controls of the new locomotives have been made as simple as, and similar to, those of a standard diesel-electric locomotive. For shunting and moving the engine and tender only, the auxiliary diesel generator provides sufficient power for speeds of up to 20 m.p.h. on level track. For control of the main power plant, a throttle handle with twenty notches is provided, rather more than for the usual diesel-electric locomotive. The additional notches enable a niceness of control to be exercised over the enormous horsepower available.

The starting sequence is controlled by the operation of a single push-button, and is entirely automatic.

Special attention has been given to the automatic prevention of wheel slip, and equipment has been provided to measure each individual axle speed and to detect and stop wheel slip at all conditions of speed and power.

The tender receives fuel oil which has been pre-heated to 200° F. The tender tank is insulated with rock wool, and this enables the fuel oil to maintain a suitable temperature for forty-five hours without further heating, in an ambient temperature of – 40° F. Electric, thermostatically controlled immersion heaters are fitted to heat the fuel oil in the tender when this is necessary, and when the turbine is operating, the fuel may be heated in a recirculating fuel line by atomizing air bled from the main turbine compressor.

The design of this locomotive has been meticulous, and since the units are permanently coupled, each detail has been deemed to be of equal importance in keeping the locomotive in service. Extensive use has been made of standard diesel-electric equipment of proven value, and many parts are therefore interchangeable from stock. Both builders and owners look forward with confidence to the future of this modern form of motive power (Plate 128, page 486).

Part II. The turbo-diesel locomotive

It has been shown in the foregoing descriptions of locomotives using the gas turbine as a prime mover, that the three principal disadvantages are:

(i) The high absorption of power by the compressor.

(ii) The low efficiency and high fuel consumption when the turbine is idling or running at low speeds.

(iii) The inability, so far, to use the excellent torque characteristics of the gas turbine by direct drive.

With the one exception noted, all gas turbine locomotives have so far employed electric transmission.

The solution of the compressor problem (i) has been largely solved in stationary and marine practice by replacing the axial flow compressor by batteries of from two to six free-piston gas generators, and the term turbo-diesel indicates a power plant comprising a free-piston gas-generator turbine set.

Sweden, France and Russia have built turbo-diesel locomotives. In the first two countries, mechanical transmission has been used, but in 1954 the Soviet Railways put into service a 3,000 H.P. locomotive with electric transmission.

SWEDISH TURBO-DIESEL LOCOMOTIVES

As has already been stated (page 478), the Swedish Götaverken Company built the first turbo-diesel locomotive in 1933. This was a small machine, of the 1–B–1 wheel arrangement. It developed 650 b.h.p., and had a starting tractive effort of 13,500 lb. The maximum speed was 45 m.p.h., and the drive from the turbine shaft was through reduction gearing, jack shaft and side rods. A simple reverse gear was fitted but no change-speed gearing was incorporated in the design. The locomotive weighed fifty-three tons of which twenty-eight tons was adhesive.

In 1955 the same company, in collaboration with the Motala Verkstad locomotive works, constructed a larger and much more advanced locomotive of 1,300 H.P., and with the 1–C–1 wheel arrangement. The starting tractive effort is 30,000 lb., maximum speed 90 Km/h. (56 m.p.h.) and the total weight is 63 tons of which $40\frac{1}{2}$ tons is adhesive. The diesel part of the power unit is unusual. A vertical opposed piston two-stroke engine has its upper pistons directly connected to those of the compressor. The lower pistons drive through connecting rods, on to a crankshaft which is direct-coupled to an alternator of 125 kW. capacity. This in turn supplies power for brake compressors, cooling fans, and electric train heating.

Power-gas from the compressors drives a two-stage gas turbine. The drive to the wheels is then transmitted through reduction gearing, friction clutches, a jack shaft and side rods. There is no change-speed gearing.

This locomotive has been in service with the Swedish State Railways, and has shown remarkable economy in operation and an over-all efficiency of 31·5 per cent when working at between 75 and 100 per cent of maximum output.

FRENCH EXPERIMENTAL TURBO-DIESEL LOCOMOTIVE, NO. 040. GA. I

Since 1951 the Renault Company of France in close collaboration with the French Railways have been experimenting with a turbo-diesel locomotive. After many bench trials of the power plant, the locomotive made its first track run in March 1952, and the subsequent results obtained have been so encouraging that two further locomotives are being constructed.

The power plant of the turbo-diesel consists of a single Pescara free-piston power-gas generator feeding a gas turbine having an output of 1,000 H.P. The turbine drives the locomotive through, first, a 6·15 : 1 reduction gear and then mechanically through shafts and a gearbox to the four driving axles of the locomotive. Each shaft has a universal joint, and the gearbox provides for two speeds forward and a reverse gear. For the sake of simplicity in this first experimental locomotive, gear changing is possible only with the locomotive at rest, and the forward gears have been arranged to give a slow speed suitable for freight working and a higher speed for passenger train operation. In practice it has been found that in future designs it will be essential to have facilities for gear change to be effected at speed, as with the fixed gear the high torque characteristics of the turbine running at low speed have not been fully available. This is largely due to the fact that the coefficients of friction in the mechanical transmission are much higher when being started from rest than they are when running, and the acceleration of the locomotive from rest has thus not come up to expectation, although it has been improved by providing a means of obtaining up to 15 per cent over-load from the turbine on starting.

As the free-piston gas generator has no rotating shaft a separate 100 H.P. diesel generator of conventional design is necessary to supply power for auxiliaries. In future locomotives, however, it is intended to replace this diesel by a compressed-air driven turbine. The free-piston gas generator will supply the compressed air. Such an arrangement will reduce both fuel consumption and noise.

The fuel used for the turbo-diesel is a light-grade

residual fuel oil. High-grade diesel oil is, however, used for starting and for about the first two minutes of running. During the period of warming up, the fuel oil is being heated by the cooling water of the gas generator, and it continues to be so during the operation of the power plant. Diesel oil of course is used exclusively for the auxiliary diesel plant.

The turbo-diesel has proved to be considerably more economical in fuel consumption than the orthodox gas turbine, particularly at idling speeds. While idling fuel consumption is still greater than that of the conventional diesel, the difference is now so small that it may disappear altogether with the introduction of the new locomotives.

The locomotive weighs fifty-four tons and has a starting tractive effort of 25,000 lb.

The experimental turbo-diesel went into regular service on the S.N.C.F. a little more than a year after its first track run. During the subsequent year it ran 65,250 miles hauling trains of 160–200 tons in express passenger service and attaining a maximum speed of 68 m.p.h. The turbine has given no serious trouble and has proved beyond all doubt that a gas turbine can be used for locomotive work in conjunction with mechanical transmission systems. The success of this locomotive using a form of propulsion so completely new to railway practice is an outstanding achievement by the engineers concerned in its design and construction. The decision to proceed with the construction of two further turbo-diesel locomotives, one of which will be rated at 2,000 H.P., is fully justified.

CHAPTER 11

Concise Biographies of Famous Locomotive Designers and Engineers

by H. M. LE FLEMING

A great variety of titles has been given to the engineers responsible for railway rolling stock departments; Locomotive Superintendent, Chief Mechanical Engineer, Chef du Matériel et de la Traction, etc., etc. For the purpose of these notes these have been condensed as follows:

CME = Chief Mechanical Engineer.
LS   = Locomotive Superintendent.
SMP = Superintendent of Motive Power.

For other titles the nearest to one of the above is substituted in order to save space and avoid unnecessary multiloquence.

Difficulty is encountered in selecting from some hundreds of well-known names, those who have had the greatest influence on the development of the locomotive.

Where other things are equal, preference has been shown for:

(*a*) the more recent engineers rather than early pioneers;

(*b*) those whose names have become part of the locomotive vocabulary;

(*c*) those who have had the greatest effect on the outward form of the locomotive.

ABT, H.C.Roman. Designed a rack-rail system in 1882, which was first installed in Switzerland in 1890. This has two or three toothed racks laid centrally on the track with the pitch of the teeth alternating for smooth working. It has become the most widely used rack system for heavy work and long stretches, including twenty-six miles of 1 in 12 gradient on the metre gauge Arica & La Paz Railway. In 1885 he introduced his combined rack and adhesion system with separate four-coupled rack engine and all cylinders in line. Compounding of such engines was introduced in the same year by A. Klose.

ADAMS, William, 1823–1904. Between 1855 and 1895 was successively LS of the North London, Great Eastern and London & South Western Railways. In 1864 he introduced the long wheelbase bogie with check springs to control the movement of the pivot. The idea of side control to overcome the oscillation set up by the short wheelbase bogies then in use was first suggested in October 1845 by Fernihough, LS of the Eastern Counties Rly., in his evidence before the Gauge Commission.

ALLAN, Alexander, 1809–91. Prominent figure in the early days of Scottish and London & North Western Rly. locomotives. His distinctive design had outside cylinders between inner and outer framing, the smokebox being united to the cylinders by gracefully curved plating. On these "Crewe"-type engines, introduced about 1842, he was associated with J. BUDDICOM, LS of the Grand Junction Railway, who later built similar engines in France. Allan pioneered many features, including balanced slide valves in 1844, and successful steel fireboxes in 1860. In 1856 his straight link valve gear first appeared and was adopted in many parts of the world.

BALDWIN, Matthias W., 1795–1866. In 1831 he opened the Baldwin Locomotive Works in Philadelphia which subsequently became the world's largest locomotive building firm developed from a single foundation. The first locomotive, *Old Ironsides*, was turned out in 1832 and the one thousandth early in 1861. Baldwin's genius was manifest in the numerous improvements in detail and methods of manufacture he introduced to prevent the frequent stoppages to which early locomotives were prone. Satisfactory pipe and tube joints were evolved and cast iron wheel centres adopted. Standardization with templates and gauges began in 1838 and by 1840 frames and hornblocks were being made in one piece and metallic packing was in use. In 1858 he initiated the casting of cylinders with half saddles, a practice which soon became standard in America. Somewhat conservative in his outlook on design, he preferred well-tried features combined with first-class workmanship as good policy. He died in 1866 when the annual output was about 120 locomotives and the first 2–8–0 was built – to become the most numerous type in America.

The first export to Europe was for Austria in 1841, since when Baldwin locomotives have been delivered to railways in every part of the world. The last steam locomotive built was for India in 1955 when the total locomotive output stood at about 75,000.

BEATTIE, Joseph H., 1814–71. LS of the London & South Western Rly. from 1850 to 1871. In 1854 he revived feedwater heating, and brought out a number of designs which were widely fitted on his locomotives. He was the first British LS to discard the single-wheeler in 1859. At the same time he was concerned with the changeover from coke to coal and evolved various types of fireboxes and combustion chambers. Later he tried long travel piston and balanced slide valves. He was succeeded on the L. & S.W.R. by his son W.G.Beattie from 1872 to 1878, who continued the developments his father had pursued.

Union Pacific Railroad Colorphoto

PLATE XVI. Union Pacific Railroad: 4,500 H.P. gas turbine-electric locomotive (1954 batch) in freight service. The tender carrying fuel oil can be seen behind the locomotive.

BELPAIRE, Alfred, 1820–93. From 1864 CME and later Administrative President of the Belgian State Rly. until his death. His flat-topped firebox with wide sloping grate was first fitted in 1864. In the previous year he had introduced a combined lever and screw reversing gear. He was the first to adopt the Stevart form of valve gear on locomotives, in which the valve motion is derived not from eccentrics but from the crosshead on the opposite side. From 1878 he brought steam railcars into extensive use.

VON BORRIES, August, 1852–1906. CME of the Hanover Division of the Prussian State Rlys. from 1875 to 1902 and pioneer of compounding. The first of his two-cylinder compounds was introduced in 1880 and his first four-cylinder compound in 1899. He was responsible for many innovations in German locomotive design, including the use of nickel steel for boilers in 1891. A great authority and writer on locomotive matters, he became a professor at the Berlin Technical School from 1902 until his death in 1906.

DU BOUSQUET, Gaston, died 1910. CME of the Northern Rly. of France from 1890 until his death in 1910. He was associated with de Glehn in the first four-cylinder compound of 1886 and in the subsequent developments with the 4–4–0, 4–4–2, 4–6–0, 4–6–2 and other types. In 1901 he brought out successful 4–6–0 tandem compound tank engines for passenger service on the Ceinture Rly. of Paris. For freight work thereon he introduced heavy articulated 0–6–2 – 2–6–0 tank engines. These "du Bousquet-Mallets" had the cylinders arranged centrally and the draw and buffing gear attached to a separate continuous frame. At the time of his death he was engaged on the first two 4–6–4 express tender engines in the world, probably the most advanced design of the day. The very large low-pressure cylinders were ingeniously staggered to fit between the frames. One engine had a water-tube firebox.

BROTAN, Johann, 1843–1923. Between 1868 and 1912 held many posts in boiler and railway works in Austria and Hungary. Inventor of the most successful semi water-tube locomotive boiler first fitted in 1902. Over a thousand were built in the next twenty-five years. He is believed to have died in 1923 but trace of him was lost in the disorganization after World War I.

BROWN, Charles, 1827–1905. Born in Uxbridge, England, and died in Basle, Switzerland. He started his career as a mechanical engineer by serving his apprenticeship with Maudslay & Field, in London. Before he had completed the seven years of his apprenticeship he left and started his own modest workshop. In 1851, a relative of Sulzer, in Winterthur, gave him the opportunity to start the building of steam engines at their factory. After great difficulties he succeeded.

In 1871 he left Sulzer and started his own firm, now known as the Swiss Locomotive Works in Winterthur. Although he was quite successful in building interesting locomotives, especially for narrow-gauge railways, he often had financial difficulties. He took an interest, at a very early date, in the Swiss electrical engineering industry, and he already had the idea that electric traction would be of great importance to Switzerland. When he was 58 he established an armaments factory near Naples. In 1890 he returned to Basle as a consulting engineer and died there in his 79th year.

BROWN, Charles Eugene (son of Charles Brown), 1863–1924. Born in Winterthur, died in Lugano.

Charles E. Brown came, at an early age, to the Oerlikon Engineering Works, where at the age of 23 he took over the management of the electrical department. He started to build powerful generators, which were among the largest in the world at that time, and in 1889 turned his attention to alternating current. He developed the first single-phase generators and motors, and, together with Dolivo-Dobrowolski, he developed the standards of multi-phase engines. After several years, Brown resigned from his position and started, together with W. Boveri, the famous firm of Brown-Boveri & Company. From 1900 onwards, Messrs Brown-Boveri started to build and improve steam turbines, following the patents of Parsons. Brown continued to manage his own firm until 1911 and then retired to Lugano.

BURY, Edward, 1794–1858. Early locomotive builder and LS of the London & Birmingham Rly., from 1836 to 1845. The first to use horizontal inside cylinders and crank axles. Characteristics were bar frames and haystack fire boxes of D-shape in plan. A number were exported to America where these features were widely adopted.

CAPROTTI, Arturo, 1881–1938. Spent his earlier years in the motor industry and in 1915 invented rotating cam valve gear for locomotives. First applied on the Italian State Rlys., where it has been widely adopted. Since used in various parts of the world and is still being developed in England.

CHAPELON, André, 1892–. After distinguished War service in the French Artillery he joined the P.L.M., and in 1925 the Paris-Orleans Rly., becoming Chief Experimental Engineer in 1936. His first success was the perfection of the "Kylchap" double blast-pipe in 1926. On this Railway he achieved the remarkable feat of doubling the power output of steam locomotives by scientific modification. The first "Pacific" rebuild appeared in 1929 and the first 4–8–0 in 1932. For further details see Illustrated Survey of Modern Steam Locomotives No. 41. In 1938 on the formation of the S.N.C.F., he was appointed to the Department of Steam Locomotive Studies, retiring as its Chief in July 1953. During this time he produced a series of most advanced designs, the execution of which was greatly curtailed by the War and subsequent rapid electrification. He is regarded as the foremost authority on the steam locomotive of his period.

CHURCHWARD, G.J., 1857–1933. CME of the Great Western Rly. from 1902 to 1921 and responsible for some earlier developments. His 1901 plan for standard locomotives was so farsighted that it was little modified when the G.W.R. was nationalized in 1948. For ten years his investigations covered every component of the locomotive. Appreciative of both American and Continental practice, three de Glehn compounds were bought for detailed study. Features of his design are combined plate and bar framing, long piston stroke, all coupled wheels individually machine balanced, and semi-plug piston valves with long travel operated by inside Stephenson valve gear giving rapid acceleration. The boiler with tapered barrel and Belpaire firebox was specially designed for rapid circulation and the narrow grate for Welsh coal. In May 1906 he fitted the first British main-line engine with Schmidt fire-tube superheater, perfecting his own type in 1909. His top feed system (1911) has been widely adopted. For express work the four-cylinder simple system with divided drive has remained unaltered since 1907. New types introduced into Britain were the 2–8–0 (1903), the 2–8–0 tank (1910) and a solitary 4–6–2, *The Great Bear* (1908). The outstanding British locomotive engineer of this century, his influence on design has become increasingly apparent.

CRAMPTON, Thomas Russell, 1816–88. In 1842 patented the type of locomotive that bears his name, the first being built in 1846. At that time when a low centre of gravity was considered essential, he combined a low-pitched boiler with large driving wheels by placing them behind the fire-box. The front end was borne by independent carrying axles and the outside cylinders were placed near the centre of the engine to minimize swaying. On the Continent he was far better known and had many more engines running than in England. In the 1850's his locomotives were probably the fastest in existence, being capable of 90 m.p.h. Years later, with four-coupled engines on the Continent, his influence persisted with central cylinders driving the rear axle. On the French Eastern Region a "Crampton" is still kept in full working order for special occasions.

DOLIVO-DOBROWOLSKI, M., 1862–1919. He was born in St Petersburg and died in Heidelberg. After studying in Germany he joined the A.E.G., and pioneered the work on multi-phase alternating currents. He is the inventor of the expression used on the Continent, *drehstrom*, meaning "revolving current" for three-phase A.C. Most of the basic principles of the three-phase A.C. technique were developed by him. He spent most of his life with the firm of A.E.G., finally as one of their technical directors.

DRIPPS, Isaac. In 1831 erected the *John Bull* imported for the Camden & Amboy R.R., although never having seen a locomotive. He added a rudimentary leading truck, forerunner of both pony truck and cowcatcher. In 1832–33 he experimented with coal burning, inventing the smoke-box deflector plate and the spark-arresting chimney with deflecting cone. His boiler design of 1832 shows the first combustion chamber and a large, wide firebox with sloping back plate, a remarkable forecast of the boiler of a century later. The *Monster* of 1836 was a 0–8–0 with coupled wheels in two groups connected by gearing. Since the C. & A.R.R. later merged into the Pennsylvania R.R. he is sometimes regarded as the first Motive Power Chief of that famous system.

ENGERTH, Wilhelm, 1814–84. Chief Engineer of the Austrian Southern Rly. None of the contestants at the Semmering trials of 1851 was considered wholly satisfactory and he was instructed to produce a design. Patented in 1852 the *Engerth* had the frames in two parts, of which the rear enclosed the firebox and was pivoted just in front of it. The cylinders drove the coupled wheels of the front unit which were connected to those of the rear unit by gearing and thus the "tender" weight was used for adhesion. However, the gears of those days and alternative forms of transmission proved unsatisfactory and were later abandoned. Use of part of the tender weight for adhesion was later revived in the Continental "Stutz-tender" locomotives.

FAIRLIE, Robert F., 1831–1885. In 1863 patented an articulated locomotive with the wheels arranged as two bogies, one or both being powered. The boiler could be single or double. The first, built in 1865, had a double boiler with the two fireboxes back to back partitioned but built as one unit. The next stage included one common firebox between the two barrels and in 1901 the Vulcan Foundry introduced two separate boilers. All these had both bogies powered, but "Single Fairlies" with one boiler and one steam bogie were built. In 1890 a few were built as compounds for the Saxon State Rly. "Fairlies" were used in many parts of the world before the advent of more modern articulated engines.

FORNEY, Matthias N., 1835–1908. A promoter of the New York Elevated Rly., who brought out in 1872 the type of locomotive for use thereon. This was an 0–4–4 back tank with vertical boiler. Most American tank engines carried their tank at the back and the term "Forney" became general for tank engines in the U.S.A.

GARBE, Dr Robert, 1847–1932. CME of the Berlin Division of the Prussian State Rlys. from 1895 to 1917. He carried out exhaustive experiments on the locomotive and was largely responsible for the rapid introduction of Schmidt superheaters. The success of his standard designs can be gauged by the numbers built not only for Germany, but elsewhere, e.g., 3,850 Class P8 4–6–0's, 5,260 G8 0–8–0's and 3,000 Class G 10, 0–10–0's. He was one of the greatest authorities on the locomotive.

GARRATT, Herbert William, 1864–1913. Patented the Garratt articulated locomotive developed by Beyer Peacock & Co., of Manchester. The first, built for Tasmania in 1909, was a compound, but all subsequent engines have been simples. The boiler is carried on a girder frame and its design is unrestricted by wheels, etc. This frame is pivoted towards the inner ends of the steam units which carry the tanks and fuel at their outer ends. Thus the engine follows the track curvature closely in three tangents. As the best type of articulated locomotive so far evolved some 2,000 have been build for many parts of the world.

DE GLEHN, Alfred G., 1848–1936. Born in England he became CME of the Alsacian Engine Works. In 1886 he designed and built a trial four-cylinder compound for the Northern Rly. of France. De Glehn compounds proper dated from 1890 and these most efficient and economical engines ran in many countries, and in France their possibilities were developed to a very advanced stage.

GOLSDORF, Dr Karl, 1861–1916. (Son of Adolf Golsdorf, CME of Austrian Southern Railway from 1885 to 1907.) CME of the Austrian State Rlys. from 1893 to his death. Few men have left such an unmistakable stamp on the locomotives of a country. Noted for the elegance and ingenuity of his designs, which numbered over sixty. He introduced the first two-cylinder Austrian compound in 1893 and developed an extremely simple automatic system whereby the engine worked semi-compound at long cut-offs. His first four-cylinder compound appeared in 1901 and both types were standard for many years. In 1900 his first ten-coupled engine was designed with adequate side-play for a long coupled wheelbase and this system superseded the cumbersome linkages of Klose and Hagans. At the same time he was the first to fit the Brotan boiler with water-tube firebox and separate steam drum. His 2–6–2 tender engines of 1904 were the first in Europe, whilst the famous 2–6–4 express engines of 1908 onwards and the 2–12–0 of 1911 were entirely new wheel arrangements. An 0–12–0 tank design for the Abt rack system was built in 1912. His valve gear, largely used on shunting engines, dispenses with the use of an expansion link. Not least amongst his many improvements were the very handy arrangement of the footplate controls and the first scientific numbering system for the locomotive stock, since widely copied.

GOOCH, Sir Daniel, Bart., 1816–89. First LS of the Great Western Rly. from 1837 to 1864 and chairman of the company from 1866 to 1889. A great protagonist of the broad gauge (7 ft), the famous 8-ft singles (4–2–2 type) first appeared in 1846, and by attaining speeds of 60 m.p.h. in the 1840's, earned for their designer the sobriquet of "father of express trains". *Lord of the Isles*, of this class, was one of the star attractions of the Great Exhibition of 1851 in London. The stationary link valve gear was brought out in 1843 and engines with condensing apparatus for underground work in 1862. Gooch was one of the earliest to insist upon interchangeability and standardization of engine parts and built the first dynamometer car for road testing in 1846. In 1865–66 he directed the laying of the first transatlantic cables in the great steamship *Great Eastern*. During this time the G.W.R. ran into difficulties and the Board elected him chairman on his return. Of his brothers, John V.Gooch was LS of the London & South Western Rly. from 1841 to 1850, and the Eastern Counties Rly. from 1851 to 1856.

GRESLEY, Sir Nigel H., 1876–1941. LS of the Great Northern Rly. from 1911 until 1923 and thereafter CME of the L. & N.E.R. until his death. He used three cylinders for all except small engines, the inside valves being operated by the simple form of conjugated gear associated with his name, and also applied in America. The first of his "Pacifics" came out in 1922 and the type was steadily developed during his regime. One of them, *Mallard*, attained a record speed of 126 m.p.h. on a trial run in 1938. New types introduced into Britain were the 2–8–2 for freight (1925) and the *Cock of the North* for express passenger (1934), also the 2–6–2 for mixed traffic (1936). In 1925 the first six-cylinder Garratt was built and in 1929 an experimental 4–6–4 No. 10000 appeared. This was a high-pressure four-cylinder compound with 450 lb. pressure and water-tube boiler. Many current developments were given a thorough trial on his engines including the booster, various poppet valve gears and feed-water heaters. He was also an exponent of articulated coaching stock and tried a similar arrangement by uniting the back of the engine and tender front over a common bogie.

HACKWORTH, Timothy, 1786–1850. Appointed to the Stockton & Darlington Rly. in 1825 he was the world's first locomotive superintendent. To keep the very primitive locomotives going was a task of exceptional difficulty. As the first to encounter many defects, his experience was invaluable to the early builders. Traffic on the S. & D. Rly. was confined to slow-moving coal trains which makes a weekly locomotive mileage of 600 in 1829 the more remarkable. His locomotives were the first to compete successfully with horses and up to 1830 he was pre-eminent in promoting their practical success.

His actual inventions were many but unfortunately they later became the subject of bitter controversy. His *Royal George* of 1827 was the first 0–6–0 and the most powerful locomotive of its day. He was amongst the first to use coupling rods, the coned blast pipe, inside cylinders with crank axle, short-stroke pumps, spring-loaded safety valves and a rudimentary feed-water heater.

His son, John W.Hackworth, patented his radial valve gear in 1859 and the name has become generic for gears of this type.

HALL, Joseph. An Englishman who went to Munich in 1839 and two years later built the first locomotive for J.A. Maffei. In 1856 he patented his arrangement of cranks with outside framing, which was widely adopted. In 1858 he moved to Austria and was engaged in various branches of railway engineering.

HAMMEL, Anton, 1812–1925. Entered the drawing office of the Maffei locomotive works in 1812, later becoming chief of the design department until his death. He re-introduced bar frames into Europe in 1905 and was responsible for many types on the Bavarian and Baden State Rlys. Most of these were four-cylinder compounds with bar frames, the Bavarian "Pacifics" and 0–8–8–0 tanks being amongst the very few non-standard designs built after formation of the Reichsbahn. His engines were always distinctive with their elegant design and beautiful finish.

HASWELL, John, 1812–97. Born in Glasgow he was appointed manager of the first locomotive works in Austria in 1837. In 1851 he produced the *Vindebona* for the Semmering trials, the first eight-coupled engine on the Continent. Introducing Stephenson's valve gear he incorporated many ideas of his own including rudimentary forms of the Belpaire firebox, thermic syphons and counter-pressure braking. The *Wien-Raab* of 1855, a large long-boiler 0–8–0 with all parts accessible, was the pattern for the Continental heavy freight locomotive for many years. The *Duplex* of 1861 was the first four-cylinder locomotive although the arrangement was unorthodox.

VON HELMHOLZ, Dr Richard, 1854–1934. Chief designer of the Krauss Locomotive Works in Munich from 1884 to 1917. In 1884 he brought out a form of Walschaert valve gear with straight expansion link, and in 1888 the Krauss-Helmholz truck combining one carrying and one coupled axle as a bogie. The latter's influence on Continental design has been enormous. Engines he designed for the Palatinate Rly. in 1894, and the Bavarian State Rly. in 1900, incorporated "boosters" in an early form. He was a great authority on the steam locomotive and its history.

HENRY, A. CME of the Paris, Lyons & Mediterranean Rly. from 1882 to 1892. In 1888 he introduced four-cylinder compounds on the P.L.M. and was the first to use considerably higher boiler pressure (213 lb.) on them. His 1889 compounds had the inside high-pressure cylinders *behind* the outside low-pressure cylinders, an arrangement which remained peculiar to the P.L.M.

HOLDEN, James, 1837–1925. LS of the Great Eastern Railway from 1885 to 1907. In 1893 introduced oil firing on British locomotives. In 1902 built the *Decapod*, a remarkable three-cylinder 0–10–0 tank engine capable of accelerating a 300-ton train from rest to 30 m.p.h. in thirty seconds to compete with electric traction. He was succeeded on the G.E.R. by his son S.D.Holden.

INGERSOLL, Howard L. Of the New York Central Railroad. Introduced the modern locomotive "booster" in 1919.

JERVIS, John B. CME of the Mohawk & Hudson R.R., was the first to apply a leading bogie to a locomotive, the 4–2–0 *Brother Jonathan* in 1832.

JOHNSON, Samuel W., 1831–1912. LS of the Great

Eastern Rly. from 1866 to 1873 and the Midland Rly. from 1873 to 1903. Designer of some of the most graceful engines, particularly the 4–2–2– type. ("Singles" having acquired a new lease of life by the invention of the steam sanding gear in 1886 by F.Holt.) In 1901, he introduced Smith compound 4–4–0's on the M.R. which remained the principal express type for over twenty years.

JOY, David, 1825–1903. Engineer and inventor in various fields. Patented his radial valve gear in 1879 and the first conjugated gear for three-cylinder engines in 1884. In this the motion for the centre valve is derived from the other two valves.

KETTERING, Charles F., 1887–1958. Responsible for General Motors diesel engines, he developed high speed two-stroke types of up to 3,000 B.H.P. His engines were specially developed for rail traction and in 1935 he prophesied that in 20 years no more steam locomotives would be built in the U.S.A.

LENTZ, Dr Hugo, 1859–1944. Ingenious inventor who will always be associated with the introduction of poppet valves on locomotives. Three principal systems were widely used:

(i) The Vertical type. Four vertical valves above the cylinder operated by a longitudinal slotted valve spindle and normal valve gear. First applied in 1905 it became standard practice on the Oldenburg State Rly.

(ii) The Oscillating Cam type with horizontal valves operated by a central camshaft and normal valve gear. First applied in Prussia in 1907 it became standard on the Austrian Federal Rly.

(iii) The Rotary Cam type also with horizontal valves and central camshaft, but rotating and operated by worm-drive. The cams are arranged in a series of steps, cut-off variation and reversal being effected by the transverse movement of the camshaft. First applied to a German industrial engine in 1921 it became standard on the Malayan Rly.

The first two systems could be fitted to existing engines with new cylinders only, but a fourth was suitable for existing piston valve cylinders. Concentric valves with oscillating cams were inserted at each end of the valve chamber, and connected by an outside link operated by normal valve gear. Another ingenious invention was his variable eccentric by which cut-off and reversal were effected in a single eccentric usually fitted outside.

LOMONOSSOF, Prof. George V., 1876–1952. In 1905 was appointed Professor of Railway Engineering at Kiev and later St Petersburg. Held many important positions, including CME of the Nicolas Rly., Director-General of Russian Railways and in 1917–18 was President of the Russian War Railway Mission to the U.S.A., being responsible for the design and ordering of 2,000 locomotives. In 1921 he was appointed High Commissioner for railway orders and fostered the development of large diesel locomotives, on which he was the first great authority. Later resident in England, Germany, U.S.A., and Canada, he was engaged in various researches and read his classic paper "Diesel Locomotives" to the Institute of Mechanical Engineers in 1933. He was also an authority on condensing locomotives.

MALLET, Anatole, 1837–1919. Patented a two-cylinder compound system in 1874, which was first used two years later. In 1884 he patented his four-cylinder compound system with the two low-pressure cylinders mounted on a front articulated frame. First adopted for light railways in 1887 it spread to European main lines by 1890 and was introduced into the U.S.A. on the Baltimore & Ohio R.R. in 1904. For many years the most widely used articulated type it reached its culmination in the Virginian Rly. 2–10–10–2's of 1918 which had 48-in. L.P. cylinders, the largest ever used on a locomotive. A few years earlier a variant type, the Triplex, was tried out in America. In these a steam tender formed a second L.P. unit, all six cylinders (2 H.P. and 4 L.P.) being of the same diameter. Three 2–8–8–2's were built for the Erie R.R. in 1914 and one 2–8–8–4 for the Virginian Rly. in 1916.

Non-compound engines on the Mallet plan were first used on the Trans-Siberian Rly. in 1902, very much against the inventor's will. The Pennsylvania R.R. introduced a simple 2–8–8–2 in 1911 and, after wartime interruption of experiments, a very powerful 2–8–8–0 in 1919. From 1923 the type became increasingly popular for the largest American freight locomotives and from 1930 for classes designed for much higher speeds. The world's largest steam locomotives were built on this plan, the Northern Pacific 2–8–8–4's, the Chesapeake & Ohio 2–6–6–6's and the Union Pacific 4–8–8–4 "Big Boys" all exceeding 500 long tons weight in working order.

MASON, William, died 1883. Builder of locomotives at Taunton, Mass., from 1853 to 1883. One of the first to adopt horizontal cylinders with longer bogie wheelbase, his engines were noted for their fine lines. He introduced the Walschaert valve gear into America in 1876 but his example was not followed. He built Fairlie articulated locomotives in the U.S.A., and adopted wedge horn-blocks, hollow-spoked wheels and the use of lead for balance weights.

MUHLFELD, James E. SMP of the Baltimore & Ohio R.R., responsible for the introduction of the first large Mallet-type engines in America in 1904 and for the adoption of Walschaert valve gear. Consulting Engineer to the Delaware & Hudson during the introduction of four high-pressure experimental compounds in 1927 to 1933.

NORRIS, William, and brothers. Began constructing locomotives in Philadelphia in 1832 and by 1859 had built 1,000 thereby ranking as the largest of the early builders. The first American firm to export locomotives to Europe, they delivered about 100 of the 4–2–0 and 4–4–0 types between 1837 and 1847. Of these at least fourteen 4–2–0's were supplied to the Birmingham and Gloucester Rly. in 1840–41. The Norris engines had a great influence in Central Europe where they introduced the bogie and the 4–4–0 wheel arrangement. The first engines built by a number of Continental firms were of the Norris type.

RAMSBOTTOM, John, 1814–97. LS of the London & North Western Rly. from 1857 to 1871. His inventions and improvements did much to advance the locomotive from a primitive to a modern machine. Most important were piston rings (1852), the displacement lubricator and improved safety valves (1856), screw reverse (1858), double-beat regulator (1859), water pick-up apparatus (1860) and solid-eye coupling rods (1866). He was also the first in England to fit injectors invented by H.Giffard in France in 1858. He re-organized Crewe works and included a steel plant. Under him reduction of locomotive classes to the minimum and standardization were developed to a high degree. Nine hundred and forty-three of his DX class 0–6–0 goods engines were built between 1858 and 1872, by far the largest class up to that time.

RAVEN, Sir Vincent L., 1859–1934. CME of the North Eastern Rly. from 1910 to 1923. On all his larger engines he adopted three cylinders with three independent sets of valve gear between the frames. Two engines were fitted with uniflow cylinders. In 1922 he designed a large electric passenger locomotive. No. 13 had the 2–Co–2 wheel arrangement and was of very advanced design for its day.

RICOUR, Théophile L., 1831–1916. Between 1854 and 1903 held a number of important engineering posts. CME of the Spanish Northern Rly. from 1861–67; he introduced counter-pressure braking in 1865. CME of the Etat Rly. of France from 1878 to 1886, he was the first to adopt extensively, modern piston valves in conjunction with relief valves.

RIGGENBACH, Nikolaus, 1817–99. In 1847 brought the first locomotive into Switzerland and in 1862 built his first rack locomotive. In this system a channel is laid down the track centre with the flanges upwards to hold a series of pins forming a ladder. The first rack-railway was opened in 1871 and the locomotives were equipped with his counter-pressure braking system, similar to the Le Chatelier brake, but using air and water in the cylinders with regulator and blast pipe closed. On his engines one set of cylinders drive the rack only or rack and adhesion coupled together.

ROGERS, Thomas, 1792–1856. Founded the Rogers Locomotive Works in 1837. In 1849 he was the first to adopt generally the link motion in America, and one of the first to use balance weights. In 1850 he brought out the "Wagon-top" boiler, the first to break away from the Bury type. In 1854 he introduced I-section coupling rods. He died in 1856 but his Works turned out some 6,300 locomotives up to 1905, by which time it had been absorbed into the American Locomotive Company.

SCHMIDT, Wilhelm, 1859–1924. Introduced modern high-degree superheating into locomotive practice, perhaps the most important single improvement in the present century. Three types were involved:

(i) A large central flue with horizontal elements, first applied on the Prussian State Rlys. in 1898.

(ii) A smokebox apparatus with annular tubing to which the firebox gases were conducted by a large central flue at the bottom of the barrel. As the large flue resulted in only 10 per cent heat loss, high-degree superheat was obtained. First used in Prussia in 1899.

(iii) The fire-tube type with elements housed in larger diameter flue tubes, first applied in Belgium in 1901 and within a decade was being fitted to nearly all large locomotives all over the world.

SEGUIN, Marc, born 1786. The first French locomotive engineer. In 1827 he rebuilt a second-hand Stephenson locomotive with a fire-tube boiler, an improvement he had already fitted on a steam launch. Independent inventor of the fire-tube boiler and blast pipe.

SHAY, Ephraim. Inventor of the Shay Patent Geared Locomotive. Since 1880 about 2,800 have been built, their manufacture being associated throughout with the Lima Locomotive Works, Ohio.

VON SIEMENS, Werner, 1810–92. He was born in Lenthe in Hanover and died in Charlottenburg.

Siemens' life is well known, though his activities in connection with the first electric railways are not so widely appreciated. In 1834 he entered the army as an artillery engineer and worked on ballistics. After the early death of his parents, he started in business to help his younger brothers. He continued to work first of all with his brother Wilhelm (later Sir William Siemens) and in 1879 he started the very first electric railway in the world at an exhibition in Berlin.

SMITH, Walter M., 1842–1906. First LS of the Imperial Japanese Government Rly. of Japan, 1874 to 1883, thereafter joining the North Eastern Rly. in England. His piston valve design came out in 1887 and three-cylinder compound system in 1899. Four-cylinder compounds were introduced on the N.E.R. in 1906, the year of his death.

STEPHENSON, George, 1781–1848. From 1803 to 1813 Superintendent Engineer of the Killingworth Collieries where locomotives were brought into successful use for the first time. After 1813 he was engaged extensively on the introduction of locomotives and the construction of new railways, including the Liverpool and Manchester. His great gifts, foresight and unshaken belief in the immense possibilities of railways made him their outstanding champion against opposition in the years before 1830 and he did more than anyone else to introduce them into general use.

In 1823, with his son Robert, with whom he was closely associated, he founded Robert Stephenson & Co., the first firm to be primarily concerned with locomotive building. As the world's greatest authority on railways at their inception he was the first President of the Institution of Mechanical Engineers in 1847 until his death in 1848.

STEPHENSON, Robert, 1803–59. Like his father was widely engaged on the development of early railways. Amongst his achievements were the construction of the London & Birmingham Rly., and the high-level bridge at Newcastle-on-Tyne. In 1829, assisted by Henry Booth, he produced the *Rocket* which is generally accepted as the first steam locomotive to combine the essential features of the orthodox type. Thereafter locomotive development by Robert Stephenson & Co. was rapid as the following brief summary of advances will show: 1830, smokeboxes, inner and outer firebox fully integrated with the boiler, inside cylinders under smokebox, crank axles. 1832, copper inner firebox and tubes, single-slide valves and a design for piston valves. 1833, six-wheeled engines of the 2–2–2, 2–4–0 and 0–6–0 types, brass and iron boiler tubes and engine steam brakes. 1835, firebox roof girders. 1841, long boiler engines which greatly improved boiler performance without increasing the wheelbase and had considerable influence on the Continent; common steam chests between the cylinders and eccentric-driven pumps. 1842, William Williams and William Howe evolved the link motion at his works. 1845, rear-driven long boiler engines of the 4–2–0 type for steadier running at speed, and the 4–2–2 type designed. 1846, three-cylinder locomotives, two built.

The genius of both Stephensons lay in their instinctive selection of sound ideas from the flood of proposals submitted. Robert Stephenson followed his father as second President of the Institution of Mechanical Engineers from 1848 to 1853.

STIRLING, Patrick, 1820–95. LS of the Glasgow & South Western Rly., 1853–66, and the Great Northern Rly.,

1866-95. On the latter his famous 8-ft singles (4-2-2's) were the principal express passenger engines from 1870 until his death, and achieved world-wide fame. His brother James followed as LS of the G. & S.W.R. from 1866 to 1878, originating the steam reversing gear in 1874. From 1878 to 1898 he was LS of the South Eastern Rly. Patrick's son Matthew was LS of the Hull & Barnsley Rly. from 1885 to 1922. Stirling engines all bore a family likeness with domeless boilers, simple lines and rounded cabs.

STRONG, George S. Introduced in America new locomotive types much in advance of their time, but remarkable forecasts of the shape of things to come. In 1885-86 he introduced the first of the 4-4-2, 4-6-2 and 2-10-2 types. Amongst novel features were twin circular corrugated fireboxes in a wide casing and vertical grid-iron valves operated by valve gear of the Hackworth type. His engines suffered from the combination of too many new and untried devices.

STROUDLEY, William, 1833-89. LS of the London, Brighton & South Coast Rly. from 1870 until his death. His locomotives were small but of excellent design and beautifully finished in a livery of yellow ochre elaborately lined out, and crimson framing. Some of the twenty-four-ton "Terrier" tank engines have lasted until the present time.

STUMPF, Johann. Introduced the uniflow principle into more recent locomotive practice. Three types were evolved:

(i) with vertical poppet admission valves in 1908;
(ii) with piston valves in 1912;
(iii) with the exhaust passage designed as a Venturi tube in 1920. Developments were cut short by the prior adoption of superheating and World War I.

THURY, René, 1860-1938. Swiss. Built in 1884 an experimental rack railway in Territet, a suburb of Montreux, to connect a hotel, several hundred feet up the mountain slope, with the town. Went to the United States where he worked for T.A.Edison. He was responsible for many inventions in electrical engineering, especially those concerned with the series coupling of electric motors. For many years he worked at Dick, Kerr & Co., of Preston, which today are part of the English Electric Company.

TREVITHICK, Richard, 1771-1833. His first attempts to construct a steam locomotive date from 1796 and the first practical road locomotive was built in 1801. The first true railway locomotive ran on Penydarran tram-road in Wales in February 1804. The "rails" were formed of cast iron plates and a train of trucks conveying ten tons of iron bars and seventy persons was successfully hauled nine miles. The engine used high-pressure steam exhausting to the atmosphere and the smooth tread wheels worked by adhesion (although geared transmission was used between the single-cylinder engine and the driving axles) which entitled Trevithick to the title of "father of the locomotive". In 1808 he laid down a circular railway in what is now the Bloomsbury district of London for exhibition purposes. Tickets for admission were five shillings and it may be regarded as the world's first passenger railway. In addition to the use of plain adhesion and the first practical steam locomotive, Trevithick is credited with other inventions including the multi-tubular boiler, blast pipe, lead plugs and direct drive with crank axles. He also fore-

saw the possibilities of feed-water heating and high degree superheat. He was, however, engaged in many other activities and his restless and adventurous nature unfortunately prevented his concentration on locomotive development. His eldest son, Francis, became an early LS of the London & North Western Rly. and other sons and nephews locomotive chiefs in various parts of the world.

VANDERBILT, Cornelius. In 1899 patented in America a boiler with tapered barrel and circular corrugated firebox, resembling that introduced by G.Lentz in Germany in 1888. A few years later he brought out a form of tender with circular water tank. The name Vanderbilt has become generic for boilers and tenders of these patterns.

VAUCLAIN, Samuel M., 1852-1940. Son of Andrew Vauclain, one of Baldwin's original workmen, he entered the Baldwin Locomotive Works in 1883 and became President in 1919. Of the many patents to his name those concerning his compounds are best known. In 1889 he brought out the system whereby the high- and low-pressure piston rods are connected to common cross-heads on each side of the engine. Later he adopted four-crank arrangements and up to 1907 over two thousand of his compounds were built. He was closely connected with Baldwin locomotive design during five decades.

VOGT, Axel S., died 1922. Joined the Pennsylvania R.R. in 1874 and was Mechanical Engineer from 1887 to 1919. Responsible for many of the beautifully proportioned and elegantly designed Pennsylvania classes which in turn had considerable influence on modern American locomotive design.

WAGNER, Dr R.P., 1882-1953. Chief of Design of the German State Rly. from its inception in 1922 to 1942. Responsible for the standard locomotive designs produced from 1924 onwards for all classes of traffic. The majority were two-cylinder simples, but the three-cylinder arrangement was adopted for the larger engines from 1935.

WALSCHAERT, Egide, 1820-1901. Foreman of the Brussels depot of the Belgian State Rlys. from 1844 to 1885. His valve gear, which has become universal, was patented in an early form in 1844 and first applied in 1848. A similar gear was independently invented in Germany by Heusinger in 1848 where it is still known by his name. It was introduced into America by Mason in 1876, but was forgotten until 1904 when it began to come into use both there and in England.

WEBB, Francis, W. 1835-1904. CME of the London & North Western Rly. from 1871 to 1903, is unfortunately remembered more for the erratic performances of his three-cylinder compounds than for his lasting and constructive improvements. These compounds introduced in 1882 with divided drive were, however, the forerunners of such types as the de Glehn, which eventually ranked amongst the most successful and economical ever built. A pioneer in the use of steel in Great Britain for frames, boilers and engine parts generally. His 500 0-6-0 coal engines were probably the simplest and cheapest ever built for a main line. One of these was erected in twenty-five and a half hours. In 1876 his radial axlebox was introduced and in 1880 he was the first to adopt Joy's valve gear. His experiments on compounding started in 1878 and his trials and patents covered a wide field. His autocratic manner and resentment of suggestion mitigated against many of his ideas being brought to a more successful conclusion.

WESTINGHOUSE, George, 1846–1914. Great American inventor and industrialist whose interests covered almost the entire field of engineering. His invention of the air-brake revolutionized safer rail travel. The first set was made in 1868 and from 1869 to 1873 he patented twenty-one improvements, including the triple valve. Introduced into Europe in 1871 the Westinghouse brake and its derivatives has become the most widely used continuous braking system.

WINANS, Ross. A prolific inventor and builder of early American locomotives, much of his work being done at the Mount Clare shops of the Baltimore & Ohio R.R., which were leased to him and his partners. He was concerned largely with freight engines amongst which were the 0–4–0 "Crabs" and 0–8–0 "Mud-diggers", with vertical boilers and horizontal outside cylinders driving through layshafts with outside coupling rods. In 1847 he was the first to burn anthracite successfully using a long firebox and in the following year brought out the "Camel", the first engine to have the cab on the boiler top, the fireman being stationed on the tender. In appreciation of the need of large engines he was ahead of his time: the *Centipede*, of 1854, was a 4–8–0 with eight-wheel tender. As early as 1829 he pointed out the advantages of coned wheel treads. His valve gear incorporated additional cams to secure almost instantaneous opening and closing of the ports. He was a pioneer in coal-burning and introduced the petticoat pipe

Appendix I. Wheel arrangement: steam power

The wheel notation of steam motive power by the Whyte system

This form of classification is generally accepted in Great Britain and the British Commonwealth and in North and South America. In Europe the French system is often used, i.e. the axles, not wheels, are indicated; so a Pacific 4–6–2 becomes 2–3–1. Another Continental system is to indicate the number of coupled axles by means of capital letters: thus a Pacific 4–6–2 becomes 2–C–1.

o o	0–4–0	Four-wheel shunter.
o o o	0–4–2	
o o o	0–6–0	Six coupled.
o o o o	0–8–0	Eight coupled.
o o o o o	0–10–0	Decapod (English).
o o o	2–4–0	
o o o o	2–4–2	Columbia.
o o o o	2–6–0	Mogul.
o o o o o	2–6–2	Prairie.
o o o o o o	2–6–4	
o o o o o	2–8–0	Consolidation.
o o o o o o	2–8–2	Mikado ("Mike").
o o o o o o o	2–8–4	Berkshire.
o o o o o o	2–10–0	Decapod (American).
o o o o o o o	2–10–2	Santa Fé.
o o o o o o o o	2–10–4	Texas.
o o o o	4–4–0	American.
o o o o o	4–4–2	Atlantic.
o o o o o o	4–4–4	Double-ender.
o o o o o	4–6–0	Ten-wheeler.
o o o o o o	4–6–2	Pacific.
o o o o o o o	4–6–4	Baltic (European). Hudson (American).

o o o o o	4–8–0	Twelve-wheeler.
o o o o o o	4–8–2	Mountain. Mohawk (New York Central).
o o o o o o o	4–8–4	Northern. Niagara (New York Central).
o o o o o o	4–10–0	Mastodon.
o o o o o o o	4–10–2	Southern Pacific.
o o o o o o o o	4–12–2	Union Pacific.

o o o o	0–4–4
o o o o	0–6–2
o o o o o	0–6–4
o o o o o	0–8–2
o o o o o o	0–8–4
o o o o o o	2–6–6

While many of the above types are seen in both tender and tank engine form, these (*left*) are almost entirely tank engine wheel formations. A tank engine is indicated by a "T" after the wheel notation thus: 4–4–2T.

Articulated types fall into two main groups. (1) Those in which both engines are placed under the boiler, as in the Mallett. These are nearly always tender engines. (2) Those in which the boiler is slung between the two engine units as in the Garratt. In the first group it is customary to show the wheel arrangement as having four groups only thus: 2–6–6–2.

In the second group there may be six groups of wheels and the notation is shown thus: 4–8–2+2–8–4. These engines are always tank engines in that they do not have a separate tender, but the suffix "T" is never used in the case of Garratt type engines.

Some examples of both types of articulated engine are given below:

o o o o o o o o	2–6–6–2	
o o o o o o o o o o	2–6–6–6	
o o o o o o o o o o	2–8–8–2	Mallett type.
o o o o o o o o o o o o	4–8–8–4	
o o o o o o o o o o	4–6–0+0–6–4	
o o o o o o o o o o o o	4–6–2+2–6–4	
o o o o o o o o o o o o	2–8–2+2–8–2	Garratt type.
o o o o o o o o o o o o o o o o	4–8–4+4–8–4	

Appendix II. Wheel arrangement: diesel and electric wheel notation

WHEEL ARRANGEMENT	NOTATION
INDIVIDUALLY DRIVEN AXLES COUPLED WHEELS CARRYING WHEELS	
	B
	Bo
	Bo-Bo
	Bo+Bo
	B-B
	C
	Co
	C-C
	Co-Co
	Co+Co
	A1A-A1A
	1Co-Co1
	1-Co-Co-1
	2-Co-Co-2
	1-Do-1
	2-Do-2
	1A-Do-A1

As the original system of Whyte notation evolved for steam locomotives is not suitable to indicate the wheel arrangement of diesel and electric locomotives, a different form has been devised in which letters are used for driving axles, and numbers for carrying (i.e., non-driving) axles. In this system "A" stands for one driving axle, "B" for two, "C" for three and "D" for four. When, however, the axles are in groups, each axle being individually driven, a suffix "o" is added after the letter (except for "A" which can of course only be individually driven). A hyphen is used to break up separate wheel groups, such as bogies, although a + sign replaces this when an articulated joint links the wheel groups together.

In the case of shunting locomotives having two or three driving axles only, the Whyte notation is often used, and these may be indicated as 0–4–0 or 0–6–0 locomotives instead of type "B" or type "C".

The accompanying table sets out some of the wheel notations for diesel electric and gas turbine locomotives at present in existence – evolution of new types may add to the variety, but this system will cover all arrangements.

Reference should be made to Chapter II, Wheel Arrangements, page 128, for American usage of this system. e.g. B–B (U.S.A.)=Bo–Bo.

Appendix III. Glossary of Locomotive Terms
English–American

English	American	English	American
Alligator crosshead	Two-bar or Laird crosshead	Firebar	Grate bar
Ashpan hopper	Ashpan dump	Fog signal	Torpedo
		Foundation ring	Mud ring
Backplate (of boiler)	Back head	Fusible plug	Drop plug
Back buffer beam	Back bumper, tail piece or end sill	Gudgeon pin	Wrist pin
Banking engine	Helper		
Belpaire throat plate	Hip joint	Hand hold or grip	Grab iron
Bogie	Four-wheel truck	Hornblock	Pedestal
Bracket bearing	Box	,, stay	,, binder, tie or brace
Brake block	Brake shoe	,, ,, (circular section)	,, thimble
,, ,, carrier	,, head		
,, shaft	,, mast	Injector	Inspirator
Breakdown crane	Wrecking crane	Inside cylinder loco.	Inside connected loco.
Buffer beam	Bumper or pilot beam		
		Keep (of axlebox)	Cellar
Cab fall plate	Cab apron		
Cab floor	Deck	Laminated spring	Semi-elliptic spring
Cast iron (best quality)	Gun iron	Leading bogie	Four-wheel lead truck
Camber (of springs etc.)	Arch	Liner	Bushing
Carrier bracket	Yoke	Longitudinal frame (of tender)	Sill
Centre pin	King bolt		
Chimney	Smokestack or stack	Main steam pipe	Dry pipe
Clack box	Check valve case	,, ,, tee-piece	The head or nigger head
Coach screw	Lag screw	Motion plate	Guide yoke
Compensating beam	Equalizer		
Coned or taper boiler	Wagon top boiler	Name plate	Badge plate
Connecting rod	Main rod		
Countersunk footsteps	Box steps	Outside axle box	Journal box
Couplers	Knuckles		
Coupling rod	Side rod or parallel rod	Petrol-electric	Gas-electric
Cowcatcher	Pilot	Petticoat pipe	Draft pipe
Crown sheet (of firebox)	Roof sheet	Piston tail rod	Extended piston rod
Cylinder cover	Cylinder head	Platform	Run board or running board
		Pricker	Slash bar
Diamond frame bogie	Arch bar truck		
Drawbar	Draft iron	Radial axle	Two-wheel idler axle
		Ramp (for re-railing)	Frog
Expansion bracket (of boiler)	Expansion pad	Regulator	Throttle
Extension rod (of valve gear)	Transmission bar	,, rod	,, rod, stem or reach rod

English	American	English	American
Release valve	Bleeding cock	Spindle (valve, etc.)	Stem
Reversing lever	Johnson bar	Spring buckle	Spring band
„ rod	Reach rod	Stay	Tie or brace
„ shaft	Tumbling shaft	Stay (firebox)	Staybolt
Rocking grate	Finger grate or shaking grate	Swan neck	Goose neck
		Swashplate	Splash plate
„ shaft bearing	Rocker box		
Rod bushes and brasses	Stubs	Top member (of diamond bogie frame)	Arch box
Rubbing block	Chafing plate or chafing iron	Train heating valve	Heater cock
Shovel	Scoop	Tube plate	Tube sheet
Shunting	Switching		
„ engine	Switcher	Water troughs	Track pans
Silencer	Muffler	Westinghouse brake	Air brake
Sliding joint	Slip joint	„ brake radiator	Aftercooler
Solid wheel centre	Plate wheel	„ brake reservoir	Air drum

American–English

American	English	American	English
Aftercooler	Westinghouse brake radiator	Deck	Cab floor
		Draft iron	Drawbar
Air brake	„ brake radiator	„ pipe	Petticoat pipe
Air drum	„ brake reservoir	Drop plug	Fusible plug
Arch	Camber (of springs, etc.)	Dry pipe	Main steam pipe
„ bar truck	Diamond-frame bogie		
„ box	Top member (of diamond frame)	Equalizer	Compensating beam
		Expansion pad	Expansion bracket (of boiler)
Ashpan dump	Ashpan hopper	Extended piston rod	Piston tail rod
Back bumper, tail piece or end sill	Back buffer beam	Finger grate and shaking grate	Rocking grate
„ head	Backplate (of boiler)	Four-wheel truck	Bogie
Badge plate	Name plate	Frog	Ramp (for re-railing)
Bleeding cock	Release valve		
Box	Bracket bearing	Gas-electric	Petrol-electric
„ steps	Countersunk footsteps	Goose neck	Swan neck
Brake head	Brake block carrier	Grab iron	Hand hold or grip
„ mast	„ shaft	Grate bar	Firebar
„ shoe	„ block	Guide yoke	Motion plate
Bumper or pilot beam	Buffer beam	Gun iron	Cast iron (best quality)
Bushing	Liner		
		Heater cock	Train heating valve
Cab apron	Cab fall plate	Helper	Banking engine
Cellar	Keep (of axlebox)	Hip joint	Belpaire throat plate
Chafing plate or chafing iron	Rubbing block	Inside connected loco	Inside cylinder loco
Check valve case	Clack box	Inspirator	Injector
Cylinder head	Cylinder cover		

GLOSSARY OF LOCOMOTIVE TERMS

American	English	American	English
Johnson bar	Reversing lever	Slash bar	Pricker
Journal box	Outside axle box	Slip joint	Sliding joint
		Smokestack or stack	Chimney
King bolt	Centre pin	Splash plate	Swashplate
Knuckles	Couplers	Spring band	Spring buckle
		Staybolt	Stay (firebox)
Lag screw	Coach screw	Stem	Spindle (valve, etc.)
		Stubs	Rod bushes and brasses
Main rod	Connecting rod	Switcher	Shunting engine
Mud ring	Foundation ring	Switching	Shunting
Muffler	Silencer		
		Tee head or nigger head	Main steam tee-piece
Pedestal	Hornblock	Throttle	Regulator
„ binder, tie or brace	„ stay	„ rod, stem or reach rod	„ rod
„ thimble	„ „ (circular section)	Tie or brace	Stay
		Torpedo	Fog signal
Pilot	Cowcatcher	Track pans	Water troughs
Plate wheel	Solid wheel centre	Transmission bar	Extension rod (of valve gear)
Reach rod	Reversing rod	Tube sheet	Tube plate
Rocker box	Rocking shaft bearing	Tumbling shaft	Reversing shaft
Roof sheet	Crown sheet (of firebox)	Two-bar or Laird cross-head	Alligator crosshead
Run board or running board	Platform	Two-wheel idler axle	Radial axle
Scoop	Shovel	Wagon top boiler	Coned or taper boiler
Semi-elliptic spring	Laminated spring	Wrecking crane	Breakdown crane
Side rod or parallel rod	Coupling rod	Wrist pin	Gudgeon pin
Sill	Longitudinal frame (of tender)	Yoke	Carrier bracket

For Further Reading

DIESEL LOCOMOTIVES

Diesel Engine Design, by H. P. F. PURDAY, Constable & Co., Ltd (1948).

Diesel Locomotives and Railcars, by BRIAN REED, The Locomotive Publishing Co., Ltd.

The Diesel Locomotive, by R. L. ASTON, Thames & Hudson, Ltd.

Electric and Diesel-Electric Locomotives, by D. W. HINDE, Macmillan & Co., Ltd (1948).

Car Builders' Cyclopedia, Simmons-Boardman Publishing Corporation, New York, N.Y. (1931 and 1953 editions).

Diesel-Electric Locomotive Handbook (Electrical and Mechanical Equipment volumes), by GEORGE F. McGOWAN, Simmons-Boardman Publishing Corporation, New York, N.Y., 1951.

Locomotive Cyclopedia, Simmons-Boardman Publishing Corporation, New York, N.Y. (1930, 1941 and 1956 editions).

Moody's Transportation Manual, Moody's Investors Service, New York, N.Y. (1957 edition).

On Time, by FRANKLIN M. RECK, Electro-Motive Division, 1948. General Motors Corporation, La Grange, Ill. (Privately published).

Railway Age (first issue 1921), Simmons-Boardman Publishing Corporation, New York, N.Y.

Super-Railroads For A Dynamic American Economy, by JOHN WALKER BARRIGER, Simmons-Boardman Publishing Corporation, New York, N.Y., 1956.

Trains Magazine, Kalmbach Publishing Company, Milwaukee, Wis. (volumes 1–19, 1940–59).

World Railways (5th edition), Rand McNally & Company, New York, N.Y.

Papers published by the Institution of Locomotive Engineers:
Journal No. 161: 'Diesel Shunting Locomotives', by C. E. FAIRBURN.
  „ „ 174: 'Some Experiences with Railcar Oil Engines in the Argentine', by C. R. PARKER.
  „ „ 179: 'The Maintenance of Diesel-Electric Shunting Locomotives on the L.M.S.', by C. E. FAIRBURN.
  „ „ 188: 'Diesel-Electric Locomotives – Running and Maintenance on the Buenos Aires Great Southern Railway', by H. M. MACINTYRE.
  „ „ 222: 'Practice and Trend in Development of Diesel Engines with Reference to Traction', by G. JENDRASSIK.
  „ „ 227: 'The Fell Diesel-Mechanical Locomotive', by L. F. R. FELL.
  „ „ 233: 'Running Tests of a 500-H.P. Diesel Mechanical Locomotive', by B. REED.
  „ „ 241: 'A Modern Hydraulic Drive for Locomotives', by R. H. FETT.

Journal No. 245: 'The Uerdingen Rail bus', by E. KREISSIG.
  „ „ 251: 'Evolution of the Internal Combustion Locomotive', by J. M. DOHERTY.
  „ „ 253: 'Experiences with Diesel Railcars', by M. J. DEVEREUX.
  „ „ 256: 'Diesel Locomotive Building and Maintenance', by T. F. B. SIMPSON.

Papers published by the Institution of Mechanical Engineers:
Volume 124: 'The Compression Ignition Engine and its applicability to British Railway Traction', by L. F. R. FELL.
  „ 130: 'Recent Developments in Hydraulic Couplings', by H. SINCLAIR.
  „ 130: 'The Lysholm-Smith Torque Converter,' by H. F. HAWORTH & A. LYSHOLM.
  „ 130: 'Voith Turbo Transmission', by WILHELM HAHN.
  „ 139: 'Some Problems in the Transmission of Power by Fluid Couplings', by H. SINCLAIR.
  „ 140: 'Control of Diesel Railcars with Particular Reference to Transmissions', by W. G. WILSON.

ELECTRIC LOCOMOTIVES

The literature on electric railways is somewhat limited. There are a number of excellent text-books but many of the older ones are out of print. Recently, however, there have appeared a number of good publications which enable the student of electric traction to gain excellent and reliable information. A number of first-class books exist in the French and German languages, and some of these are mentioned below:

British Electric Trains, by H. W. A. LINECAR, Ian Allan Ltd, 1947.

Electric and Diesel-Electric Locomotives, by D. C. and M. HINDE, Macmillan & Co., Ltd, 1947.

Electric Railways, by S. ASHE and J. W. KEILEY, Constable & Co., Ltd, 1905.

Electric Railway Engineering, by T. FERGUSON, Macdonald & Evans Ltd, 1955.

Electric Traction, by E. P. BURGH, McGraw-Hill Inc., 1911.

Electric Traction, by A. T. DOVER, (Sir Isaac) Pitman & Sons Ltd, 1937 and 1956.

Electric Traction, by PHILIP DAWSON, 'The Electrician,' 1909.

Electric Traction, by R. H. SMITH, Harper Bros, 1900.

Electric Traction, by E. WILSON and F. LYDALL, Edward Arnold, 1907.

Electric Traction Engineering, by E. A. BINNEY, Cleaver-Hume Press Ltd, 1955.

Electric Traction Jubilee, by J. H. CANSDALE, BTH., 1946.

Electric Trains, by Agnew, Virtue & Co., 1937.

Electric Trains and Locomotives, by B. K. COOPER, Hill (Leonard) Ltd, 1953.

Electric Triebfahrzeuge, by K. SACHS, Huber, 1953.

Electric Zugfoerderung, by E. E. SEEFEHLNER, Springer, 1924.

Individual Axle Drive, by A. HUG, Congress Journal, 1949–50.

Individual Axle Drive, by A. HUG, O. Fuessli, 1931.

Single-Phase Railways, by E. AUSTIN, Constable & Co., Ltd, 1915.

The Early History of the Electric Locomotive, by F. J. G. Haut, Newcomen Society, 1952.

Periodicals

Electric Railway Traction. Supplement to Railway Gazette (ended 1940).

Engineering. The vols. 1880 to date contain many illustrated references to Electric Locomotives.

STEAM LOCOMOTIVES

The design and construction of steam locomotives is well documented and there are few, if any, aspects of it which have not received adequate treatment.

In addition to the many books, some of which are listed hereafter, there are additional sources of information available in the form of articles in the technical press and especially papers and addresses presented to the Engineering Institutions. The proceedings of the Institution of Locomotive Engineers (28 Victoria Street, London, S.W.1), contain nearly 700 papers and addresses, and the subject-index issued by that Institution contains no less than sixty-two headings. Practice the world over is considered and copies of these papers (together with the discussions), the majority of which are available for purchase, are particularly recommended to those wishing to make a more detailed study of matters necessarily treated somewhat sketchily in this book.

In the case of some periodicals, current numbers may not contain any reference to steam practice, but much information has appeared in former volumes.

Many excellent publications have been issued by railways, locomotive builders, and the manufacturers of components; generally speaking, the distribution of these publications is necessarily limited to those professionally concerned, and for this reason they are not mentioned here. Literature which is now mainly of historic interest is also excluded with the exception of a few older books which cover practice on locomotives still extant.

Periodicals:

Bulletin of the International Railway Congress Association, Brussels.

Organ fur die Fortschritte das Eisenbahnweseus, 1846 to date.

The Engineer, London, 1856 to date.

Engineering, London, 1866 to date.

The Railway Age, New York, 1876 to date.

Glasers Annalen, Berlin, 1877 to date.

**The Railway Engineer*, London, 1880–1935.

Revue Generales des Chemins de Fer, Paris, 1881 to date.

The Locomotive, London, 1896 to date.

The Railway Gazette, London, 1904 to date.

*Die Locomotive, Vienna, 1904–44.

Railway Mechanical Engineer, New York, 1916–52.

Railway and Locomotive Historical Society Bulletins, 1921 to date.

Railway Locomotives and Cars, New York, 1952 to date.

*Publication discontinued.

Transactions of Engineering Institutions, etc.:

American Society of Mechanical Engineers, New York; Association of American Railroads, Chicago; Institution of Civil Engineers, London; Institution of Locomotive Engineers, London; Institution of Mechanical Engineers, London; The Royal Society, London; University of Illinois Bulletin, Urbana.

Books:

Application of Highly Superheated Steam, by R. GARBE, Crosby Lockwood & Son, London, 1908.

Articulated Locomotives, by L. WIENER, Constable & Co., Ltd, London, 1930.

Balancing of Locomotives, by W. E. DALBY, Edward Arnold & Co., Ltd, London, 1930.

British Locomotive Types, Tothill Press Ltd, London, 1946.

Berechnung und Konstruktion von Dampflokomotiven, by W. BAUER und X. STURZER. C. W. Kreidel's Verlag, Berlin, 1923.

British Steam Locomotive 1825–1925, by E. L. AHRONS, Locomotive Publishing Co., Ltd, London, 1927.

Century of Locomotive Building by Robert Stephenson & Co., 1823–1923, by J. G. H. WARREN, Reid (Andrew) & Co., 1923.

Das Eisenbahn-Maschinenwesen der Gegenwart, various authors. Kreidel's Verlag, Wiesbaden, 1920.

Der Dampfbetrieb der Schweizerischen Eisenbahnen 1847–1936, by A. MOSER, Birkhauser, 1936.

Design for Welding, The James F. Lincoln Arc Welding Foundation, Cleveland, Ohio, 1948.

De Locomotieven van der Hollandsche Yseren Spoorweg Maatschappij, 1839–1921, by J. J. KARSTENS.

Development of British Locomotive Design, by E. L. AHRONS, Locomotive Publishing Co., Ltd, London, 1914.

Development of the Railcar, by R. W. KIDNER, Oakwood Press, 1949.

Die Dampflokomotive, by J. JAHN, J. Springer, Berlin, 1924.

Die Dampflokomotiven der Gegenwart, by R. GARBE, J. Springer, Berlin, 1921.

Dynamics of Energy Movement, by L. K. SILLCOX, Massachusetts Institute of Technology, 1952.

Essays of a Locomotive Man, by E. A. PHILLIPSON, Locomotive Publishing Co., Ltd, London, c. 1930.

Four Thousand Miles on the Footplate, by O. S. NOCK, Ian Allan, Ltd, London, 1952.

Fundamentals of the Locomotive Machine Shop, by F. M. A'HEARN, Simmons-Boardman Publishing Co., New York, 1926.

German Locomotive Experience with Pulverised Fuels, H. M. Stationery Office, London, 1946.

Halting High Speed Horses, by L. K. SILLCOX, Massachusetts Institute of Technology, 1952.

History of the Railway Locomotive down to the end of the Year 1831, by C. F. DENDY MARSHALL, Locomotive Publishing Co., Ltd, London, 1953.

Horsepower of Locomotives – its Calculation and Measure-

ment, by E. L. DIAMOND, The Railway Gazette, London, 1936.

25 Jahre Deutsche Einheitslokomotive, 1925–1950, Stockklausner & Weinstottee, Miba-Verlag, Nurnberg, 1950.

Injectors, their Theory, Construction, and Working, by W. W. F. PULLEN, The Technical Publishing Co., Ltd, Manchester, 1900.

La Locomotova (Spanish), Hereter & Nieguel, Apolo, Barcelona, 1923.

Lectures on Superheating on Continental Locomotives, by E. SAUVAGE, University of London Press, Ltd, London, 1911.

Leitfaden fur den Dampflokomotivdienst, by L. NIEDERSTRASSER, Verkehrswissenschaftliche Lehrmittelgesellschaft, Frankfort-on-Main, 1954.

Locomotive, by U. LAMELLE ET F. LEGEIN, Ramlot, Bruxelles, and Dunod, Paris.

Locomotive Actuelle (3rd edn. 1948), by M. DEMOULIN, Librairie Polytechnique, Paris.

Locomotive Actuelle à Vapeur, by E. DEVERNAY, Dunod, Paris, 1954.

Locomotive Catechism, by R. GRIMSHAW, The Norman W. Henley Publishing Co., New York, 1923.

Locomotive Compounding and Superheating, by J. F. GAIRNS, Griffin (Charles) & Co., Ltd, London, 1907.

Locomotive Cyclopaedia, Simmons-Boardman, New York.

Locomotive Engineers Pocket Book, 42nd Edition, Locomotive Publishing Co., Ltd, London.

Locomotive Firing, The New York State Vocational and Practical Arts Association, Buffalo, N.Y., 1944.

Locomotive Injectors, Locomotive Publishing Co., Ltd, London, 1913.

Locomotive Maintenance and Operation, by R. E. BRINKWORTH, 1938, Thacker, Spink & Co., Ltd, Calcutta, 1933.

Locomotive Management: Cleaning, Driving, Maintenance. (1st Edn. 1909, 10th Ed. 1954.) Tenth edition revised by W. R. Oaten, 1954, earlier editions Hogdson and J. W. Williams, 1909, by J. T. HODGSON and C. S. LAKE, Tothill Press, Ltd, London.

Locomotive Operation, by G. R. HENDERSON, The Railway Age, Chicago, 1904.

Locomotive Practice and Performance in the Twentieth Century, by C. J. ALLEN, Heffer, Cambridge, 1949.

Locomotive Performance, by W. F. M. GOSS, J. Wiley, New York, 1907.

Locomotive Running Department, by J. G. B. SAMS, Locomotive Publishing Co., London, 1939.

Locomotive in Service, Locomotive Publishing Co., Ltd, London.

Locomotive Sparks, by W. F. M. GOSS, J. Wiley, New York, 1914.

Locomotive of Today (Various editions), Locomotive Publishing Co., Ltd, London.

Locomotive Valve Gears and Valve Setting, Locomotive Publishing Co., Ltd, London, 1924.

Locomotive Valves and Valve Gears, by J. H. YODER and G. B. WHAREN, D. van Nostrand Co., New York, 1917.

Locomotive Valves and Valve Gears, by C. S. LAKE and A. REIDINGER, Percival Marshall & Co., Ltd, London, 1940.

Locomotive à Vapeur (2nd Edn. 1952), by ANDRE CHAPELON, J. B. Bailliere et Fils, Paris.

Locomotives (Many Editions 1935–1949), by A. M. BELL, Virtue & Co., Ltd, London.

Locomotive on Curved Track. Mechanics of a, by S. M. PORTER, The Railway Engineer, London, 1935.

Locomotives Simple, Compound and Electric, by H. C. REAGAN, John Wiley, New York, 1907.

Locomotives and their Working, by C. R. H. SIMPSON and F. B. ROBERTS, Virtue & Co., Ltd, London, 1951.

Lokomotivy (Russian), by V. A. RAKOV, Gostrausizdat, Moscow, 1955.

L.M.A. Handbook. The Locomotive Manufacturers' Association of Great Britain, London, 1949.

Lubrication of Locomotives, by E. L. AHRONS, Locomotive Publishing Co., Ltd, London, 1921.

Machine Locomotive (10th Edn.), by E. SAUVAGE et A. CHAPELON, Ch. Beranger, Paris.

Manual of Locomotive Engineering, by W. J. PETTIGREW and A. F. RAVENSHEAR, Charles Griffin & Co., Ltd, London, 1899.

Metallurgy of a High-speed Locomotive. The Railway Gazette, London, 1938.

Modern Locomotive Practice, by C. E. WOLFF, The Scientific Publishing Co., Cheshire, 1903.

Modern Railway Motive Power, by BRIAN REED, Temple Press, Ltd, London, 1951.

Oil Fuel Equipment for Locomotives and Principles of Application, by A. H. GIBBINGS, Constable & Co., Ltd, London, 1915.

On Engines in Britain and France, by P. RANSOME-WALLIS, Ian Allan Ltd, 1957.

Osterreichs Lokomotiven und Treibwagen, Stockklausner, Ployer, 1954.

Pacific Locomotive Committee Report. Manager of Publications, Delhi, 1939.

Practical Evaluation of Railroad Motive Power, by P. W. KIEFER, Steam Loco Research Inst., Inc., New York, 1947.

Practical Locomotive Operating, by C. ROBERTS, and R. M. SMITH, J. B. Lippincott Company, Philadelphia, 1913.

Principles of Locomotive Operation, by A. J. WOOD, McGraw-Hill Publishing Co., Ltd, London, 1925.

Railroad Shop Practice, by F. A. STANLEY, McGraw-Hill Book Co., Inc., New York, 1921.

Railway Locomotive, by VAUGHAN PENDRED, Archibald Constable & Co., Ltd, London, 1908.

Railway Fuel, by E. McAULIFFE, Simmons-Boardman Publishing Co., New York, 1927.

Repairing of Locomotives, by E. L. AHRONS, Locomotive Publishing Co., Ltd, London, 1920.

Resistance of Express Trains, by C. F. DENDY MARSHALL, The Railway Engineer, London, 1925.

Soft Water for Loco Boiler Feed, by B. D. FOX, Locomotive Publishing Co., Ltd, London, 1946.

Springs. A Miscellany. Locomotive Publishing Co., Ltd, London.

Stabilité de route des Locomotives. Herman et Cie, Paris, 1935.

Steam Locomotion, by E. C. POULTNEY, The Caxton Publishing Co., Ltd, London, 1951.

Steam Locomotive, by R. P. JOHNSON, Simmons-Boardman Publishing Corporation, New York, 1945.

Steam Locomotive, by O. S. NOCK, Allen (George) and Unwin, Ltd, 1956.

Steam Locomotive in America, by A. W. BRUCE, W. W. Norton & Co., Inc., New York, 1952.

Steam Locomotive Design; Data and Formulae, E. A.

PHILLIPSON, Locomotive Publishing Co., Ltd, London, 1936.

Steam Locomotive of Today, by M. P. SELLS, Locomotive Publishing Co., Ltd, London, 1951.

Steam Locomotive in Traffic, by E. A. PHILLIPSON, Locomotive Publishing Co., London, 1949.

Study of the Locomotive Boiler, by L. H. FRY, Simmons-Boardman Publishing Co., New York, 1924.

Superheating on Locomotives, by J. F. GAIRNS, Locomotive Publishing Co., Ltd, London.

Tandem Compound Locomotives, by P. M. KALLA-BISHOP, Kalla-Bishop Books Ltd, London, 1949.

Traité Pratique de la Machine Locomotive, by M. DEMOULIN, Librairie Polytechnique, Paris, 1898.

Traité de Stabilité du Material des Chemins de Fer, Ch. Berenger, Paris, 1924.

Vacuum Brake. Locomotive Publishing Co., Ltd, London, 1921.

Vacuum Brake and Related Appliances. R. E. BRINKWORTH. Locomotive Publishing Co., Ltd, London.

Westinghouse Brake. Locomotive Publishing Co., Ltd, London, 1921.

World Locomotives, by C. S. LAKE, Marshall (Percival) & Co., Ltd, London, 1905.

Reference:
20,000 Schriftquellen zur Eisenbahnkunde. K. EWALD, Henschel, 1941.

Notes on Contributors

G. FREEMAN ALLEN. Son of a famous father, Cecil J. Allen, and a director of the railway publishing firm of Ian Allan, Ltd., Editor of *Trains Illustrated Magazine*, which has the largest circulation of any independent railway journal in the World. Writer of many articles on railway topics and author of *Railways the World Over*, 1956 and of *British Railways To-day and Tomorrow*, 1959, both Ian Allan, Ltd.

J. M. DOHERTY, A.M.I.LOCO.E., is a mechanical engineer who is now working on the design and manufacture of diesel engine components. He is an authority on the development of the diesel railway locomotive, and he has made a special study of methods of transmission. He has presented a number of papers to the Institution of Mechanical Engineers and the Institution of Locomotive Engineers.

S. O. ELL, Assistant (Testing) to the Mechanical and Electrical Engineer, British Railways, Western Region, is an authority on the draughting of steam locomotives. He is a leading member of the Swindon "team" who have so successfully modified various older types of locomotives, and have brought their performance up to modern standards.

H. M. LE FLEMING, M.A., A.M.I.Mech.E., M.I.LOCO.E., M.N.E.Coast Inst., Retired Mechanical Engineer, author and artist. He was formerly assistant to the Chief Mechanical Engineer of the Malayan Railways. He has had experience of steam locomotive techniques all over the world. A well-known authority on steam locomotives and warships, he has also a very wide knowledge of marine engineering. He has written many articles on locomotives in technical and semi-technical journals, and he is the author of *ABC of Ocean Liners; Ocean Freighters; Ocean Tankers; British Warships; British Tugs; British Trawlers; Coastal Passenger and Cargo Ships;* all Ian Allan Ltd, 1953–58 and he is part author of *The Locomotives of the Great Western Railway*, R.C.T.S., 1954–58.

F. J. G. HAUT, B.SC.(Eng.), A.M.I.Mech.E., M.I. and S.Inst., is a consulting Mechanical Engineer and Metallurgist. He is an international authority on electric traction and a regular contributor to technical journals both in the U.K. and overseas. He was born in Austria and is a member of the Newcomen Society. He is the author of *The Early History of the Electric Locomotive*, The Newcomen Society, 1952.

DAVID P. MORGAN, Editor-in-Chief of *Trains Magazine*, Milwaukee, U.S.A., has made a special study of the diesel locomotive in North America, and has written extensively on the subject. His publications include: *Louisville and Nashville's Pacifics and Mountains*, Bulletin 87 of Railway and Locomotive Historical Society, Boston, 1952; *True Adventures of Railroaders*, Little, Brown & Company, Boston, 1954.

O. S. NOCK, B.SC.(London), M.I.C.E., M.I.Mech.E., M.I.R.S.E., F.P.W.Inst., is chief draughtsman to the Signal and Colliery Division of the Westinghouse Brake and Signal Company, and as such is in charge of all design work in connection with many of the new signalling schemes for British Railways. He is an authority on locomotive practice and performance and is now the author of those well known surveys in the *Railway Magazine*. He is a frequent contributor to *The Engineer* and *The Locomotive Magazine*. He is the author of twenty-two books on railways and locomotives, the most important of which are *Scottish Railways*, Nelson, 1950; *The Premier Line*, Ian Allan, 1952; *The Railway Engineers*, Batsford, 1955; *Fifty Years of Western Express Running*, Everard, 1954; *The Railway Race to the North*, Ian Allan, 1959.

C. R. H. SIMPSON, A.M.I.LOCO.E., was for many years Editor of *Locomotive Magazine*. He has an exceptionally wide knowledge of locomotive construction and practice throughout the world, and is an expert practical engineer. He has written more than 1,500 articles for the technical press on every aspect of locomotive design and construction. He is the author of *The Locomotive Engineer's Pocket Book* (42nd ed.), The Locomotive Publishing Co., main author of *Locomotives and their Working*, Virtue & Co, 1951 and editor of *Steam Locomotion*, by E. C. Poultney, Caxton Press, 1951.

P. RANSOME-WALLIS, M.B., CH.B., is a physician, audiologist and author. He has made a life-long study of railway motive power and is well known as a writer on locomotives in Britain, Continental Europe and the United States. Publications include: *On Railways at Home and Abroad*, Betchworth, 1951; *Men of the Footplate; On Engines in Britain and France; ABC of Ships and the Sea; The Royal Navy*, all Ian Allan Ltd, 1954–58.